Tributes
Volume 17

Logic without Frontiers
Festschrift for Walter Alexandre Carnielli on the occasion of his 60th Birthday

Volume 8
Logos and Language.
Essays in Honour of Julius Moravcsik.
Dagfinn Follesdal and John Woods, eds.

Volume 9
Acts of Knowledge: History, Philosophy and Logic.
Essays dedicated to Göran Sundholm
Giuseppe Primiero and Shahid Rahman, eds.

Volume 10
Witnessed Years. Essays in Honor of Petr Hájek
Petr Cintula, Zuzana Haniková and Vítězslav Švejdar, eds.

Volume 11
Heuristics, Probability and Causality. A Tribute to Judea Pearl
Rina Dechter, Hector Geffner and Joseph Y. Halpern, eds.

Volume 12
Dialectics, Dialogue and Argumentation. An Examination of Douglas Walton's Theories of Reasoning and Argument
Chris Reed and Christoher W. Tindale, eds.

Volume 13
Proofs, Categories and Computations. Essays in Honour of Grigori Mints
Solomon Feferman and Wilfried Sieg, eds.

Volume 14
Construction. Festschrift for Gerhard Heinzmann
Solomon Feferman and Wilfried Sieg, eds.

Volume 15
Hues of Philosophy. Essays in Memory of Ruth Manor
Anat Biletzki, ed.

Volume 16
Knowing, Reasoning, and Acting. Essays in Honour of Hector J. Levesque
Gerhard Lakemeyer and Sheila A. McIlraith, eds.

Volume 17
Logic without Frontiers. Festschrift for Walter Alexandre Carnielli on the occasion of his 6^{0th} Birthday
Yean-Yves Béziau and Marcelo Esteban Coniglio, eds.

Tributes Series Editor
Dov Gabbay   dov.gabbay@kcl.ac

Logic without Frontiers
Festschrift for Walter Alexandre Carnielli on the occasion of his 60th Birthday

edited by

Jean-Yves Béziau

and

Marcelo Esteban Coniglio

© Individual author and College Publications 2011. All rights reserved.

ISBN 978-1-84890-055-4

College Publications
Scientific Director: Dov Gabbay
Managing Director: Jane Spurr
Department of Computer Science
King's College London, Strand, London WC2R 2LS, UK

http://www.collegepublications.co.uk

Original cover design by Laraine Welch
Illustration by Kathryn Finter
Printed by Lightning Source, Milton Keynes, UK

All rights reserved. No part of this publication may be reproduced, stored in a retrieval system or transmitted in any form, or by any means, electronic, mechanical, photocopying, recording or otherwise without prior permission, in writing, from the publisher.

WAC006: For the sake of logic

JEAN-YVES BÉZIAU AND MARCELO E. CONIGLIO

This book is dedicated to Walter Alexandre Carnielli colleague and friend for his 60th birthday. Walter has been working in logic since four decades. As a globetrotter and polyglot speaking more than 10 languages (including *Latin sine flexion*, also known as *Peano's interlingua* in the name of Peano who invented it), he has managed to develop many contacts around the world. The papers in this book reflect the wideness of his personal contacts and correlated logical interests. Walter was always very active, developing relations with many people and promoting logic in Brazil and in the world.

Walter was born on January 11 in 1952 in Campinas, Brazil — town which was the world capital of coffee at some point — from Italian origin. He studied mathematics at the State University of Campinas (UNICAMP), the second highest ranked university of Brazil. His PhD (1984) was directed by the legendary Brazilian logician Newton da Costa and he did a post-doc at the University of Berkeley in California with Leon Henkin. He developed his professional career at UNICAMP becoming full professor in 1996. He was Director of the Center for Logic, Epistemology and the History of Science at UNICAMP for three terms, and also President of the Brazilian Logic Society, participating to the organization of many events, among them: the 6th Workshop on Logic, Language Information and Computation (WoLLIC99, Itatiaia, 1999), the 2nd World Congress on Paraconsistency (WCP2, Juquehy 2000), the 9th Model-Based Reasoning in Science and Technology (MBR09, Campinas, Brazil, 2009) and of course most of the Brazilian logic meetings.

He also was visiting professor and/or researcher in many different countries: Germany (Alexander von Humboldt Grant), Portugal, Luxembourg, Chile, France, Poland, Italy, USA and Australia. He has been very active as an editor and/or member of editorial boards of major journals: *Studia Logica*, *Logic and Logical Philosophy*, *Journal of Applied Logic*, *Reports on Mathematical Logic*, *Journal of Applied Non-Classical Logics*, *Logica Universalis*.

His work touches many different areas of logic. He started to work on the systematization of many-valued logics through tableaux. He has been working quite a lot on topics related with paraconsistency and other non-classical logics. His research activities encompass also a general perspective on logical systems, in particular his work on multimodal logics and combination of logics. He has developed original ideas such as polynomial calculus, modulated quantifiers, society semantics and possible-translationd semantics. He also has been working on finite and infinite combinatorics (with A.M.Sette, P.A.Veloso and C. di Prisco) and on critical thinking with Richard Epstein.

His published works are listed in the following bibliography:

Bibliography of Walter Carnielli

1 Papers

(1) W.A.Carnielli and L.P.de Alcantara, "Transfinite induction on ordinal configurations", *Zeitschrift für mathematische Logik und Grundlagen der Mathematik*, **27**, 1981, 531–538.

(2) W.A.Carnielli and O.M.Torres, "Completitud estructural", *Bulletin del Departamiento de Cencias*, **23**, 1983, 17–29.

(3) W.A.Carnielli and L.P.de Alcantara, "Paraconsistent algebras" *Studia Logica*, **43**, 1984, 79–88.

(4) W.A.Carnielli, "On coloring and covering problems for rook domains" *Discrete Mathematics*, **57**, 1985, 9–16.

(5) W.A.Carnielli, "An algorithm for axiomatization and theorem proving in finite many-valued logics", *Logique et Analyse*, **112**, 1985, 363–368.

(6) W.A.Carnielli, "The problem of quantificational completeness and the characterization of all perfect quantifiers in three-valued logics", *Mathematical Logic Quarterly*, **33**, 1987, 19–29.

(7) W.A.Carnielli, "Methods of proof for relatedness and dependence logics", *Reports on Mathematical Logic*, **21**, 1987, 35–46.

(8) W.A.Carnielli, "Systematization of the finite many-valued logics through the method of tableaux", *The Journal of Symbolic Logic*, **52**, 1987, 473–493.

(9) W.A.Carnielli and N.C.A. da Costa, "Paraconsistent deontic logics", *Philosophi - The Philos. Quarterly of Israel*, **16**, 1988, 293–305.

(10) W.A.Carnielli and N.C.A. da Costa, "Kantian and non-Kantian logics", *Logique et Analyse*, **16**, 1988, 293–305.

(11) W.A. Carnielli, "Many-valued logics and plausible reasoning", in *Proceedings of the XX International Congress on Many-Valued Logic*, IEEE Computer Society, Los Alamitos, 1990, pp.328–335.

(12) W.A.Carnielli, "Hyper-rook domain inequalities" *Studies in Applied Mathematics*, **82**, 1990, 59–69.

(13) W.A.Carnielli, L.Fariãs del Cerro and M.Lima-Marques, "Contextual negations and reasoning with contradictions", in J.Mylopoulos and R.Reiter (eds), *Proceedings of the International Joint Conference on Artificial Intelligence (IJCAI'91)*, Morgan Kaufmann, Sidney, 1991, pp.532–537.

(14) W.A. Carnielli and M. Lima-Marques, "Reasoning under inconsistent knowledge", in *The Journal of applied non-classical logics*, **2**, 1992, 49–79.

(15) W.A.Carnielli and C.A. Di Prisco, "Some results on polarized partition relations of higher dimension", *Mathematical Logic Quarterly*, **39**, 1993, 461–474.

(16) W.A.Carnielli and J.C.Cifuentes, "Is there logic behind fuzzyness?", in W.A.Carnielli (ed), Logics, sets and Information: Proceedings Of The XI Brazilian Conference On Mathematical Logic, CLE, Campinas, 1995, pp.73–91.

(17) W.A.Carnielli and A.M.Sette, "Maximal weakly-intuitionistic logics", *Studia Logica*, **55**, 1995, 181–203.

(18) W.A.Carnielli, I.M.L. D'Ottaviano and E.H.Alves, "The centre for logic in Campinas and the development of logic in Brazil", *Logique et Analyse*, **154**, 1996, 15–29.

(19) W.A.Carnielli and I.M.L. D'Ottaviano, "Translations between logical systems: a Manifesto", *Logique et Analyse*, **157**, 1997, 67–81.

(20) P.A.S. Veloso and W.A.Carnielli, " Ultrafilter logic and generic reasoning", in *Computational Logic and Proof Theory - Vienna, 1997*, Springer, 1997, pp.34–53.

(21) A.M.Sette, W.A. Carnielli and P.A.S.Veloso, "An alternative view of default reasoning", in E.H.Haueusler and L.C.Pereira (eds), *Proofs, Types and Categories*, PUC, Rio de Janeiro, 1999, pp.127–158.

(22) W.A. Carnielli and M. Lima-Marques, "Society semantics and multiple-valued logics", in *Advances in contemporary logic and computer science*, AMS, Providence, 1999, pp.33–52.

(23) W.A.Carnielli, "On the Ramsey problem for multicolor bipartite graphs", Advances In Applied Mathematics, **22**, 1999, 48–59.

(24) W.A.Carnielli and M.E.Coniglio, "A categorical approach to the combination of logics", in *Manuscrito*, **22**, 1999, 69–74.

(25) W.A.Carnielli, " Possible-translations semantics for paraconsistent logics", in D.Batens et al. (eds), *Frontiers in Paraconsistent Logic: Proceedings of the I World Congress on Paraconsistency, Ghent, 1998*, Research Studies Press, Baldock, 2000, pp.159–172.

(26) W.A.Carnielli and E.L.Monte Carmelo, "K2,2-K1, n and K2, n-K2, n bipartite Ramsey numbers", *Discrete Mathematics*, **223**, 2000, 83–92.

(27) S.Rahman and W.A.Carnielli, "The dialogical approach to paraconsistency", *Synthese*, **125**, 2000, 201–221.

(28) W.A.Carnielli, J.Marcos and S.de Amo, "Tableaux for logics of formal inconsistency", *Logic and Logical Philosophy*, **8**, 2000, 115–152.

(29) W.A.Carnielli and J.Marcos, "Ex Contradictione Non Sequitur Quodlibet", in *Bulletin of Advanced Reasoning and Knowledge*, **1**, 2001, 89–109.

(30) W.A.Carnielli and J.Marcos, "Tableaux for logics of formal inconsistency", in *Proceedings of the 2001 International Conference on Artificial Intelligence*, CSREA Press, Athens, 2001, pp.848–852.

(31) W.A.Carnielli and J.Marcos, "A taxonomy of C-systems", in *Paraconsistency - the Logical Way to the Inconsistent*, Marcel Dekker, New York, 2002, pp.1–94.

(32) W.A.Carnielli, C.Sernadas and J.Rasga, "Modulated fibring and the collapsing problem", *The Journal of Symbolic Logic*, **67**, 2002, 1541–1569.

(33) W.A.Carnielli and M.E.Coniglio, "Transfers between logics and their applications", *Studia Logica*, **72**, 2002, 367–400.

(34) S.de Amo, W.A.Carnielli and J.Marcos, "A logical framework for integrating inconsistent Information in multiple databases", in T.Eiter and K.-D.Schewe (eds), *Lecture Notes in Computer Science, vol 2284* , Springer, Berlin, 2002, pp.67–84.

(35) W.A.Carnielli, C.Caleiro, M.E.Coniglio, A.Sernadas and C.Sernadas, "Fibring non-truth-functional logics: completeness preservation", *Journal of Logic, Language and Information*, **12**, 2003, 183–211.

(36) J.Bueno-Soler, M.E.Coniglio and W.A.Carnielli, "Finite algebraizability via possible-translations semantics", in W.A.Carnielli et al. (ed), *Proceedings of CombLog'04 - Workshop on combination of logics: Theory and applications*, IST, Lisbon, 2004, pp.79–86.

(37) M.M.D.Cunha, M.E.Coniglio and W.A.Carnielli, "An incoherence in the AGM theory?", in H.Feitosa and F.Sautter (eds), *Lógica: Teoria, Aplicações e Reflexões*, CLE-UNICAMP, Campinas, 2004, p.107–117.

(38) P.A.S.Veloso and W.A.Carnielli, "Logics for qualitative reasoning", in S.Rahman and J.Symons (eds), *Logic, Epistemology and the Unity of Science*, Kluwer, Dordrecht, 2004, pp.487–526.

(39) C.Caleiro, W.A.Carnielli, M.E.Coniglio and J.Marcos, "Two's company: The humbug of many logical values", in J.-Y.Béziau (ed), *Logica Universalis*, Birkhäuser, Basel, 2005, pp.169–189.

(40) J.Bueno-Soler and W.A.Carnielli, "Possible-translations algebraization for paraconsistent logics", *Bulletin of the Section of Logic*, **34**, 2005, 77–92.

(41) A.B.M.Brunner and W.A.Carnielli, "Anti-intuitionism and paraconsistency", *Journal of Applied Logic*, **3**, 2005, 161–184.

(42) W.A.Carnielli and M.E.Coniglio, "Splitting Logics", in S.Artemov et al. (eds), *We Will Show Them: Essays in Honour of Dov Gabbay - vol. 1*, College Publications, London, 2005, pp.389–414.

(43) W.A.Carnielli, C.Caleiro, J.Rasga and C.Sernadas, "Fibring of logics as a universal construction", in D.M.Gabbay and F.Guenthner (eds), *Handbook of Philosophical Logic, vol. 13*, Springer, Dordrecht, 2005, pp.123–187.

(44) W.A.Carnielli, "Polynomial ring calculus for many-valued Logics", in *36th International Symposium on Multiple-Valued Logic*, IEEE Computer Society, Los Alamitos, 2005, pp.20–25.

(45) W.A.Carnielli, "Surviving abduction", *Logic Journal of the IGPL*, **14**, 2006, 237–256.

(46) J.Bueno-Soler, M.E.Coniglio and W.A.Carnielli, "Possible-translations algebraizability", in J-Y-Beziau et al. (ed), *Handbook of Paraconsistency*, College publications, London, 2007, pp.321–340.

(47) W.A.Carnielli, "Polynomizing: logic inference in polynomial format and the legacy of Boole", *Studies in Computational Intelligence*, **64**, 2007, 349–364.

(48) W.A.Carnielli and M.E.Coniglio, "Combining Logics", in *Stanford Encyclopedia of Philosophy*, 2007.

(49) W.A.Carnielli, M.E.Coniglio and J. Marcos, "Logics of formal inconsistency", in D.Gabbay and F.Guenthner (eds), *Handbook of Philosophical Logic, vol. 14*, Springer, Dordrecht, 2007, pp.15–107.

(50) W.A.Carnielli and M.E.Coniglio, "Bridge principles and combined reasoning", in T.Müller and A,Newen (eds), *Logik, Begriffe, Prinzipien des Handelns*, Mentis, Paderborn, 2007, pp.32–48.

(51) J.C.Agudelo and W.A.Carnielli, "Unconventional models of computation through non-standard logic circuits", in *Lecture Notes in Computer Science, vol 4618*, Springer, Berlin, 2007, pp.29–40.

(52) W.A.Carnielli and F.A.Doria, "Are the foundations of computer science logic dependent?", in C.Dégremont et al (eds), *Dialogues, Logics and Other Strange Things- Essays in Honour of Shahid Rahman*, College publications, London, 2008, pp.87–107.

(53) W.A.Carnielli, J.Rasga and C.Sernadas, "Preservation of interpolation features by fibring", *Journal of Logic and Computation*, **18**, 2008, 123–151.

(54) W.A.Carnielli, "The tyranny of knowledge", *Manuscrito*, **31**, 2008, 511–518.

(55) W.A.Carnielli and M.C.E. Coniglio, "Aristóteles , paraconsistentismo e a tradição budista" *O Que nos Faz Pensar*, **23**, 2008, 163–175.

(56) W.A.Carnielli and M.C.C. Grácio, "Modulated logics and flexible reasoning", *Logic and Logical Philosophy*, **17**, 2008, 211–249.

(57) I.C.Oliveira and W.A.Carnielli, "The Ricean objection: An analogue of Rice's theorem for first-order theories", *Logic Journal of the IGPL*, **16**, 2008, 585–590.

(58) W.A.Carnielli, "Meeting Hintikka's challenge to paraconsistentism", *Principia*, **13**, 2009, 283–297.

(59) W.A.Carnielli, "Uma lógica da modalidade económica?", *Revista Brasileira de Filosofia*, **232**, 2009, 209–225.

(60) W.A.Carnielli, M.C.E. Coniglio and I.M.L. D'Ottaviano, "New dimensions on translations between logics" *Logica Universalis*, **3**, 2009, 1–18.

(61) J.Rasga, W.A.Carnielli and C.Sernadas, "Interpolation via translations", *Mathematical Logic Quarterly*, **55**, 2009, 515–534.

(62) W.A.Carnielli, "Formal polynomials and the laws of form", in J.Y.Beziau and A.Costa Leite (eds) *Dimensions of logical concepts*, CLE-UNICAMP, Campinas, 2009, pp.201–212.

(63) W.A.Carnielli and L.Magnani, "Years of reasoning. In honor of the 65th birthday of Claudio Pizzi", in L.Magnani et al (eds), *Model-Based Reasoning in Science and Technology - Abduction, Logic, and Computational Discovery*, Springer, Berlin, 2010, pp.1–15.

(64) W.A.Carnielli, "On a theoretical analysis of deceiving: how to resist a bullshit attack", in L.Magnani et al (eds), *Model-Based Reasoning in Science and Technology - Abduction, Logic, and Computational Discovery*, Springer, Berlin, 2010, pp.237–244.

(65) J.C.Agudelo and W.A.Carnielli, "Paraconsistent machines and their relation to quantum computing", *Journal of Logic and Computation*, **20**, 2010, 573–595.

(66) W.A.Carnielli, "Formal polynomials, heuristics and proofs in logic", *Logical Investigations*, **16**, 2010, 280–294.

(67) W.A.Carnielli and M.C.E. Coniglio, "On discourses addressed by infidel logicians", in K.Tanaka et al, (eds), *Paraconsistency: Logic and Applications*, Springer, Berlin, 2011, p.1–14.

(68) W.A.Carnielli, "The single-minded pursuit of consistency and its weakness", *Studia Logica*, **97**, 2011, 81–100.

(69) J.C.Agudelo and W.A.Carnielli, "Polynomial ring calculus for modal logics: a new semantics and proof method for modalities", *The Review of Symbolic Logic*, **4**, 2011, 150–170.

(70) W.A.Carnielli, "Paul Bernays and the eve of non-standard models in logic", in J.-Y.Beziau (ed), *Universal Logic: an Anthology*, Birkhäuser, Basel, 2012, pp.33–42.

2 Books

(1) R.L.Epstein and W.A.Carnielli, *Computability: computable functions, logic and the foundations of mathematics, with the timeline Computability and Undecidability*, Second edition. Wadsworth/Thomson Learning, Belmont, CA, 2000.

(2) W.A.Carnielli and C.Pizzi, *Modalità e multimodalità*, Franco Angeli, Milan, 2001.

(3) W.A.Carnielli and R.L. Epstein, *Computabilidade: Funções Computáveis, Lógica e os Fundamentos da Matemática*, Winner of 2007 Jabuti Award, the most prestigious literary prize in Brazil.

(4) W.A.Carnielli and C.Pizzi, *Modalities and Multimodalities*, Springer-Verlag, 2008.

(5) W.A.Carnielli, M.E.Coniglio, D.M.Gabbay, P.Gouveia and C.Sernadas, *Analysis and Synthesis of Logics. How to Cut and Paste Reasoning Systems*, Springer, Berlin, 2008.

(6) W.A.Carnielli and R.L.Epstein, *Pensamento Crítico: o poder da lógica e da argumentação*, Rideel, São Paulo: 2009.

(7) W.A.Carnielli, *Possible-translations semantics: their scope, limits and capabilities*, Birkhäuser, Basel, 2012.

3 Edited Books

(1) W.A.Carnielli and L.P. de Alcantara (eds), *Methods and Applications of Mathematical Logic: Proceedings of the Seventh Latin American Symposium on Mathematical Logic, held July 29–August 2, 1985, at the University of Campinas in Brazil*, AMS, Providence, 1988.

(2) W.A.Carnielli (ed), *Logic, Sets And Information: Proceedings Of The XI Brazilian Conference On Mathematical Logic*, CLE, Campinas, 1995.

(3) W.A.Carnielli and I.M.L.D'Ottaviano (eds), *Advances in Contemporary Logic and Computer Science: Proceedings of the Eleventh Brazilian Conference on Mathematical Logic, May 6-10, 1996, Salvador da Bahia, Brazil*, AMS, Providence, 1999.

(4) W.A.Carnielli, M.E.Coniglio and I.M.L.D'Ottaviano (eds), *Paraconsistency: The Logical Way to Inconsistency*, Marcel Dekker, New York, 2002.

(5) J.-Y.Béziau, W.A.Carnielli and D.M.Gabbay (eds), *Handbook of Paraconsistency*, College publications, London, 2007.

(6) W.A.Carnielli, M.E.Coniglio and I.M.L.D'Ottaviano (eds), *The Many-Sides of Logic*, College publications, London, 2009.

(7) L.Magnani, W.A.Carnielli and C.Pizzi (eds), *Model-Based Reasoning in Science and Technology: Abduction, Logic, and Computational Discovery*, Springer, Berlin, 2010.

4 Editions of special issues of journals

(1) W.A.Carnielli (ed), Many-valued logics, *Journal of Applied Non Classical Logics*, **9(1)**, 1999.

(2) W.A.Carnielli, M.E.Coniglio and I.M.L.D'Ottaviano (eds), An Event on Brazilian Logic Part I. Proceedings of the XIII Brazilian Conference on Mathematical Logic, *Logic Journal of the IGPL*, **12(6)**, 2004.

(3) W.A.Carnielli, M.E.Coniglio and I.M.L.D'Ottaviano (eds), An Event on Brazilian Logic Part II. Proceedings of the XIII Brazilian Conference on Mathematical Logic, *Logic Journal of the IGPL*, **13(1)**, 2005.

(4) W.A.Carnielli, J.Marcos and D.Batens (eds), A Paraconsistent Decagon, *Journal of Applied Logic*, **3(1)**, 2005.

(5) W.A.Carnielli and J.J.da Silva (eds), Logic, Language, and Knowledge. Essays on Chateaubriand's Logical Forms, *Manuscrito*, **31**, 2008.

The present volume contains 28 articles of renowed logicians, philosophers, mathemathicians and computer scientists all around the world, dedicated to a great logician and friend: Walter Alexandre Carnielli. Feliz aniversário, Walter!

Jean-Yves Béziau
Department of Philosophy Federal University of Rio de Janeiro
Rio de Janeiro, Brazil
E-mail:

Marcelo E. Coniglio
Department of Philosophy and Centre for Logic, Epistemology and the History of Science (CLE)
State University of Campinas
Campinas, Brazil
E-mail: coniglio@cle.unicamp.br

The Authors

Michael Abraham had his PhD in Physics at Bar-Ilan University (BIU), Israel. Now teaches talmud and is a member of the Talmudical logic group at BIU. He published many books on philosophy, logic, law, and talmud, and especially on the relations beteen these areas.

Arnon Avron is a Professor in the school of computer science of Tel-Aviv University. He has a Ph.D in Mathematics from Tel-Aviv university. His main areas of research are proof theory, non-classical logics, logic in computer science, and foundations of mathematics.

Israel Belfer is a PhD Candidate in the Science, Technology and Society program (STS) in Bar-Ilan University, Israel. He is a research assistant at the Edelstein Center for the Philosophy of Science, Hebrew University, Jerusalem. Belfer participates in several research groups dealing with bioethics, Israeli society, technology, as well as the Talmudic Logic research group at Bar-Ilan University under Professor Dov Gabbay.

Jean-Yves Béziau is professor of logic at the Federal University of Rio da Janeiro, PhD in mathematical logic (University of Paris 7) and PhD in philosophy (University of São Paulo). He has been working in France, Brazil, Poland, Switzerland and California (UCLA and Stanford). He is editor-in-chief of the journal Logica Universalis (Birkhäuser, Springer Basel).

Ross Brady is a Reader and Associate Professor in Philosophy at La Trobe University in Melbourne, having worked there for 40 years, after a short appointment as a Lecturer in the Mathematics Department at University of Western Australia. He completed a PhD in Logic from the University of St. Andrews, Scotland, in 1971, after completing a course-work MA in Formal Logic at the University of New England and a BSc in Pure Mathematics at the University of Sydney. He is the editor and a major contributor to the book, Relevant Logics and their Rivals, Volume 2, Ashgate, 2003, and the author of the book, Universal Logic, CSLI, 2006. He has researched in the area of set-theoretic and semantic paradox solution, using many-valued and relevant logics, and has developed the relevant logic MC of meaning containment.

Oswaldo Chateaubriand is Professor of Philosophy at the Pontifical Catholic University of Rio de Janeiro where he has taught since 1978. He obtained his PhD in Philosophy at the University of California at Berkeley and taught at the

University of Washington, Cornell University, and Fundao Getlio Vargas. He held visiting appointments at the Universty of So Paulo and at Harvard University. He is a founding member of the Sociedade Brasileira de Lgica (SBL), of which he was president for two terms. He is an external member of the Centro de Lgica, Epistemologia e Histria da Cincia (CLE) and is a member of the Institut International de Philosophie (IIP). His research areas are philosophy of logic, philosophy of mathematics, and philosophy of language. His main publication is the book *Logical Forms, Part I: Truth and Description* (2001) and *Part II: Logic, Language, and Knowledge* (2005).

Sandra de Amo is an Associate Professor in the Faculty of Computer Science at the Federal University of Uberlândia, Brazil. She received her PhD in Computer Science from the University of Paris 13, France, in 1995 and her MSc in Mathematics from the Institute or Pure and Applied Mathematics (IMPA-Rio de Janeiro, Brazil). Her research interests include data mining, query languages, preference modeling and reasoning, temporal databases, applications of logic in databases and inconsistent information management in databases.

Anderson de Araújo is a FAPESP Postdoctoral Fellow at University of São Paulo (IME-USP), after having completed his PhD in Philosophy at State University of Campinas (CLE-UNICAMP).

Milton A. de Castro obtained his PhD from the State University of Campinas (UNICAMP) in 2004. He is author of several articles on proof-systems for non-classical logics.

Valeria de Paiva is a is a mathematician, logician and computer scientist based in Cupertino, CA. She works as senior research scientist at Rearden Commerce, Foster City, CA. She was a search analyst at Cuil, Inc. in Menlo Park, CA, from May 2008-Sept 2010 and before that she was a research scientist at the Intelligent Systems Laboratory of PARC (Palo Alto Research Center), California (2000-2008). She received her PhD in Mathematics from Cambridge University in 1988 for work on "Dialectica Categories", under Martin Hyland's supervision, and has ever since worked on logical approaches to computation, especially using Category Theory. She is on the editorial board of *Theoria and Application of Categories*, *Logical Methods in Computer Science* and of *Logica Universalis*, as well as in the editorial board of the Springer series of books *Logic, Language and Information*. She's an Honorary Research Fellow at the School of Computer Science, University of Birmingham, UK and taught recently at Stanford University and Santa Clara University.

Max Dickmann is Directeur de Recherche (Emeritus since 2007) with the Centre National de la Recherche Scientifique (CNRS), France, where he has been a researcher since 1974. He works at the research groups in Mathematical Logic and in Algebraic Geometry and Topology of the Institut de Mathématiques de Jussieu, Universities Paris 6 and 7. His present areas of research concern the interaction between mathematical logic, real algebraic geometry and algebra, with special emphasis on the algebraic theory of quadratic forms and on spectral spaces.

Carlos A. Di Prisco is Emeritus Professor at Instituto Venezolano de Investigaciones Científicas and professor at Universidad Central de Venezuela. He obtained his Ph.D. in mathematics at the Massachussetts Institute of Technology. His main research interests are set theory and the foundations of mathematics, and in particular combinatorial set theory. He has been visiting professor at University of Paris VII, Centre de Recerca Matemática, Barcelona, Smith College, University of Evora; research associate at University of California, Berkeley; Fellow of the Guggenheim Foundation. He is a member of the Center of Logic, Epistemology and the History of Science (CLE), University of Campinas.

Itala M.L. D'Ottaviano is full professor of Logic and Foundations of Mathematics at the State University of Campinas (UNICAMP), Brazil. She received her MSc in Mathematics in 1974 and her PhD in Mathematics in 1982, both from the State University of Campinas. She is founding member of the Centre for Logic, Epistemology and the History of Science (CLE) at UNICAMP, and of the Brazilian Logic Society (SBL). She has been working in non-classical logics, foundations of mathematics, history and philosophy of science, algebraic logic, and theory of self-organization. She is member of the Académie Internationale de Philosophie des Sciences (AIPS).

Richard L. Epstein is the head of the Advanced Reasoning Forum in Socorro, New Mexico, USA. He received his B.A. from the University of Pennsylvania in 1969 and his Ph.D. from the University of California, Berkeley in 1973. He has published books on both formal logic and logic as the art of reasoning well.

Manuel Fidel is full professor in the Department of Computer Science of the Universidad Nacional del Sur (Bahía Blanca, Argentina). He has an Doctorate in Mathematics. His areas if interest are Foudations of Mathematics, Algebraic Logic and Software Development.

Aldo V. Figallo is professor of Foundations of Mathematics at the Universidad Nacional del Sur and professor of Topology at the Universidad Nacional de La Pampa. He directs the research institutional program Development in Mathematics and Logic at the Instituto de Ciencias Básicas, Universidad Nacional de San Juan. He got his Ph.D. in Mathematics in 1990 from the Universidad Nacional del Sur and the subject of his thesis was an algebraic treatment of implicative fragments of Lukasiewicz logics. He started his studies in Algebraic Logic under the supervision of Antonio Monteiro in the 1970s, being one of his latest disciples. He has been working in Algebraic Logic, Universal Algebra and Ordered Algebraic Structures the last thirty years producing several articles in these areas.

Martín Figallo is assistant professor of Calculus at the Department of Mathematics of Universidad Nacional del Sur. He received his MSc in Mathematics (2005) from the same university. He has been working in Classical and non–Classical logic, Algebraic Logic and Proof-Theory.

Marcelo Finger is professor of Computer Science at the University of São Paulo. He received his PhD in Computing (1994) and his MSc (in Foundations of Information Technology) from the Department of Computing at the Imperial College of Science and Technology of the University of London. He has been working in classical and non-classical Logics for Computer Science, with several applications to Artificial Intelligence, Databases and Computational Linguistics.

Melvin Fitting is a professor at the City University of New York, in the undergraduate department of Mathematics and Computer Science at Lehman College, and in the graduate departments of Mathematics, of Computer Science, and of Philosophy at the CUNY Graduate Center. He received his PhD in mathematical logic from Yeshiva University in 1968 and has had a long and productive career that is reaching it's formal, but not creative, end.

Dov Gabbay, FRSC FAvH FRSA FBCS is Augustus De Morgan Professor of Logic (Emeritus) at King's College London, Special professor at Bar-Ilan University israel and a Visiting Professor at the University of Luxembourg. He is one of the world's most active and influential researchers in logic, is the author of over four hundred research papers and twenty five research monographs, and has initiated several new and active research areas. He is editor of several international Journals, and over 50 Handbooks of Logic. Dov Gabbay is Chairman and founder of several international conferences, executive of the European Foundation of Logic Language and Information and is President of The International IGPL Logic Group as well as one of the four founders - and council member for many years - of FoLLI, the Association of Logic, Language and Information. Now retired and Life Member. He is also Founder, Executive and Vice President of the International Federation of Computational Logic, (UK Charity, Number 1112512).

Edward Hermann Haeusler is associate professor of Computer Science at Pontifical Catholic University of Rio de Janeiro. He received his DSc in Computing from PUC-Rio. He has been working in Logic and Proof-Theory, with applications to Computer Science.

Jaakko Hintikka was born in Vantaa, Finland and studied mostly at the University of Helsinki under G.H. von Wright. He has held different academic appointments, most recently at Boston University. Jaakko Hintikka's scholarly and scientific interests comprise philosophy of language and theoretical linguistics, philosophical and mathematical logic, epistemology, philosophy of science and philosophy of mathematics, as well as history of philosophy and history of ideas, especially Aristotle, Descartes, Leibniz, Kant, and Wittgenstein. He is known as the main architect of game-theoretical semantics, independence-friendly logic, and of the interrogative approach to inquiry, and as one of the architects of epistemic logic, distributive normal forms, possible-worlds semantics and the logical theory of inductive generalization.

Norihiro Kamide received his Ph.D. in Information Science from Japan Advanced Institute of Science and Technology in 2000. From 2008 to 2009, he was an Alexander von Humboldt Fellow at Dresden University of Technology. He is an editorial board member of Far East Journal of Applied Mathematics. He is now a visiting researcher of Waseda Institute for Advanced Study and a visiting associate professor of Cyber University.

Beata Konikowska is a professor of mathematical sciences at the Institute of Computer Science of the Polish Academy of Sciences in Warsaw. She has Ph.D. and habilitation (Polish post-doctoral degree) in mathematical sciences, discipline: computer science. Since 2006, she has been the Deputy Director for Science of the above Institute.

Frederick Kroon is associate professor of philosophy at the University of Auckland, New Zealand. He has a PhD in philosophy from Princeton University, and is an associate editor for the Australasian Journal of Philosophy and subject editor in 20th century philosophy for the Stanford Encyclopedia of Philosophy. Following early work in computability theory, and, more recently, on semantic paradox and paradoxes of rationality, he now works mainly in the philosophy of language and philosophical logic. He has authored papers in these various areas for a broad range of journals, including *Ethics*, *Analysis*, *The Journal of Philosophy*, *Philosophical Review*, and *Noûs*.

Mamede Lima-Marques is full professor of Information Science at University of Brasilia. He received his Doctorat in Informatique (1992) from Université Paul Sabatier, Toulouse III, France. He has been working in Architecture of Information.

Jennie Louise received her undergraduate philosophy degree from the University of Queensland, and her PhD from the Research School of Social Sciences, Australian National University in 2004. She has been at the University of Adelaide since 2005. Her research interests are in ethical theory and applied ethics (especially medical and health ethics), rationality and reasoning, decision theory and philosophy of logic.

João Marcos is professor of Logic at the Federal University of Rio Grande do Norte, and a founding member of its Group for Logic, Language, Information, Theory and Applications (LoLITA). He holds a PhD in Mathematics from TU-Lisbon (Portugal) and a PhD in Philosophy from State University of Campinas (Brazil). Walter Carnielli was his supervisor and co-author in numerous papers.

Paulo Mateus is associate professor of logic and computation at the Department of Mathematics of Instituto Superior Técnico (IST), Universidade Técnica de Lisboa (UTL), and coordinator of the Security and Quantum Information Group of Instituto de Telecomunicações. He graduated at IST where he got his PhD in mathematics, and obtained his habilitation in mathematics from UTL. Most of his research has been focused on reasoning about probabilistic and quantum systems.

Francisco Miraglia is Professor of Mathematics at the University of São Paulo, having obtained his Master of Philosophy and PhD in Mathematics at Yale University (1979). He has worked in Model Theory and its interaction with Sheaf Theory, as well as, more recently, in the algebraic theory of quadratic forms. He has been a visting professor at Oxford, Verona, Maryland and the University of Paris VII.

Chris Mortensen was born in Rockhampton, Australia. He studied philosophy and mathematics at the University of Queensland before taking a PhD in the philosophy of mind at the University of Adelaide. He was for some years a research fellow at the Australian National University, working in non-classical logics with Routley, Meyer and notable visitors including da Costa. He then took a lecturing position at the University of Adelaide, where he spent the remainder of his career, retiring as professor of philosophy. His research interests include logic, metaphysics, philosophy of mathematics and science, and Buddhism. He is known for his work on the theory of inconsistency, and his books *Inconsistent Mathematics* (1995) and *Inconsistent Geometry* (2010).

Daniele Mundici is professor of mathematical logic in the University of Florence. He has served as a professor of computer science in the University of Milan, and has given courses in universities of Europe, America and Africa. His degrees are in physics. Former president of the Italian Association of Logic, and of the Gödel Society of Vienna, he is now Secretary of the GNSAGA group of the Istituto Nazionale di Alta Matematica, Rome. He is a member of the Academy of Exact and Natural Sciences, Buenos Aires, and of the International Academy of Philosophy of Science, Bruxelles. He is managing editor of various journals and book series, including the *Journal of Algebra and its Applications*, the Chinese journal *Studies in Logic*, and *Studia Logica*.

Inés Pascual is professor of Topology and Functional Analysis at Universidad Nacional de San Juan, Argentina. She is co-director of the institutional research project *Lógicas y Álgebras* within the program *Desarrollos en Matemática y Lógica* at Instituto de Ciencias Básicas, Universidad Nacional de San Juan, Argentina. At present she is preparing her doctoral thesis on θ-valued Łukasiewicz–Moisil algebras at Universidad Nacional del Sur, Bahía Blanca, Argentina.

Tarcísio Pequeno has a PhD in Computer Science from the Catholic University of Rio de Janeiro and did a post-doc at the University of New Hampshire (USA). He was full professor of the Federal University of Ceará (UFC) and president of the Research Foundation of Ceará (FUNCAP); he is presently professor at the University of Fortaleza (UNIFOR). He has been working in foundations of computer science, non monotonic logic, paraconsistent logic and philosophy of science.

Luiz Carlos Pereira is assistant professor of Philosophy at Pontifical Catholic University of Rio de Janeiro (PUC-Rio) and associate professor of Philosophy at the State University of Rio de Janeiro (UERJ). He received his PhD in Philosophy from the University of Stockholm. His main areas of interest are Logic and Analytic Philosophy.

João Rasga is assistant professor of logic and computation at the Department of Mathematics of Instituto Superior Técnico (IST), Universidade Técnica de Lisboa, and vice-coordinator of the Security and Quantum Information Group of Instituto de Telecomunicações. He graduated at IST where he got his PhD in mathematics. Most of his research has been focused on combinations of logics, proof theory and abstract deductive systems.

Eike Ritter is a Lecturer in Computer Science at the School of Computer Science, University of Birmingham, UK. His research interests include security, applications of mathematical logic and category theory to computer science, type theory and its applications to functional programming, proof theory and automated theorem proving.

Uri J. Schild is a professor of Computer Science at the Ashkelon Academic College and Bar-Ilan University, PhD in Computer Science (Imperial College, London). He has been working in Israel, Australia and England.

Amílcar Sernadas is professor of logic and computation at the Department of Mathematics of Instituto Superior Técnico, Universidade Técnica de Lisboa, and chair of the scientific council of the Security and Quantum Information Group of Instituto de Telecomunicações. He got his PhD in computer science from the University of London and obtained his habilitation in computational mathematics from the University of Lisbon. Most of his research has been focused on applications of logic to computer science. He belongs to the editorial board of the journal Formal Aspects of Computing.

Cristina Sernadas is professor of logic and computation at the Department of Mathematics of Instituto Superior Técnico, Universidade Técnica de Lisboa (UTL), and senior researcher at the Security and Quantum Information Group of Instituto de Telecomunicações. She got her PhD in mathematics from the University of London and obtained her habilitation in mathematics from UTL. Most of her recent research has been focused on the topic of combined logics.

Raymond Smullyan got his PhD in 1959 from Princeton University under the supervision of Alonzo Church. He is famous for his work on refutation trees and his generalization of Gödel's theorem. He has also published many popular books on logical puzzles, appearing in several languages. Some of these lead the reader through proofs of the Gödel completeness and incompleteness theorems, combinatory logic, boolean algebra, and other fundamental topics. He has worked as a magician, and as a pianist, with piano videos available through YouTube.

Paulo Veloso is professor of Computer Science at the Federal University of Rio de Janeiro. He received his PhD in Computer Science (1975) from the University of California, Berkeley. He has been working in algebraic and logical aspects of Foundations of Computing and Programming.

Sheila Veloso is professor of Computer Science at the University of Estado do Rio de Janeiro since 2005, after retiring from the Federal University of Rio de Janeiro as Adjoint Professor. She received her D.Sc in Computing (1985) from the Programa de Engenharia de Sistemas e Computao, COPPE of the University of Rio de Janeiro and her M .A (in Mathematics) from the Department of Mathematics of the University of California, Berkeley. She has been working in classical and non-classical Logics, with applications to Foundations of Computing and Artificial Intelligence.

Heinrich Wansing is professor of logic and epistemology at the Ruhr-University Bochum (Germany). He obtained his PhD in Philosophy at the Free University of Berlin in 1992 and defended his Habilitation in Logic and Analytical Philosophy at the University of Leipzig in 1997. From 1998 until 2010 he was a professor of Philosophy of Science and Logic at the Technical University of Dresden. Heinrich Wansing is a managing editor of Studia Logica and editorial board member of several other journals.

Anna Zamansky is a Marie Curie Postdoctoral Fellow at Vienna Technical University, after having completed her PhD in Computer Science at Tel-Aviv University.

Alicia Ziliani is associate professor of Foundations of Mathematics at the Departament of Mathematics at the Universidad Nacional del Sur. She received her Ph.D. in Mathematics at the same university. She was one of the latest disciples, together with Aldo V. Figallo, of Professor Antonio Monteiro.

Contents

WAC006: For the sake of logic vii
Jean-Yves Béziau and Marcelo E. Coniglio

The Authors xv

1. Duplex Diagonalization 1
   Raymond M. Smullyan

2. Perfect Set Properties and large cardinals 17
   Carlos A. Di Prisco

3. Why is the $P=?NP$ question so difficult? 25
   Newton C. A. da Costa and Francisco A. Doria

4. Reflections on the Church-Turing Thesis 43
   Frederick Kroon

5. On the logical relativity of quantum computability 65
   Marcelo Finger and Anderson de Araújo

6. Delegation, Count as, and Security in Talmudic Logic, a preliminary study 73
   Michael Abraham, Israel Belfer, Dov Gabbay, and Uri J. Schild

7. A Fibonacci estimation for the height of Schütte-like cut-free proofs 97
   Edward H. Haeusler and Luiz C. Pereira

8. Valid Inferences 105
   Richard L. Epstein

9. Rules of the Game 131
   Tarcísio Pequeno and Jean-Yves Béziau

10. Proving Completeness for Nested Sequent Calculi 145
    Melvin Fitting

11. Real Semigroups and Rings 155
    Max Dickmann and Francisco Miraglia

12. Once More on the Unexpected Examination 173
    Chris Mortensen and Jennie Louise

13. Revisiting 'Generally' and 'Rarely' 183
    Paulo A. S. Veloso and Sheila R. M. Veloso

14. Rational measure of rational simplexes 20
    Daniele Mundici

15. Principal and Boolean congruences on θ-valued Łukasiewicz–Moisil algebras . . . 21
    Aldo V. Figallo, Inés Pascual and Alicia Ziliani

16. Analytical Tableaux for da Costa's Paraconsistent Predicate Calculi C_n^* . . . 23
    Itala M. Loffredo D'Ottaviano and Milton Augustinis de Castro

17. Systematic Construction of Analytic Calculi for Basic Logics of Formal Inconsistency . . . 26
    Arnon Avron, Beata Konikowska and Anna Zamansky

18. The Value of the Two Values 27
    João Marcos

19. A Note On The Logic Of *Una Tantum* Truth 29
    Claudio Pizzi

20. Outline of a theoretical framework of Architecture of Information: a School of Brasilia proposal . . . 31
    Mamede Lima–Marques

21. Non-deterministic combination of connectives 32
    Amílcar Sernadas, Cristina Sernadas, João Rasga and Paulo Mateus

22. Metavaluations, Naive Set Theory and Inconsistency 33
    Ross T. Brady

23. A Deductive Language for Querying Inconsistent Databases 36
    Sandra de Amo and Mônica Sakuray Pais

24. Connexive Modal Logic Based on Positive S4 38
    Norihiro Kamide and Heinrich Wansing

25. Basic Constructive Modality 41
    Valeria de Paiva and Eike Ritter

26. On the theory of Dynamic Sets 42
    Manuel Fidel and Martín Figallo

27. Truth-value gaps and the minimalist conception of truth 44
    Oswaldo Chateaubriand

28. On Skolem Functions in Proof Theory 45
    Jaakko Hintikka

Duplex Diagonalization[1]

RAYMOND M. SMULLYAN

Introduction

We introduce the notion of a *duplex diagonalization system*, all results of which simultaneously provide proofs of results in three areas—recursion theory, incompleteness theorems and decision machines (called universal machines in [Smullyan, 1994, p. 39].) In Part I, we define a type of calculating machine that I call a *decision machine* and state several facts about them.

I do not prove these facts but pose them as problems which I believe the readers should enjoy trying to solve on their own. However, the solutions are all special cases of the theorems of Part IV, whose proofs are given in full.

In Part II is presented a type of system called *R-Systems*, all results of which are generalizations of results of recursion theory and in fact indexed relational systems in general. Again, a series of problems is given all solutions of which are special cases of results proven in Part IV. In Part III we similarly treat some generalizations of incompleteness theorems of [Gödel, 1931] and [Rosser, 1947] as well as some related results.

All the systems of Parts I, II and III are synthesized in the concluding part– Part IV, in which we introduce the notion of a Duplex Diagonalization System and show that the three previous systems are indeed duplex diagonalization systems. We then state and prove eighteen theorems about duplex diagonalization systems in general, which simultaneously solves all the problems of the three previous systems.

Throughout this article I use the word "number" to mean positive integer and I consider a 1-1 function from $N \times N$ onto N, where N is the set of positive integers. This function is to be constant throughout this whole article, and it assigns to each ordered pair $\langle x, y \rangle$ of numbers a number denoted xy such that for every number n, there is one and only one x and one and only one y such that $n = xy$.

Also, I often use an abbreviation due to the late mathematician Paul Halmos, namely "iff" for "if and only if".

Part I How to Stump a Decision Machine

We consider a calculation machine M with infinitely many registers R_1, R_2, ..., R_n, ... We refer to each number n as the *index* of R_n. Each register is, so to speak, in charge of a property of numbers and its function is to try and determine which numbers have the property and which ones don't. To find out

---

[1]Dedicated to Walter Carnielli on the happy occasion of his sixtieth birthday. Keep up the good work, Walter!

whether a given number x has a certain property one goes to the register in charge of the property and feeds the number into that register. The machine then goes into operation and one and only one of three things happens:

(1) The machine eventually halts and flashes a signal (say a green light) signifying that it *does* have the property, in which case we say that the register *affirms* x.

(2) The machine eventually halts and flashes a different signal (say a red light) to indicate that the number doesn't have the property, in which case we say that the register *denies* x.

(3) The machine runs on forever and never halts. In that case we say that x *stumps* the register, or that the register is stumped by x.

We are given that the decision machine M satisfies the following two conditions:

M1: To each register R is assigned a register $R^\sharp$ called the *diagonalizer* of R such that for any number x, the register $R^\sharp$ affirms x iff R affirms xx, and $R^\sharp$ denies x iff R denies xx.

M2: To each register R is assigned a register R' called the *opposer* of R, which affirms those and only those numbers that R denies, and denies those and only those numbers that R affirms.

For any number n, we let n' be the index of the opposer of R_n. Thus $R_{n'}$ is the opposer of R_n.

We shall call a register U *universal* if for all numbers x and y, the register U affirms xy if and only if R_x affirms y.

PROBLEM 1.1 Prove that if U is a universal register then there is a number that stumps U.

PROBLEM 1.2 Prove that if one of the registers is universal, then some register is stumped by its own index—in other words there is a number a such that R_a is stumped by a.

We shall call two registers R and S *similar* if for every number x, R affirms (denies) x iff S respectively affirms (denies) x.

PROBLEM 1.3 Given a register R, the registers $R'^\sharp$ and $R^{\sharp\prime}$ are usually not the same, but are they necessarily similar?

We shall call a register V *contra-universal* if for all numbers x and y, V affirms xy iff R_x denies y.

PROBLEM 1.4 Prove that any contra-universal register can be stumped.

PROBLEM 1.5 Prove that if some register is contra-universal then some register is stumped by its own index.

PROBLEM 1.6 Prove that if some register is universal, then there is a register that is stumped by the index of its opposer—in other words that there is a number n such that R_n is stumped by n'.

We will call a register K *creative* if for every register R there is at least one number n such that K affirms n if and only if R affirms n.

PROBLEM 1.7 Prove that any creative register can be stumped.

PROBLEM 1.8 Prove that any universal register is creative and any contra-universal register is creative.

NOTE: Problems 1.7 and 1.8 jointly provide additional and simple ways of solving Problems 1.1 and 1.4

Fixed Points In what follows, we let A be the set of all numbers xy such that x affirms y, and we let B be the set of all numbers xy such that x denies y.

We shall call a number n a *fixed point* of a register R to mean that R affirms n iff $n \in A$, and R denies n iff $n \in B$.

PROBLEM 1.9 Prove that every register has a fixed point.

The use of fixed points provide neat and simple proofs of several results, as we will see.

PROBLEM 1.10 - Suppose that U is universal and V is contra-universal. Let R be any register.
  (a) Prove that if n is any fixed point of R then U affirms n iff R affirms n (which proves again that U is creative).
  (b) Prove that if n is any fixed point of R' then V affirms n iff R affirms n (which proves again that V is creative).

PROBLEM 1.11 Suppose that U is universal and V is contra-universal.
  (a) Prove that any fixed point of the opposer U' of U will stump U.
  (b) Prove that any fixed point of V will stump V.

NOTE: This proves alternate solutions to problems 1.1 and 1.4.

Call a register R a *Gödel* register if for every number x, R denies x iff R_x affirms x.

PROBLEM 1.12
  (a) Prove that any Gödel register can be stumped.
  (b) Prove that if some register is universal then some register is a Gödel register.

We shall call a register *s-special* if for all x, the register affirms x iff R_x denies x.

PROBLEM 1.13
  (a) Prove that any s-special register can be stumped.
  (b) Prove that if some register is contra-universal, then some register is an s-register.

We recall the set A of all numbers xy such that x affirms y, and B, the set of all numbers xy such that x denies y.

PROBLEM 1.14 Suppose that R is a register that affirms all numbers in A and no numbers in B. Prove that R can be stumped.

PROBLEM 1.15 Suppose that R is a register that affirms all numbers in A and denies all numbers in B. can R be stumped?

PROBLEM 1.16 Suppose there is a register K that affirms all numbers in B and denies all numbers in A.
   (a) Prove that K can be stumped.
   (b) Prove that there exists a register R that affirms all numbers x such that R_x denies x and denies all numbers x such that R_x affirms x. (Such a register might aptly be called a *Rosser* register, for reasons that will be manifest later.)
   (c) Prove that such a register R can be stumped (which is pretty obvious!).

PROBLEM 1.17 Let us say that a register *halts* at x if it either affirms or denies x.

Call a register a *stump detector* if it halts at all and only those numbers xy such that y stumps R_x. Is it possible that some register is a stump detector?

By the *affirmation set* of a register R we shall mean the set of all numbers n such that R affirms n. We shall call a set of numbers an *affirmation set* if it is the affirmation set of some register.

PROBLEM 1.18 Prove that the complement $\neg A$ of A is not an affirmation set.

Part II R-Systems

We consider a denumerable collection Σ of sets of numbers and an enumeration $A_1, A_2, \ldots, A_n, \ldots$ of them in some order and another enumeration $B_1, B_2, \ldots, B_n, \ldots$ in a different order such that for each number n, the set A_n is disjoint from B_n (contain no common element). we shall call these two sequences an *R-system* if the following two conditions hold.

Condition R1 Each number n is assigned a number $n^\sharp$ such that for every number x:
   (1) $x \in A_n^\sharp$ iff $xx \in A_n$
   (2) $x \in B_n^\sharp$ iff $xx \in B_n$

Condition R2 Associated with every number n is a number denoted n' such that $A_{n'} = B_n$ and $B_{n'} = A_n$.

Remarks For Σ the collection of all recursively enumerable sets, it is possible to arrange them in two sequences $A_1, \ldots, A_n, \ldots$ and $B_1, \ldots, B_n, \ldots$ such that Conditions *R1* and *R2* hold, for the funtion xy to be recursive (usually taken to be $J(x,y)$). Thus all results of this section are generalization of results of recursion theory.

We shall call an element U of Σ *universal* if for all numbers x and y, the numbers $xy \in U$ iff $y \in A_x$.

PROBLEM 2.1 Prove that if A_n is universal then some number lies outside both A_n and B_n.

PROBLEM 2.2 Prove that if some element of Σ is universal, then there is a number n such that n itself lies outside both A_n and B_n.

PROBLEM 2.3 Prove that for any number n, the sets $A_{n'}^\sharp$ and $A_n^{\sharp\prime}$ are the same, and the sets $B_{n'}^\sharp$ and $B_n^{\sharp\prime}$ are the same..

We shall call an element V of Σ *contra-universal* if for all numbers x and y, the number $xy \in A$ iff $y \in B_x$.

PROBLEM 2.4 Prove that if A_n is contra-universal then A_n is not the complement of B_n.

PROBLEM 2.5 Prove that if some element of Σ is contra-universal then there is some n such that $n \notin A_n \cup B_n$ (A_n union B_n).

PROBLEM 2.6 Prove that if some element of Σ is universal then there is a number n such that $n' \notin A_n \cup B_n$.

We shall call an element C of Σ *creative* if for every element M of Σ there is at least one number n such that $n \in C$ iff $n \in M$.

PROBLEM 2.7 Prove that if A_n is creative then A_n is not the complement of B_n.

We let A be the set of all numbers xy such that $y \in A_x$ and B be the set of all numbers xy such that $y \in B_x$. Obviously A is universal and B is contra-universal (but we are not given that A and B are members of Σ).

PROBLEM 2.8 Prove that the sets A and B are both creative.

Given an ordered pair $\langle A_k, B_k \rangle$ we shall call a number n a *fixed point* of the pair if $n \in A_k$ iff $n \in A$, and $n \in B_k$ iff $n \in B$.

PROBLEM 2.9 Prove that every pair $\langle A_k, B_k \rangle$ has a fixed point.

PROBLEM 2.10 Prove that for any number k:
(a) If n is a fixed point of $\langle A_k, B_k \rangle$ then $n \in A$ iff $n \in A_k$ (which, with Problem 2.9, proves again that A is creative).
(b) If n is a fixed point of $\langle B_k, A_k \rangle$ then $n \in A$ iff $n \in B_k$ (which with Problem 2.9 proves again that B is creative).

PROBLEM 2.11
(a) Prove that if A_k is universal then any fixed point of $\langle B_k, A_k \rangle$ lies outside both A_k and B_k.
(b) Prove that if A_k is contra-universal then any fixed point of $\langle A_k, B_k \rangle$ will lie outside both A_k and B_k.

PROBLEM 2.12 Prove that if $A \in \Sigma$ then there is a number g such that B_g is the set of all numbers x such that $x \in A_x$, and for such a number g, the set A_g is not the complement of B_g.

PROBLEM 2.13 Prove that if $B \in \Sigma$, then there is a number s such that A_s is the set of all numbers x such that $x \in A_x$ and for such a number s, the set A_s is not the complement of B_s.

PROBLEM 2.14 Prove that if A is a subset of A_n and B is disjoint from A_n, then A_n is not the complement of B_n. [This is a stronger result than Problem 2.1. Can you see why?]

PROBLEM 2.15 Prove that if A is a subset of A_n and B is a subset of B_n then A_n is not the complement of B_n.

PROBLEM 2.16 Suppose A is a subset of B_n and B is a subset of A_n. Prove:
(a) A_n is not the complement of B_n.
(b) There is a number r such that for all numbers x, if $x \in A_x$ then $x \in B_r$, and if $xinB_x$ then $x \in A_r$.
(c) For such a number r, the set A_r is not the complement of B_r.

PROBLEM 2.17 Call a number m a *magic number* if for all x and y, the number $xy \in A_m \cup B_m$ iff $y \notin A_x \cup B_x$. Can there be a magic number?

PROBLEM 2.18 Consider the set of all numbers xy such that $y \notin A_x$. Prove that this set cannot be an element of Σ.

DISCUSSION Theorem 2.18 generalizes a fundamental result in recursion theory—namely, that there is a recursively enumerable set whose complement is not recursively enumerable:

For Σ the collection of all recursively enumerable sets, the universal set A (as well as the set B) is indeed recursively enumerable (element of Σ), but by Theorem 2.18 its complement is not.

Another basic result of recursion theory is that there exists a disjoint pair M_1, M_2 of recursively enumerable sets that are *recursively inseparable*—i.e., for any disjoint pair S_1, S_2 of recursively enumerable sets such that M_1 is a subset of S_1 and M_2 is a subset of S_2, the set S_1 is not the complement of S_2.

Now, for Σ the collection of recursively enumerable sets, every disjoint pair $\langle S_1, S_2 \rangle$ of recursively enumerable sets is $\langle A_n, B_n \rangle$ for some n. Hence the pair $\langle A, B \rangle$ must be recursively inseparable, for suppose S_1 and S_2 are disjoint recursively enumerable sets such that A is a subset of S_1 and B is a subset of S_2. For some number n, $S_1 = A_n$ and $S_2 = B_n$. Thus A is a subset of A_n and B is a subset of B_n and so by Theorem 2.15, A_n is not the complement of B_n, hence S_1 is not the complement of S_2.

Part III - Generalized Incompleteness Theorems

We now turn to some generalizations of incompleteness theorems of Gödel and Rosser and some related results.

In the mathematical systems under consideration there is a denumerable sequence $H_1, H_2, \ldots, H_n, \ldots$ of expressions called *predicates*, and to each predicate H and each number n is assigned an expression denoted $H(n)$, called a *sentence*. [Informally we can think of H as the name of a property of numbers, and the sentence $H(n)$ as expressing the proposition that n has the property named by H.]

We assume that every sentence is $H_x(y)$ for some x and y. We arrange all the sentences in a sequence $S_1, S_2, \ldots, S_n, \ldots$ in such a manner that for all numbers x and y, the sentence S_{xy} is the sentence $H_x(y)$. We call n the *index* of S_n. Thus xy is the index of the sentence $H_x(y)$.

There is a well-defined subset P of the set of sentences whose members are called *provable* sentences, and a well-defined set R of sentences called *refutable* sentences. The system is called *consistent* if no sentence is both provable and refutable. We shall assume that the systems under consideration are consistent.

A sentence is called *decidable* if it is either provable or refutable, otherwise *undecidable*. The system is called *complete* if every sentence is decidable, and *incomplete* if some sentence is undecidable.

We will call two sentences X and Y *equivalent* if either one is provable, so is the other, and if either one is refutable, so is the other (and hence if either one is undecidable, so is the other). Thus to say that X and Y are equivalent is to say that they are either both provable, both refutable, or both undecidable.

The systems under consideration are required to have the following two features:

F1: To each predicate H is assigned a predicate $H^\sharp$ called the *diagonalizer* of H such that for every number n, the sentence $H^\sharp(n)$ is equivalent to $H(nn)$.

F2: To each predicate H is assigned a predicate H', called the *negator* of H such that for every number x, the sentence $H'(x)$ is provable iff $H(x)$ is refutable, and is refutable iff $H(x)$ is provable.

For any set W of sentences, we let W_0 be the set of its indices. Thus P_0 is the set of all numbers n such that S_n is provable, and R_0 is the set of all numbers n such that S_n is refutable.

We define a set A of numbers to be *representable* if there is a predicate H such that for all numbers n, the sentence H_n is provable iff $n \in A$, and such a predicate H will be said to *represent* A. Thus H represents the set of all numbers n such that $H(n)$ is provable.

PROBLEM 3.1 Prove that if H represents P_0 then there is a number n such that $H(n)$ is undecidable.

PROBLEM 3.2 Prove that if the set P_0 is representable then there is a number n such that the sentence $H_n(n)$ is undecidable.

PROBLEM 3.3 Let us call two predicates H and K *similar* if for all numbers n, the sentences $H(n)$ and $K(n)$ are equivalent.

Given a predicate H, the predicate $H^{\sharp'}$ and $H'^\sharp$ are not necessarily the same, but are they necessarily similar?

PROBLEM 3.4 Prove that if H represents R_0 then there is a number n such that $H(n)$ is undecidable.

PROBLEM 3.5 Prove that if R_0 is representable then there is a number n such that $H_n(n)$ is undecidable.

PROBLEM 3.6 Prove that if P_0 is representable then there is a number n such that $H_n(n')$ is undecidable.

Call a predicate K *creative* if for every predicate H there is at least one number n such that $K(n)$ is provable iff $H(n)$ is provable.

PROBLEM 3.7 Prove that if K is creative then there is a number n such that $K(n)$ is undecidable.

PROBLEM 3.8 Prove that any predicate that represents P_0 is creative, and any predicate that represents R_0 is creative.

A sentence S_n is called a *fixed point* of a predicate H if S_n is equivalent to $H(n)$.

PROBLEM 3.9 Prove that every predicate has a fixed point.

PROBLEM 3.10
(a) Prove that if H represents P_0 then for any predicate K, if n is a fixed point of K then $H(n)$ is provable iff $K(n)$ is provable. [This proves again that any predicate that represents P_0 is creative.]
(b) Prove that if H represents R_0 then for any predicate K, if n is a fixed point of K', then $H(n)$ is provable iff $K(n)$ is provable. [This proves again that any predicate that represents R_0 is creative.]

PROBLEM 3.11
(a) Prove that if H represents P_0 then for any fixed point n of H', the sentence $H(n)$ is undecidable.
(b) Prove that if H represents R_0, then for any fixed point n of H, the sentence $H(x)$ is undecidable.

For any set W of sentences, define W^* as the set of all numbers n such that $H_n(n) \in W$. Thus P^* is the set of all n such that $H_n(n)$ is provable, and R^* is the set of all n such that $H_n(n)$ is refutable.

We shall call a predicate H a *Gödel predicate* if H' represents the set P^*.

PROBLEM 3.12 [after Gödel]
(a) Prove that if H_a is a Gödel predicate then $H_a(a)$ is undecidable.
(b) Prove that if P_0 is representable then there is a Gödel predicate.

In [Smullyan, 1961, p. 45] I stated and proved what might aptly be called a "dual" form of Gödel's construction, by introducing a predicate that represents R^* instead of P^*. I will call such a a predicate an *s-predicate*.

PROBLEM 3.13
(a) Prove that if H is an s-predicate, then there is a number n such that $H(n)$ is undecidable.
(b) Prove that if R_0 is representable, then there is a an s-predicate.

DISCUSSION Using a Gödel predicate, Gödel constructed his famous sentence which, so to speak, asserts its own non-provability. It is true but not provable (and also not refutable) in the system. By contrast, using my predicate, the undecidable sentence I got is not one which says "I am not provable," but one which says "I am refutable." The sentecne is false but not refutable, hence its negation is true but not provable (and also not refutable). This sentence was later independently discovered by Jerislaw [Jerislaw, 1973] and is known as the *Jerislow sentence*. As shown by Jerislaw, it could accomplish certain interesting things that could not be accomplished with the Gödel sentence.

SEPARABILITY Consider two disjoint sets A and B of numbers. We shall say that a predicate H *weakly separates* A from B if $H(n)$ is provable for every number n in A, but not provable for any number n in B.

We shall say that H *strongly* separates A from B if $H(n)$ is provable for every number n in A, and refutable for every number n in B.

PROBLEM 3.15 Suppose that H strongly separates P_0 from R_0. Does it necessarily follow that there must be a number n such that $H(n)$ is undecidable?

J.B. Rosser [Rosser, 1947] constructed a predicate J that strongly separates R^* from P^*. I will call such a predicate a *Rosser* predicate.

PROBLEM 3.16 - Suppose that there is a predicate K that strongly separates R_0 from P_0.
(a) Prove that $K(n)$ is undecidable for some n.
(b) Prove that there is a Rosser predicate J.
(c) Prove that for such a predicate J, $J(n)$ is undecidable for some n.

DISCUSSION For Gödel to show that the negator G' of his predicate G represented P^*, he had to assume that the system in question had a stronger property than mere consistency, a property known as omega-consistency. Thus Gödel did not show outright that the systems were incomplete, but only that if they were omega-consistent then they were incomplete. By contrast, Rosser could show that his predicate J strongly separates R^* from P^* without having to assume omega-consistency.

PROBLEM 3.17 Prove that there is no predicate H such that for all numbers n, the sentence $H(n)$ is decidable iff S_n is undecidable.

PROBLEM 3.18 Prove that the complement $\neg P_0$ of the set P_0 is not representable.

Part IV - A Unification

We now unify the material of Parts I, II and III by means of a *duplex diagonalization system*, which we now define.

We consider two disjoint sets A and B of numbers. We shall call the ordered pair $\langle A, B \rangle$ a *duplex diagonalization system*—more briefly a *duplex system* if the following two conditions hold:

D1: Assigned to every number n is a number denoted $n^\sharp$ such that for all numbers x, the number $n^\sharp x \in A$ iff $xx \in A$, and $n^\sharp x \in B$ iff $xx \in B$.

D2: Assigned to every number n is a number denoted n' such that for every number x, the number $n'x \in A$ iff $n \in B$, and $n'x \in B$ iff $nx \in A$.

APPLICATIONS Duplex systems will be applied as follows:

1. For the decision machine of Part I, we take A to be the set of all numbers xy such that the register R_x affirms y, and B to be the set of all xy such that R_x denies y. For any number n we take $n^\sharp$ to be the index of the diagonalizer of the register R_n and n' to be the index of the opposer of R_n. By the given conditions *M1* and *M2*, the conditions *D1* and *D2* hold respectively, and so $\langle A, B \rangle$ is a duplex system. Thus all results we prove about duplex systems in general, hold for the particular duplex system of decision machines.

2. For the R systems of Part II, we take A to be the set of all numbers xy such that $y \in A_x$, and B to be the set of all xy such that $y \in B_x$. By the given conditions *R1* and *R2*, the pair $\langle A, B \rangle$ is a duplex system.

3. For the mathematical systems of Part III, we take A to be the set of all numbers xy such that $H_x(y)$ is provable, which in fact is the set P_0, and B to be the set of all numbers xy such that $H_x(y)$ is refutable, which is the set R_0. For any number n, we take $n^\sharp$ to be the index of the predicate $H_n^\sharp$—the diagonalizer of H_n, and we take n' to be the index of the negator H_n' of H_n. By the given conditions *F1* and *F2*, it follows that the pair $\langle P_0, R_0 \rangle$ is a duplex system.

We now state and prove eighteen theorems about duplex systems in general. For each n less than or equal to 18, Theorem n simultaneously solves Problem 1.n of Part I, Problem 2.n of Part II and Problem 3.n of Part III.

We shall call a number n *undecidable* if it is in neither A nor B. For Part III, this means that the sentence S_n is undecidable (n is in neither P_0 nor R_0). For Part I, a number xy is undecidable iff the register R_x is stumped by y. For Part II, xy is undecidable iff y is in neither A_x nor B_x.

We shall call a number w *universal* if for all numbers n, the number $wn \in A$ iff $n \in A$. For Part I, this means that the register R_n is universal. For Part II, this means that the set A_n is universal. For Part III, this means that the predicate H_n represents the set P_0.

We call a number v *contra-universal* if for all numbers n, the number $vn \in A$ iff $n \in B$. This means for Part I (Part II) that the register R_n (set A_n) is universal, and for Part III, the predicate H_n represents R_0.

THEOREM 1 If w is universal then there is a number n such that wn is undecidable.

Proof. - Suppose w is universal. To reduce clutter, let a be the number $w'^\sharp$ and n the number aa (which is $w'^\sharp w'^\sharp$). We will see that wn is undecidable.

For any number x, the number $w(xx) \in B$ iff $w(ax) \in A$, iff $w'^\sharp x \in A$, iff $ax \in A$ (since $a = w'^\sharp$), iff $w(ax) \in A$ (since w is universal). Thus for *all* numbers x, we see that $w(xx) \in B$ iff $w(ax) \in A$. We now take a for x, and thus $w(aa) \in B$ iff $w(aa) \in A$. Thus $wn \in B$ iff $wn \in A$, and hence wn is undecidable. [In general, any number x that is in A iff it is in B must be undecidable, because such an x is either outside both A and B, or inside both A and B, but the latter cannot be, since A is disjoint from B, hence x is outside both A and B]. ∎

THEOREM 2 *If one of the numbers is universal, then there is a number a such that aa is undecidable.*

Proof. Suppose w is universal. We take a to be $w'^{\sharp}$, as in the proof of Theorem 1.

For any number x, the number $ax \in B$ iff $w'^{\sharp}x \in B$, iff $w'(xx) \in B$ iff $w(xx) \in A$ iff $xx \in A$ (since w is universal). Thus for all x, we see that $ax \in B$ iff $xx \in A$. We take a for x, and so $aa \in B$ iff $aa \in A$, and thus aa is undecidable (since A is disjoint from B). ∎

REMARK The number $w^{\sharp\prime}w^{\sharp\prime}$ is also undecidable, as the reader can verify.

We will call two numbers *equivalent* provided if either one belongs to A, so does the other, and if either belongs to B, so does the other. We shall call two numbers m and n *similar* if mx is equivalent to nx for every x.

THEOREM 3 *For every number n, the numbers $n'^{\sharp}$ and $n^{\sharp\prime}$ are similar.*

Proof. $n'^{\sharp} \in A$ iff $n'(xx) \in A$ iff $n(xx) \in B$ iff $n^{\sharp}x \in B$ iff $n^{\sharp\prime}x \in A$.
Thus $n'^{\sharp}x \in A$ iff $n^{\sharp\prime}x \in A$. Similarly, $n'^{\sharp}x \in B$ iff $n^{\sharp\prime}x \in B$. ∎

THEOREM 4 *If v is contra-universal then for some number n, the number vn is undecidable.*

Proof. Take $v^{\sharp}v^{\sharp}$ for n. Then $v(v^{\sharp}v^{\sharp}) \in A$ iff $v^{\sharp}v^{\sharp} \in B$ iff $v(v^{\sharp}v^{\sharp}) \in B$ and so vn is undecidable. ∎

THEOREM 5 *If some number is contra-universal then there is a number n such that nn is undecidable.*

Proof. Suppose v is contra-universal. Now take n to be $v^{\sharp}$. Then $v^{\sharp}v^{\sharp} \in A$ iff $v(v^{\sharp}v^{\sharp}) \in A$ iff $v^{\sharp}v^{\sharp} \in B$. Thus $v^{\sharp}v^{\sharp} \in A$ iff $v^{\sharp}v^{\sharp} \in B$, and so $v^{\sharp}v^{\sharp}$ is undecidable. ∎

THEOREM 6 *If some number is universal then there is a number n such that nn' is undecidable.*

Proof. Suppose w is universal. We take n to be the number $w^{\sharp}$. Then $w^{\sharp}w^{\sharp\prime} \in A$ iff $w(w^{\sharp\prime}w^{\sharp\prime}) \in A$ iff $w^{\sharp\prime}w^{\sharp\prime} \in A$ (since w is universal), iff $w^{\sharp}w^{\sharp\prime} \in B$. Thus $nn' \in A$ iff $nn' \in B$, and so nn' is undecidable. ∎

Call a number k *creative* if for every number h there is a number n such that $kn \in A$ iff $hn \in A$.

[For Part I (Part II, Part III) the number k is creative iff the register R_k (set A_k, predicate H_k) is respectively creative.]

THEOREM 7 If k is creative, then there is a number n such that kn is undecidable.

Proof. Suppose k is creative. Then for every number h there is a number n such that $kn \in A$ iff $hn \in A$. Take k' for h and so there is a number n such that $kn \in A$ iff $k'n \in A$, but $k'n \in A$ iff $kn \in B$, hence $kn \in A$ iff $kn \in B$, and so kn is undecidable. ∎

THEOREM 8 Any universal number is creative and any contra-universal number is creative.

Proof.
(a) Suppose w is universal. Given a number h, we take n to be the number $h^\sharp h^\sharp$. Then $w(h^\sharp h^\sharp) \in A$ iff $h^\sharp h \notin A$ iff $h(h^\sharp h^\sharp) \in A$. Thus $wn \in A$ iff $hn \in A$.
(b) Suppose v is contra-universal. Given a number h, take n to be $h'^\sharp h'^\sharp$. Then $v(h'^\sharp h'^\sharp) \in A$ iff $h'^\sharp(h'^\sharp) \in B$, iff $h'(h'^\sharp h'^\sharp) \in B$ iff $h(h'^\sharp h'^\sharp) \in A$. Thus $vn \in A$ iff $hn \in A$. ∎

NOTE If k is universal or contra-universal then by Theorem 8 it is creative, hence by Theorem 7, there is a number n such that kn is undecidable, which again proves Theorems 1 and 4.

We shall call a number n a *fixed point* of a number b if n is equivalant to bn. [For Part I (Part II, Part III) this means that n is a fixed point of the register R_b (pair $\langle A_b, B_b \rangle$, predicate H_b) respectively.]

THEOREM 9 [Fixed Point Theorem] Every number b has a fixed point.

Proof. By condition D1, for every number x, the number $b^\sharp x$ is equivalent to $b(xx)$. We take $b^\sharp$ for x and we see that $b^\sharp b^\sharp$ is equivalent to $b(b^\sharp b^\sharp)$. Thus $b^\sharp b^\sharp$ is a fixed point of b. ∎

As we will see, the use of fixed points provide relatively swift proofs of several theorems and sometimes yields more information than is given in the statement of the theorems.

The following is a sharpening of Theorem 8.

THEOREM 10 Suppose w is universal and v is contra-universal. Then for any number h:
(a) If n is a fixed point of h then $wn \in A$ iff $hn \in A$ (proving again that w is creative).
(b) If n is a fixed point of h' then $vn \in A$ iff $hn \in A$ (proving again that v is creative).

Proof. Assume hypothesis
(a) Suppose n is a fixed point of h. Then $wn \in A$ iff $n \in A$ iff $hn \in A$ (since n is equivalent to hn). Thus $wn \in A$ iff $hn \in A$.
(b) Suppose n is a fixed point of h'. Then $vn \in A$ iff $n \in B$ iff $h'n \in B$ (since n is equivalent to $h'n$), iff $hn \in A$. Hence $vn \in A$ iff $hn \in A$. ∎

THEOREM 11 Suppose w is universal and v is contra-universal. Then
(1) Any fixed point n of w' is undecidable, and so is wn.
(2) Any fixed point n of v is undecidable, and so is vn.

Proof.
(1) Suppose n is a fixed point of w'. Then $n \in B$ iff $w'n \in B$ (since n is equivalent to $w'n$), iff $wn \in A$ iff $n \in A$. Thus $n \in B$ iff $n \in A$, and so n is undecidable. Since n is equivalent to $w'n$, then $w'n$ is also undecidable, and so of course wn is undecidable.

(2) Suppose n is a fixed point of v. Then $n \in A$ iff $vn \in A$ iff $n \in B$. Thus $n \in A$ iff $n \in B$, hence n is undecidable, and vn, being equivalent to n, is also undecidable. ∎

Call a number k *g-special* (after Gödel) if for all numbers x, the number $kx \in B$ iff $xx \in A$. [For Part I, k is g-special iff the register R_k is a Gödel register. For Part II, k is g-special iff it is a number denoted "g" in Problem 2.12. For Part III, k is g-special iff H_k is a Gödel predicate.]

THEOREM 12
(a) If k is g-special then kk is undecidable.
(b) If some number is universal then some number is g-special.

Proof.
(a) Suppose k is g-special. Since $kx \in B$ iff $xx \in A$ for every x, then for $x = k$, it follows that $kk \in B$ iff $kk \in A$, and so kk is undecidable.
(b) Suppose w is universal. Then $w'^\sharp x \in B$ iff $w'(xx) \in B$ iff $w(xx) \in A$, iff $xx \in A$. Thus $w'^\sharp x \in B$ iff $xx \in A$ holds for every x, and so $w'^\sharp$ is g-special. We note that since $w^{\sharp\prime}$ is similar to $w'^\sharp$ (Theorem 3) then $w^{\sharp\prime}$ is also g-special. ∎

REMARK It follows from (a) and (b) above that if w is universal, then if a is either of the numbers $w'^\sharp$ or $w^{\sharp\prime}$, the number aa is undecidable, which of course again proves Theorem 2.

A number k shall be called *s-special* if for all x, $kx \in A$ iff $xx \in B$. [For Part I, k is s-special iff the register R_k is s-special. For Part II, k is s-special iff k is a number denoted "s" in Theorem 2.13. For Part IV, k is s-special iff H_k represents R^*.]

THEOREM 13
(a) If k is s-special then kk is undecidable.
(b) If some number is contra-universal then some number is s-special.

Proof.
(a) Suppose k is s-special. Then for all x, the number $kx \in A$ iff $xx \in B$. We take k for x, and so $kk \in A$ iff $kk \in B$, hence kk is undecidable.
(b) Suppose v is contra-universal. Then for every number x, the number $v^\sharp x \in A$ iff $v(xx) \in A$ iff $xx \in B$. Hence $v^\sharp$ is s-special. ∎

NOTE From (a) and (b) it follows that if v is contra-universal, then $v^\# v^\#$ is undecidable, which again proves Theorem 5.

We now turn to some theorems that are closely related to Rosser's improvement of Gödel's theorem.

The next theorem is a strengthening of Theorem 1.

THEOREM 14 Suppose h is a number such that for all numbers x, the following two conditions hold:
 (1) If $x \in A$ then $hx \in A$.
 (2) If $x \in B$ then $hx \notin A$.
Then there is a number n such that hn is undecidable.

Proof. We could prove this from scratch, but it will be quicker to take advantage of the Fixed Point Theorem already proved.

Assume the hypothesis. Let n be a fixed point of h'. We will see that hn is undecidable.
 (1) $n \in A$ iff $hn \in B$ (since $n \in A$ iff $h'n \in A$ iff $hn \in B$). Similarly,
 (2) $n \in B$ iff $hn \in A$.
 (3) If $n \in A$ then $hn \in A$ (by hypothesis).
 (4) If $n \in B$ then $hn \notin A$ (by hypothesis). From (1) and (3) it follows that if $n \in A$ then hn is in both A and B, contrary to the fact that A is disjoint from B. Hence $n \notin A$.

From (2) and (4) it follows that if $n \in B$, then $hn \in A$ and $hn \notin A$, which is logically impossible. Therefore $n \notin B$. Since $n \notin A$ and $n \notin B$, then n is undecidable. Since $h'n$ is equivalent to n, then $h'n$ is also undecidable, and so hn is undecidable. ∎

NOTE The reason that the above theorem is stronger than Theorem 1 is that the hypothesis is weaker—that is, if h is a universal number, then h must satisfy conditions (1) and (2) of the hypothesis, because: (1) if $n \in A$ then of course $hn \in A$, since $h \in A$ iff $hn \in A$; (2) If $n \in B$, then $n \notin A$. (since A is disjoint from B), hence $hn \notin A$ (since $hn \in A$ iff $n \in A$) and thus if $n \in B$ then $hn \notin A$.

THEOREM 15 Suppose that for all x, the number h satisfied the following two conditions:
 (1) If $x \in A$ then $hx \in A$ (same as (1) of Th. 14)
 (2) If $x \in B$ then $hx \in B$.
Then for some n, the number hn is undecidable.

Proof. Immediately from Theorem 14, since (2) of the present hypothesis implies (2) of the hypothesis of Theorem 14, since B is disjoint from A. ∎

We will call a number k *r-special* (after Rosser) if for all numbers x,
 r1: If $xx \in B$ then $kx \in A$.
 r2: If $xx \in A$ then $kx \in B$.

THEOREM 16 Suppose there is a number h such that: for all numbers x:
(1) If $x \in B$ then $hx \in A$.
(2) If $x \in A$ then $hx \in B$.
Then;
(a) hn is undecidable for some n.
(b) There is an r-special number k.
(c) If k is r-special, then kn is undecidable for some n.

Proof.
(a) Let n be a fixed point of h. By (1) (taking n for x) if $n \in B$ then $hn \in A$, hence $n \in A$ (since hn is equivalent to n) Thus if $n \in A$, then $n \in B$. Similarly, using (2) instead of (1), if $n \in B$ then $n \in A$. Thus n is undecidable, and so is hn, which is equivalent to n.
(b) In (1) and (2) we take xx for x, and so
(1) If $xx \in B$ then $h(xx) \in A$, and hence $h^\sharp x \in A$.
(2) If $xx \in A$ then $h(xx) \in B$, and hence $h^\sharp x \in B$. Therefore $h^\sharp$ is r-special.
(c) Suppose k is r-special. In the defining conditions, we take k for x, and so by $r1$, if $kk \in B$ then $kk \in A$, and by $r2$, if $kk \in A$ then $kk \in B$. Thus kn is undecidable for n the number k. ∎

THEOREM 17 There exists no number k such that for all n, the number kn is decidable if and only if n is undecidable.

Proof. Take a fixed point n of k. Since kn is equivalent to n, then of course kn is decidable iff n is decidable, hence it cannot be that kn is decidable iff n is undecidable. ∎

We define a number set S to be *representable* if there is a number k such that S is the set of all numbers n such that $kn \in A$, and such a number k will be said to represent S. [For Part I, this means that S is the affirmation set of the register R_k. For Part II, this means that $S = A_k$. For Part III this means that the predicate H_k represents S.]

THEOREM 18 The complement $\neg A$ of A is not representable.

Proof. Given a number k, let n be a fixed point of k. then $kn \in A$ iff $n \in A$, hence it is not the case that $kn \in A$ iff $n \in \neg A$, and so k does not represent $\neg A$. ∎

Let us note that Theorem 1 follows almost from Theorem 18, because if w is universal, it represents A, and since w' does not represent $\neg A$ (no number does), then there must be at least one number n such that it is not the case that $w'n \in A$ iff $n \in \neg A$, hence $w'n \in A$ iff $n \in A$, but also $n \in A$ iff $wn \in A$ (since w represents A). Therefore $wn \in A$ iff $w'n \in A$ iff $wn \in B$, hence wn is undecidable.

[More generally, it should be noted that for any number k, if k represents a set whose complement is not representable, then kn is undecidable for some n.]

One might wonder why such a simple proof of Theorem 1 is possible, until one realizes that the proof of Theorem 18 uses the Fixed Point Theorem!

Acknowledgment

I wish to express my heartfelt thanks to Prof. Melvin Fitting for his invaluable advice in the preparation of this article.

BIBLIOGRAPHY

[Gödel, 1931] K. Gödel. Über formal unentscheidbare Sätze der Principia Mathematica und verwandter systeme, i. *Monatshefte für Mathematik und Physik*, 38(173-198), 1931.

[Jerislaw, 1973] R. K. Jerislaw. Redundancies in the Hilbert-Bernays derivability conditions for Gödel's second incompleteness theorem. *Journal of Symbolic Logic*, 38:359–367, 1973.

[Rosser, 1947] J. B. Rosser. Extensions of some theorems of Gödel and Church. *Journal of Symbolic Logic*, 2:129–137, 1947.

[Smullyan, 1961] R. M. Smullyan. *Theory of Formal Systems*. Princeton University Press, 1961.

[Smullyan, 1994] R. M. Smullyan. *Diagonalization and Self-Reference*. Oxford University Press, 1994.

Raymond M. Smullyan
Professor Emeritus of Philosophy
Indiana University
Bloomington, USA

Perfect Set Properties and large cardinals

CARLOS A. DI PRISCO

ABSTRACT. Regularity properties of sets of real numbers such as Lebesgue measurability, Baire property and the perfect set property are shared by all analytic sets. Using the axiom of choice it is possible to show that there are sets of real numbers without these properties. Results of Solovay and Shelah unveiled deep connections between the consistency of the axioms of Zermelo-Fraenkel together with the axiom of dependent choices and all sets of real numbers having these properties, and the consistency of the existence of inaccessible cardinals. Mathias's analysis of the Ramsey property in this context establishes further results and also leaves some open questions. We present a concise survey of results relating perfect sets properties and Ramsey properties with large cardinals.

1 Introduction

The main objective of this expositive article is to present some results that relate large cardinals with properties of sets of real numbers, with special emphasis on perfect set properties. Also, we present some open problems in this area of research.

Probably the result that brought up the interest on these connections between large cardinals and sets of real numbers is Solovay's construction of a model of set theory where all sets of real numbers are Lebesgue measurable [Solovay, 1970]. Solovay showed that in this model every set of real numbers has also the property of Baire, and the perfect set property. Mathias [Mathias, 1977] proved some years later that in this model every set of real numbers has the Ramsey property. Some of these properties of sets of reals will be defined below for the benefit of the reader.

Clearly, the axiom of choice does not hold in Solovay's model, since it permits to prove that there are non-measurable sets. Nevertheless, a weaker version, the axiom of dependent choices (DC) is true in the model. It is pertinent to note that DC is enough for most of the usual constructions in analysis that require some choice, and for this reason Solovay's result indicates that it is unlikely that a non-measurable set will appear in the normal practice of mathematicians.

To obtain his model, Solovay started assuming the existence of an inaccessible cardinal. Then proceeded to collapse the inaccessible to $\aleph_1$, the first uncountable cardinal, using Levy's forcing notion to obtain a generic extension where every cardinal below the inaccessible becomes countable. Solovay showed that in the generic extension , every "definable" set of reals is Lebesgue measurable, has the property of Baire and the perfect set property; and then defined an inner model of the generic extension in which all sets of reals have these nice properties.

Clearly, it is enough to assume the consistency of the existence of an inaccessible cardinal: starting from a model where there is an inaccessible, one gets the appropriate generic extension and the corresponding inner model. In his paper, Solovay asks if the assumption regarding inaccessible cardinals is really necessary. This problem remained open for some time until Shelah [Shelah, 1984] discovered that it is necessary to get a model where all sets of reals are Lebesgue measurable, but it is not necessary to obtain a model where all sets of reals have the property of Baire.

The necessity of an inaccessible to get a model where all sets of reals have the perfect set property had already been noticed by Specker years before [Specker, 1957].

In this note we will also consider some variations of the Ramsey property and their relation to large cardinals.

We will use conventional set theoretic notation. The set of natural numbers will be identified with ω, the first infinite ordinal. The letter aleph is used to denote de cardinals, $\aleph_0 (= \omega)$ is the first infinite cardinal, $\aleph_1$ is the first uncountable cardinal. If A, B are sets, A^B denotes de set of all functions from B to A.

2 Some properties of sets of real numbers.

We will define several properties of sets of real numbers. Sometimes, instead of working with the real numbers themselves, we will work with other closely related spaces, such as the Baire space ω^ω of all the infinite sequences of natural numbers, or the collection $[\omega]^\omega$ of all infinite sets of natural numbers. Each of these two collections, equipped with the corresponding product topology is homoeomorphic to $\mathbb{R} \setminus \mathbb{Q}$ the irrational numbers. For this reason, the elements of ω^ω or $[\omega]^\omega$ are also considered real numbers.

A subset P of the set of real numbers $\mathbb{R}$ is perfect if it is non-empty, closed and contains no isolated points. It can be easily shown that if P is perfect, then $|P| = 2^{\aleph_0}$. The same definition can be used to define perfect subsets of ω^ω or of $[\omega]^\omega$.

The theorem of Cantor and Bendixon states that every closed set $C \subseteq \mathbb{R}$ can be decomposed in the form $C = P \cup N$ where P is perfect (or empty) and N is at most countable. This is shown by a transfinite procedure starting with the set C and taking at each successor step the Cantor-Bendixon derivative which consists in removing the isolated points, and taking intersections at limit stages. By the separability of the space, only at most countable many points are removed at each application of the Cantor Bendixon derivative, and the process stops at a countable ordinal. If at the end of the process nothing remains of the original set, then that set is countable. Otherwise, what remains is a perfect set.

This result indicates that no closed set can be a counter example to the Continuum Hypothesis. The argument of Cantor-Bendixon can be generalized to analytic sets, which are images of Borel sets under continuous functions (and thus it is also true for Borel sets).

As was shown by Bernstein [Bernstein, 1908], using a well ordering of the real numbers one can produce a set which does not contain any perfect subset and has non-empty intersection with every perfect set. Such a set is usually called a totally imperfect set, or a Bernstein set. Notice that given a totally imperfect set, it or its complement must be uncountable, thus we have an uncountable set which does not contain a perfect subset.

A weaker consequence of the axiom of choice suffices to show that existence of an uncountable set with no perfect subset. We will write $\aleph_1 \leq 2^{\aleph_0}$ to express that there is an injective function from $\aleph_1$ into $2^{\aleph_0}$. Since (under the Axiom of Choice) the cardinality of $\mathbb{R}$ is $2^{\aleph_0}$, the hypothesis $\aleph_1 \leq 2^{\aleph_0}$ implies that there is an injection from $\aleph_1$ into the real numbers. Obviously, such an injection exists if the real numbers can be well ordered (which follows from the Axiom of Choice). If $\aleph_1 = 2^{\aleph_0}$ then we have a well ordering of the reals (in type $\aleph_1$), and as indicated above, this is enough to get a totally imperfect set, and thus an uncountable set without perfect subsets. On the other hand, if $\aleph_1 < 2^{\aleph_0}$ then the image A of an injection from $\aleph_1$ into $\mathbb{R}$ is an uncountable subset of $\mathbb{R}$ which contains no perfect subset since if P is perfect and contained in A, then $|A| = 2^{\aleph_0}$, contradicting that A has cardinality $\aleph_1$ and $\aleph_1 < 2^{\aleph_0}$.

It is customary to say that a set of real numbers has the perfect set property if it is countable or it contains a perfect subset.

Summarizing, if $\aleph_1 \leq 2^\omega$, there is an uncountable set of real numbers without perfect subsets. This contrasts with the observation from [Di Prisco and Galindo, 2010] that $\aleph_1 \leq 2^\omega$ does not imply the existence of a Bernstein set.

A question immediately arises: Is the statement "every uncountable subset of $\mathbb{R}$ contains a perfect subset" consistent with ZF? Or, perhaps, even in the absence of the axiom of choice it is possible to refute that statement. As we mentioned in the introduction, Solovay in [Solovay, 1970], assuming the consistency of the existence of an inaccessible cardinal, obtained a model of ZF where DC, a weak form of choice holds, and every set of real numbers is Lebesgue measurable, has the property of Baire, and if uncountable, contains a perfect subset.

The following section will be devoted to the perfect set property and Specker's result of 1957 about the perfect set property and inaccessibility. Then, in other sections we will define the Ramsey property and explore the necessity of large cardinals to prove the consistency of all sets of reals having the Ramsey property in each of various of its versions.

3 The perfect set property and inaccessible cardinals

First we recall the definition of inaccessible cardinal.

DEFINITION 1. We say that an infinite cardinal κ is regular if for every $\alpha < \kappa$ and every function $f : \alpha \to \kappa$, the image of f is bounded in κ.

An uncountable cardinal κ is inaccessible if it is regular and for every $\alpha < \kappa$, $2^\alpha < \kappa$.

Let L denote Gödel's universe of constructible sets. Remember that the continuum hypothesis holds in L, as well as the axiom of choice. If A is a set, then $L[A]$ denotes the class of all sets constructible from A; $L[A]$ satisfies ZFC.

If $a \in \omega^\omega$, then $L[a]$ satisfies the generalized continuum hypothesis.

We will follow closely the exposition of Chapter 11 of [Kanamori, 1994] for the rest of the section.

LEMMA 2. *If ω_1 is regular and for every real $a \in [\omega]^\omega$, $\omega_1^{L[a]} < \omega_1$, then for every $a \in \omega^\omega$, ω_1 is inaccessible in $L[a]$.*

Proof. Suppose, to get a contradiction, that there is a real $a \in \omega^\omega$ such that ω_1 is not inaccessible in $L[a]$. Then, since ω_1 is regular and $L[a]$ satisfies the generalized continuum hypothesis, ω_1 must be a successor cardinal in $L[a]$. Let thus $\omega_1 = \gamma^{+L[a]}$, for some cardinal γ of $L[a]$. Since γ is countable, there is a real $b \in \omega^\omega$ that codes a well ordering of ω of type γ. Now, if $c \in \omega^\omega$ codes both a and b, we get that $\omega_1^{L[c]} = \omega_1$ which contradicts our hypothesis. ∎

Putting together results of Solovay, Specker and Levy, we have the following.

THEOREM 3. *Each of the following theories are equiconsistent.*

(i) $ZF + DC +$ Every uncountable set of reals contains a perfect subset.

(ii) $ZF + \omega_1$ is regular $+ \aleph_1 \not\leq 2^{\aleph_0}$.

(iii) $ZF + \omega_1$ is regular $+ \forall a \in \omega^\omega (\aleph_1^{L[a]} < \aleph_1)$.

(iv) $ZFC + \exists \kappa (\kappa \text{ is inaccessible})$

Proof. As we saw in the previous section, if $\aleph_1 \leq 2^{\aleph_0}$, there is an uncountable set of reals without perfect subsets; and DC implies that $\aleph_1$ is a regular cardinal. Therefore, (i) implies (ii).

Given $a \in \omega^\omega$, $\mathcal{P}(\omega) \cap L[a]$ has a well ordering in $L[a]$ of order type $\omega_1^{L[a]}$. Therefore, if for some real $a \in \omega^\omega$ $\omega_1^{L[a]} = \omega_1$, we get that $\aleph_1 \leq 2^{\aleph_0}$. This shows that (ii) implies (iii).

That the consistency of (iii) implies the consistency of (iv), follows from Lemma 2. And finally, The consistency of (iv) implies the consistency of (i) is part of Solovay's result. ∎

4 The Ramsey property

The Ramsey property refers to subsets of the collection $[\omega]^\omega$ of infinite sets of natural numbers. If A is such an infinite set, then $[A]^\omega$ denotes the collection of all infinite subsets of A. If a and A are sets of natural numbers, a finite and A infinite, then
$$[a, A] = \{B \in [\omega]^\omega : a \sqsubset B \subseteq A\},$$
the collection of all infinite subsets B of A that have a as an initial segment.

A subset $\mathcal{X}$ of $[\omega]^\omega$ is **Ramsey** if for every $[a, A] \neq \emptyset$ with $A \in [\omega]^\omega$ there exists $B \in [a, A]$ such that $[a, B] \subseteq \mathcal{X}$ or $[a, B] \cap \mathcal{X} = \emptyset$. Sometimes the term "completely Ramsey" is used to denote this property, reserving the term "Ramsey" for the weaker property "there is an infinite set A such that $[A]^\omega \subseteq \mathcal{X}$ or $[A]^\omega \cap \mathcal{X} = \emptyset$".

Galvin and Prikry [Galvin and Prikry, 1973] showed that all Borel subsets of $[\omega]^\omega$ are Ramsey. Silver [Silver, 1970] extended this to all analytic sets.

Ellentuck [Ellentuck, 1974] characterized the Ramsey sets as those having the Baire property with respect to the exponential topology of $[\omega]^\omega$, i.e. the topology generated by the basic sets of the form $[a, A]$.

Mathias proved that in Solovay's model every set of real numbers is Ramsey, obtaining the following theorem.

THEOREM 4. *[Mathias, 1977] Con(ZF+ $\exists$ an inaccessible cardinal) implies Con(ZF+DC+ every set of real numbers is Ramsey).*

Whether the hypothesis if inaccessibility is necessary in this case remains an open problem (see [Kanamori, 1994] 11.16).

5 Selective and semiselective coideals

DEFINITION 5. A family $\mathcal{H} \subset \wp(\omega)$ is a **co-ideal** if it satisfies:

(i) $A \subseteq B$ and $A \in \mathcal{H}$ implies $B \in \mathcal{H}$, and

(ii) $A \cup B \in \mathcal{H}$ implies $A \in \mathcal{H}$ or $B \in \mathcal{H}$.

Given a decreasing sequence $A_0 \supseteq A_1 \supseteq A_2 \supseteq \cdots$ of infinite subsets of ω, a set B is a **diagonalization** of the sequence (or B **diagonalizes** the sequence) if and only if $B/n \subseteq A_n$ for each $n \in B$. A co-ideal $\mathcal{H}$ is **selective** if and only if every decreasing sequence of elements of $\mathcal{H}$ has a diagonalization in $\mathcal{H}$.

Mathias introduced the concept of a selective co-ideal (or a happy family), which has turned out to be of wide interest. He considered sets that are $\mathcal{H}$-Ramsey with respect to a selective co-ideal $\mathcal{H}$, and generalized Silver's result to this context.

DEFINITION 6. Let $\mathcal{H} \subseteq [\omega]^\omega$ be a co-ideal. $\mathcal{X} \subseteq [\omega]^\omega$ is $\mathcal{H}$-**Ramsey** if for every $[a, A] \neq \emptyset$ with $A \in \mathcal{H}$ there exists $B \in [a, A] \cap \mathcal{H}$ such that $[a, B] \subseteq \mathcal{X}$ or $[a, B] \cap \mathcal{X} = \emptyset$.

Mathias showed that if Solovay's model is obtained collapsing a Mahlo cardinal, then in this model every set $\mathcal{A} \subseteq [\omega]^\omega$ is $\mathcal{H}$-Ramsey for every selective coideal $\mathcal{H}$.

Recall that κ is Mahlo if it is inaccessible and the set of inaccessible cardinals below κ is stationary (in particular there are κ many inaccessible cardinals below κ)[1].

THEOREM 7. *[Mathias, 1977] Con(ZF+ $\exists$ a Mahlo cardinal) implies Con(ZF+DC+ for every selective coideal $\mathcal{H}$, every set of real numbers is $\mathcal{H}$-Ramsey).*

Eisworth [Eisworth, 1999] Showed that the hypothesis of the consistency of a Mahlo cardinal is necessary. In fact, if every set of reals is $\mathcal{U}$-Ramsey for every selective ultrafilter $\mathcal{U}$, then ω_1 is Mahlo in L.

---

[1] See [Kanamori, 1994] for the definition of stationary and other notions related to large cardinals

Given a co-ideal $\mathcal{H}$, a set $\mathcal{D} \subseteq \mathcal{H}$ is **dense open** in the ordering $(\mathcal{H}, \subseteq)$, if (a) for every $A \in \mathcal{H}$ there exists $B \in \mathcal{D}$ such that $B \subseteq A$ and, (b) for every $A, B \in \mathcal{H}$, if $B \subseteq A$ and $A \in \mathcal{D}$ then $B \in \mathcal{D}$.

Semiselective coideals were introduced in [Farah, 1997]

DEFINITION 8 (Farah). A coideal $\mathcal{H}$ is semi-selective if for every sequence $D_0, D_1, \ldots,$ of open dense subsets of $\mathcal{H}$ there is $B \in \mathcal{H}$ such that for every n, $B/n \in D_n$.

It is easy to verify that if a coideal $\mathcal{H}$ is selective then it is semi-selective.

To illustrate the interest of the concept of semiselectiveness, we recall that $\mathcal{A}$ is Ramsey if and only of it has the property of Baire with respect to Ellentuck's topology. We say that $\mathcal{A}$ has the abstract Baire property with respect to a coideal $\mathcal{H}$ if for every $[a, A] \neq \emptyset$ with $A \in \mathcal{H}$ there exists $[b, B] \subseteq [a, A]$, with $B \in \mathcal{H}$, such that $[b, B] \subseteq \mathcal{X}$ or $[b, B] \cap \mathcal{X} = \emptyset$. Clearly if $\mathcal{A}$ is Ramsey, it has the abstract Baire property.

If $\mathcal{H}$ is selective then $\mathcal{A}$ is $\mathcal{H}$-Ramsey if and only if it has the abstract property of Baire with respect to $\mathcal{H}$.

Semiselective ideals capture exactly this property, in the sense that a coideal $\mathcal{H}$ is semi-selective if and only if the following statement holds:

$\mathcal{A}$ is $\mathcal{H}$-Ramsey if and only if $\mathcal{A}$ has the abstract Baire property with respect to $\mathcal{H}$.

In [Matet, 1993] Matet gave a characterization of the property of being Ramsey with respect a selective coideal in terms of games. These games coincide with those of Kastanas [Kastanas, 1983] when the coideal is $[\omega]^\omega$, and with the games of Louveau [Louveau, 1976] when the coideal is in fact a Ramsey ultrafilter. A similar characterization of the property of being $\mathcal{H}$-Ramsey when $\mathcal{H}$ is semiselective coideal was given in [Di Prisco et al., to appear], and moreover, it is shown there that a coideal $\mathcal{H}$ is semiselective if and only if the sets that are $\mathcal{H}$-Ramsey are exactly those for which the corresponding game is determined.

Before stating the next result, we give the definition of a stronger large cardinal notion. A cardinal κ is $\mathbf{\Pi}_1^1$-indescribable if for every $\mathbf{\Pi}_1^1$ formula ϕ with n unary relational symbols, and $U_1, U_2, \ldots, U_n \subseteq V_\kappa$, if

$$\langle V_\kappa, \in, U_1, \ldots, U_n \rangle \vDash \phi,$$

then there is $\beta < \kappa$ such that

$$\langle V_\beta, \in, U_1 \cap V_\beta, \ldots, U_n \cap V_\beta \rangle \vDash \phi.$$

PROPOSITION 9. *If κ is $\mathbf{\Pi}_1^1$-indescribable, then κ is Mahlo and the set of Mahlo cardinals below κ is stationary (in particular, there are κ many Mahlo cardinals below κ).*

In Solovay's model obtained collapsing a $\mathbf{\Pi}_1^1$-indescribable cardinal, every set of reals is $\mathcal{H}$-Ramsey for every semi-selective coideal $\mathcal{H}$.

THEOREM 10. *[Di Prisco et al., to appear]* $Con(ZF+\exists \ \Pi_1^1\text{-}indescribable)$ implies
$Con(ZF+DC+\text{for every semi-selective coideal } \mathcal{H}, \text{ every } \mathcal{A} \text{ is } \mathcal{H}\text{-}Ramsey)$.

It is not known if the hypothesis about Π_1^1-indescribable cardinals is necessary.
The concept of semiselective coideal was introduced by Farah in [Farah, 1997] to answer the following question posed by Todorcevic: what are the combinatotrial properties a family in the ground model of infinite sets of natural numbers must have so that in any perfect-set forcing extension any Borel map $f : [\omega]^\omega \to \{0,1\}$ is constant on a set of the form $[A]^\omega$ with $A \in \mathcal{H}$?

Under stronger large cardinal hypothesis it can be shown that all definable sets of real numbers have the regularity properties mentioned above.

Shelah and Woodin [Shelah and Woodin, 1990] showed that if there is a supercompact cardinal all sets of reals in $L(\mathbb{R})$ are Lebesgue measurable, have the Baire property, have the perfect set property, and are Ramsey. Todorcevic extended this showing that every set of reals in $L(\mathbb{R})$ are $\mathcal{U}$-Ramsey for any selective ultrafilter $\mathcal{U}$. Farah proves in his article mentioned before that if there is a supercompact cardinal and $\mathcal{H}$ is a semiselective coideal then all sets of real numbers in $L(\mathbb{R})$ a $\mathcal{H}$-Ramsey.

BIBLIOGRAPHY

[Bernstein, 1908] F. Bernstein, *Zur Theorie der Trigonometrischen Reihe*. Berichte über die Verhandlungen der Königlich Sächsischen gesellschaft der Wissenschaften zu Leipzig Mathematische-Physiche Klasse. 60 (1908) 325-338.

[Di Prisco, 2005] C. A. Di Prisco, *Mathematics versus metamathematics in Ramsey theory of the real numbers*. In *Logic, Methology and Phylosophy of Science. Proccedings of the Twelfth International Congress*. Petr Hajek, Luis Valdes-Villanueva, Dag Westerstalhl. Eds. Kings College Publications. London. 2005.

[Di Prisco and Galindo, 2010] C. A. Di Prisco and F. Galindo, *Perfect Set Properties in Models of ZF*. Fundamenta Mathematicae 208 (2010) 249-262.

[Di Prisco et al., to appear] C. A. Di Prisco, J.G. Mijares and C. E. Uzcátegui, *Ideal games and Ramsey sets*. Proc. Amer. Math. Soc. (to appear).

[Eisworth, 1999] Eisworth, T., *Selective ultrafilters and* $\omega \to (\omega)^\omega$, Proc. Amer. Math. Soc. 127 (1999) 3067-3071.

[Elentuck, 1974] Elentuck, E., *A new proof that analytic sets are Ramsey*, J. Symbolic Logic, 39 (1974), 163-165.

[Farah, 1997] Farah, I., *Semiselective co-ideals*. Mathematika, 45 (1997), 79-103.

[Galvin and Prikry, 1973] Galvin, F. and K. Prikry, *Borel sets and Ramsey's theorem*, J. Symbolic Logic, 38 (1973), 193-198.

[Kanamori, 1994] Kanamori, A., *The higher infinite*. Springer-Verlag, 1994.

[Kastanas, 1983] Kastanas, I., *On the Ramsey Property for Sets of Reals*. J. Symbolic Logic, 48 (1983), 1035-1045.

[Louveau, 1976] Louveau, A., *Une méthode topologique pour l'étude de la propriété de Ramsey*. Israel J. Math. 23 (1976) 97-116.

[Matet, 1993] Matet, P. *Happy Families and Completely Ramsey Sets*. Archive for Mathematical Logic, 38 (1993), 151-171.

[Mathias, 1977] A.R.D. Mathias, *Happy families*. Annals of Mathematical Logic 12 (1977) 59-111.

[Shelah, 1984] S. Shelah, *can you take Solovay's inaccessible away?* Israel Journal of Mathematics 48 (1984) 1-47.

[Shelah and Woodin, 1990] S. Shelah and W.H. Woodin, *Large cardinals imply that that eery reasonably definable set of reals is Lebesgue measurable* Israel Journal of Mathematics 70 (1990) 381-394.

[Silver, 1970] Silver, J., *Every analytic set is Ramsey*, J. Symbolic Logic, 35 (1970) 60-64.

[Solovay, 1970] R. Solovay, *A model of set theory in which every set of reals is Lebesgue measurable*. Annals of Mathematics 92 (1970) 1-56.

[Specker, 1957] E. Specker, *Zur Axiomatik der Mengenlehre (Fundierungs und Auswahlaxiome)*. Zeitschrift fur Mathemastische Logik 3 (1957) 173-210.

Carlos Augusto Di Prisco
Departamento de Matemáticas
Instituto Venezolano de Investigaciones Científicas
Caracas, Venezuela
E-mail: cdiprisc@ivic.gob.ve

Why is the $P=?NP$ question so difficult?[1]

NEWTON C. A. DA COSTA AND FRANCISCO A. DORIA

ABSTRACT. We argue in the present paper that the $P=?NP$ question is difficult because one should perhaps approach it as a metamathematical question, and not as a question that might be settled with the tools of ordinary, informal mathematics. This paper is essentially an elaboration on the fact that the concept of total recursive function is meaningful in intuitive mathematics, but presents difficulties when considered within some formal systems. We suggest an approach to deal with $P=?NP$ which is based on the differences between Turing machine theory as seen with respect to an intuitive framework and within the bounds of a formal system as ZFC or PA.

1 Introduction

> The $P=?NP$ question looks so innocent.
> Why is it so difficult?

(Overheard during a talk on computational complexity.)

The recent failed attempt by V. Deolalikar at a proof of $P < NP$[1] raises the question of its difficulty: why is such an innocent–looking question so difficult? We argue here that it looks so because it has to be dealt with from the metamathematical viewpoint, not as if it were a problem of ordinary mathematics or computer science. However this definitively isn't the majority viewpoint in the mathematical community. The prevailing ideas about metamathematics and the $P=?NP$ question have been recently summarized by Feferman [Feferman, 1999]:

> Finally we must take note of the fact that up to now no previously open problem from number theory or finite combinatorics, such as the Goldbach conjecture, or the Riemann Hypothesis, or the twin prime conjecture, or the $P=?NP$ problem is known to be independent of the kinds of formal systems we have been talking about, not even of Peano Arithmetic [...] I think it is more likely, as it has been demonstrated in the case of Fermat's Last Theorem, that the truth of these will essentially be settled — if at all —

---

[1] Partially supported by CNPq, Philosophy Section.
[1] See the Wikipedia entry for "Vinay Deolalikar."

by ordinary mathematical reasoning without any passage through metamathematics [...]

With those more than sobering words as a necessary caveat, we now propose to explore the out–of–the–mainstream possibility discarded by Feferman in the preceding comment: that there actually is an important metamathematical side to the $P =?NP$ question, and that it bears on its solution.

Just for starters: see below Example 5, which presents a simple undecidable sentence with respect to Zermelo–Fraenkel set theory (ZFC) that directly matters to a crucial concept used in the formulation of the $P =?NP$ question: that of a time–polynomial (poly) Turing machine.

On the concept of "difficult problem"

The $P =?NP$ conjecture has a long history of leading astray both professional researchers and laymen who think that they can tackle it. But before we consider possible reasons for that, we must say a word about "difficult problems." Why is a problem seen as difficult?

The first characterization for "difficult problem" stems from Gödel's length of proofs theorem (see references and related results in [da Costa and Doria, 1991]). In a nutshell, a theorem which has a short statement may have an arbitrarily long proof, that is to say, the length of a proof in PA (Peano Arithmetic) or ZFC (Zermelo–Fraenkel set theory, plus the axiom of choice) isn't a recursive function of the length of the theorem's statement). Now if we add adequate axioms, the proof may be considerably shortened.

It can be shown that there are infinitely many sentences in ZFC that have short formalizations and arbitrarily long proofs, in all usual domains of mathematics [da Costa and Doria, 1991].

Beyond that we have undecidability and incompleteness.

We may take this characterization as a working definition for "difficult problem." If so, our question becomes: can we settle $P =?NP$ within PA? Can we do it in ZFC? That means, can we use the axioms of PA to prove, say, $P < NP$. Or do we require a stronger sytem, like ZFC, to do it? Or even have to go beyond ZFC?

2 The $P =?NP$ question

Assuredly the $P =?NP$ problem arises out of concrete situations (traveling salesman problem, allocation problems). We know that problems in the NP–class are sketchily characterized as follows:

- They have solutions that are "hard" to find by the known methods in the general case.

- Such solutions are always "easy" to check once one has correctly guessed or discovered them.

(We have added the quotation marks to "easy" and "hard" because their meaning hasn't been made precise yet.)

Such is their motto: easy to check a solution, hard to find them.

As it is well–known, "easy" stands for *time–polynomial in the length of the binary input*, that is, if a binary word x of length $|x|$ is input to algorithm M_m, this algorithm will be *time–polynomial in the length of the input*, or just a *poly algorithm* if the operation time $t_m(x)$ of M_m over any such x satisfies: $t_m(x) \leq |x|^a + b$, a, b nonzero positive integers. Then "hard" is everything that isn't easy; in the present case, at least a time–exponential operation time in the binary input's length.

Granted the above considerations, we can formulate our question in an informal vein: we know that there are infinitely many "hard" ways to find a solution for, say, the traveling salesman problem; but — *is there an easy way to obtain a solution?* Such is the $P =?NP$ question, in a first approximation.

Thus $P =?NP$ seems at first to be a very concrete, commonplace kind of question. We have procedures that settle the traveling salesman problem after lenghty exponential–size calculations [Machtey and Young, 1979]; can we speed them up? Can we shorten those procedures?

One would expect a yes or no answer here. After all, common wisdom tells us that ordinary, concrete problems are expected to have ordinary, concrete answers, or at least nearly so. Yet we know that common wisdom in general only enters the picture to misguide our efforts. This is even truer of ordinary problems that are made into mathematical questions.

Another informal characterization for $P =?NP$

Due to its origin in problems of everyday life, this class of problems has several loose, very informal characterizations as a kind of first or immediate approach to it. Another viewpoint is: the $P =?NP$ question has to do with the power of nondeterministic computation. Can some nondeterministic computation [Machtey and Young, 1979] be simulated without extra resources by a corresponding deterministic computation? More specifically, given a nondeterministic, time–polynomial, program that (ideally) settles some arbitrary question known to be in the so–called NP class of problems, can we find a deterministic, time–polynomial program that also solves the same question?

An example

Let's be more specific: consider the satisfiability question for Boolean expressions in conjunctive normal form (cnf), which is usually taken as the model problem in the area [Machtey and Young, 1979].

Recall that Boolean expression x is *satisfiable* if there is a line of truth–values that makes x true. We in general restrict our interest to cnf Boolean expressions and the set of all adequately satisfiable binarily–coded cnf expressions is noted SAT. SAT is a primitive recursive subset of the set of all Boolean cnf expressions.

A nondeterministic computer may try several alternatives out of a previously prescribed, finite, set at each computational step. Therefore, given one such cnf Boolean expression, the nondeterministic program might scan it and proceed by simultaneously testing truth values "true" and "false" for each new propositional variable. When the scanning of the expression is concluded, some branch or branches in the nondeterministic computation will also have produced a sequence (or several sequences) of truth values that satisfy our original Boolean expression. Time spent will be (roughly) a linear function of the scanning time required to go through the expression we started from.

Then the question arises: can we simulate that time–polynomial, nondeterministic procedure, by some *deterministic* algorithm, which is again time–polynomial in the input's length? This is of course the same as asking for a polynomial algorithm that simulates the known nonpolynomial ones for a problem in the NP class.

To add some extra detail to the preceding picture, suppose that a satisfiable Boolean expression x in conjunctive normal form has p propositional variables. It will be satisfied by a binary line s of truth–values of length p. An obvious algorithm which looks for an s that satisfies x would list all such s and try them in the Boolean expression x. That would make at worst 2^p trials.

Now for each value of p, we can easily check that the shortest Boolean expression in cnf with p variables has a length that is roughly proportional to p under some reasonable binary coding. Therefore the operating time of such an algorithm would blow up in an exponential way.

(Here the nondeterministic algorithm enters the picture again. Think of a nondeterministic machine that operates as follows: for x a Boolean cnf expression as its input, at each node in the computation the procedure bifurcates and simultaneously tests "true" in one branch, and "false" in another; therefore computation of a satisfying line of truth values by a nondeterministic algorithm along some branches will be approximately linear on the length of x.)

Informal vs. formal mathematics

Before we proceed, we must ask a preliminary question: do we really need to formalize in a rigorous way the above concepts to deal with the $P =?NP$ question? Can't we just try to settle $P =?NP$ within informal, intuitive mathematics, even if one has to deal with situations like the one described in Example 5 ?

We must give more precision to our concepts:

- *Informal mathematics* is to be understood as the domain of naïve set theory and everything that proceeds out of it through the usual mathematical techniques.

- *Formal mathematics* means Zermelo–Fraenkel set theory (with the Axiom of Choice, ZFC), or some extensions of it; or perhaps Peano Arithmetic (PA). Context will make explicit the axiom system we are dealing with.

Which is the main difference between both approaches? In our opinion, and in reference to the $P =?NP$ question, a crucial one: the concept of *total recursive function* is meaningful in intuitive mathematics but presents difficulties when considered within a given axiomatic system like PA or ZFC or similar theories based on the classical predicate calculus wih enough arithmetic and with a recursively enumerable set of theorems.

This paper is essentially an elaboration on this theme, and contains the suggestion of an approach to deal with $P =?NP$ which is based on the differences between Turing machine theory as seen with respect to an intuitive framework and within the bounds of a formal system as ZFC or PA.

Informal vs. formal approaches to $P =?NP$

The choice of an informal vs. formal approach essentially depends on our intuition about the problem's nature.

If we really think that this is just a concrete (albeit very difficult) problem, that will *per force* have some concrete, straightforward answer, say, some kind of miracle poly algorithm that settles all instances of the satisfiability problem, then any extra rigor will surely be a burden. For if that's our starting viewpoint, there will just be two possible answers to our query: either one will exhibit a poly algorithm that settles all instances of the satisfiability problem, or will find some no–go argument that proves that such a program cannot exist. In that case one usually proceeds by a detailed examination of a particular problem in the NP–class. This is the underlying approach to the well–known Razborov partial results (see e.g. [Razborov and Rudich, 1997]) on the "hardness" of particular cases of problems in the NP–class, and is certainly the approach favored by most active researchers in the field.

But suppose that the reader feels that there might be some extra subtlety involved in our apparently pedestrian question. For example, even if one believes that there might actually exist a poly algorithm with the desired characteristics, one might feel that the correct approach to focus on it would be to split the problem into two steps:

- First, prove an existential result: there is an algorithm with the desired properties.

  This portion of the presumed proof can be nonconstructive, and no hint on how to obtain the desired algoritm may eventually be found at this step.

- Then, we must show how to obtain or recognize the desired poly program among all algorithms.

(If $P = NP$ holds then we can recursively do that in the present case; see on that Remark 6 and comments close to it.)

Let's elaborate on that: this isn't an unusual situation as it happens e.g. with Shannon's main coding theorem in information theory [Bartholo et al., 2011; Shannon, 1948]. Shannon's proof is nonconstructive; it doesn't lead to the construction of any kind of coding. However many efficient codings are available in the literature. The same is true of the very simple, very elegant Nash proof of the existence of a solution for the game of Hex [Milnor, 2002]: the Nash proof of the advantage of the first player is nonconstructive; it gives no information on strategies, and yet there are several known winning strategies.

The matter may sometimes require an extra machinery if, say, after a nonconstructive existential proof one still cannot find an instance of the desired poly algorithm; that is, if the desired algorithm whose existence had been already proved still eludes us after many efforts to construct it. Then one might wonder whether some deep, elusive phenomenon of metamathematical character might be at work here, such as for instance ω–inconsistency.

However this shouldn't be the case with the $P =?NP$ question, as the truth of formal sentence $[P = NP]$ implies that it can proved in a reasonable axiomatic theory; see Remark 6.

What does that mean for the actual $P =?NP$ question?

Remarks far from the mainstream

Let's go back a while: which preliminary evidence do we have that makes one think that the $P =?NP$ question *cannot* be settled with the help of informal mathematics?

Let's sketch a possible setting for the metamathematical approach. The following are the standard definitions for $P < NP$:

Recall that SAT is the set (adequately coded through binary words) of all satisfiable Boolean expressions in cnf [Machtey and Young, 1979]. SAT $\subset \omega$ is a primitive recursive subset of the set ω of all positive integers. Thus we can code SAT by ω through a primitive recursive coding, which is supposed here.

REMARK 1. We can informally state $P = NP$ as:

> There is a time–polynomial Turing machine M_m that correctly "guesses" a satisfying line of truth–values for every input $x \in$ SAT. □

More precisely:

REMARK 2. STANDARD FORMALIZATION, INTUITIVE:

> There is a Turing machine M_m of Gödel number m, and there are positive integers a, b so that for every $x \in$ SAT, the output $\mathsf{M}_m(x)$ is a satisfying line for x, and the number of cycles of M_m over x, $t_m(x) \leq |x|^a + b$. □

Rigorous presentation of the standard formalization

Within ZFC:

DEFINITION 3. (STANDARD FORMALIZATION FOR $P = NP$.)

$$[P = NP] \leftrightarrow_{\text{Def}} \exists m, a, b \in \omega \, \forall x \in \omega \, [(t_m(x) \leq |x|^a + b) \wedge R(x, m)].$$

($R(x, y)$ is a polynomial predicate; it subsumes a kind of "verifying machine" that checks whether or not x is satisfied by the output of Turing machine m.) □

DEFINITION 4. $[P < NP] \leftrightarrow_{\text{Def}} \neg [P = NP]$. □

$[P < NP]$ is a Π_2 sentence.

Consider $P = NP$ and its negation $P < NP$ the way done in Definitions 3 and 4, of this paper (those are the standard formalizations). Of course we may argue that there may exist different formalizations for the intuitive concepts expressed in the sentences $P = NP$ and $P < NP$ as we present them here, but we use the one which we now describe and which is sort of consensual.

We see that the formal sentence that we choose to translate the assertion $P < NP$ has the following characteristics: it is,

- A purely arithmetical sentence (that is, it "fits" easily within Peano Arithmetic and even within weaker theories).

- It is at most a Π_2 arithmetic sentence, that is, of a reasonably low level within the arithmetical hierarchy.

So, it has a simple and clearcut formalization. It also has a well–known place within a large body of research on the foundations, beginning with the landmark 1939 paper by Turing on what he called ordinal logics [Feferman, 1960; Feferman, 1962; Turing, 1939].

The main questions about such sentences are:

1. For a first–order classical theory T that includes arithmetic and whose theorems form a recursively–enumerable set, given a Π_2 sentence ϕ in the language of T, does $T \vdash \phi$?

2. Given T, if T doesn't prove a Π_2 sentence ϕ, then how strong must be a theory T' that includes T in some prescribed way, in order that $T' \vdash \phi$?

The second question was dealt with in the Turing and Feferman papers. The first question was given an answer (for recursive extensions of arithmetic) in two papers by Kreisel [Kreisel, 1951-52]. In a nutshell the Kreisel results (proved for a theory T which is a version of formalized arithmetic) are: if a given total recursive function, defined out of the predicate in the Π_2 sentence ϕ, grows fast enough, then ϕ cannot be proved from the axioms of T; the function related to the predicate fits within a well–studied hierarchy of functions [Schwichtenberg, 1989]. Kreisel's result directly depends on a Gentzen–like consistency proof [Schwichtenberg, 1989].

An earlier, more general result appears in the 1936 paper by Kleene [Kleene, 1936] on general recursive functions.

Can we make precise the concept of "infinite set of poly machines" within ZFC ?

In order to prove that $P < NP$ one somehow has to list all poly machines and prove that, for each such machine, the right answer to an instance of a NP–complete problem fails infinitely many times.

But — if we can show that *some infinite sets of poly machines depend on the underlying axiomatic framework?* As we will now see, ZFC has (in a certain sense) more subsets of the set of all poly machines than does PA. That's an apparent paradox that stems from an undecidable sentence which can be formulated within PA, ZFC, and recursive extensions therein [da Costa and F. A. Doria, 2007].

We describe here that undecidable sentence with respect to the ZFC:

EXAMPLE 5. Consider the following picture for a poly machine (it's the so–caled BGS set) [Baker et al., 1975]: it is a coupled system $\langle \mathsf{M}_m, \mathsf{C}_p \rangle$, where M_m is a Turing machine of Gödel number m, and C_p is a clock. The coupled system operates as follows: after some binary word x is input to M_m, the clock C_p observes its operation, and interrupts it after $p(a, b, x) = |x|^a + b$ cycles. If $\mathsf{M}_m(x)$ has already stopped, the output of the coupled system is precisely

$M_m(x)$; if the operation of M_m over x has to be interrupted, we agree that the coupled system outputs 0 and stops.

Now consider the set $\{\langle M_m, C_p\rangle\}$, where $p(F, a, b, x) = |x|^{F(a)} + F(b)$, all a, b. Each couple in this set is still a poly machine, if F is total in the framework where we operate. However for an adequate naïvely total recursive function F, if F is naïvely total, as for example in Kleene's 1936 undecidable sentence [da Costa and F. A. Doria, 2007; Kleene, 1936]:

> We can neither prove nor disprove in consistent ZFC the formal counterpart to the following sentence: "for all a, b, each couple $\langle M_m, C_p\rangle$ is a poly machine."

For particular numerals in the place of a and b we can of course determine that the corresponding machine is a poly machine, but we cannot prove or disprove in ZFC that the *whole* set $\{\ldots \langle M_m, C_p\rangle \ldots\}$ is a set of poly machines, as we cannot prove in ZFC that F converges for all values of its argument. This imposes some haziness on the definition of the set of all poly machines, and makes it still harder for the conventional approach, where one has to prove *for all* poly machines, that each one will fail somewhere when dealing with a problem in the NP–class, when trying to argue that $[P < NP]$ holds. □

Goals of the paper

We argue here that the metamathematical viewpoint as sketched above may be one of the possible effective ways to envision $P =?NP$, and that this lies at the root of its difficulty: one simply isn't looking in the right direction.

3 Paris–Harrington–like results

We now briefly recall a specific example of a Π_2 sentence with a nontrivial metamathematical behavior, namely the Paris–Harrington result [Paris and Harrington, 1977]. That result exhibits an arithmetic sentence which can be formulated as a Π_2 sentence which cannot be proved within Peano arithmetic. Still, the more recent results by H. M. Friedman [Friedman, 1998; Smoryński, 1985; Smoryński, 1985], together with similar results by Dougherty and Jech [Dougherty and Jech, 1997] show that analogous sentences are implied by the introduction of large cardinals to a strong system like Zermelo–Fraenkel set theory.

These examples (Paris–Harrington, Friedman) have the following characteristics:

- They can be formalized as Π_2 sentences.

- None of these sentences can be proved from the axioms of Peano Arithmetic (PA); in the case of Friedman's examples, they cannot be proved in much stronger theories [Smoryński, 1985; Smoryński, 1985], which go up towards Zermelo–Fraenkel set theory [Friedman, 1998].

- Their unprovability is related to fast-growing recursive functions that cannot be proved (or disproved) to be total in Peano arithmetic (for the Paris–Harrington result) or for some stronger theory (for the Friedman examples).

More precisely: if ϕ is unprovable within theory T, usually a large piece of ZFC, then for an adequate fast growing total recursive function F_T, $T \vdash \phi \leftrightarrow [\mathsf{F}_T \text{ is total}]$.

F_T is obtained out of a diagonalization procedure over all T–provably recursive functions.

- For the Friedman [Friedman, 1998] (and Dougherty–Jech [Dougherty and Jech, 1997]) examples: if we add a large cardinal to the theory, the corresponding sentences become provable.

The Paris–Harrington unprovability result arises from a simple modification of the finite Ramsey theorem: one requires that some sets be "large," in a given particular sense. This is enough to make the modified sentence unprovable in Peano arithmetic, supposed consistent. The Friedman examples arise out of a theorem of Kruskal on the embedding of trees: again a simple additional requirement makes something provable into an unprovable sentence, this time in theories well beyond Peano Arithmetic.

So, there is a possibility that a simple, intuitive Π_2 sentence may have a highly nontrivial metamathematical behavior.

What happens if we have an independence result?

We may now quote a preliminary result which is part of the folklore in the area:

REMARK 6. Suppose that $[P < NP]$ holds. Let $G(m,x)$ be a primitive recursive predicate that intuitively means: "poly machine coded by m correctly guesses a satisfying line for x." (See the definition of $P < NP$ above.)

Then $[P < NP]$ is formalized as the Π_2 sentence $\forall m \, \exists x \, \neg G(m,x)$. For:

$$\mathsf{f}_{\neg G}(m) = \mu_x [\neg G(m,x)]$$

($\mathsf{f}_{\neg G}$ is the so-called counterexample function to $P = NP$) we have that:

$$[\mathsf{f}_{\neg G} \text{ is total}] \leftrightarrow [P < NP].$$

Now given a total recursive function h so that for every m, $\mathsf{h}(m) > \mathsf{f}_{\neg G}(m)$, we have that $[P < NP]$ becomes a Π_1 sentence, $\forall m \, \exists x \leq \mathsf{h}(m) \, \neg QG(m,x)$. Thus, if $[P < NP]$ is independent of ZFC, then it holds of the standard model for arithmetic.

Thus, if $P < NP$ is proved in some theory T, then it is provably Π_1 in that theory.

Conversely, if $[P = NP]$ is true of the standard model then it is provable in ZFC. For given each code m for a poly machine, compute $\mathsf{h}(m)$ and test in that machine all inputs $x \leq \mathsf{h}(m)$. If $G(m,x)$ holds for all those x, then m is the "miracle" poly algorithm that settles all of SAT. $\square$

Then, if $[P = NP]$ is independent, it will only hold true of a nonstandard model for arithmetic.

To stress the point:

- If $[P = NP]$ is true, then we can prove it in ZFC (actually in a much weaker system; it is enough to extend PA with a true Π_1 sentence).

Moreover we can recursively find the "miracle algorithm" that settles all of SAT in polynomial time.

- However if $[P = NP]$ and $[P < NP]$ are independent of ZFC, then $[P < NP]$ is true of the standard model for arithmetic.

That is, it holds in the real world of concrete computers.

4 The counterexample function as a fast growing function

Let's now formalize a few of those ideas.

DEFINITION 7. For $f, g : \omega \to \omega$,

$$f \text{ dominates } g \leftrightarrow_{\text{Def}} \exists y \, \forall x \, (x > y \to f(x) \geq g(x)).$$

We write $f \succ g$ for f dominates g. □

Quasi–trivial machines

Recall that the operation time of a Turing machine is given as follows: if M stops over an input x, then the operation time over x,

$$t_M = |x| + \text{number of cycles of the machine until it stops.}$$

EXAMPLE 8.

- **First trivial machine.** Note it O. O inputs x and stops.

$$t_O = |x| + \text{moves to halting state} + \text{stops.}$$

So, operation time of O has a linear bound.

- **Second trivial machine.** Call it O$'$. It inputs x, always outputs 0 (zero) and stops.

Again operation time of O$'$ has a linear bound.

- **Quasi–trivial machines.** A *quasi–trivial machine* Q operates as follows: for $x \leq x_0$, x_0 a constant value, Q = R, R an arbitrary total machine. For $x > x_0$, Q = O or O$'$.

This machine has also a linear bound. □

REMARK 9. Now let H be any fast–growing, superexponential total machine. Let H$'$ be another such machine. Form the following family of quasi–trivial Turing machines with subroutines H and H$'$:

1. If $x \leq \text{H}(n)$, $\mathsf{Q}^{\text{H},\text{H}',n}(x) = \text{H}'(x)$;
2. If $x > \text{H}(n)$, $\mathsf{Q}^{\text{H},\text{H}',n}(x) = 0$. □

PROPOSITION 10. *There is a family* $\mathsf{R}_{g(n,|\text{H}|)}(x) = \mathsf{Q}^{\text{H},\text{H}',n}(x)$, *where* g *is primitive recursive, and* $|\text{H}|$ *denotes the Gödel number of* H.

Proof. By the composition theorem and the s–m–n theorem. ∎

We first give a result for the counterexample function when defined over all Turing machines (with the extra condition that the counterexample function $= 0$ if M_m isn't a poly machine). We have:

PROPOSITION 11. *If $N(n) = \mathsf{g}(n)$ is the Gödel number of a quasi–trivial machine as in Remark 9, then $f(N(n)) = k(n) + 1 = \mathsf{H}(n) + 1$.*

Proof. Use the machines in Proposition 10. ∎

Proof of non–domination

Our goal here is to prove the following result:

PROPOSITION 12. *For no total recursive function h does $\mathsf{h} \succ f$.*

Proof. Suppose that there is a total recursive function h such that $\mathsf{h} \succ f$.

REMARK 13. Given such a function h, obtain another total recursive function h' which satisfies:

1. h' is strictly increasing.

2. For $n > n_0$, $\mathsf{h}'(n) > \mathsf{h}(\mathsf{g}(n))$.

∎

LEMMA 14. *Given a total recursive h, there is a total recursive h' that satisfies the conditions in Remark 13.*

Proof. Given h, obtain out of that total recursive function by the usual constructions a strictly increasing total recursive h^*. Then if, for instance, F_ω is Ackermann's function, $\mathsf{h}' = \mathsf{h}^* \circ \mathsf{F}_\omega$ will do. (The idea is that F_ω dominates all primitive recursive functions, and therefore h^* composed with it dominates $\mathsf{g}(n)$.) ∎

We have that the Gödel numbers of the quasi–trivial machines Q are given by $\mathsf{g}(n)$. Choose adequate quasi–trivial machines, so that $f(\mathsf{g}(n)) = \mathsf{h}'(n)+1$, from Proposition 11. From Remark 13 and Lemma 14 we conclude our argument. If we make explicit the computations, for $\mathsf{g}(n)$ (as the argument holds for any strictly increasing primitive recursive g):

$$f(\mathsf{g}(n)) = \mathsf{h}'(n) + 1 = \mathsf{h}^*(\mathsf{F}_\omega(n)) + 1,$$

and

$$\mathsf{h}^*(\mathsf{F}_\omega(n)) > \mathsf{h}^*(\mathsf{g}(n)).$$

For $N = \mathsf{g}(n)$,

$$f(N) > \mathsf{h}^*(N) \geq \mathsf{h}(N), \text{ all } N.$$

Therefore no such h can dominate f. □

COROLLARY 15. *No total recursive function dominates f.* □

Now the counterexample function f is nonrecursive: can we obtain a similar fast–growing result for its recursive avatars?

5 The main conjecture: if $P < NP$ is true then it cannot be proved by reasonable axiomatic systems

(Contents of this section summarize the discussion in [Chaitin et al., 2011].) We must stress at this point what we understand by "reasonable axiomatic system." It is a system like PA (Peano Arithmetic, axiomatized arithmetic with finite induction) or ZFC (Zermelo–Fraenkel set theory, strengthened with the axiom of choice). These systems share the following characteristics:

- If consistent, we may suppose that they have a model with standard (usual) arithmetic.

- They have a set of theorems which may be generated by a Turing machine (we say that the theorems are *recursively enumerable*).

These reasonable theories, so to say, have the following already mentioned property:

> Given a reasonable theory S there is a total recursive function F_S which cannot be proved total by the axioms of S.

We can construct F_S in such a way that it dominates all provably recursive total functions in S. So F_S grows faster than any S–provably total recursive function. Now the conjecture:

> No consistent reasonable theory S proves that the counterexample function f is total.

Could we then argue that it would then prove F_S total recursive, as a segment of f? If so, no such theory S would then prove $P < NP$ (modulo our hand waving; for details and loopholes see the references).

Construction of F_S is easy as we now review it: we enumerate with the help of a computer program all theorems of S, pick up those that say f_i is total, for some i that codes our enumeration, and diagonalize over those functions, that is, we define:

$$\mathsf{F}(i) = \mathsf{f}_i(i) + 1,$$

where by construction F is different from any f_i. There are other constructions where one immediately sees the fast–growing property of that function.

But do we really have independence?

We now use a trick that first appeared in a famous 1975 paper by Baker, Gill and Solovay. Pick up an arbitrary Turing machine and plug to it a polynomial clock that counts the steps in the computation and shuts down the machine the moment it is about to exceed the bound computed by the clock. That arrangement is also a Turing machine, one that we can ensure is a poly machine, from its construction. The set of all those machines is the BGS set; it contains representatives of every conceivable poly machine (that is, it contains machines that emulate all known poly machines).

So far the arguments exhibited point towards the strong possibility that a theory like S, which subsumes PA and ZFC, does not prove $P < NP$. To prove independence we require another hypothesis:

If S proves that $P = NP$, then the set of BGS poly machines that decide every instance in a NP-problem is recursive in theory S.

(Recall that a set is recursive if there is a computer program that allows us to decide whether some element x is in that set or not; the BGS set is a convenient way to represent all poly machines, as it is a recursive set — however it doesn't include all poly machines, only sort of representatives for each collection of machines that perform the same calculations; see the next section.) The argument exhibited by Costa–Doria in support of that hypothesis is simple, but again incomplete: one turns on the machine that lists the theorems of S and separates all theorems of the form "integer n codes an algorithm for a machine in the BGS set that solves all instances of a NP-problem." One then should conclude that the listing is exhaustive.

Then, given that hypothesis, we can concoct a version f' of the counterexample function which again grows faster than any total recursive function, and the conclusion follows: S cannot prove $P = NP$ too.

More precisely: if $P = NP$ is true, then it can be proved so by a reasonable theory, as we have seen. Then the counterexample function will be undefined at the values of n that code algorithms that settle, for such machine will never fail. The conjecture asserts that the set of undefined values of f, given that $P = NP$ holds, is recursive. We can therefore easily fill in the holes and obtain a fast–growing function f' that grows faster than any total recursive function etc.

And we get the independence of $P = NP$ and of $P < NP$ from strong, reasonable, axiomatic systems at the end of our discussion.

Still more on the counterexample function

$P < NP$ is given by a Π_2 arithmetic sentence, that is, a sentence of the form "for every x there is an y so that $P(x, y)$," where $P(x, y)$ is a very simple kind of relation.[2] Now given a theory S with enough arithmetic in it, S proves a Π_2 sentence ξ if and only if the associated Skolem function f_ξ is proved to be total recursive by S. For $P < NP$, this function is what we have been calling the counterexample function.

However there are infinitely many counterexample functions we may consider, an *embarras de choix*, as they say in French. Why is it so? For many adequate, reasonable theories S, we can build a recursive (computable) *scale of functions*[3] $F_0, F_1, \ldots, F_k, \ldots$ with an infinite set of S–provably total recursive functions so that F_0 is dominated by F_1 which is then dominated by $F_2, \ldots$, and so on.

Given each function F_k, we can form a BGS–like set BGS^k, where clocks in the time–polynomial Turing machines are bounded by a polynomial:

$$|x|^{F_k(n)} + F_k(n),$$

where $|x|$ denotes the length of the binary input x to the machine. We can

---

[2] It is a primitive recursive predicate.
[3] We thank L. Gordeev for this construction.

then consider the recursive set:

of all such sets.

Each BGS^k contains representatives of all poly machines (time polynomial Turing machines). Now, what happens if:

- There is a function g which is total provably recursive in S and which dominates all segments f_k of counterexample functions over each BGS^k?

- There is no such an g, but there are functions g_k which dominate each particular f_k, while the sequence $g_0, g_1, \ldots$ is unbounded in S, that is, grows as the sequence $F_0, F_1, \ldots$ in S?

In the first case, S proves $P < NP$, and we are done. However the second case leads to an interesting nontrivial situation which is so far unexplored. We believe we can use here techniques similar to those that we applied in the study of the full counterexample function.[4]

6 Concluding remarks

We notice two main obstacles to the perception of metamathematical aspects in the $P =?NP$ question. (We might even consider them to be actual conceptual blocks.) We may call them the concreteness prejudice, already mentioned in the Introduction, and the doubts that might exist about the so–called "correct," "standard" formal statement of the problem.

The concreteness prejudice

The concreteness prejudice can be stated as: if a problem arises out of concrete situations, then it should have an intuitive, concrete solution. The $P =?NP$ question certainly arises out of some very concrete, everyday situations, like the traveling salesman problem or allocation problems. So, it should be expected to have a concrete, intuitive solution.

Another version of the same conceptual block is: if a problem has a simple arithmetic formulation, and deals with finite objects, then it should be decidable within arithmetic. However we know that this certainly isn't true, as the discussion in Section 3 has clearly indicated.

Formalizing $P = NP$

The "standard" (or, one might say, "orthodox") formulation for $P = NP$ is the one in Remark 2 and in Definition 3. However we strongly believe that it is too restrictive and misleading; our exotic formulations give a clearer idea of the concepts and difficulties involved.

Recall the definition for a poly machine, all of them. They allow us to say that a polynomial Turing machine is a Turing machine whose operation time is bounded by *some* polynomial on the input's length. Emphasis is on "some."

We now add more detail to our viewpoints.

---

[4]See F. A. Doria, *Chaos, Computers, Games and Time*, E–PAPERS/PEP/COPPE (2011).

Total recursive functions and axiomatic theories

First, it doesn't matter whether a growing sequence of polynomial bounds has very close or widely separated bounds; it doesn't matter whether two such bounds encompass just a few or many poly machines.

To phrase it differently: as "the size of gaps doesn't matter," we may code the exponents in the polynomial bounds by any computable, total, fast–growing function. The usual axiomatic theories can be roughly seen as convenient artifacts that allow us to organize our knowledge about a given domain, but which are severely restricted in many respects; in particular these theories cannot fully "see" which recursive functions are naïvely total or not.

In other words, there is in fact some incompatibility between axiomatic theories as usual (PA, ZFC, for instance) and naïve computation theory — and naïve computation theory could be characterized as the kind of computation theory that immediately applies to the world of Turing machines. (We recall a remark at the beginning of Rogers' textbook where the author makes explicit that the domain of his discourse is intuitive mathematics [Rogers, 1967].)

Then, why should we restrict ourselves, say, to total recursive functions within Primitive Recursive Arithmetic (PRA) just because bounds depending on Ackermann's Function don't fit within PRA, even if the Ackermann Function is a perfectly reasonable, interesting, mathematical function? Or why can't we use F_{ϵ_0} in our bounds, just because it can't be proved total in Peano Arithmetic? Or even F, in the case of ZFC? After all, in order to define $[P = NP]^f$ we are only interested in polynomial bounds, together with some convenient, reasonable way of describing them. We see that there is no "natural" axiomatic theory to place them inside.

This fact, that "total recursive function" is a concept that has only meaning with respect to some axiomatic system as the ones above, has as we have already seen, important consequences for the matter we discuss here.

ω–incompleteness, Σ_1–unsoundness

Second, and as a complement to the preceding discussion, when we talk about "*some* bound," we will be dealing with an existential quantifier when we move into an axiomatic system. And due to that quantifier we may eventually stumble upon counterintuitive properties like ω–incompleteness and Σ_1–unsoundness. We must now make room for this "some" in our theory, that is, we cannot use a definition that would restrict the possibilities implicit in those two properties, Σ_1–unsoundness and ω–incompleteness. We didn't elaborate on this aspect of the question in the present paper; we just want to note it here. And this is certainly allowed by the use of functions like F in the bounds for poly Turing machines.

Conclusion

We think that the results stated in this paper make a good case for the metamathematical viewpoint, when dealing with the $P =?NP$ question. We wouldn't even be surprised if some kind of large cardinal turned out to be essential for the proof of $P < NP$ within ZFC, as in the already quoted Dougherty–Jech and Friedman examples.

7 Acknowledgments

M. Guillaume helped the authors with many detailed criticisms and suggestions which were essential for the present work. The construction of the scale of functions used here is due to L. Gordeev.

This paper was supported in part by CNPq, Philosophy Section. It is part of the research efforts of the Advanced Studies Group, Production Engineering Program, at COPPE–UFRJ and of the Logic Group, HCTE–UFRJ. We thank Profs. R. Bartholo, C. A. Cosenza, S. Fuks, S. Jurkiewicz, R. Kubrusly, L. Ventura and F. Zamberlan for support.

BIBLIOGRAPHY

[Baker et al., 1975] T. Baker, J. Gill, R. Solovay, "Relativizations of the $P =?NP$ question," *SIAM Journal of Computing* **4**, 431 (1975).

[Bartholo et al., 2011] R. Bartholo, C. A. Cosenza, F. A. Doria, M. Doria, *Allocation Problems, Economics, Fuzzy Sets, Information*, GAE/PEP–COPPE, UFRJ (2011).

[Chaitin et al., 2011] G. J. Chaitin, N. C. A. da Costa, F. A. Doria, *Gödel's Way*, CRC Press/Taylor & Francis (2011).

[da Costa and Doria, 1991] N. C. A. da Costa and F. A. Doria, "Undecidability and incompleteness in Classical Mechanics," *International J. of Theor. Physics* **30**, 1041 (1991).

[da Costa and F. A. Doria, 1994] N. C. A. da Costa and F. A. Doria, "Gödel incompleteness in analysis, with an application to the forecasting problem in the social sciences," *Philosophia Naturalis* **31**, 1 (1994).

[da Costa and F. A. Doria, 2003] N. C. A. da Costa and F. A. Doria, "Consequences of an exotic definition for $P = NP$," *Applied Mathematics and Computation* **145**, 655–665 (2003).

[da Costa and F. A. Doria, 2002] N. C. A. da Costa and F. A. Doria, "Informal and formal mathematics," preprint 07–RGC–IEA (2002).

[da Costa and F. A. Doria, 2002] N. C. A. da Costa and F. A. Doria, "Barebones," working notes privately circulated, IEA–USP (2002).

[da Costa and F. A. Doria, 2007] N. C. A. da Costa, F. A. Doria, and E. Bir, "On the metamathematics of the P vs. NP question," *Applied Mathematics and Computation* **189**, 1223–1240 (2007).

[da Costa et al., 2000/2005] N. C. A. da Costa, F. A. Doria and M. Guillaume, e–mail exchanges (2000/2005).

[Doria, 2011] F. A. Doria, *Chaos, Computers, Games and Time*, GE/PEP==COPPE, UFRJ (2011).

[Dougherty and Jech, 1997] R. Dougherty and T. Jech, "Left–distributive embedding algebras," *ERA Amer. Math. Soc.* **3**, 28 (1997).

[Feferman, 1960] S. Feferman, "Arithmetization of metamathematics in a general setting," *Fund. Math.* **49**, 35 (1960).

[Feferman, 1962] S. Feferman, "Transfinite recursive progressions of axiomatic theories," *J. Symbolic Logic* **27**, 259 (1962).

[Feferman, 1999] S. Feferman, "Does mathematics need new axioms?" *Amer. Math. Monthly* **106**, 99 (1999).

[Fortnow, 2009] L. Fortnow, "The status of the P versus NP problem," *Communications of the ACM* **52**, 78 (2009).

[Friedman, 1998] H. M. Friedman, "Finite functions and the necessary use of large cardinals," *Ann. Math.* **148**, 803 (1998).

[Kleene, 1936] S. C. Kleene, "General recursive functions of natural numbers," *Math. Annalen* **112**, 727 (1936).

[Kleene, 1967] S. C. Kleene, *Mathematical Logic*, Wiley (1967).

[Kreisel, 1951-52] G. Kreisel, "On the interpretation of non–finitist proofs," I and II, *J. Symbol. Logic* **16**, 241 (1951) and **17**, 43 (1952).

[Machtey and Young, 1979] M. Machtey and P. Young, *An Introduction to the General Theory of Algorithms*, North–Holland (1979).

[Milnor, 2002] J. Milnor, "The game of Hex," in H. W. Kuhn and S. Nasar, *The Essential John Nash*, Princeton U. P. (2002).

[Paris and Harrington, 1977] J. Paris and L. Harrington, "A mathematical incompleteness in Peano arithmetic," in J. Barwise, ed., *Handbook of Mathematical Logic*, North–Holland (1977).
[Razborov and Rudich, 1997] A. Razborov and S. Rudich, "Natural proofs," *J. Computer and System Science* **55**, 24 (1997).
[Rogers, 1967] H. Rogers Jr., *Theory of Recursive Functions and Effective Computability*, McGraw–Hill (1967).
[Schwichtenberg, 1989] H. Schwichtenberg, "Proof theory: some applications of cut–elimination," in J. Barwise, ed., *Handbook of Mathematical Logic*, North–Holland (1989).
[Shannon, 1948] C. E. Shannon, "A mathematical theory of communication," I and II, *Bell System Tech. J.* **27**, 379 and 623 (1948).
[,] C. Smoryński, "The incompleteness theorems," in J. Barwise, ed., *Handbook of Mathematical Logic*, North–Holland (1989).
[Smoryński, 1985] C. Smoryński, " 'Big' news from Archimedes to Friedman," in L. A. Harrington, ed., *Harvey Friedman's Research on the Foundations of Mathematics*, North–Holland (1985).
[Smoryński, 1985] C. Smoryński, " Some rapidly growing functions," in L. A. Harrington, ed., *Harvey Friedman's Research on the Foundations of Mathematics*, North–Holland (1985).
[Turing, 1939] A. M. Turing, "Systems of logic based on ordinals," Proc. London Math. Society, Ser. 2 **45**, 161 (1939).

Newton C. A. da Costa and Francisco A. Doria
Institute for Advanced Studies
University of São Paulo
São Paulo, Brazil
E-mail: ncacosta@terra.com.br, fadoria@gmail.com

Reflections on the Church-Turing Thesis
FREDERICK KROON

ABSTRACT. The Church-Turing Thesis is regarded by most philosophers, logicians, and mathematics as clearly correct, although Epstein and Carnielli point out in their classic text Computability that debates about the nature of mathematics have the potential to affect one's view of the thesis. But this consensus disappears when computer scientists and physicists in particular enter the picture. The present paper revisits the Church-Turing Thesis in the light of ongoing debate about whether work in physics and other sciences provides grounds for thinking that the thesis is in fact false, a question that Epstein and Carnielli don't directly consider in their text.

1 Introduction

I have known Walter since the time of the first ARF conference (September, 1999) when we were both inducted as founding members. ARF, short for *Advanced Reasoning Forum*, is a loosely circumscribed international group of philosophers and logicians who study reasoning and argument and the way these notions are articulated and applied in logic and philosophy, especially, but by no means solely, using the thematization provided in the work of Richard ('Arf') Epstein. The latter – a prominent recursion theorist in the 1970s and, later, a prominent and influential author of texts in formal and mathematical logic as well as in critical thinking – is the main-stay of ARF. He has held the group together for the last 15 years, generously supporting the intellectual enterprises of its members through conferences, occasional publications, and a web-site, while continuing to promote his own ambitious views of logic, philosophy and critical thinking through papers and monographs. I am glad he too is contributing to this volume in honor of Walter on his 60th birthday.

Epstein and Walter have been the resident logicians on ARF, while I have been one of the resident philosophers, albeit a philosopher with a teaching and research interest in logic. Indeed this teaching and research interest makes for a connection to Walter that long preceded the founding of ARF. In the 1970s, in the course of giving advice about some material I was teaching on subrecursive hierarchies, Epstein provided me with some marvelous notes on computability theory that were later revised and expanded with the help of another young logician, Walter Carnielli. The notes were subsequently published as the Epstein / Carnielli text *Computability*, first published in 1989 and now in its third edition.

My contribution to the present volume in honor of Walter concerns an old issue. One of the fundamental *philosophical* issues in computability theory is the status of the Church-Turing thesis – the thesis that a function is effectively computable if and only if it is lambda-definable or (equivalently) Turing-computable. In their text, Epstein and Carnielli are unusually even-handed, giving arguments for and against the thesis (pp. 231ff), and then providing a sympathetic account of this disagreement that traces it to different ways of understanding the nature of mathematics (p. 237). The lack of dogmatism is surprising as well as refreshing, because in my experience logicians and philosophers rarely display uncertainty about the Church-Turing thesis: most are avid believers. But this consensus in favor of the thesis disappears when computer scientists and physicists in particular enter the picture. My plan in the present paper is to revisit the Church-Turing thesis in the light of ongoing debate about whether work in physics and other sciences provides grounds for thinking that the thesis is in fact false. This is a debate that Epstein and Carnielli more or less bypass in their text, although it is one to which Carnielli has made his own contribution (in some joint work with Agudelo). What I have to say will be partly historical and expository, partly evaluative, and partly speculative.

2 The origins of the Church-Turing Thesis

It is a cliché, although no doubt correct, to say that we live in a time when few of the old verities retain their old hold on us. It should come as no surprise that the Church-Turing Thesis – regarded at one time as about as certain a hypothesis you could expect to get in logic or mathematics – has also become a casualty of this falling away from certainty. I want to begin this essay by briefly reminding readers how the thesis began, why it was regarded by most as evidently true (even if by its nature strictly unprovable), and what has happened to make so many lose faith in this certainty.

The Church-Turing Thesis began its life in the 1930s as a thesis concerning the notion of an effective or mechanical method. The latter was understood in roughly the following sense [Copeland, 2008]:

> (E) A method M for achieving some desired kind of result (e.g., calculating the values of a specific function **R** on the basis of its arguments, or deciding whether given numbers have a specific property **B**, etc) is *effective* or *mechanical* just in case 1) M is set out in terms of a finite number of exact instructions; 2) M will, if carried out properly, always produce the desired result in a finite number of steps (assuming there is a result); 3) M can (in practice or in principle) be carried out by a human being unaided by any machinery save paper and pencil; 4) M demands no insight or ingenuity on the part of the human being carrying it out.

Notice that 'mechanical' so understood has nothing to do with machines. Understanding it as applying to any procedure that is somehow mechanizable or implementable on computing machines in the modern sense would not only betray an appalling ignorance of the meaning of a certain term of art but would also betray an ignorance of the kind of issues that were important at the time. These issues are well known to philosophers of mathematics. Hilbert's formalist program of mathematics had thrown out a challenge to mathematicians: it was one thing to develop intriguing and important mathematics; it was quite another thing to justify that mathematics. Some of the most important parts of mathematics such as Cantor's set theory were in need of justification in this sense, since on their earliest and perhaps most elegant formulations they were inconsistent (Cantor himself already realized this, but Russell's paradox, which was just one of the many antinomies that were developed in the years following Cantor's work, brought the issue to a head), and it wasn't clear which ways of revising these formulations allowed one to escape the antinomies. But how then could we justify using them, given that their subject matter involved objects that could never in any sense be presented to perception and so be modeled in the physical world (and thereby proved consistent)? The answer that Hilbert gave was that we needed finitary methods that themselves stood in no need of justification. Hilbert had proposed formalizing mathematics so that formally presented mathematics could then become the objects of mathematical study (*metamathematics*), using effective, finitary and hence ultra-reliable methods. Doing so should, among other things, help to solve the *Entscheidungsproblem*, the decision problem for predicate logic, and help to show that formalized Zermelo-Fraenkel set theory, for example, is consistent. But even to contemplate such questions required having a precise notion of what an effective, finitary method was. These questions and their history are elegantly covered in the Epstein-Carnielli text, which also includes essential readings such as relevant sections of Hilbert's *On the Infinite*.

The task of delivering such a notion was thus of the utmost mathematical and philosophical importance. But of course the *idea* of an effective method, presented above as (E), is an informal one, since certain key requirements, e.g., that the method demand no insight or ingenuity, are left unexplicated. In his famous 1936 paper *On Computable Numbers, with an Application to the Entscheidungsproblem*[Turing, 1936], Turing presented a formally exact concept – *computable by Turing machine* – with which the informal concept *calculable by means of an effective method* could be replaced. Church did the same in his *An Unsolvable Problem of Elementary Number Theory* [Church, 1936], using the concept of *lambda-definability*. It turned out that the two replacement notions were equivalent: each picks out the same set of mathematical functions, as does Godel's formal notion of *recursiveness*. (The latter was proved equivalent to lambda-definability by Church and Kleene, while Turing proved lambda-definability equivalent to Turing computability in the appendix to *On Computable Numbers* ....) The Church-Turing Thesis is the assertion that

> A (total) function from the positive integers to the positive integers is calculable by means of an effective method (or effectively calculable) iff it is recursive (equivalently, Turing computable, lambda-definable, etc.).

When the thesis is expressed in terms of the formal concept proposed by Turing, it is usually referred to as 'Turing's thesis'; and *mutatis mutandis* in the case of Church. (Sometimes the Church-Turing Thesis is presented as the left-to-right conditional; the claim that every Turing-computable function is effectively calculable is then presented as the *converse* of the thesis. I will not follow that usage here. In any case, the right-to-left conditional is usually regarded as unproblematic on the grounds that a Turing machine program is itself a specification of an effective method: a human being can understand the instructions, and carry out the operations called for, without the exercise of any ingenuity or insight.)

One other point. Church actually suggested that we *define* the notion of an effectively calculable function of positive integers as the notion of a recursive function of positive integers or of a lambda-definable function of positive integers ([Church, 1936], 356). Famously, Post demurred, and criticized Church for masking what was essentially a 'working hypothesis' as a definition:

> [T]o mask this identification under a definition ... blinds us to the need of its continual verification. ([Post, 1936], 105.)

Post's reading of the situation has become more or less standard, and for most commentators the use of the word 'thesis' captures the fact that we should never expect to find a rigorous proof.

Still, this wasn't always the standard. Both Church and Gödel thought that Turing's explication of the notion of effectiveness was convincing, and Church even said that Turing's analysis had the advantage of making the identification with

> effectiveness in the ordinary (not explicitly defined) sense evident immediately. ([Church, 1937], 43.)

From this standpoint, it is not at all clear that Church, Gödel, and Turing would have agreed that the Thesis stands in 'need of ... continual verification'.

3 The Post-Church debate and its aftermath

This difference between the attitudes of Post and Church represents a suggestive model of the on-going debate about the Thesis. Post wanted to see 'wider and wider formulations' of the notion of effectiveness, and then thought these would reduce to his own seemingly restrictive notion, thereby showing that Church's Thesis was tantamount to a *natural law*. Church thought that Turing had produced an utterly persuasive formalization of something that was inherently informal, a formalization that captured in explicit terms something that is only implicit in the informal notion. While many seem to think that there is no real difference between these two understandings, it is not hard to see how they encourage different ways of approaching the question of the plausibility of the Thesis.

After all, Post's understanding in principle encourages the idea that we might have evidence *against* the thesis. Such evidence might take us in one of two directions. It might cause problems for the claim that any Turing-computable function is effectively calculable (the supposedly unproblematic part of the thesis), since it might involve philosophical evidence in favor of tightening the notion of effectively calculability in a constructivist and even finitist direction, leading to the thought that only a restricted version of the thesis can be correct

(*Computability*, pp. 234–5)[1]. Epstein and Carnielli spend some time in their text emphasizing the concerns of constructivists, but research in this direction has turned out to be a minority pursuit. By and large, such concerns have tended to be seen as purely philosophical, of no interest to the working logician, mathematician or computer scientist (and, increasingly, the philosopher, who might well be suspicious of Platonism but no less suspicious of metaphysically inspired points of view aimed at trimming the sails of classical mathematics).

Of wider interest are concerns that go the other way. These challenge the claim that we have overwhelming evidence that effectively calculable functions are all Turing-computable. One such complaint comes from those who question the force of what Epstein and Carnielli call *the Most Amazing Fact*: the fact that the various attempts to formalize the notion of effective calculability by what appears to be radically different means have all been shown to be equivalent on the basis of effective translations. The complaint is not that these equivalences fail, but that they provide us with less support for the Church-Turing Thesis than we might think. After all, we might have been focusing on too narrow a range of means of formalization. To quote Georg Kreisel:

> What excludes the case of *systematic* error? (c.f. the overwhelming empirical support from ordinary mathematics for: if an arithmetic identity is provable at all, it is provable in classical first-order arithmetic; they all overlook the principle involved in, for example, consistency proofs.) ([Kreisel, 1965], p. 144; quoted in *Computability*, p. 237)

And, rather more cuttingly, Richard Sylvan and Jack Copeland opine that:

> There are grounds for suspecting that the famous convergences may have been, in part at least, contrived. Those attempting later analyses knew where they were going, and what equivalences would be nice to prove and would be duly applauded. The vaunted independence of these analyses is something of an illusion. Several of the analyses were pursued even in the one small place, Princeton. ([Sylvan and Copeland, 2000], p. 192)[2]

So the *degree* of evidential support that the Most Amazing Fact is supposed to give the thesis is one source of concern. This is independent of another concern, which is whether the notion of an effective procedure is precise enough *on its own* to discourage the thought that there are computational models with instructions we can understand, and procedures for executing instructions we can follow, but that permit the computation of non-recursive, non-Turing

---

[1] The relevant notion of constructivity can be understood in different ways. [Kroon and Burkhard, 1990] describe a complexity-based way, using an idea of Hao Wang.

[2] This seems more than a little unkind. Even if these results don't look very deep to us now, they surely did when first presented. For Post, such equivalences were initially only a working hypothesis. And some of them don't even involve models based on symbolic manipulation on the part of humans. Thus, virtually any modern textbook will mention the equivalence of Turing machines and Shepherdson's and Sturgis's register machines (even Copeland does so in his Stanford Encyclopedia Entry on the topic), but register machines are not really designed to mimic the way humans work.

computable functions. Kreisel thought that it was (more on this later); Sylvan and Copeland thought that it wasn't. Post would clearly have agreed with Sylvan and Copeland on this latter point, but he would have disagreed with Sylvan and Copeland as well as Kreisel that the Most Amazing Fact has little evidential bearing on the Thesis. Post, after all, directly stated that Church's Thesis is a 'working hypothesis' to the extent that the logical equivalence of his model and 'wider and wider formulations' of the notion of effectiveness is a working hypothesis.[3] This suggests that he might well have regarded the matter as close to settled once the Most Amazing Fact was in place.

Despite this, it is clear how Post's comments can lead to a rather different picture. For suppose we look at conditions (3) and (4) in (E) and decide that the reference to 'humans' is rather parochial if we have in mind 'wider and wider formulations' of the kind Post wanted, especially if we are allowed to take into account what has happened since the heyday of early attempts at formalizing the notion of effectiveness (and why not?). Suppose, in short, that we no longer emphasize the *actual* way humans mechanically compute or calculate, and instead emphasize *possibilities* for human computation. Then we might find it arbitrary to concentrate on the capacities of individual human computing agents; we might instead stress the way humans can work in combination, for example. Or we might find it arbitrary to concentrate on what human minds can do, when in practice we also rely on the power of electronic computers.

In the end, then, our search for 'wider and wider formulations' might lead us (with Post's tacit blessing) to take the required informal notion to be something rather broad: say, the notion of whatever is *intuitively* computable, no matter how. And that in turn might lead us to understand the Church-Turing Thesis as saying that a function is intuitively computable just when it is Turing computable (and lambda-definable, and recursive, ...).

4 The physical Church-Turing Thesis

Here, then, we have a deep potential source of disagreement about the Church-Turing Thesis, a disagreement rooted in conflicting views about which kind of computing models count as relevant, and why. To reiterate: If we understand the Church-Turing Thesis in terms of the way humans computed prior to the advent of the on-going revolution in computing, this threatens to turn it into something rather parochial, not worthy of serious debate, so why not turn it into something that more grandiosely talks about functions computable in any way whatsoever, using whatever physical resources the world have to offer, and see if the thesis thus construed continues to demand assent? And this, of course, is where current talk of hypercomputation and 'breaking the Turing barrier' gets its grip. [4] For thus construed the thesis – call it CTT_{PHYS} when

---

[3]To quote Post: '...our aim will be to show that all such [wider and wider formulations] are logically reducible to [Post's notion of finite combinatory processes]. We offer this conclusion at the present moment as a working hypothesis. And to our mind such is Church's identification of effective calculability with recursiveness.' ([Post, 1936], 105)

[4]Hypercomputation is computation in the course of computing a function that is not Turing computable. (The term was invented by Jack Copeland.) For a particularly dismissive perspective on hypercomputation from a well known computability theorist, see [Davis, 2004], [Davis, 2006]. [Cotogno, 2003] is another who dismisses the possibility of hypercomputation

construed this way, and CCT construed on its classic reading as a thesis about effective calculability – invites us to contemplate all the physical resources that physics, chemistry, and biology in particular have to offer when tied to humans' increasingly sophisticated ability to find *uses* for these resources, including their use as aids to computing. What reason do we have for thinking that computability thus understood can't get us beyond the Turing computable?

Before we get to this point, it is important to be clear about the modality involved in claims about the comput*able*. We need to ask whether CTT_{PHYS} is supposed to be in trouble because there are logically possible machines, working on the basis of logically possible physical resources, that can compute non-recursive functions, or whether something stronger is being claimed. Perhaps, for example, the claim is that there are *physically* possible computing devices (ones relying on what is theoretically possible, given current physics) that can do more than Turing machines can do; or perhaps the claim is that such machines are not merely theoretically possible, but practically possible (they can actually be built).[5] Following common practice, it is the second version of the claim that we shall (initially) focus on. Most theorists think that practical possibility makes hypercomputation too dependent on what is available at a particular time, while logical possibility makes it hard to see why we should care (why care about what is the case in worlds whose physical laws are vastly different from ours?) At a first approximation, then, CTT_{PHYS} simply maintains that a function from the positive integers to the positive integers is physically computable (in the sense that it is physically possible for it to be computed) if and only if it is Turing computable. The challenge to the thesis would come from evidence that it is physically possible for certain non-Turing computable functions to be computed.

But this still seems to leave the content of CTT_{PHYS} underdescribed. For even as we talk about using the resources of physics, chemistry, and biology in the aid of computing, we still need to know what counts as computing in a physical environment. One influential suggestion is due to Robin Gandy [Gandy, 1980], who understands physical computing in terms of what he calls discrete deterministic mechanical devices. A discrete deterministic mechanical device is a pair $\langle S, F \rangle$ where S is a potentially infinite set of states consisting of atomic parts whose number is unbounded (although there is only a finite number of kind of atomic parts), and F a structural operation from S to S. Two minimal physical constraints that Gandy imposes on these devices are that "there is a lower bound on the linear dimensions of every atomic part of the device and that there is an upper bound (the velocity of light) on the speed of propagation of changes" ([Gandy, 1980], 126). Together these constraints imply that an atom or smallest constituent can transmit and receive information only in its bounded neighborhood, and in bounded time – a principle of Local Causation. There are two other seemingly reasonable principles, all at a high level of physical abstraction, that can be imposed on such machines: the Principle of the Limitation of Hierarchy (the set-theoretic rank of the states is bounded) and

---

outright, although by seeming to confuse Turing uncomputability with *logical* uncomputability.

[5]See [Copeland, 2008].

the Principle of Unique Assembly (any state can be assembled from parts of bounded size). Gandy thought that devices regulated by such principles gave us a sufficiently general, suitably abstract, notion of physical computing.

They don't, however, show that physical computing beyond the Turing barrier is possible. Indeed, Gandy demonstrated that these principles allow discrete deterministic mechanical devices to be simulated by Turing machines, yielding his more precise version of CTT_{PHYS} (Gandy's Thesis, or G):

(G) A function is calculable by a discrete deterministic mechanical device iff it is Turing computable.

Far from supporting the idea of computable non-recursive functions, therefore, it seems that once we explicate what computation in a physical medium involves, understanding this notion at an appropriately high level of abstraction in the manner of Gandy, the prospects of computing beyond the Turing barrier disappears.

But such a pessimistic conclusion would be premature. Earlier I cited Copeland and Sylvan's remark about the unsurprising nature of what Epstein and Walter call the Most Amazing Fact ('There are grounds for suspecting that the famous convergences may have been, in part at least, contrived'). While their remark surely exaggerated the situation, something like this reaction has merit when applied to Gandy's proof that even physical computability is subject to the Church-Turing Thesis. The reason is straightforward. By his own admission, Gandy wasn't trying to invent a new notion of computation informed by the physical sciences. Turing had presented an informal argument for the claim what can be done by human computors in a routine, "mechanical" way can be accomplished by a Turing machine, where the mechanical operations in question were assumed to obey locality conditions and operate deterministically on states that satisfy boundedness conditions. What Gandy did was to appeal to physical limitations of mechanisms in order to capture, at a high level of physical abstraction, similar restrictive conditions on the operations of mechanisms. To that extent he was following Turing's own argument for Turing's Thesis. The convergence is well described by Sieg and Byrnes, whose 1999 paper gave an easier, improved version of Gandy's argument:

> [According to Gandy] any discrete deterministic mechanical device (can be represented as a discrete dynamical system and its operation) must satisfy the restrictive principles of local causation and unique assembly. This is completely parallel to [Turing's] claim that any human computor (can be represented as a discrete dynamical system and its operation) must obey appropriate locality conditions. [Sieg and Byrnes, 1999]

In short, while Gandy's demonstration of (G) is compelling, what has *not* been demonstrated by Gandy's argument is that (G) is the only or best way of understanding CTT_{PHYS}. Where the notion of physical computing is concerned, there is nothing amiss with a Post-like plea for 'wider and wider formulations' so that we have a better understanding of the prospects of the truth of CTT_{PHYS}.

5 The role of finitude

What might such 'wider and wider formulations' look like? One suggestion, following the work of David Deutsch [Deutsch, 1985], is to allow any universal model computing machine operating "by finite means". Deutsch argues that this is less restrictive than a proposal like Gandy's. While the class of Turing machines satisfies the conditions of 'computing by finite means', so does a class of model computing machines that is the quantum generalization of the class of Turing machines; and so in particular does the universal quantum computer Q, which Deutsch shows to be capable of simulating every finitely realizable physical system. But Deutsch also shows that while such a model quantum computer has important properties not reproducible by any Turing machine, it is still not capable of computing non-recursive functions on the positive integers, and hence of computing beyond the Turing barrier.

Deutsch's work suggests that the prospects of breaking the Turing barrier look bleak indeed, especially as quantum computing is often held up as the most important new development in computing. Could there be even *wider* formulations of CTT$_{PHYS}$ that might offer more hope? Here the computing literature has offered a number of proposals. This section briefly reviews some of them.

An obvious first proposal is to find physically meaningful ways of relaxing conditions that define Gandy's machines as well as Deutsch's "universal model computing machines", including the universal quantum computer Q. And the obvious condition to relax is the background condition of finitude, in particular the condition, which occurs implicitly in Gandy[6] and explicitly in Deutsch, that only a finite number of operations can be performed in a finite interval of time. But before we look at such machines, let us first give a working account of physical computability that would, in principle at least, allow for such non-finite machines:

> (PC) A function, f, is physically computable iff there is a machine 'blueprint' M for computing f such that it is physically possible for there to be a machine instantiating M that physically outputs $f(x)$ on input x for any x.

Perhaps the best known example of non-finite machines are accelerated Turing machines or *Zeno* machines (c.f. [Copeland, 1998a]): Turing machines that are capable of computing an infinite sequence of steps in a finite amount of time (by, say, performing one step in 1 unit of time, the next step in half that amount of time, and, in general, the nth step in $1/2^{n-1}$ units of time). Such machines are capable of solving the halting problem for standard Turing machines (not surprisingly, the halting problem for Zeno machines is not decidable by any Zeno machine).[7] On the surface, there is nothing *conceptually* incoherent about

---

[6]See [Shagrir and Pitowsky, 2003], who complain that unless this is acknowledged as a background condition (G) is demonstrably false.

[7]Here is an informal presentation of an accelerated Turing machine that solves the halting problem for Turing machines: Given Turing machine T and input w:

> begin program
> write 0 on the first position of the output tape;

this kind of acceleration. What counts as a single step surely depends on how a step is carved up, and that is plausibly done in terms of an elementary task undertaken, not on how long a task takes. The claim that the halting function for ordinary Turing machines is physically computable might thus be based on the claim that physical manifestations of Zeno machines are physically possible.

A common response to the suggestion of Zeno machines is that they violate ordinary stress-energy conditions and need an infinite amount of space, so that they remain at best *notional* machines: logically possible, perhaps, but not physically possible. One suggestion that appears to escape this objection is to think instead of a computing machine that, after performing some fixed number of operations, builds a smaller and faster copy of itself which will then perform a number of operations before it too builds a faster and smaller copy, ... and so on [Davies, 2001]. Given appropriate assumptions, the resulting series of infinitely shrinking machines will be able to complete an infinite number of steps within a finite time, thereby surpassing the power of Turing machines and doing so without the individual machines requiring an infinite energy supply or an infinite amount of space. But while the existence of such a series of infinitely shrinking machines appears to be consistent with Newtonian mechanics, Davies acknowledges it is not consistent with the atomic and quantum mechanical nature of matter in our universe, and in that case they are not physically possible after all.

Perhaps the best known proposal for a hypercomputer that uses the resources of the *physics* of time rather than relying on the kind of conceptual possibilities for time described above is the relativistic hypercomputer described by Mark Hogarth (Hogarth machines, for short).[8] This kind of machine exploits the properties of Malament-Hogarth spacetime, which constitutes a solution to Einstein's field equations for General Relativity. Malament-Hogarth spacetimes contain regions with an infinite time-like trajectory A that can be circumvented by a finite time-like trajectory B, so that B contains a point p such that A lies entirely in p's chronological past even though it is infinite. In principle, if an observer then launches a Turing machine along A while she travels along B, she may, in a finite amount of time, find herself in the future of an infinitely long computation performed by the Turing machine. If the Turing machine can also send signals to the observer, the observer would be able to know the

---

set $i = 1$;
begin loop simulate the first i steps of T on w;
if $T(w)$ has halted, then write 1 on the first position of the output tape;
$i = i + 1$;
end loop
end program

One familiar problem for such machines is that a simple modification turns them into paradoxical machines: after '...write 1 on the first position of the output tape' add 'if (i is odd) then print 1; else (print 0)'. If $T(w)$ does not halt, the resulting machine will be in no determinate state 'after' it has finished the ω-sequence of steps (for this version of Thomson's Lamp paradox, see [Earman and Norton, 1993], 28).

[8]C.f. [Hogarth, 1994], [Hogarth, 2004], and [Shagrir and Pitowsky, 2003]. Hogarth himself uses the acronym 'SAD', for 'arithmetical sentence deciding' – somewhat misleadingly, as it turns out, since there are SAD machines that can decide *hyperarithmetical* sentences (see Welch 2008).

outcome of such a computation, thus giving her greater computational powers than ordinary Turing machines on their own. In particular, the observer may be able to determine whether an arbitrary machine T halts given some input w, and so compute the halting function for Turing machines.

But this suggestion too faces enormous problems. For one thing, the mere fact that Malament-Hogarth spacetimes are solutions to the field equations does not seem enough for genuine physical possibility. Earman and Norton, for example, have argued that while such spacetimes are consistent with the equations of general relativity, they violate strong cosmic censorship, 'which states that naked singularities do not develop in physically reasonable models of general relativity theory' ([Earman and Norton, 1993], 35). Furthermore, it seems that any machine that runs for an infinite length of time will malfunction with probability 1 ([Button, 2009], 779), which must surely render such machines useless. (Button notes that a version of this complaint may apply equally to infinitely shrinking machines. I say more about this kind of complaint later, arguing that it reveals a very different kind of difficulty from the others.)

Hogarth machines constitute one of the most widely discussed physical models with an alleged hypercomputation capability. As mentioned earlier, another frequently mentioned possibility is quantum computing, which involves the manipulation of qubits (more generally, qudits) in accordance with the laws of quantum mechanics. But recall Deutsch's argument that his universal quantum computer Q can simulate any quantum computer with arbitrary precision, yet cannot compute beyond the Turing barrier (although it can in certain cases achieve extremely good speed-up). So it is unclear how to understand this possibility. There are a number of lines of enquiry in the literature. [Calude and Pavlov, 2002], for example, propose breaking the Turing barrier through a semi-quantum method, tailored for the so-called infinite Merchant Problem which is assumed equivalent to the halting problem. The authors argue that the method should detect a solution to the problem with a 'tiny but not-zero probability' that improves as more time is spent on the search. In a recent paper, Calude and some other co-workers have recommended the use of quantum oracles that function as random number generators [Abbott et al., 2011]. Such proposals face a range of difficulties. Why, in the case of the first proposal, should we count this as a method for computing the halting function if the method doesn't even make it *probable* that our answer is correct? In the case of the second proposal, even if the quantum oracle produces a random uncomputable string, how could this capability possibly help us solve some given unsolvable problem like the halting problem? More fundamentally, how does this count as *computing* an uncomputable function rather than just generating the course-of-values of a function that we know to be uncomputable on physical grounds?[9]

---

[9]It is interesting, in this connection, to point out that Kreisel distinguished between number-theoretic functions for which there were mechanical rules and that satisfy CTT, and physically realizable number-theoretic functions. With regard to the latter, Kreisel says it 'seems to be open whether there are (finitely specified) physical systems whose most probable behavior is non-recursive' ([Kreisel, 1970], 143, note 2). He then briefly considers one possibility involving the three body problem, suggests the problem may well be unsolvable, and then wonders whether, if so, 'one could use this situation for an analog computation of a

Possibly the most prominent proposal for a quantum hypercomputer is due to Tien Kieu (see, e.g., [Kieu, 2005]), who argues that an appropriately constructed quantum system can decide the Turing-unsolvable problem whether an arbitrary Diophantine equation has an integral solution. But this proposal too faces formidable problems. It requires infinite precision in setting up and maintaining the system, and the problem of determining when the solution state is reached is unsolvable by Turing machines (see, e.g., [Smith, 2006]) so would need its own hypercomputer. In essence, Kieu assumes a form of non-effective oracle computation, one in which an infinite search through the integers is accomplished in a finite amount of time. Showing how such a search is physically possible remains a deep problem for the proposal. (Oracles more generally are one of the main tools suggested by Copeland for hypercomputation; see, for example, [Copeland, 1998b]. For a critique, see [Davis, 2004].)

My final example of an argument for the possibility of hypercomputation is one that indirectly returns us to the work of Carnielli. The argument – in many ways the most radical of the arguments encountered so far – is due to the late Richard Sylvan, well known as a proponent of paraconsistent logic as well as dialetheism (the view that there are true contradictions). In a paper co-authored with Copeland, Sylvan argues for the physical possibility of D-machines (dialethic machines), which are described as Turing machines acting under a dialethic, or paraconsistent, logic. A D-machine of 'type 1' is described as a machine that, when it encounters a contradiction, 'can proceed with its computation satisfactorily' on the basis of this dialethic logic. Sylvan and Copeland claim that D-machines of this type 1 'are able to compute diagonal functions f that are classically regarded as non-computable'; in particular, when, during the usual 'proof by contradiction' that f is computable the machine encounters a contradiction of the typical '$d = d+1$' kind, 'the computation continues, there being no danger that the machine will infer absurdities from this esoteric contradiction' ([Sylvan and Copeland, 2000], p. 198).

Note that such D-machines, like the other models we have looked at, again involve a relaxation of the demand of finitude. The computation of the 'non-computable' diagonal function first involves a conception of the infinite totality of all algorithmic functions, and then a diagonalizing out of this infinite class in order – so the argument goes – to obtain a diagonal function that is 'both algorithmic and not' (p. 195), just as the Liar sentence 'This very sentence is not true' is for Sylvan both true and not true. But Sylvan and Copeland have virtually nothing else to say about the computation of classically non-computable functions, so it is difficult to know what to make of their suggestion. (It is significant that an arch-dialetheist like Graham Priest has not taken it up.)

But Sylvan/Copeland also briefly consider D-machines of 'type 2', which are defined as 'machines whose metalogic is dialethic: for such a machine, M, one of whose states is x, 'M is in x' and '$\neg(M$ is in $x)$' may both be the case'

---

non-recursive function (by repeating collision experiments sufficiently often)' (*ibid.*) Without more detail, it is difficult to know whether this suggestion is supposed to escape the problem I identify above. See also Piccinini's discussion of the need for process-independent rules in [Piccinini, 2011].

(p. 197). They then propose two questions to the paraconsistent community: 'whether D-machines of type 2 compute classically uncomputable functions and, if so, which', and 'whether or not the halting theorem holds within a paraconsistent framework' (p. 197, 199). Interestingly, there are physically meaningful ways of modeling D-machines of type 2. Carnielli and his student Juan Agudelo have shown how to define a paraconsistent Turing machine model that allows a partial simulation of superposed states of quantum computing, while a sharpened version of this model allows them to represent such notions as entangled state and relative phase. But they also show that the computing power of such models of computation is no greater than that of Deutsch's quantum computers; these models do not break the Turing barrier, and the halting problem continues to hold ([Agudelo and Carnielli, 2010], §3.2). The two questions posed by Sylvan/Copeland therefore have a negative answer, at least with respect to this way of modeling the notion of a paraconsistent Turing machine. It will be interesting to see if Sylvan's own vastly more ambitious hopes for paraconsistent machines will survive in the paraconsistent community in the face of this result. The prospects do not look good.

6 Epistemic considerations

What conclusions can we draw from of all this? The first point to make on the basis of this brief survey is that the prospects of hypercomputation and breaking the Turing barrier do not look particularly promising, even with physical computability understood on the model of (PC) so that the modality of our choice is mere *physical* possibility rather than the stronger notion of practical possibility. All the various proposals face difficulties under this construal of physical computability. Hypercomputation on the basis of notional, merely logically possible, machines is a different matter, of course. Zeno machines and infinitely shrinking Turing machines in a world unencumbered by the sorts of physical forces and "noise" that make their operation unreliable seem at least *logically* possible. So do Hogarth machines. The kinds of problems that have been raised for such machines are contingent problems, after all. Still, this is scant comfort for devotees of hypercomputation, who clearly think that the *actual* laws of physics are on their side.

But I believe the matter is rather more complicated than this. To see why, it is important to go back to CTT itself. I began my reflections on CTT_{PHYS} by pointing out that Post was looking for 'wider and wider formulations' of the idea of an effective method of computation, and suggested that such a search might lead one to de-emphasize the way humans calculate in favor of a more general account. Post is frequently cited by those in the hypercomputation community as (at least tacitly) in favor of such a search.

But it is important to stress that this was unlikely to have been Post's intention. There is every reason to think that Post would have approved of Turing's attempt to analyze human computability in terms of various constraints on operations and states. Post, after all, reminds us that he had tried to produce

> a system of a certain logical potency but also, in its restricted field,
> of *psychological fidelity* (Post 1936, 105; the emphasis is mine).

While there is a way in which CTT_{PHYS} can be regarded as the result of generalizing CTT along important dimensions (more on this later), this has virtually nothing to do with what concerned these early workers on computability. As I emphasized earlier, CTT arose in the context of a debate about the correct way to capture the notion of an effective, finitary, procedure: the kind of procedure that would show that existing mathematics was consistent, for example, or that would permit mathematicians to decide the truth of an arbitrary statement of mathematics or the validity of an arbitrary formula of first-order logic (all questions that had been raised in the context of Hilbert's formalist program of mathematics). The goal here was *epistemological*. The kind of procedure in question would yield *certain* knowledge, the kind of knowledge that mathematics had always aspired to. Once logicians and mathematicians had presented their formalizations of an effective method, the questions that Hilbert asked could be given definitive answers in terms of these formalizations – negative answers, in the case of the most fundamental of these questions. And with the mounting evidence, as they saw it, in favor of CTT, these answers acquired the status of absoluteness: there can be no effective general method for settling logical validity or mathematical truth, and (as Gödel showed) there can be no effective proof of the consistency of arithmetic (assuming it is consistent), let alone an effective proof of the consistency of existing systems of set theory.[10]

Armed with this clarification, let us briefly return to the question we began with: is CTT *true*? That is, with effective computability or calculability understood in the manner of (E), and assuming idealized circumstances (in particular, enough time and space) is it true that a function from the positive integers to the positive integers is effectively calculable iff it is Turing computable / recursive? To this question the answer is almost certainly 'Yes'. I am persuaded by Kreisel that the evidential role played by the Most Amazing Fact is less significant than it is often taken to be. What in fact impressed Kreisel far more were Turing's informal analysis of the notion of an effective procedure (later put in a physical setting by Gandy) and his proof sketch that such procedures can be simulated by Turing machines. As he put it, the support for CTT 'consists above all in the analysis of machine-like behavior and in a number of closure conditions, for example diagonalization' ([Kreisel, 1965], p. 144). And there are other possibilities, such as Peter Smith's use of a kind of 'squeezing' argument to show that the notion of an algorithmic computable function is identical to that of a Turing computable function ([Smith, 2007] [Ch. 35], 2010).[11] There is also Kripke's argument that CTT can be viewed as a corollary of Gödel's completeness theorem [Kripke, 2011]. We have more reason to be confident of CTT, in fact, than of many mathematical results that have supposedly received formal proof – and this in spite of the informal nature of the notion of an effective process. (Think, for example, about the Hales-Ferguson proof of Kepler's Conjecture or Grigori Perelman's proof

---

[10]Despite this, many writers confuse CTT and CTT_{PHYS}. Hogarth, for example, has this to say: "But is [CTT] true? Well, it will be if there is at least one physically possible Turing machine and if there are no physically possible non-Turing computers." ([Hogarth, 1994], 134). For more examples, see [Copeland, 2008].

[11]"Squeezing" arguments of this type were initially introduced by Kreisel in the context of fixing the formal content of the informal notion of (first-order) validity-in-virtue-of-form.

of Poincaré's Conjecture: proofs that were initially resisted by many, and even after checking are still not fully believed by many.)

Now return to CTT$_{\text{PHYS}}$. I have already suggested, largely following what others have written, that current attempts to describe machines that are able to break the Turing barrier do not look particularly convincing, with the possible exception of Hogarth machines. The kind of barriers that have been raised to the physical realizability of these machines (see especially Earman/Norton 1993) have been addressed by a number of authors (see, for example, [Etesi and Németi, 2001] and [Shagrir and Pitowsky, 2003]), although it is far from clear that they can all be overcome. What is clear, however, that is that even if they *can* be overcome, actually overcoming them will be an utterly prohibitive task, well beyond the capacity of current humans or their descendents. These machines seem a paradigm of the practically impossible. No one could realistically ever use such machines.

As I noted earlier, as a complaint against hypercomputation the comment that such machines are practically or technologically impossible is usually countered with the response that we are interested in technologies permitted by the laws of physics and hence in the *limits* of what is technologically possible. The fact that they are impossible in practical terms is usually considered of little consequence. But there is reason to think that this may be a mistake. It may be that CTT$_{\text{PHYS}}$ is best formulated in terms of a notion of physical computability that emphasizes practical possibility or feasibility rather than mere physical possibility. And, surprisingly enough, this may to some extent *improve* the chances of hypercomputation rather than lessen them. Or so I shall argue.

To this end, recall the way we defined physical computability earlier:

(PC) A function, f, is physically computable iff there is a machine 'blueprint' M for computing f such that it is physically possible for there to be a machine instantiating M that physically outputs $f(x)$ on input x for any x.

Note that this definition is set out in purely ontological terms. If, for any x, it is physically possible for the computation to proceed with x as input and $f(x)$ as output then the function is physically computable. A little reflection should convince us that this is a surprisingly undemanding requirement. Suppose that in the actual world, although perhaps not in every physically possible world, such a machine has the property that its users are unable to detect the output $f(x)$ in a range of cases where we should expect to detect the output (assume the machine does in fact produce an output but energy/mass restrictions make it unreadable). Or suppose that in the actual world the machine's hardware is notoriously unreliable when operating on a range of arguments where we expect to be able to use the machine (it keeps breaking down). Both kinds of cases suggest a sense – a "for all practical purposes" sense – in which the machine fails to compute f.

Anomalies of this kind have inspired some to revise an account like (PC). Gualtiero Piccinini, for example, has recently argued for the following account of physical computability in terms of constraints of usability [Piccinini, 2010], [Piccinini, 2011]:

(PC$_U$) A function, f, is physically computable iff there is a machine 'blueprint' M for computing f that can be implemented by a physical system P satisfying constraints (1)-(6) below such that, for any x, P can be set to undergo a process that results in P's producing output $f(x)$ on input x.

The constraints are these: (1) *Readable inputs and outputs* (the inputs and outputs must be readable); (2) *Process-independent rule* – there must be a fixed rule, specifiable independently of the physical process, that links the outputs to the inputs; (3) *Repeatability* – the process must 'in principle be repeatable by any competent observer who wishes to obtain its results' ([Piccinini, 2011], 742-3); (4) *Settability* – the system undergoing the process must be settable, so that a user can perform different computations on the same system; (5) *Physical constructability* – the system must be physically constructible in the sense that 'the relevant physical materials can be arranged to exhibit the relevant [computational] properties' (744); and (6) *Reliability* – the system 'must operate correctly long enough to yield (correct) results at least some of the time ...[and so] the system's components must not break too often' (745).

One difficulty for (PC$_U$) as it stands is that the constraints are rather loosely formulated; in particular, although (PC$_U$) is replete with occurrences of the modal 'can' little is said about how to understand this expression (if 'can' is the 'can' of physical possibility there is little to distinguish (PC$_U$) from (PC)). But the discussion around the constraints make it clear what is intended. Piccinini wants a formulation of (PC) that makes it the physical parallel of the notion of effective computability. The latter property holds of a function if there is an effective or finitary procedure that properly functioning human calculators have the ability to use here and now, allowing them to determine with complete certainty the value $f(x)$ of function f on argument x (assuming no limits of time and space, etc). As we emphasized earlier (and as Gödel stressed in an oft-quoted comment on recursiveness or Turing computability),[12] such a notion of computability is *epistemic* in orientation. So what Piccinini wants is an epistemically loaded version of (PC). He is explicit on the point: after presenting his *Usability Constraint* ('if a physical process is a computation, it can be used by a finite observer to obtain the desired values of a function'), he adds that '[t]he usability constraint is an epistemic constraint', one that makes CTT$_{PHYS}$ 'analogous' to CTT (740). He also makes it clear that constraints (1)-(6) above simply highlight what is needed to make the physical processes employed in the course of computation 'usable'.

Earlier I argued for moderate skepticism about the prospects of breaking the Turing barrier on the basis of various proposed methods of computation. These appealed, by and large, to the apparent physical impossibility of the proposed methods, and were based on understanding CTT$_{PHYS}$ in terms of mere physical computability (PC). (PC$_U$) is meant to be stronger than (PC), and so we should expect matters to look even bleaker for the prospects of

---

[12]'It seems to me that [the great importance of the concept of general recursiveness / Turing computability] is largely due to the fact that with this concept one has for the first time succeeded in giving an absolute definition of an interesting epistemological notion, i.e., one not depending on the formalism chosen' ([Gödel, 1965], p. 84).

Turing-barrier breaking if (PC$_U$) is our preferred account. Is this really so? The answer is: yes, so long as we keep the intended epistemic parallel in mind. In the first place, the modal 'can' should be construed as practical possibility or feasibility. Secondly, 'usability' should be understood in partly epistemic terms. On one interpretation, simply being successful in achieving the desired values when the observer uses the process on different arguments suffices for making the process usable. But that is not the intended meaning. The finite observer must be epistemically *justified*, although she need not be *certain*, in taking the output values to be the values on the given arguments. (That is why Repeatability is important, for example.) Indeed, Piccinini slates the appeal to Hogarth machines on just such justificatory grounds. He cites with evident approval Tim Button's point that any Hogarth machine that runs for an infinite length of time will malfunction with probability 1 ([Button, 2009], 778), so that they 'are genuinely useless and cannot be made useful' ([Button, 2009] 2009, 779). But of course there are physically possible worlds where such machines do run for an infinite length of time without malfunction, so this complaint would make little sense unless we wanted our computing devices to give us *reason* to believe that they do what they are supposed to do and not simply to *do* what they are supposed to do.

Subject to these clarifications, (PC$_U$) does indeed make the prospects of Turing-barrier breaking look even bleaker. For a start, it rules out one suggestion found in the quantum computing literature: the idea of using a (physical) Turing machine working with a quantum oracle. [Abbott *et al.*, 2011] describe a device that, under specified assumptions involving measurements of observables, produces an incomputable sequence. They then ask 'what is the computational power of a Turing machine working with [such] a quantum oracle?' But if this simply means what functions are Turing computable relative to, or in, this incomputable sequence, (PC$_U$) denies that this has anything to do with the question of what functions can be physically computed by relying on the computational power of such an oracle-dependent machine. To answer the latter question requires functions to be described using a process-independent rule (constraint (2)). Given that the string generated by the quantum process is genuinely random, there is no reason to think that appealing to a quantum oracle in this way allows the physical computation of new incomputable functions – functions, that is, able to be specified using a rule that is independent of this quantum process.[13]

(PC$_U$) also allows a quicker route to the conclusion that Hogarth machines, for example, do not falsify Modest Physical CT (Piccinini's term for CTT$_{PHYS}$ when physical computability is understood in terms of (PC$_U$)). Remember that the relevant modality for (PC$_U$) is practical or technological possibility. Piccinini points out a range of problems with making workable Hogarth machines, including the likelihood of machine breakdown and the difficulty of accessing distant spacetime regions with the Malament-Hogarth property. He agrees that this is a matter of practical possibility rather than physical possibility, but in-

---

[13]Of course, if it could be established that the string (probably) corresponds to the halting function for Turing machines, say, it would be a different matter; but showing that is the hard part.

sists such problems are enough to ensure that Hogarth machines can't compute beyond the Turing computable (757, fn. 28) – at least for now:

> ...at the moment, [Hogarth machines] are not even close to being technologically practical. It is extremely unlikely that [they] will ever be built and used successfully. So for now and the foreseeable future, [Hogarth machines] do not falsify Modest Physical CT. (759)

Indeed, Piccinini thinks he can make a bolder statement. Recall some of the other challenges that have been posed for proposed hypercomputers. A number of these computers seemed on the cusp of physical impossibility, although there might have been some lingering doubt. With (PC_U) in place, it is easier to place them all on the side of the (practically) impossible, allowing us to find in favor of Modest Physical CT – at least for now:

> Modest Physical CT is true if and only if genuine hypercomputers are impossible in the sense that they do not satisfy our usability constraints. ...As yet, there is no hard evidence against Modest Physical CT. Instead, there are good reasons to believe Modest Physical CT (763).

7 Reconsidering the role of the epistemic

Piccinini's reminder of the importance of computability's epistemic dimensions is salutary. But his proposal contains an obvious yet significant flaw. Understanding the flaw allows us to see why there remains hope for a reasonable sense in which CTT_{PHYS} might after all be false. The flaw is simply this: unlike both logical and physical possibility, practical possibility is not an absolute but a graded notion. This is evident from the fact that constraints like Reliability and Repeatability clearly admit of degrees (How reliable? Repeatable how often, and at what cost?); so do (virtually) all the other constraints.[14] Even constructibility is not an absolute notion. Suppose a hypercomputer could be constructed but only if the physical and intellectual resources needed were so great that putting them to this use would mean the end of civilization, leaving only a robot or two to run the machine; such a hypercomputer is surely *barely* constructible. More generally, the central epistemic notion that informs all the other constraints – usability – is clearly a graded rather than an absolute notion. This severely compromises Piccinini's (PC_U) and hence also his understanding of CTT_{PHYS} (Modest Physical CT), since it is never made clear which precisifications of these various graded notions we should opt for, and why. Piccinini's proposal harbors a tacit relativism.

I do not propose to offer a solution to this problem, but a suggestion – one that is not only in the spirit of Piccinini's laudable emphasis on an epistemically informed notion of physical computability but also in the spirit of the increased

---

[14]That even Repeatability admits of degrees, for example, is clear from Piccinini's admission that Hogarth machines operating in spacetime regions with a large number of edges which, if accessible one after the other, can each be exploited to run a distinct computation ([Piccinini, 2011], 756-757). In that case, the more (accessible) edges there are, the better the degree of Repeatability.

emphasis on fallibilism in modern epistemology, including mathematical epistemology. No longer do we think, as philosophers in the past thought, and as even Hilbert thought, that to trust a mathematical result we need, in the end, to have a super-reliable way of securing such a result, one where we can say with certainty that the result holds. Here (as elsewhere in life) a reasonably high degree of probability is enough if it is too hard to do much better with available resources. One stark reminder of this came in 1976 with the discovery of a proof of the Four-Color theorem based on an enormous computer computation, but even more interesting in some ways are proofs that are simply too complicated to be checked by a single mathematician. Probability is here often best construed as intersubjective probability; roughly speaking, we achieve high intersubjective probability if the mathematical community agrees, on the basis of checking and re-checking by experts, and on the inability of experts to find mistakes even when the stakes are high, that a given proof does indeed establish its result and that the result can stand. Such, broadly speaking, was the case with the Hales-Ferguson proof of Kepler's Conjecture or Grigori Perelman's proof of Poincaré's Conjecture, examples I mentioned earlier in connection with our very high degree of confidence in CTT.[15]

But now suppose that even high probability is impossible to get. Perhaps the checks needed are just too difficult and time-consuming (and too wasteful of resources; legions of mathematicians working together could perform the task, but it would not be worth it). In that case, we might well judge that considerably less suffices for us to accept the result, at least for now and even if only tentatively. Now turn to the subject of physical computing again. If the stakes are high enough, we might well be prepared to accept a solution to a computing problem even where our degree of credence is not very high at all. In such cases, we will insist on repeated checks (how many will depend on a number of factors), but we will not require certainty or even high probability. In my view, that ought to be the response of the hypercomputation community to Piccinini's views.

We might put it as follows. Let us agree that usability is an important, even essential, criterion for processes that might be used in physical computation. But if the stakes are high and the difficulty of achieving a high degree of credence based on a high probability of correctness low, we might well set our expectations for what is required for usability rather low. Hypercomputers face difficulties that digital computers and even classical quantum computers don't have to face: they have to deal with the problem of executing infinitely many "operations" in a finite amount of time. It might be argued that they deserve some corresponding slack.

Two examples will suffice to give a flavor of the idea. Consider first the suggestion of [Calude and Pavlov, 2002] that a semi-quantum method they describe should detect a solution to a particular unsolvable problem with a 'tiny but not-zero probability' that improves as more time is spent on the

---

[15]There are many other examples. Marston Conder points out, for example, that the proof of the theorem giving the typology of the finite simple groups required about fifteen thousand pages, spread over five-hundred separate articles written by about three-hundred different authors [Conder, 1994].

search. Assume the details of the study are correct. Rather than dismiss such a method on the basis of the fact that the probability of success is tiny, we should applaud the fact that there is *some* reason to think that a solution has been detected, and that there are ways of increasing our level of confidence. If the stakes are high enough, we should regard such a process as "usable", even as we hope for better methods that allow us to improve our level of confidence.

My second example involves Hogarth machines. Recall Button's argument that when completing an infinite number of stages any such machine will malfunction with probability 1 ([Button, 2009], 778). This is not a case where we get a tiny but non-zero probability of success; this is a case where we get *zero* probability of success. Piccinini cites this as proof that such machines are unusable. But even more so than in the case of Calude's and Pavlov's semi quantum method, this strikes me as a clear *non sequitur*. Here we have a case where we are owed a more relaxed view of what suffices for usability. Suppose that we test for malfunction at various stages, and nothing shows up. Suppose we repeat the process a number of times and achieve the same result each time.[16] Even if we do both things often, this will not change the probability of success, which remains 0. But it will, or at any rate should, strongly increase our preparedness to rely on the result. At some point, we might even rationally come to *believe* the result.[17]

8 Conclusion

This paper has traversed the recent debate about the Church-Turing Thesis. Like many others, I have argued for a strict divide between the mathematical and physical forms of the thesis, decrying the attempt to conflate the two. But I have also argued that there is more in common between the two forms of the thesis than is often acknowledged, and in particular that there is merit in adding an epistemic element to the thesis, to match the (strikingly) epistemic nature of the thesis as it was first formulated by Church, Turing, Post, Gödel, and others. The question is how this is best done. In the final part of the paper, I suggested that one recent proposal, by Piccinini, is beset by a kind of tacit relativism that threatens to derail the proposal. I must confess that I have no firm ideas of how to fix this problem, but I also think that this should not take us away from trying to understand the physical form of the Church-Turing thesis in partly epistemic terms. The final section has offered a tentative suggestion; it develops the tacit relativism I see in Piccinini's own proposal in a direction that is rather more sympathetic to the viability of hypercomputation than Piccinini's own proposal, partly by appealing to some recent trends in the epistemology of mathematics.

---

[16]See [Shagrir and Pitowsky, 2003], 91-3, and [Piccinini, 2011], 756-7, on how Hogarth machines satisfy Repeatability.

[17]Such a claim will strike some as paradoxical. It is important, however, to separate two things that are often conflated: rational degree of credence or subjective probability, which many think should follow the objective chances (as in David Lewis's Principal Principle), and doxastic possibility. If you think that something is doxastically possible only if its degree of credence is greater than 0, my suggestion will make little sense. There is ample reason to resist the latter claim, however (a version of the Bayesian constraint of Regularity). See, for example, [Williamson, 2007] and [Hàjek, 2011].

I am rather pleased to leave matters here. This is not only because pursuing such a strategy would lead us too far afield (and would require a great deal more intellectual effort on my part) but also because we have, in a way, come full circle. I began this paper by talking about Walter's and Epstein's book *Computability*, and what it had to say about the Church-Turing Thesis. They didn't say much about physical forms of the thesis but they did spend some time exploring ways in which views about the nature of mathematics, especially views that emphasize our epistemic limitations (constructivism and intuitionism, for example) might condition our views about the thesis. In my opinion, some newer trends in the epistemology of mathematics equally have the potential to teach us about physical forms of the Church-Turing Thesis.

BIBLIOGRAPHY

[Abbott et al., 2011] A.A. Abbott, C.C. Calude, and K. Svozil. On Demons and Oracles. Forthcoming, 2011.

[Agudelo and Carnielli, 2010] J.C. Agudelo and W. Carnielli. Paraconsistent Machines and their Relation to Quantum Computing. *Journal of Logic and Computation*, 20(2):573–595, 2010.

[Button, 2009] T. Button. SAD Computers and Two Versions of the Church-Turing Thesis. *British Journal for the Philosophy of Science*, 60(4):765–792, 2009.

[Calude and Pavlov, 2002] C.S. Calude and B. Pavlov. Coins, Quantum Measurements, and Turing's Barrier. *Quantum Information Processing*, 1(1-2):107–127, 2002.

[Church, 1936] A. Church. An Unsolvable Problem of Elementary Number Theory. *American Journal of Mathematics*, 58:345–363, 1936.

[Church, 1937] A. Church. Review of Turing 1936. *Journal of Symbolic Logic*, 2:42–43, 1937.

[Conder, 1994] M. Conder. Pure Mathematics: an Art? or an Experimental Science? *NZ Science Review*, 51(3):99–102, 1994.

[Copeland, 1998a] B. J. Copeland. Turing's O-Machines, Penrose, Searle, and the Brain. *Analysis*, 58:128–138, 1998.

[Copeland, 1998b] B.J. Copeland. Even Turing Machines Can Compute Uncomputable Functions. In C. Calude, J. Casti, and M. Dinneen, editors, *Unconventional Models of Computation*, pages 150–164. Springer, London, 1998.

[Copeland, 2008] B. J. Copeland. The Church-Turing Thesis. In E.Z. Zalta, editor, *The Stanford Encyclopedia of Philosophy (Fall 2008 Edition)*, http://plato.stanford.edu/archives/fall2008/entries/church-turing/, 2008.

[Cotogno, 2003] P. Cotogno. Hypercomputation and the Physical Church-Turing Thesis. *British Journal for the Philosophy of Science*, 54:181–223, 2003.

[Davies, 2001] E. B. Davies. Building Infinite Machines. *British Journal for the Philosophy of Science*, 52:671–682, 2001.

[Davis, 2004] M. Davis. The Myth of Hypercomputation. In C. Teuscher, editor, *Alan Turing: Life and Legacy of a Great Thinker*, pages 195–212. Springer, Berlin, 2004.

[Davis, 2006] M. Davis. Why there is no such discipline as hypercomputation. *Applied Mathematics and Computation*, 178:4–7, 2006.

[Deutsch, 1985] D. Deutsch. Quantum Theory, the Church-Turing Principle and the Universal Quantum Computer. *Proceedings of the Royal Society*, A 400:97–117, 1985.

[Earman and Norton, 1993] J. Earman and J. D. Norton. Forever is a Day: Supertasks in Pitowsky and Malament-Hogarth Spacetimes. *Philosophy of Science*, 60:22–42, 1993.

[Epstein and Carnielli, 2008] R. Epstein and W. Carnielli. *Computability: Computable Functions, Logic, and the Foundations of Mathematics*. Advanced Reasoning Forum, Socorro, New Mexico, USA, 2008.

[Etesi and Németi, 2001] G. Etesi and I. Németi. Non-Turing computations via Malament-Hogarth Space-times. *International Journal of Theoretical Physics*, 41:341–370, 2001.

[Gandy, 1980] R. Gandy. Church's Thesis and Principles for Mechanisms. In J. Barwise, H. J. Keisler, and K. Kunen, editors, *The Kleene Symposium*, pages 123–148. North-Holland, Amsterdam, 1980.

[Gödel, 1965] K. Gödel. Remarks Before the Princeton Bicentennial Conference on Problems in Mathematics. In M. Davis, editor, *The Undecidable*, pages 84–88. Raven, Ewlett, NY, 1965.

[Hàjek, 2011] A. Hàjek. Staying Regular. Forthcoming, 2011.

[Hogarth, 1994] M. L. Hogarth. Non-Turing Computers and Non-Turing Computability. *PSA: Proceedings of the Biennial Meeting of the Philosophy of Science Association*, 1:126–138, 1994.

[Hogarth, 2004] M. L. Hogarth. Deciding Arithmetic Using SAD Computers. *British Journal for the Philosophy of Science*, 55:681–691, 2004.

[Kieu, 2005] T. D. Kieu. An Anatomy of a Quantum Adiabatic Algorithm that Transcends the Turing Computability. *International Journal of Quantum Information*, 3(1):177–183, 2005.

[Kleene, 1936] S.C. Kleene. Lambda-Definability and Recursiveness. *Duke Mathematical Journal*, 2:340–353, 1936.

[Kreisel, 1965] G. Kreisel. Mathematical Logic. In T.L. Saaty, editor, *Lectures in Modern Mathematics vol. III*, pages 95–195. Wiley, York, 1965.

[Kreisel, 1970] G. Kreisel. Church's Thesis: a Kind of Reducibility Axiom for Constructive Mathematics. In A. Kino, J. Myhill, and R.E. Vesley, editors, *Intuitionism and Proof Theory, Studies in Logic and the Foundations of Mathematics*, pages 121–150. North-Holland, Amsterdam, 1970.

[Kripke, 2011] S.A. Kripke. Another Approach: The Church-Turing 'Thesis' as a Special Corollary of Gödel's Completeness theorem. In B.J. Copeland, C. Posy, and O. Shagrir, editors, *Computability: Gödel, Turing, Church, and Beyond*, pages 53–67. MIT Press, Cambridge, 2011. In print.

[Kroon and Burkhard, 1990] F. Kroon and W. A. Burkhard. On a Complexity-Based Way of Constructivizing the Recursive Functions. *Studia Logica*, 49:133–149, 1990.

[Piccinini, 2010] G. Piccinini. Computation in Physical Systems. In E.Z. Zalta, editor, *The Stanford Encyclopedia of Philosophy (Fall 2010 Edition)*, http://plato.stanford.edu/archives/fall2010/entries/computation-physicalsystems/, 2010.

[Piccinini, 2011] G. Piccinini. The Physical Church-Turing Thesis: Modest or Bold? *The British Journal for the Philosophy of Science*, 62(4):733–769, 2011.

[Post, 1936] E.L. Post. Finite Combinatory Processes – Formulation 1. *Journal of Symbolic Logic*, 1:103–105, 1936.

[Shagrir and Pitowsky, 2003] O. Shagrir and I. Pitowsky. Physical Hypercomputation and the Church-Turing Thesis. *Minds and Machines*, 13:87–101, 2003.

[Sieg and Byrnes, 1999] W. Sieg and J. Byrnes. An Abstract Model for Parallel Computations: Gandy's Thesis. *The Monist*, 82:150–164, 1999.

[Smith, 2006] W. D. Smith. Three Counterexamples Refuting Kieu's Plan for Quantum Adiabatic Hypercomputation; and Some Uncomputable Quantum Mechanical Tasks. *Applied Mathematics and Computation*, 178(1):184–193, 2006.

[Smith, 2007] P. Smith. *An Introduction to Gödel's Theorems*. Cambridge University Press, Cambridge, 2007.

[Smith, 2010] P. Smith. Squeezing Arguments. *Analysis*, 71(1):22–30, 2010.

[Sylvan and Copeland, 2000] R. Sylvan and J. Copeland. Computability is Logic-Relative. In G. Priest and D. Hyde, editors, *Sociative Logics and Their Applications: Essays by the Late Richard Sylvan*, pages 189–199. Ashgate, London, 2000.

[Turing, 1936] A. Turing. On Computable Numbers, with an Application to the Entscheidungsproblem. *Proceedings of the London Mathematical Society*, 42:230–265, 1936.

[Welch, 2008] P. D. Welch. The Extent of Computation in Malament-Hogarth Spacetimes. *British Journal for the Philosophy of Science*, 59:659–674, 2008.

[Williamson, 2007] T. Williamson. How Probable Is an Infinite Sequence of Heads? *Analysis*, 67:173–180, 2007.

Fred Kroon
University of Auckland
Auckland, New Zealand
E-mail: f.kroon@auckland.ac.nz

On the logical relativity of quantum computability

Anderson de Araújo, Marcelo Finger

ABSTRACT. We analyze the thesis sustained by Agudelo and Carnielli that some crucial aspects of quantum computations are indeed properties of paraconsistent computations. To evaluate the plausibility of this thesis, we generalize Agudelo and Carnielli's paraconsistent Turing machines, obtaining the multiple Turing machines. These machines satisfies, in a certain sense, an aspect of quantum computing that was lacking in the paraconsistent Turing machines, notably, reversibility. As the notion of reversibility used in quantum computing is restricted to deterministic computations, we leave, however, as an open problem the logical relativity of quantum computability.

1 Introduction

In [Sylvan and Copeland, 2000], Sylvan and Copeland have argued that the concept of computation is relative to logic, that is to say, the properties of computability depends on the logical system adopted in the formalization of our intuitive notion of efectiveness. We can call this position by *logical relavitity thesis*. In particular, Sylvan and Copeland have suggested a 'theory of paraconsistent computability' [Sylvan and Copeland, 2000, p.196], showing how it is possible to conceive a Turing machine that performs computations on contradictions printed in its tape.

In [Carnielli and Dria, 2008], Carnielli and Dria have endorsed the logical relativity thesis and, more, in [Agudelo and Carnielli, 2010] Agudelo and Carnielli have carried out the idea of paraconsistent computability, by proposing a logical formalization of Turing machines through the logics of formal inconsistency [Carnielli *et al.*, 2007]. Agudelo and Carnielli also have sustained a strong version of the logical relativity thesis: crucial features of quantum computability could be thought as special characteristics of paraconsistent computations.

The possible connections between paraconsistent logic and the logic of quantum mechanics is not a new theme. The main motivation relies on the fact that the logic associated to the observables of quantum systems, Hermitian operators in the Dirac-Neumann axiomatization of quantum mechanics (Cf. [Cohen-Tannoudji *et al.*, 1977]), can be thought as an ortholatticce (Cf. [Birkhoff and von Neumann, 1936]). In this way, paraconsistent quantum logics are characterized by the class of all realizations based on an involutive bounded lattice, in which the non-contradiction principle does not hold (Cf. [Chiara and Giuntini, 2002]). Agudelo and Carnielli's perspective is, however, more strong than that traditional trend relating paraconsistent and quantum logics: they argue that characteristic quantum phenomena are in fact paraconsistent phenomena. To defend that, they have tried in [Agudelo and Carnielli, 2010] to show how some quantum algorithms such as that defined by Deutsch [Deutsch, 1985] and Deutsch-Jozsa [Deutsch and Jozsa, 1992] as well as some well-known quantum phenomena such as quantum entanglement can be simulated through the paraconsistent Turing machines defined in [Agudelo and Carnielli, 2010].

A crucial feature of quantum computations is that they are *reversible*: each configuration of a computation can be reached via exactly one path so that at every step the computation can be retraced unambigously all the way back to the initial configuration (Cf. [Nielsen and Chuang, 2000, p.17-22]). Nevertheless, the paraconsistent Turing machines defined by Agudelo and Carnielli are nondeterministic Turing machines, i.e., their computations are not reversible. From this point of view, Agudelo-Carnielli thesis as stated in [Agudelo and Carnielli, 2010] is implausible. In this paper, we will show, however, that in a certain restrict case it may be made tenable.

In Section 2 we generalize we define the concept of *multiple Turing machines*, without using any logical system. In Section 3 we prove that Agudelo and Carnielli's paraconsistent Turing machines are indeed special cases of multiple Turing machines. In Section 4 we show that these machines are reversible in the sense that is necessary to quantum computing. In particular, this shows that paraconsistent Turing machines, when considered as multiple Turing machines, also satifies the required reversibility of quantum systems. Despide this improvement, we conclude that the logical relativity of the concept of quantum computation is open problem, because in principle the notion of reversibility used in quantum computing is restricted to deterministic computation.

2 Multiple Turing machines

Following [de Arajo, 2011], let us jointly define deterministic and nondeterministic Turing machines[1].

DEFINITION 1. A *Turing machine* is a (partial) function $M : Q \times S \to \wp(Q \times S \times V) - \{\varnothing\}$ such that for some $k, m \in \mathbb{N}$:

- $Q = \{q_0, q_1, \ldots, q_k\}$ is the set of *states* of M; q_0 is the *initial state* and q_k is the *final state*;

- $S = \{s_0, s_1, \ldots, s_m\}$ is the set of *symbols* of M; s_0 is the symbol *empty*;

- $V = \{\triangleleft, \triangleright\}$ is the set of *moves* of M; $\triangleleft$ is the move *to the left* and $\triangleright$ is the move *to the right*.

The triples in M are called *instructions* of M. If each set in the range of a Turing machine M has only one triple, M is called *deterministic*; otherwise, M is called *nondeterministic*. Given a Turing machine $M : Q \times S \to \wp(Q \times S \times V)$, it will be denoted by $M = (Q, S)$.

As paraconsistent Turing machines have been defined in [Agudelo and Carnielli, 2010], they are nondeterministic machines. This means that in principle paraconsistent Turing machines are not reversible, since they are not even functions.

---
[1]We will use the set-theoretical concepts as they are defined in [Jech, 2002].

Fortunatelly, we can overcome this difficulty, because it is possible to put every Turing machine in the form of a function without loosing its nondeterministic feature. For this end, consider the next definition.

DEFINITION 2. Given a Turing machine $M = (Q, S)$, the *multiple partition* of M is the (partial) function $f_M : \wp(Q \times S \times V) - \{\varnothing\} \to \{\circ, \bullet\}$ such that $f(q', s', \diamond) = \circ$ if there are q and s for which $M(q, s) = (q', s', \diamond)$ and $|M(q, s)| = 1$, but $f(q', s', \diamond) = \bullet$ if there are q and s such that $M(q, s) = (q', s', \diamond)$ and $|M(q, s)| > 1$, where $\diamond \in \{\triangleleft, \triangleright\}$; otherwise, f_M is undefined.

The idea behind this definition is, given a Turing machine M, to assign the symbol $\circ$ to the instructions of M that have a single triple associated to a specific argument, but $\bullet$ to the instructions of M that have more than one triple associated to a specific argument of M. Intuitively, this means we are labelling the deterministic and nondeterministic "parts" of M.

DEFINITION 3. Let $M = (Q, S)$ be a Turing machine, $\diamond$ be a move in $\{\triangleleft, \triangleright\}$ and f_M the multiple partition of M. Then, the *multiple Turing machine* associated to M is the (partial) function

$$\hat{M} : Q \times S \to \{\circ, \bullet\}^{\wp(Q \times S \times V) - \{\varnothing\}}$$

such that $\hat{M}(q, s) = f_M(q', s', \diamond)$ if, and only if, $M(q, s) = (q', s', \diamond)$; otherwise, $\hat{M}(q, s)$ is not defined. If $\hat{M} = (Q, S)$ is the multiple Turing machine associated to the Turing machine M, it will be said that M is the *generator* of $\hat{M}$. Besides, the generator $\hat{M}$ of M is called *deterministic* or *nondeterministic* according M is deterministic or nondeterministic.

It is clear that a multiple Turing machine always is a function, even when its generator is nondeterministic.

3 Computations of multiple Turing machines

In this section, we will indicate how multiple Turing machines can be viewed as a generalization of paraconsistent Turing machines.

DEFINITION 4. Let $\hat{M} = (Q, S)$ be a multiple Turing machine and $\bar{s} = s_0^i, \ldots, s_m^i$ be a string in S^*, where S^* is the Kleene closure of S. A *computation* of $\hat{M}$ with *input* $\bar{s}$ is a (partial) function $C_{\hat{M}}^{\bar{s}} : \mathbb{N} \to \wp(S^* \times Q \times S^*) - \{\varnothing\}$ defined by the following clauses:

- $C_{\hat{M}}^{\bar{s}}(0) = \{(\epsilon, q_0, s_1^i, \ldots, s_m^i)\}$, where ϵ is the empty string in S^*;

- If $C_{\hat{M}}^{\bar{s}}(t) = \{(r_0^0, \ldots, r_l^0, q_p^0, s_0^0, \ldots, s_u^0), \ldots, (r_0^n, \ldots, r_{l'}^n, q_{p'}^n, s_0^n, \ldots, s_{u'}^n)\}$ and there is an j with $0 \leq j \leq n$ for which $\hat{M}(q_{p''}^j, s_0^j)$ is defined, then $C_{\hat{M}}^{\bar{s}}(t+1)$ is the smaller set $\wp(S^* \times Q \times S^*) - \{\varnothing\}$ that satisfies these conditions:
  (2.1) For each j wherein $\hat{M}(q_w^j, s_0^j)$ is not defined, $\diamond \in \{\triangleleft, \triangleright\}$, $(r_0^j, \ldots, r_{l''}^j, q_w^j, s_0^j, \ldots, s_{p''}^j) \in C_{\hat{M}}^{\bar{s}}(t+1)$;
  (2.2) For each j wherein $\hat{M}(q_w^j, s_0^j)$ is defined, the following holds, where M is the generator of $\hat{M}$:

- If $(q_a^j, s_b^j, \triangleleft) \in M(q_w^j, s_0^j)$ and $r_0^j, \ldots, r_{l''}^j \neq \epsilon$, then
  $(r_0^j, \ldots, r_{l''-1}^j, q_a^j, s_b^j, s_1^j, \ldots, s_{p''}^j) \in C_{\hat{M}}^{\bar{s}}(t+1)$;
- If $(q_a^j, s_b^j, \triangleleft) \in M(q_w^j, s_0^j)$ and $r_0^j, \ldots, r_{l''}^j = \epsilon$, then
  $(\epsilon, q_a^j, s_b^j, s_1^j, \ldots, s_{p''}^j) \in C_{\hat{M}}^{\bar{s}}(t+1)$;
- If $(q_a^j, s_b^j, \triangleright) \in M(q_w^j, s_0^j)$ and $s_1^j, \ldots, r_{p''}^j \neq \epsilon$, then
  $(r_0^j, \ldots, r_{l''}^j, s_b^j, q_a^j, s_1^j, \ldots, s_{p''}^j) \in C_{\hat{M}}^{\bar{s}}(t+1)$;
- If $(q_a^j, s_b^j, \triangleright) \in M(q_w^j, s_0^j)$ and $s_1^j, \ldots, r_{p''}^j = \epsilon$, then
  $(r_0^j, \ldots, r_{l''}^j, s_b^j, q_a^j, s_0) \in C_{\hat{M}}^{\bar{s}}(t+1)$.

- If $C_{\hat{M}}^{\bar{s}}(t) = \{(r_0^0, \ldots, r_l^0, q_p^0, s_0^0, \ldots, s_u^0), \ldots, (r_0^n, \ldots, r_l^n, q_{p'}^n, s_0^n, \ldots, s_{u'}^n)\}$ and, for all i such that $0 \leq i \leq n$, $\hat{M}(q_{p''}^j, s_0^j)$ is not defined, then $C_{\hat{M}}^{\bar{s}}(t')$ is not defined for every $t' > t$. In this case, $C_{\hat{M}}^{\bar{s}}$ is called a *(Turing) computation of $t+1$ steps*.

The sequences in $C_{\hat{M}}^{\bar{s}}$ are called *configurations* of $C_{\hat{M}}^{\bar{s}}$. In particular, $C_{\hat{M}}^{\bar{s}}(0)$ is called the *initial configuration* and $C_{\hat{M}}^{\bar{s}}(t)$ is called the *final configuration* if $C_{\hat{M}}^{\bar{s}}$ is a computation of $t+1$ steps. The computation $C_{\hat{M}}^{\bar{s}}$ is called *deterministic* or *nondeterministic* according the generator M of $\hat{M}$ is deterministic or nondeterministic.

Note that computations of multiple Turing machines can be nondeterministic, although this machines are indeed functions. This is the desired preservation of the nondeterministic features, mentioned above. Bellow, we define the computability of multiple Turing machines in such a way that it incorporates the paraconsistent characteristics delimited in [Agudelo and Carnielli, 2010].

DEFINITION 5. Let $\hat{M} = (Q, S)$ be a multiple Turing machine. It is said that $\hat{M}$ *halts* for (the input) $\bar{s}$ in S^* if the computation $C_{\hat{M}}^{\bar{s}}$ has a final configuration; otherwise, it is said that $\hat{M}$ *does not halts* for $\bar{s}$. In this way, $\hat{M}$ *accepts* $\bar{s}$ in S^* with probability n/m if $\hat{M}$ halts for ϵ in a final configuration $C_{\hat{M}}^{\bar{s}}(t)$ of $C_{\hat{M}}^{\bar{s}}$ such that $|C_{\hat{M}}^{\bar{s}}(t)| = n$ and the number of tuples $(r_0^i, \ldots, r_l^i, q_u^i, s_0^i, \ldots, s_p^i)$ in $C_{\hat{M}}^{\bar{s}}(t)$ for which q_u^i is the final state of $\hat{M}$ and $\bar{s} = s_0^i, \ldots, s_p^i$ is equal to m. In this case, it will also be said that $\hat{M}$ *rejects* the string $\bar{s}$ in S^* with probability $1 - n/m$. Finally, $\hat{M}$ *decides* a language $L \subseteq S^*$ with error probability at most $1 - n/m$ if, for every $\bar{s} \in L$, $\hat{M}$ accepts $\bar{s}$ with probability at least n/m and, for every $\bar{r} \notin L$, $\hat{M}$ rejects $\bar{s}$ with probability at least n/m.

In [Agudelo and Carnielli, 2010], it was defined two models of paraconsistent Turing machines. The common feature of both models is that a single computation can accept more than one string. It is clear that the same is true with respect to multiple Turing machines.

PROPOSITION 6. *Let $M = (Q, S)$ be a paraconsistent Turing machine in the sense of [Agudelo and Carnielli, 2010]. Thus, if M decides a language L in polynomial-time, then there is a multiple Turing machine $\hat{M}$ such that M is a generator of M and $\hat{M}$ also decides L in polynomial-time.*

Proof. We will show this result with respecto to the model of paraconsistent computation defined in [Agudelo and Carnielli, 2010, p.589]. Let $M = (Q, S)$ be a paraconsistent Turing machine in the sense of [Agudelo and Carnielli, 2010, p.589] that decides the language $L \subseteq S^*$ in polynomial-time. Then, according the definition in [Agudelo and Carnielli, 2010, p.591], for every $\bar{s} \in L$, M accepts $\bar{s}$ with probability error at most $1/3$ and, for every $\bar{r} \notin L$, M rejects $\bar{s}$ with probability error at most $1/3$. Define the multiple Turing machine $\hat{M}$ whose generator is M. Thus, by definition 5, it is immediate that $\hat{M}$ also decides L in polynomial-time. For others models of paraconsistent Turing machines, the proof is similar, it is only necessary to change the definition 5 above in an appropriate way. ∎

In this way, it is now open the possibility that paraconsistent Turing machines, in the form of multiple Turing machines, to be reversible without loosing its nondeterministic feature; as desired.

4 Computational reversibility

In this section, we show that multiple Turing machines are reversible in the sense that is necessary to quantum computing. In particular, this shows that paraconsistent Turing machines, when considered as multiple Turing machines, can satisfy, in a certain sense, this fundamental property.

It is well-known that Turing machines can be represented by graphs whose vertices are the states of the machine and the edges are labeled by its instructions (Cf. [Sipser, 2006]). In this way, it can be said that a Turing machines is *logically reversible* if its graph has out-degree and in-degree one. This means that each configuration can be reached via exactly one path so that at every step the computation can be retraced unambigously all the way back to the initial configuration. As observed by Crescenz and Papadimitriou [Crescenzi and Papadimitriou, 1995], reversible computation is a weakening of determinism rather similar to the way symmetric computation weakens nondeterminism: a nondeterministic Turing machine is *symmetric* if it has a graph that is undirected.

With respect to quantum computability, the notion of reversibility is somewhat different from the logical notion explained above. Bernstein and Vazirani [Bennet and Vazirani, 1997] has proved that a deterministic Turing machine can be used to defined a quantum Turing machine if, and only if, it is computationally reversible in the following sense.

DEFINITION 7. Let M be a Turing machine. A tuple $(r_0, \ldots, r_l, q_u, s_0, \ldots, s_p)$ $\in C_M^{\bar{s}}(t)$ is a *predecessor* of the tuple $(\dot{r}_0, \ldots, \dot{r}_{l'}, \dot{q}_{u'}, \dot{s}_0, \ldots, \dot{s}_{p'}) \in C_M^{\bar{s}}(t+1)$ if $(\dot{q}_{u'}, s, \diamond) \in M(q_u, s_0)$ for $s = \dot{s}_1$ if $\diamond = \triangleleft$ but $s = \dot{r}_{l'}$ if $\diamond = \triangleright$. In this case, it can also be said that $(\dot{r}_0, \ldots, \dot{r}_{l'}, \dot{q}_{u'}, \dot{s}_0, \ldots, \dot{s}_{p'})$ is a *successor* of $(r_0, \ldots, r_l, q_u, s_0, \ldots, s_p)$. Besides, M is *computationally reversible* when, for every $\bar{s} \in S^*$ and $t \in \mathbb{N}$, if the tuple $(r_0^j, \ldots, r_{l'}^j, q_u^j, s_0^j, \ldots, s_p^j)$ is in $C_M^{\bar{s}}(t+1)$, then it has at most one predecessor in $C_M^{\bar{s}}(t)$.

Note that the notion of computational reversibility defined above also can be applied to multiple Turing machines. The next result shows under what condition a multiple Turing machine is computationally reversible.

PROPOSITION 8. *A multiple Turing machine $\hat{M}$ is computational reversible if, and only if, the generator M of $\hat{M}$ is one-to-one.*

Proof. First, we prove the sufficient condition. Suppose that $M = (Q, S)$ is a Turing machine such that $\hat{M}$ is one-to-one. Let $(r_0^j, \ldots, r_l^j, q_u^j, s_0^j, \ldots, s_p^j)$ and $(\dot{r}_0^j, \ldots, \dot{r}_{l'}^j, \dot{q}_{u'}^j, \dot{s}_0^j, \ldots, \dot{s}_{p'}^j)$ be tuples in $C_M^{\bar{s}}(t)$. Suppose that

$$(r_0^j, \ldots, r_l^j, q_u^j, s_0^j, \ldots, s_p^j) = (\dot{r}_0^j, \ldots, \dot{r}_{l'}^j, \dot{q}_{u'}^j, \dot{s}_0^j, \ldots, \dot{s}_{p'}^j).$$

Then there are tuples

$$(r_0^i, \ldots, r_{l*}^i, q_w^i, s_0^i, \ldots, s_{p*}^i) \text{ and } (\dot{r}_0^i, \ldots, \dot{r}_{l**}^i, \dot{q}_{w*}^i, \dot{s}_0^i, \ldots, \dot{s}_{p**}^i)$$

in $C_M^{\bar{s}}(t-1)$ such that the first is a predecessor of $(r_0^j, \ldots, r_l^j, q_u^j, s_0^j, \ldots, s_p^j)$ and the second is a predecessor of $(\dot{r}_0^j, \ldots, \dot{r}_{l'}^j, \dot{q}_{u'}^j, \dot{s}_0^j, \ldots, \dot{s}_{p'}^j)$. Since $q_u^j = \dot{q}_u^j$, $(q_w^i, s_0^i, \diamond_1), (\dot{q}_{w*}^i, \dot{s}_0^i, \diamond_1) \in M(q_u^i, s_0^i)$, and so $\hat{M}(q_u^i, s_0^i) = f_M(q_w^i, s_0^i, \diamond_1)$ and $M(q_u^i, s_0^i) = f_M(\dot{q}_{w*}^i, \dot{s}_0^i, \diamond_2)$. As M is one-to-ne, it must be that $q_w^i = \dot{q}_{w*}^i$, $s_0^i = \dot{s}_0^i$ and $\diamond_1 = \diamond_2$. Hence,

$$(r_0^i, \ldots, r_{l*}^i, q_w^i, s_0^i, \ldots, s_{p*}^i) = (\dot{r}_0^i, \ldots, \dot{r}_{l**}^i, \dot{q}_{w*}^i, \dot{s}_0^i, \ldots, \dot{s}_{p**}^i)$$

in $C_M^{\bar{s}}(t-1)$. Therefore, $\hat{M}$ is computationally reversible.

In this paragraph, we prove the necessary condition. Let $\hat{M}$ be a computationally reversible multiple Turing machine. Take two tuples

$$(r_0^i, \ldots, r_l^i, q_u^i, s_0^i, \ldots, s_p^i) \text{ and } (\dot{r}_0^i, \ldots, \dot{r}_{l'}^i, \dot{q}_{u'}^i, \dot{s}_0^i, \ldots, \dot{s}_{p'}^i)$$

in $C_M^{\bar{s}}(t)$. Without lost of generality, let us assume that $r_0^i, \ldots, r_l^i = \dot{r}_0^i, \ldots, \dot{r}_{l'}^i$ and $s_1^i, \ldots, s_p^i = \dot{s}_1^i, \ldots, \dot{s}_{p'}^i$. Suppose that $\hat{M}(q_u^i, s_0^i) = \hat{M}(\dot{q}_u^i, \dot{s}_0^i)$; say that $\hat{M}(q_u^i, s_0^i) = f_M(q, s, \triangleright)$. Then,

$$(r_0^i, \ldots, r_l^i, s, q, s_1^i, \ldots, s_p^i) \text{ and } (\dot{r}_0^i, \ldots, \dot{r}_{l'}^i, s, q, \dot{s}_1^i, \ldots, \dot{s}_{p'}^i)$$

in $C_M^{\bar{s}}(t+1)$ are the respective successors of $(r_0^i, \ldots, r_l^i, q_u^i, s_0^i, \ldots, s_p^i)$ and $(\dot{r}_0^i, \ldots, \dot{r}_{l'}^i, \dot{q}_{u'}^i, \dot{s}_0^i, \ldots, \dot{s}_{p'}^i)$ in $C_M^{\bar{s}}(t)$. Now,

$$(r_0^i, \ldots, r_l^i, s, q, s_1^i, \ldots, s_p^i) = (\dot{r}_0^i, \ldots, \dot{r}_{l'}^i, s, q, \dot{s}_1^i, \ldots, \dot{s}_{p'}^i)$$

and, because $\hat{M}$ is computationally reversible, it follows that

$$(r_0^i, \ldots, r_l^i, q_u^i, s_0^i, \ldots, s_p^i) \text{ and } (\dot{r}_0^i, \ldots, \dot{r}_{l'}^i, \dot{q}_{u'}^i, \dot{s}_0^i, \ldots, \dot{s}_{p'}^i)$$

also are equal. Therefore, $q_u^i = \dot{q}_u^i$ and $s_0^i = \dot{s}_0^i$, i.e., $\hat{M}$ is one-to-one. ∎

In this way, it is shown that multiple Turing machines satisfy the property of reversibility necessary to develop quantum computability. It is important to emphasize that Bernstein and Vazirani's notion of reversibility is restricted to deterministic Turing machines, as in the case of the logical reversibility. If the determinism also is a necessary condition to quantum computing, then multiple Turing machines cannot be used to define quantum models of computation. In particular, the same is true for paraconsistent Turing machines. We will not approach this point. Our concern here was in showing that there is a way to combine reversibility and nondeterminism in a sense appropriate to quantum computing.

5 Conclusion

Paraconsistent Turing machines were invented to simulate quantum computations, as an attempt to support the logical relativity thesis about computation. We have defined the same concepts without resorting to logic.

It is of concern that paraconsistent Turing machines may not be reversible, while quantum computation must be reversible. We have proposed a new class of Turing machines, the Multiple Turing machines, and shown they are reversible *and* that they simulate paraconsistent Turing machines; the latter has not been demonstrated, however. So, in principle, multiple Turing machines may be used to simulate quantum Turing machines and bridge the gap between paraconsistent and quantum computations.

This leave the path of simulating on kind of computation with another kind clear for exploration. The classes involved here may be inspired by logical properties, but the computations they generate have not been related to any logic. Moreover, in [de Arajo and Finger, 2011] we have shown that a natural quantum version called QSAT of the SAT problem does not allow a direct comparison between the complexity classes NP and QMA, for which SAT and QSAT are respectively complete (Cf. [Kitaev *et al.*, 2002]). Therefore, our conclusion is indeed a question: In which sense quantum computing is a matter of logic?

BIBLIOGRAPHY

[Agudelo and Carnielli, 2010] J.C. Agudelo and W. Carnielli. Paraconsistent machines and their relation to quantum computing. *Journal of Logic and Computation*, 20(2):573–595, 2010.

[Bennet and Vazirani, 1997] C.H. Bennet and U. Vazirani. Quantum complexity theory. *SIAM Journal of Computing*, 26(5):1411–1473, 1997.

[Birkhoff and von Neumann, 1936] G.D. Birkhoff and J. von Neumann. The logic of quantum mechanics. *Annals of Mathematics*, 37:823–843, 1936.

[Carnielli and Dria, 2008] W.A. Carnielli and F.A. Dria. Are the Foundations of Computer Science Logic Dependent? In C. Degremont, L. Keiff, and H. Ruckert, editors, *Dialogues, Logics and Other Strange Things - Essays in Honour of Shahid Rahman*, volume 1 of Londres, pages 87–107. College Publicatios, 2008.

[Carnielli *et al.*, 2007] W.A. Carnielli, M.E. Coniglio, and J. Marcos. Logics of formal inconsistency. In D. Gabbay and E. Guenthner, editors, *Handbook of Philosophical Logic*, volume 14, pages 15–107. Kluwer Academic publishers, 2007.

[Chiara and Giuntini, 2002] M.L.D. Chiara and R. Giuntini. Quantum logic. In D. Gabbay and E. Guenthner, editors, *Handbook of Philosophical Logic*, volume 6, pages 129–228. Kluwer Academic publishers, 2002. 2nd ed.

[Cohen-Tannoudji *et al.*, 1977] C. Cohen-Tannoudji, B. Diu, and F. Lalo. *Quantum Mechanics*. Wiley, New York, 1977.

[Crescenzi and Papadimitriou, 1995] P. Crescenzi and C.H. Papadimitriou. Reversible simulation of space-bounded computations. *Theoretical Computer Science*, 143:159–165, 1995.

[de Arajo and Finger, 2011] A. de Arajo and M. Finger. Classical and quantum satisfiability. In *Eletronic Proceedings of Theoretical Computer Science*, 2011. To appear.

[de Arajo, 2011] A. de Arajo. *A model-theoretical approach to classical Turing computability*. PhD thesis, State University of Campinas (UNICAMP), Brazil, 2011. In Portuguese.

[Deutsch and Jozsa, 1992] D. Deutsch and R. Jozsa. Rapid solution of problems by quantum computation. *Proceedings of the Royal Society of London*, Series A(439):553–558, 1992.

[Deutsch, 1985] D. Deutsch. Quantum theory, the church-turing principle and the universal quantum computer. *Proceedings of the Royal Society of London*, Series A(400):97–117, 1985.

[Jech, 2002] T. Jech. *Set Theory*. Spring Verlag, Berlin, 2002.

[Kitaev et al., 2002] A. Kitaev, A. Shen, and M. Vyalyi. *Classical and quantum computation*, volume 47 of *Graduate Studies in Mathematics*. American Mathematical Society, New York, 2002.

[Nielsen and Chuang, 2000] M.A. Nielsen and I.L. Chuang. *Quantum Computation and Quantum Information*. Cambridge University Press, Cambridge, 2000.

[Sipser, 2006] M. Sipser. *Introduction to the theory of computation*. Thomson Course Technology, Massachusetts, 2006. 2nd ed.

[Sylvan and Copeland, 2000] R. Sylvan and J. Copeland. Computability is logic-relative. In G. Priest and D. Hyde, editors, *Sociative Logics and Their Applications: Essays by the Late Richard Sylvan*, American Mathematical Society Proceedings of Symposia in Pure Mathematics, pages 189–199. Ashgate Publishing Company, 2000.

Anderson de Araújo and Marcelo Finger
Institute of Mathematics and Statistics (IME)
University of São Paulo (USP)
São Paulo, Brazil
E-mail: {aaraujo,mfinger}@ime.usp.br

Delegation, Count as, and Security in Talmudic Logic, a preliminary study

M. ABRAHAM, R. BELFER, D. GABBAY, AND U. SCHILD

ABSTRACT. Delegation is a commonplace feature in our society. Individuals give power of attorney to their lawyers to perform certain actions for them (e.g. buy or sell property), institutions delegate to certain employees to sign for them (human resources send letters of appointment) and owners can grant access and administrative rights to other people in relation to their servers.

The logic behind such a system has been studied by several communities.

In philosophy this is known as "count as". X counts as Y in context C.

In law there are various rules for power of attorney.

In computer science one talks about access control and delegation.

This paper examines the approach to delegation in Talmudic Logic.

The current approaches to delegation, mainly study three features

1. Dominance — if several primary sources delegate to secondary sources who carry on delegating then what is the dominance relationship among the chains of delegations

2. Revocation — if some sources revoke the delegation or some change their minds and reinstate, how does this propagate through the chains of delegations?

3. Resilience — if one source revokes delegation do we cancel other delegations from other sources on the grounds that we now do not trust the delegate?

In the literature systems have been constructed which either model or implement a calculus of Delegation-Revocation (Privilege calculus). Their purpose is to answer the question of whether the chain of delegation and revocations can allow an agent to perform an action and their models are chain update models.

The Talmudic approach is slightly different not only in the details of its model but also in its point view.

The Talmud not only examines the procedure of the actual acts of delegation and revocation and its calculus but also includes ordinary actions(not just chain update actions) which may have preconditions addressing not only facts but also delegation and revocation chains leading to the actions. The Talmud also addresses inability of delegated agents to execute the actions and the possibility of agents going mad or dying during the delegation revocation process.

1 Background and orientation

Delegation is a commonplace feature in our society. Individuals give power of attorney to their lawyers to perform certain actions for them (e.g. buy or sell property), institutions delegate to certain employees to sign for them (human resources send letters of appointment) and owners can grant access and adminisrtative rights to other people in relation to their servers.

The logic behind such a system has been studied by several communities.

In philosophy this is known as "count as". X counts as Y in context C [Searle, 1969; Searle, 1995; Jones and Sergot, 1996], and see [Grossi and Jones, to appear] for a survey.

In law there are various rules for power of attorney.

In computer science one talks about access control and delegation, see for example [Rissanen *et al.*, 2005].

This paper examines the approach to delegation in Talmudic Logic.

The following are feature to be addressed:

1. The general logical context in which delegation takes place.

2. Exactly how (by what process) does agent **a** delegate to agent **b** item φ.

3. What are the rules for making chains of delegation?

4. How can delegation be revoked in a chain?

5. What happens in a delegation chain if some of the agents in the chain become insane (i.e. irresponsible or generally break down) and how to continue if such agents become sane again? What if they die (drop out permanently)?

6. What to do if some agents exceed their remit in a chain (e.g. human resources in a University sends a letter of appointment by mistake to the wrong candidate)?

7. It may be the case that several agents a_i capable of executing action α, each delegates to the same agent **b** to do α. Meanwhile some of these agents a_i go insane, some die, and some cancel the delegation. What can **b** do?

A word on methodology. The Talmud (completed at the end of the fifth century) and its later interpreters (another 5-10 centuries) is full of debate about various cases of delegation. There is no formal logic, just argumentation involving various case studies by various scholars. So finding the logic behind it means that we have to find a logical model with some degrees of freedom and a mapping of the various scholars or views to parameters in the logical model. We then have to go to all places and cases in the Talmud where there is a debate and the model must explain each move in each argument in each debate in each place in a perfect match. This is possible to do because Talmudic debates are remarkably coherent and consistent.

The topic of delegation is the sixth topic in Talmudic logic which we are examining. We have already published five books modelling five previous topics, using the same methodology.[1]

2 Motivating the Talmudic system

Let $\mathcal{A}$ be a set of actions and $\mathbf{A}$ be a set of agents. We need a relation $\mathbb{R} \subseteq \mathbf{A} \times \mathcal{A}$ giving us for each agent $\mathbf{a}$ in $\mathbf{A}$ the set of all actions $\alpha \in \mathcal{A}$ such that $\mathbf{a}$ can

---

[1]

1. Non-deductive Inference in the Talmud (with M. Abraham and U. Schild). 350pp, College Publications, 2010.
   We analyse the three basic non-deductive rules of Talmudic inference; namely Kal Vachomer (Argumentum A Fortiori) and the two kinds of Binyan Av (Analogy and Induction). We construct a unified Matrix Abduction model that explains all the major instances of these rules in the Talmud.

2. The Textual Inference Rules Klal uPrat. How the Talmud Defines Sets (with M. Abraham, G. Hazut, Y. Maruvka and U. Schild). 300pp, College Publications, 2010.
   We analyse the Klal uPrat family of textual rules in the Talmud. We view them as common-sense practical rules for defining sets. Such methods do not exist in general common-sense logical systems, and they complement the existing common-sense (non-monotonic) deductive logics.

3. Talmudic Deontic Logic (with M. Abraham and U. Schild). 296pp, College Publications, 2010.
   In this book we study the Deontic Logic of the Talmud. We find the system is different from the formal deontic logical system currently used in the general scientific community, both in its ethical aspects as well as in its legal aspects. We show that the Talmudic distinctions between Obligations and Prohibitions are not based on the manner of execution of actions (positive action or lack of action) and offer a suitable model for such distinctions.Our model distinguishes between the normative and practical aspects of the Talmudic legal and ethical argumentation and discusses several applications and clarifications to current so called paradoxes of Deontic Logic as related to Contrary to Duties and to legal and ethical practical decision making.

4. Temporal Logic in the Talmud (with M. Abraham, I. Belfer and U. Schild). 674pp, College Publications 2011.
   This book studies Talmudic temporal logic and compares it with the logic of time in contemporary law. Following a general introduction about the logical handling of time, the book examines several key Talmudic debates involving time. The book finds that we need multi-dimensional temporal models with backward causation and parallel histories.
   It seems that two major issues are involved:

   (a) Actions conditional about future actions (Tenayim), connecting with backward causality;

   (b) Actions involving entities defined using future events (Breira), connecting with ideas from quantum mechanics.The book concludes with a general comparative discussion of the handling of time in general law and in the Talmud

5. Resolution of Conflicts and Normative Loops in the Talmud (with M. Abraham and U. Schild). 316pp, College Publications 2011.
   In this book we describe the fundamental rules for conflict resolution and address the basic Talmudic methods for resolving conflicts. We also investigate logical loops in Talmudic argumentation. It is obvious that one needs meta-level (out of the box) considerations. We also consider conflicts between Biblical Obligations and Prohibitions, a topic we studied in our third book. We conclude by comparing some features of conflict resolution with our matrix model presented in our first book.

execute α (**a** has the authority to execute α). The actions have the form $\alpha = (A_\alpha, B_\alpha)$, where A_α is the pre-conditon and B_α is the post-condition. A_α and B_α are written in some predicate language $\mathbb{L}$, to be described later, after we see what we need!

An agent **a** can delegate his authority to do any action to agent **b** (there are some restrictions on agent **b**, like he has to be sane and responsible and can perform actions similar to α). He must not be involved in the action α himself, and the action α must be legally meaningful). We need a relation $\mathbb{D}$ where $\mathbb{D}(\mathbf{a}, \mathbf{b}, \alpha)$ means that **a** delegated action α to **b**. This can be delegated further by **b**. So the relation $\mathbb{R}$ can be expanded to a relation $\mathbb{F}^*$, namely

$$x\mathbb{R}^*\alpha \text{ iff } \exists y_1, \ldots, y_k \text{ for some } k, \text{ such that } y_1\mathbb{R}\alpha \wedge \bigwedge_{i=1}^{k-1} \mathbb{D}(y_i, y_{i+1}, \alpha) \wedge y_k = x.$$

In order to model the complexities of delegation in Talmudic logic we want to realise $\mathbb{D}$ using tokens (modern papers call them certificates, see for example [Bandmann et al., 2002; Final Amendment, 1999] and [Rissanen et al., 2004]).

An agent **a** which can do α has a token $\mathbb{T}(\mathbf{a}, \alpha)$. Think of it as a copy print of $(\mathbf{a}, \alpha) \in \mathbb{R}$. If **a** wants to delegate to **b**, he signs on the token, "I authorise **b**́. We denote this by $(\mathbf{a}, \mathbf{b}, \alpha)$. Thus we can get the chain $(y_1, \ldots, y_k, \alpha)$. If y_k wants to execute an action α, $\alpha = (A_\alpha, B_\alpha)$, the language $\mathbb{L}$ must also enable A_α to ask y_k: do you have a token $(y_1, \ldots, y_k, \alpha)$?

So for example to sell a table t, we need the agent **a** to own t and then he can sell it to agent **b**. Or we can have $(\mathbf{a}, y_1, \ldots, y_k, \text{sell table})$ and y_k can sell the table on behalf of **a**. So the language $\mathbb{L}$ must contain $\mathbb{D}$, as well as the names of agents and facts about the world.

This is a language where $\alpha = (A_\alpha, B_\alpha)$ and A_α can talk about α. It is a self reflecting language.

Different delegation theories will be implemented by different properties of the tokens.

There are two main types of delegation in Talmudic logic.

1. Power of attorney view (Maimonides[2] view).

2. The long arm view (Tur[3] view).

If **a** delegates to **b** and **b** delegates further to **c**, let us refer to **a** as the master (or principal, using modern terminology) and to **b** as the agent and to **c** as the subagent.

The power of attorney view is for the master to delegate to an agent to do action α.

The action is done by the agent and the result of the action is passed on to the master.

---

[2] Moses ben-Maimon, called Maimonides Ramban (Hebrew acronym for "Rabbi Moshe ben Maimon").

[3] Jacob ben Asher, also known as Ba'al ha-Turim as well as Rabbi Yaakov ben Raash (Rabbeinu Asher), was likely born in Cologne, Germany, c. 1269 and likely died in Toledo, Spain, c. 1343.

| Owner of token: | John Smith |
| | Social Security number (SSN): |

Action	Master (who has authority over the action)	Verify nomination of owner of this token (i.e. John Smith)
1. Sell house (property ♯)	Terry Jordan owner of the house	+from Terry
2. α	a	+from a

Figure 1.

action α.
Issued by agent **a**
such that $a\mathbb{R}\alpha$.

Figure 2.

The long arm view is that the agent is an extension of the arm of the master, and the master is doing the action α by means of his arm extension — the agent.

So the agent "counts as" the master.

We model the difference between these two views through the properties of the token. The token is given from the master to the delegated agent and in the token there is a list of actions to be done by the agent.

The power of attorney view postulates a token for each agent. The long arm view postulates a token for each delegated action/job.

The token per agent view envisages the token as listing all the actions to be done. These include actions that the master has authority to do, as well as actions which he the master was recruited by a previous master to do (to whom he acts as an agent).

Figure 1 shows what this token looks like. Note also that this token allows for the master to cancel the appointment of the agent as an agent for the action (in modern terminology, the master revokes the delegation to the agent).

In the case of the long arm view, the tokens look like Figure 2

We now describe what happens when agent John Smith wants to execute an action. We check the following:

1. The power of attorney view checks whether action α shows in John Smith token (see Figure 1). Is he the master for this action? Was he appointed agent for this action by a master who has authority? Was he appointed by an agent who was himself appointed by a master? etc. All the above is supposed to be recorded in the token.

2. The long arm view would simply check if our John Smith has the token as in Figure 2.

EXAMPLE 1. To see the difference between the two views, let us assume that the master **a** appointed **b** as an agent for him to do action α, and then lost his mind.

The long arm view will say the action cannot be performed because the source of the long arm, the master, is without a sound mind. If we look at the token 2, the **a** in a$\mathbb{R}\alpha$ is no longer sane.

The power of attorney view, the owner of the token (the agent **b**) is capable and sane, his token indicates he has the authority to take action, so he can do it!

Let us take a very simple example from practice. The manager of a company delegates to a secretary to delete certain sensitive files from the server, just before a shareholders' meeting is about to take place. The secretary goes to the meeting and intended to do the action afterwards. During the stormy meeting, the manager resigned and discussions were ongoing about appointing a new manager. The long arm view would say the secretary cannot delete the files because she is the long arm of the manager who is no longer in power, he resigned. The power of attorney view says that the secretary has a power of attorney, he/she should do the action and delete the files.

EXAMPLE 2 (Cancellation and reinstatement).

1. Both views allow for cancellation (the modern term is revocation). The master cancels either the token of Figure 1 or of Figure 2, depending on the view.

2. Both views agree that if the master becomes sane again (the manager of Example 1 gets reinstated) the action can take place without the need for doing again the formal appointment of delegation (i.e. the secretary need not ask the reinstated manager to reconfirm his instructions to him/her).

EXAMPLE 3 (The delegated agent becomes insane). Suppose the agent goes crazy, and then becomes sane again (goes through a mental breakdown for a while). Can he/she continue being an agent and execute the action?

According to the long arm view he can. He got the token, he is now sane, so he can do it. Similarly according to the power of attorney view. He got the token.

There is a difference however in the view about pre-condition A_α of α.

The long arm view says A_α must check the sanity (capability) of both the master who controls the long arm and the agent, who is the arm. Both have to be functional.

The power of attorney view needs the functionality check in A_α of the agent only. The agent carries the token, he is supposed to execute the action!

We now examine how, according to each view, an agent can nominate a subsgent for himself.

The long arm view treats this very simply. The agent has a token as in Figure 2. So he just passes this token on to his subagent. All very simple. The subagent is now the long arm of the master. The agent is no longer in the picture. We may have a long chain of such nominations. So a by-product of this view is that the master can cancel the nomination of his long arm agent at the end of the chain, no matter how long the chain is.

Owner of token: John Smith. SSN:

Action	Master a	who nominated John	confirm nomination from Levy not retracted	Who nominated Levy	Confirm nomination from Terry not retracted
sell house: address	Terry Jordan	+Levy =c	+confirm	+Terry =b	+confirm
α	a	c	+	b	(negative, a retracted the nomination

Figure 3.

The power of attorney view would have to say that the subagent is delegated from the agent and not from the master. Thus the token must record this information. It must record the chain of delegations from agent to agent. A by-product of this view is that the master cannot cancel the nomination of the subagent. The subagent was nominated by the agent not by the master! The master can cancel the nomination of the agent but if the agent has already nominated a subagent then the nomination of the subagent stands! Figure 3 shows what the token looks like.

3 Technical definitions of the logical model

3.1 Preliminary discussion

The framework in which we are working involves agents and actions, with a relation $\mathbb{R}$ between agents **a** and actions α, saying whether agent **a** can execute action α ($a\mathbb{R}\alpha$). α has preconditions A_α and postconditions B_α.

To this model we add the delegation component, in which agent **a** can delegate to agent **b** the execution of action α. We wrote this as $\mathbb{D}(\mathbf{a}, \mathbf{b}, \alpha)$.

Thus any model of delegation needs to be based on a model for agents and actions. Since our primary interest is in modelling delegation we can take a simple basic model of agents and actions, without any fine refinements (coming from a sophisticated multi-agent theories), provided that this agent-action model is rich enough to allow us to express all the delegation features we need to model.

DEFINITION 4. Let **A** be a finite set of agents and $\mathcal{A}$ a finite set of actions. By a basic multi-agent system we mean a tuple of the form $\mathfrak{M} = (S, \mathbf{R})$, where S is a non-empty set of states and $\mathbf{R} \subseteq (S \times S) \times \mathbf{A} \times \mathcal{A}$.

When $(t, s, \mathbf{a}, \alpha) \in \mathbf{R}$, we draw it graphically as in Figure 4. The figure means that at state t agent **a** can execute action α which moves the system to state s.

It may be that agent **b** can also execute α at state t but not agent **c**. So we write Figure 5.

where $\mathbf{E} \subseteq \mathbf{A}$ is the set of agents which can execute α at state t.

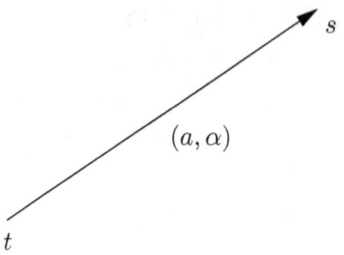

Figure 4.

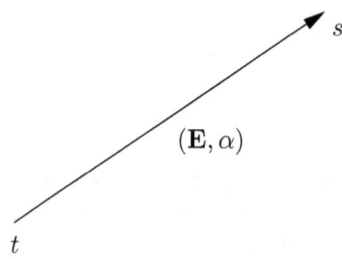

Figure 5.

Thus
$$\mathbf{E} = \{\mathbf{x} | (t, s, \mathbf{x}, \alpha) \in \mathbf{R}\}.$$
We need to require that the execution of α is deterministic, i.e.

- $(t, s, \mathbf{x}, \alpha) \in \mathbf{R}$ and $(t, s', \mathbf{y}, \alpha) \in \mathbf{R}$ implies $s = s'$.

We believe that this simple model is good enough for our purposes.

REMARK 5. Note that we ignored the representation of the preconditions A_α and postconditions B_α of actions α. This we can do because we use a set of states. If A_α does not hold at state t then the action cannot be taken. If it is taken then B_α holds at state s.

Thus to complete the model, we take $(S, \mathbf{R})$ and associate with each $t \in S$, a model $\mathbf{m}_t$ of the language $\mathbb{L}$ of the preconditions and postconditions, so that we can write $t \vDash A_\alpha$ or $s \nvDash B_\alpha$, etc.

Thus our final models have the form
$$\mathfrak{M} = (S, \mathbf{R}, \mathbf{m}_t), t \in S.$$

The above does not deal with delegation yet. We now examine our options for modelling delegation.

It is convenient to list the views and types of delegation we encountered in Section 2.

Type 1. **a** delegates to **b** action α in any context (state), e.g. "sell my house".

Type 2. **a** delegates to **b** action α only in certain contexts, e.g. "sell my house when the Euro is over 1.30 to the dollar".

Type 3. Preconditions and postconditions of actions involve delegation considerations, even though the action itself is not a delegation action.

For example, when **a** delegates to **b** action α then when **b** wants to execute α, part of the precondition of α is that agent **a** is sane and alive.

View 1. Agent **b** is the long arm of agent **a**.

In this case if we are at state t, we move to state s after the execution of the action (either by **a** or by **b**).

View 2. Agent **a** gives Talmudic power of attorney to agent **b** for example, to buy a house for him from agent **c** or to collect a debt for him from agent **c**.

In this case the execution of the action may result in an intermediate state. In the case of buying a house there is no intermediate state, but in the case of collecting a debt, where agent **c** also owes money to agent **b**, there is an intermediate state. We view the sequence of actions to be that the money first goes in the hands of agent **b**, who then passes it to agent **a**. So if we start at state t we move to t' and then to s, unlike the long arm delegation, where we move from t to s directly.

Option 1: The fibred (combined) option

This option puts a delegation program or logic next to a model $\mathfrak{M}$ for agents and actions. The program is used to update the relation $\mathbf{R}$ in $\mathfrak{M}$.

Let Δ be a database of all the delegation tokens in the system. The model becomes

$$\mathfrak{M} = (S, \mathbf{R}^\Delta, \mathbf{m}_t), t \in S.$$

When an agent delegates an action to another agent, the delegation database Δ is updated to Δ' and $\mathfrak{M}$ changes to

$$\mathfrak{M}' = (S, \mathbf{R}^{\Delta'}, \mathbf{m}_t), t \in S.$$

So the update system of Δ communicates with $\mathbf{R}$ of $\mathfrak{M}$.

Thus logically modelling such a system requires the logical modelling of communication between a program and a logic.

Since the updating system is independent of $\mathfrak{M}$, many researchers use Δ only as models of delegation and do not mention $\mathfrak{M}$ at all. This may not be possible if the systems interact. For example, part of the precondition of action α may be that it is not performed through delegation. The prime minister of a country or a King, for example, cannot freely delegate some actions associated with his position. A wife having difficulties giving a child to her husband cannot, nowadays, delegate the job to her maid, as was the custom in Biblical times.

Another example is when the delegations comes as a result of an action α in the real world is the following. If I run over the parents of a small child then by law the court becomes delegates for his interests. So here the delegation is a postcondition of my actions.

In such cases, where there is interaction between the delegation system and the preconditions and postconditions of ordinary actions, the next integrated model is a better option.

Option 2: The integrated model

This model views the delegation details (certificates, tokens, etc.) as part of the state $t \in S$, and the act of delegation is viewed as just another action modifying the state. So in this integrated model, $\mathbf{m}_t$ talks not only about facts, but includes the details of each delegation certificate/token ever issued. The precondition of actions must include that the action is executed by an agent who has the valid delegation token for the action.

A variety of models can be constructed under this option, but they all have the drawback that the delegation part gets a bit lost as a separate system. We could save the uniqueness of the delegation part by separating delegation actions from ordinary actions and add temporal like connectives to the model which can chain correctly the delegation actions, etc., etc. If we do that well, we will be back to Option 1, hidden inside Option 2 under the guise of additional connectives.

Option 3: Reactivity model for the long arm view of the Tur

We now intuitively explain the components of this model. It is an integrated model which uses reactivity to separate delegation actions from ordinary actions in a natural way. For reactive Kripke models see [Gabbay, 2008].

It is best suited when there is intereaction between delegation and post and preconditions of actions or where there are delegations which are valid only in certain contexts (states).

If delegations involve just agents and actions then Option 1 may be best.

First we decide that we keep the agent action model as it is, with the states describing pure facts, with no mention of delegation. So the relation $\mathbf{R}$ can be represented as in Figure 5. So there is no mention of delegation in this figure. We are going to add delegation into it. We need to write Figure 5 more explicitly. Consider Figure 6.

This Figure contains some information as Figure 5. We just wrote the information explicitly. We have $\mathbf{E} = \{\mathbf{a}_1, \ldots, \mathbf{a}_n\}$ and $\mathbf{A} - \mathbf{E} = \{\mathbf{b}_1, \ldots, \mathbf{b}_k\}$.

The action α can be executed by $\mathbf{a}_1, \ldots, \mathbf{a}_n$ and so we have the arrows $t \to s$ annotated by $+(\mathbf{a}_i, \alpha), i = 1, \ldots, n$. The action α cannot be taken by $\mathbf{b}_1, \ldots, \mathbf{b}_k$ and so we have the arrows $t \to s$ annotated by $-(\mathbf{b}_j, \alpha)$.

It would be easier to replace the relation $\mathbf{R}$ by its characteristic function $\mathbf{F_R}$. We have

$$\mathbf{F_R}(t, s, \mathbf{a}, \alpha) = 1 \text{ iff } (t, s, \mathbf{a}, \alpha) \in \mathbf{R}.$$

From now on we regard $\mathbf{R}$ as such a function (by abuse of notation).

We begin discussing delegation for the long arm view of the Tur:

Now suppose we perform an action of delegation. Agent $\mathbf{a}_1$, who can perform action α, wants to delegate to agent $\mathbf{b}_1$, the execution of α at state t. Agent $\mathbf{b}_1$

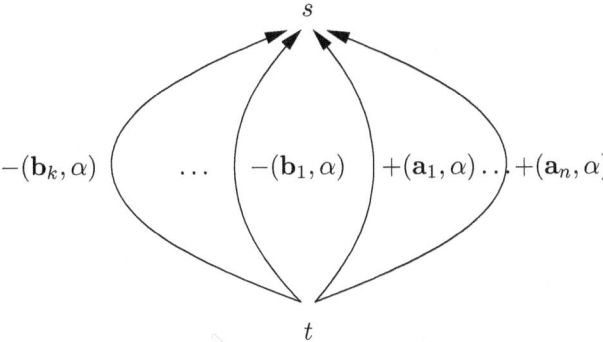

Figure 6.

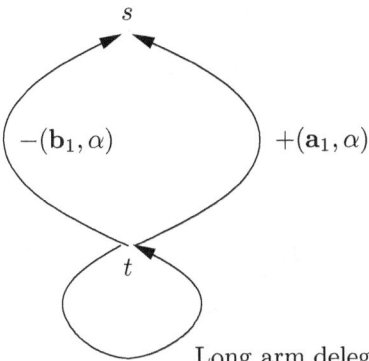

Long arm delegation of action α from $\mathbf{a}_1$ to $\mathbf{b}_1$

Figure 7.

cannot perform action α at state t before the delegation (i.e. we have $+(\mathbf{a}_1, \alpha)$ and $-(\mathbf{b}_1, \alpha)$ before the delegation) but after delegation $-(\mathbf{b}_1, \alpha)$ is updated and changed to $+(\mathbf{b}_1, \alpha)$.

How do we represent this using reactive arrows? When we perform the delegation action, we remain at state t, since in our model, the states represent facts about the world and do not contain any token/certificate delegation information.

Figure 7 represents this move for $\mathbf{a}_1$ and $\mathbf{b}_1$.

We represent Figure 7 by the reactive Figure 8.

The reactive double arrow is written as

$$(t \to_{(\mathbf{a}_1, \mathbf{b}_1, \alpha)} t) \twoheadrightarrow_\lambda (t \to_{(\pm(\mathbf{b}_1, \alpha))} s).$$

λ is a label indicating the nature of the delegation/revocation double arrow. As $\mathbf{a}_1$ moves from t to t along the arrow, he triggers the double arrow which sends a signal λ to $t \to_{\pm(\mathbf{b}_1, \alpha)} s$. Let us assume that $\lambda = $ switch. Then if the annoation of $t \to s$ is "+", it turns into "-" and if it is "-" it turns it into "+".

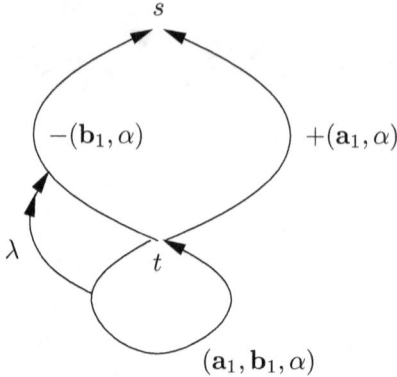

Figure 8.

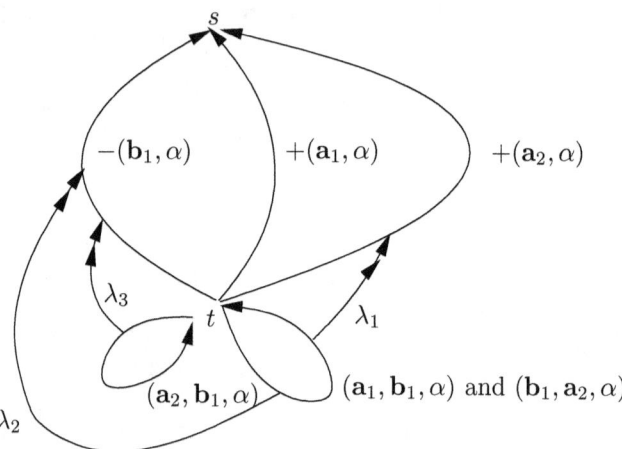

Figure 9.

thus the double arrow is a switch!

Let us look at Figure 9.

Again, let us assume that $\lambda_1 = \lambda_2 = \lambda_3 =$ switch. Suppose $\mathbf{a}_1$ goes along the path

$$t \to_{(\mathbf{a}_1,\mathbf{b}_1,\alpha) \text{ and } (\mathbf{a}_1,\mathbf{a}_2,\alpha)} t \to_{(\mathbf{a}_1,\mathbf{b}_1,\alpha) \text{ and } (\mathbf{a}_1,\mathbf{a}_2,\alpha)} t$$

and then $\mathbf{a}_2$ continues along the path $t \to_{(\mathbf{a}_2,\mathbf{b}_1,\alpha)} t$.

The total movement is

$$t \to_{(\mathbf{a}_1,\mathbf{b}_1,\alpha) \text{ and } (\mathbf{a}_1,\mathbf{a}_2,\alpha)} t \to_{(\mathbf{a}_1,\mathbf{b}_1,\alpha) \text{ and } (\mathbf{a}_1,\mathbf{a}_2,\alpha)} t \to_{(\mathbf{a}_2,\mathbf{b}_1,\alpha)} t$$

The first $t \to_{(\mathbf{a}_1,\mathbf{b}_1,\alpha) \text{ and } (\mathbf{a}_1,\mathbf{a}_2,\alpha)} t$ switches $-(\mathbf{b}_1,\alpha)$ into $+(\mathbf{b}_1,\alpha)$. We consider that legitimate because we do have in the Figure $t \to_{+(\mathbf{a}_1,\alpha)} s$.

This same first movement also revokes the right of $\mathbf{a}_2$ to execute α. So it switches $+(\mathbf{a}_2,\alpha)$ into $-(\mathbf{a}_2,\alpha)$. Now $\mathbf{a}_2$ cannot delegate α because his ability to execute α was revoked by $\mathbf{a}_1$. Fortunately, $\mathbf{a}_1$ went again through $t \to_{(\mathbf{a}_1,\mathbf{b}_1,\alpha)\ and\ (\mathbf{a}_1,\mathbf{a}_2,\alpha)} t$ and switched back to $+(\mathbf{a}_2,\alpha)$ and $-(\mathbf{b}_1,\alpha)$. Now $\mathbf{a}_2$ can go through his arc $t \to_{(\mathbf{a}_2,\mathbf{b}_1,\alpha)} t$ and switch on $+(\mathbf{b}_1,\alpha)$.

In the general case, where $\lambda_1, \lambda_2, \lambda_3$ can be general labels, not necessarily "switches", we need to collect the labels and decide whether the target arc (which is hit by several labels) is supposed to be on ("+") or not ("-").

For example, assume that

$$\lambda_1 = \text{switch}$$
$$\lambda_2 = \text{dominant delegation}$$
$$\lambda_3 = \text{switch}$$

So as we move along

$$t \to_{(\mathbf{a}_1,\mathbf{b}_1,\alpha)\ and\ (\mathbf{a}_1,\mathbf{a}_2,\alpha)} t$$

the arc $t \to_{+(\mathbf{a}_2,\alpha)} s$ is switched to $t \to_{-(\mathbf{a}_2,\alpha)} s$ and the arc $t \to_{-(\mathbf{b}_1,\alpha)} s$ is changed to dominant $t \to_{+(\mathbf{b}_1,\alpha)} s$.

As we continue along

$$t \to_{(\mathbf{a}_1,\mathbf{b}_1,\alpha)\ and\ (\mathbf{a}_1,\mathbf{a}_2,\alpha)} t$$

the arc $t \to_{-(\mathbf{a}_2,\alpha)} s$ changes back to become $t \to_{+(\mathbf{a}_2,\alpha)} s$ as it is hit by λ_1 but the arc $t \to_{+(\mathbf{b}_1,\alpha)} s$ does not change as it is hit again by $\lambda_2 = $ dominant delegation.

Now we continue along the arc

$$t \to_{(\mathbf{a}_2,\mathbf{b}_1,\alpha)} t$$

and the arc $t \to_{+(\mathbf{a}_1,\alpha)} s$ is hit by $\lambda_3 = $ switch. the $+(\mathbf{b}_1,\alpha)$ does not change because it was hit before by $\lambda_2 = $ dominant delegation and it is now hit by just a switch, which is not dominant.

Now if λ_3 were

$$\lambda_3' = \text{dominant revocation}$$

then we would need to decide whether the triple $\{\lambda_2, \lambda_2, \lambda_3'\}$ should end up with + or with -.

This means that in the general case, when we go along a path and trigger various double arrows with labels, we need to calculate using a flattening algorithm Λ, whether any given arc is "+" or "-". We collect all the albels λ_j which hit the arc along the path and let Λ "flatten" it to either "+" or "-".

We also note, see Figure 10, that we have the notation to delegate from state t to state r, where r can be anywhere in the system. However the Talmud does not allow for delegation for action which is not definite now but will be in the future. We can however put a condition into the precondition of an action but we have to delegate immediately. So I cannot say: when the Euro rate becomes over 1.4 to the Dollar you become my delegate to sell my house, but I can say: I delegate you now to sell my house on the condition that the Euro rate becomes over 1.4 to the Dollar.Thus according to the Talmud, Figure 10 cannot arise unless $r = t$.

What we learn from the above examples and discussion is the following:

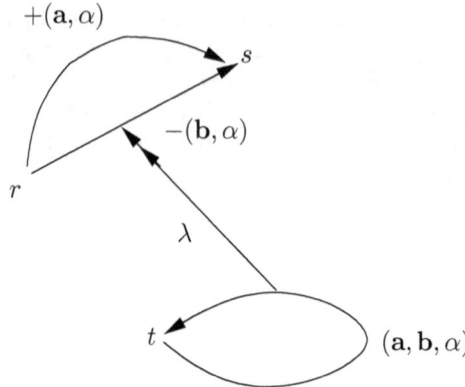

Figure 10.

1. We need to specify an annotated path π of delegation.

2. Final revocation or delegation is according to the last delegation (revocation) action in case of switch double arrows but requires a falttening function Λ if we use labels.

3. The double arrows, through their labels, give us who has power over whom to delegate or revoke.

4. We can constrain the movements along the arcs by the geometry of the arcs.

We are devoting special attention to switch labels because this is the simplest model. An agent **a** at node t who wants to delegate can go the appropriate arc $t \to t$ with the appropriate double arrow emanating from it. To revoke he just goes again through the arc. To cancel the revocation, he goes again, etc.

This is the simplest model.

Option 4: Reactivity model for the power of attorney view of Maimonides

Our starting point for modelling this view is Figure 7, which we modify for representing the case of power of attorney. We get Figure 11.

$\mathbf{a}_1$ can execute action α and move from state t to state s. $\mathbf{b}_1$ cannot do this. In Figure 11, the fact that $\mathbf{a}_1$ can go from t to s is represented by the continuous arrow

$$t \to_{(\mathbf{a}_1, \alpha)} s.$$

The fact that $\mathbf{b}_1$ cannot execute action α and go from t to s is represented by the broken arrow

$$t \not\to_{(\mathbf{b}_1, \alpha)} s$$

The big circle around t and the big circle around s are the locations (round tables for t and s respectively) where our agents are sitting ready to delegate and take action. In the long arm view, if applied to Figure 11, agent $\mathbf{a}_1$ can

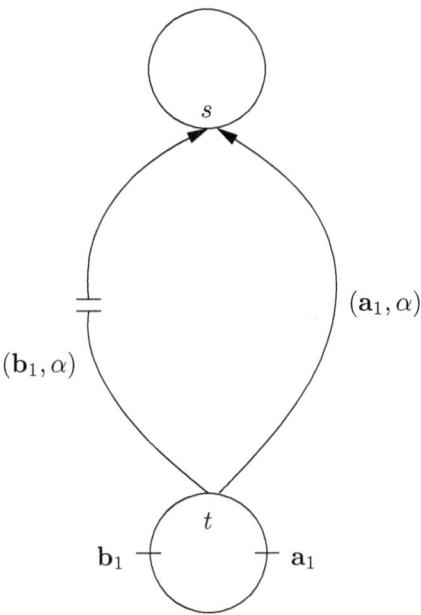

Figure 11.

delegate (long arm) to agent b_1 the action α by sending a double arrow to the gap of agent b_1 and closign the gap for him so b_1 can move from t to s along his own arrow.

Figure 12 shows this long arm delegation and it should be compared with the slightly different Figure 8.

However, in the case of the power of attorney view, a_1 delegates to b_1 by allowing for a double arrow from b_1 to a_1 inside the round talbe circle at t. Figure 13 shows what we mean.

In Figure 13, agent b_1 does have a way to move from t to s exeuting α. He moves from t to s executing α. He moves along the double arrow $b_1 \twoheadrightarrow a_1$ to a_1 position and then moves to s along the a_1 arc

$$t \to_{(a_1, \alpha)} s.$$

Here we see how a_1 truly sends b_1 along his own arc!

We note that actually there was no need to draw the broken arc $t \not\mapsto_{b_1,\alpha} s$ in Figure 11. Since b_1 cannot go from t to s by executing α, it is quite sufficient not to draw an arc for b_1 at all, and the absence of such an arc would indicate that b_1 cannot get to s. However for the sake of comparison, of the long arm view iwth the power of attorney view (as shown in Figures 12 and 13) it is advantageous to draw the broken arc in Figure 11.

Note that the Talmudic options for delegation are either the long arm view in all cases, or the power to attorney view in all cases. We do not have the mixed option, as in Figure 14. In other words, the Talmud has many cases and debates about delegtion. Maimonides takes the view that they are all power of attorney cases and the Tur takes the view that they are all long arm cases.

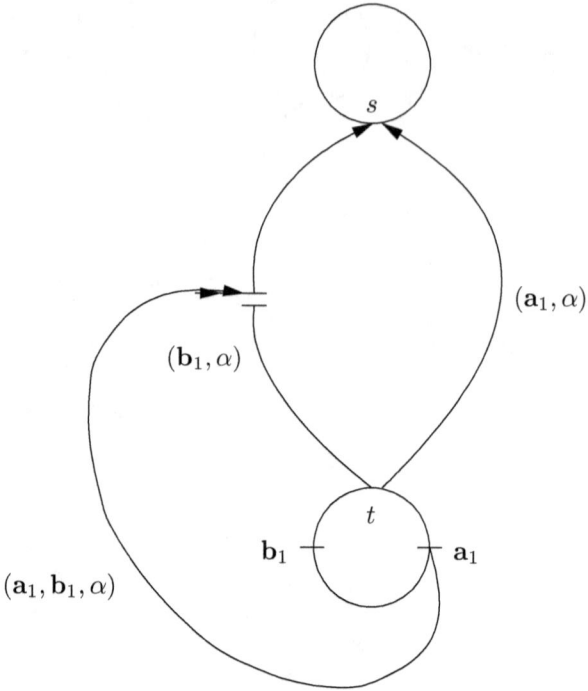

Figure 12. Long arm delegation of α from $\mathbf{a}_1$ to $\mathbf{b}_1$

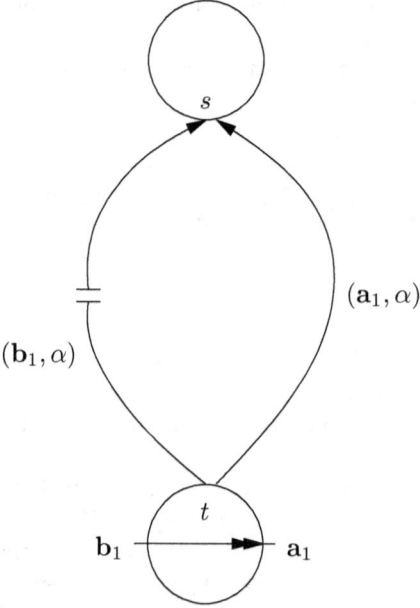

Figure 13. Power of attorney delegation of α from $\mathbf{a}_1$ to $\mathbf{b}_1$

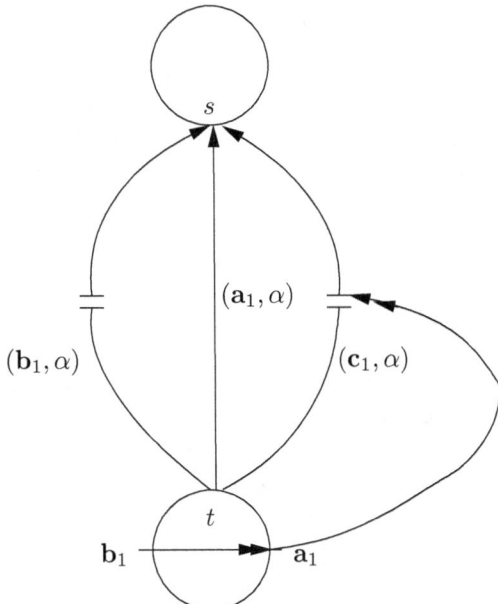

Figure 14.

In Figure 14, agent a_1 gives a power of attorney delegation to b_1 to execute α and at the same time gives a long arm delegation to c_1 to execute α.

EXAMPLE 6. To show the difference between the long arm and power of attorney view, consider the following examples.

1. Ruby and Simon delegate to Levy to dig a hole in the road (to access some pipes). An accident happened. A child fell into the hole and died. Ruby panicked and fled the country.

   According to the long arm view of delegation, we consider it as if each of Ruby and Simon dug the hole himself. So each is 100% responsible for damages. According to the power of attorney view, Levy was their joint agent and so each is 50% responsible for damages.

2. Another example is Simon delegates Levy to go to Sarah and give her a ring of engatement on behalf of himself. Simon has no money to buy the ring, so Levy (a good friend) buys the ring from his own money and gives it to Sarah.

   The Talmudic argumentation for this story goes as follows:

   (a) According to the long arm view, Levy is the long arm of Simon. So it is as if Simon himself gives the ring to Sarah, which he did not buy with his own money. The law says that the ring must be bought with the money of the person to whom Sarah is to be engaged (i.e. Simon). So the engagement action as performed is not valid. According to the power of attorney view, the engagement is valid.

Levy has a power of attorney to do a job. Levy is not a long arm of Simon. He does the job with his own money. This is fine.

(b) The Talmud could have argued differently from the above.

It could have said that according to the long arm view Levy is now Simon and so Levy's money counts as Simon's money, so the engagement is valid. While the power of attorney view would say that as an empowered agent, Levy is not Simon, so Levy's money is not Simon's money and so the engagement is not valid.

Our model supports the Talmudic acutal argument (a) and not the alternative argument (b) and thus properly models the Talmud.

In Figure 12, it is essentially $\mathbf{a}_1$ which goes through the arc of $\mathbf{b}_1$ to execute α. His long arm $\mathbf{b}_1$ goes for him through the arc

$$t \to_{(\mathbf{b}_1, \alpha)} s$$

which now has no gap.

In Figure 13, $\mathbf{b}_1$ goes thorugh the arc

$$t \to_{(\mathbf{a}_1, \alpha)} s$$

and so $\mathbf{b}_1$ is indeed a power of attorney agent for $\mathbf{a}_1$.

The difference between the cases is which arc is traversed by the agent $\mathbf{b}_1$. Is it $t \to_{(\mathbf{b}_1, \alpha)} s$, (long arm case) or is it $t \to_{(\mathbf{a}_1, \alpha)} s$ (power of attorney case)?

3.2 The reactive switch model for the long arm view of Tur

We now present formal definitions for the reactive case with switch double arrows.

DEFINITION 7 (Reactive delegation action model for switch double arrows). Let $\mathbf{A}$ be a finite set of agents and $\mathcal{A}$ a finite set of actions. Let $\mathbb{L}$ be a predicate language and assume that $\{\mathbf{m}\}$ are models for $\mathbb{L}$. We assume the actions $\alpha \in \mathcal{A}$ have preconditions A_α and postconditions B_α in the language $\mathbb{L}$. By a reactive model we mean a tuple $\mathfrak{M} = (S, \mathbf{R}_a, \mathcal{R}, \mathbb{D}, a, \mathbf{m}_t), t \in S$ where

S is a set of states

$\mathbf{R}_a$ is a function giving values in $\{0, 1\}$ to tuples of the form $(t, \mathbf{a}, \alpha)$ and to tuples of the form $(t, s, \mathbf{a}, \alpha), t, s \in S, \mathbf{a} \in \mathbf{A}, \alpha \in \mathcal{A}$,

$a \in S$ is the initial state.

$\mathbb{D} \subseteq \mathbf{A} \times \mathcal{A}$ says which agent is a master of which action. Such actions the agent can delegate. We can code $\mathbb{D}$ as part of $\mathbf{R}$ by having

$$\mathbb{D} = \{(\mathbf{a}, \alpha) | \mathbf{R}_a(\mathbf{a}, \alpha) = 1\}.$$

$\mathcal{R}$ is a set of double arrows of the form

$$(t, \mathbf{a}, \mathbf{b}, \alpha) \twoheadrightarrow (u, r, \mathbf{b}, \alpha)$$

or the form
$$(t, \mathbf{a}, \mathbf{b}, \alpha) \twoheadrightarrow (u, \mathbf{b}, \mathbf{c}, \alpha)$$
where $t, u, r \in S, \alpha \in \mathcal{A}, \mathbf{a}, \mathbf{b} \in \mathbf{A}$.

$\mathbf{m}_t$ are models of $\mathbb{L}$.

We assume that if $(t, s, \mathbf{a}, \alpha) \in$ domain $\mathbf{R}_a$ and $(t, s', \mathbf{b}, \alpha) \in$ domain $\mathbf{R}_a$ then $s = s'$.

Note the following:

1. If $\mathbf{R}_a(t, \mathbf{a}, \mathbf{b}, \alpha) = 0$ then at state t, agent $\mathbf{a}$ cannot delegate α as he cannot pass through the arc $t \to_{(\mathbf{a},\mathbf{b},\alpha)} t$.

2. Whenever $(t, \mathbf{a}, \mathbf{b}, \alpha) \twoheadrightarrow (u, r, \mathbf{b}, \alpha)$ (or respectively, $(t, \mathbf{a}, \mathbf{b}, \alpha) \twoheadrightarrow (u, \mathbf{b}, \mathbf{c}, \alpha)$) is in $\mathcal{R}$ then agent $\mathbf{a}$ by going through the arc $t \to_{(\mathbf{a},\mathbf{b},\alpha)} t$ (if he is allowed to delegate and other conditions hold) can delegate or revoke the ability of agent $\mathbf{b}$ to execute action α at state u (ending at state v if allowed by other factors), (respectively delegate or revoke the ability of agent $\mathbf{b}$ to delegate or revoke action α at state u to agent $\mathbf{c}$).

DEFINITION 8 (Legitimate path in a model for switch double arrows). Let $\mathfrak{M} = (S, \mathbf{R}_a, \mathcal{R}, \mathbb{D}, a, \mathbf{m}_t), t \in S$ be a model. We define the notion of legitimate annotated path of the form
$$\pi = (a \to_{e_1} x_1 \to_{e_2} x_2 \to \ldots \to x_{n-1} \to_{e_n} x_n)$$
where $x_i \in S$ and e_i are annotation labels of the form
$$e_i = (\mathbf{a}_i, \mathbf{b}_i, \alpha).$$

As part of the definition of legitimate path we associate by induction a function $\mathbf{R}_{\pi_i}$ with the initial path $a \to_{e_1} x_1 \to \ldots \to_{e_i} x_i$.

Step 0 $\pi = (a)$ and $\mathbf{R}_{(a)} = \mathbf{R}_a$.

Step $i + 1$ Assume we have $\mathbf{R}_{\pi_i}$. We define $\mathbf{R}_{\pi_{i+1}}$, where $\pi_{i+1} = \pi_1 \cup \{x_{i+1} \to_{e_{i+1}} x_{i+1}\}$.

Subcase 1 Delegation. In this case $x_i = x_{i+1}$ and $\mathbf{R}_{\pi_i}(x_i, \mathbf{a}_i, \mathbf{b}_i, \alpha) = 1$ and
$$\mathbf{R}_{\pi_{i+1}}(u, v, \mathbf{b}, \alpha) = \begin{cases} 1 - \mathbf{R}_{\pi_i}(u, v, \mathbf{b}_i, \alpha) \\ \text{if } (t, \mathbf{a}, \mathbf{b}_i, \alpha) \twoheadrightarrow (u, v, \mathbf{b}_i, \alpha) \\ \text{is in } \mathcal{R} \\ \text{and } \mathbf{R}_{\pi_i}(u, v, \mathbf{b}_i, \alpha) \text{ otherwise} \end{cases}$$

Similarly
$$\mathbf{R}_{\pi_{i+1}}(u, \mathbf{b}, \alpha) = \begin{cases} 1 - \mathbf{R}_{\pi_i}(u, \mathbf{b}_i, \mathbf{c}_i, \alpha) \text{ if} \\ (t, \mathbf{a}, \mathbf{b}_i, \alpha) \twoheadrightarrow (u, \mathbf{b}_i, \mathbf{c}_i, \alpha) \text{ is in } \mathcal{R} \\ \text{and } \mathbf{R}_{\pi_i}(u, \mathbf{b}_i, \mathbf{c}_i, \alpha) \text{ otherwise.} \end{cases}$$

Subcase 2 Action. In this case $x_i \neq x_{i+1}$ and $\mathbf{R}_{\pi_i}(x_i, x_{i+1}, \mathbf{a}_i, \alpha) = 1$ and $\mathbf{m}_{x_i} \vDash A_\alpha$. Let $\mathbf{R}_{\pi_{i+1}} = \mathbf{R}_{\pi_i}$.

Note that if the action α does not change the state x_i, then we still go to $x_{i+1} \neq x_i$, but we will have $\mathbf{m}_{x_i} = \mathbf{m}_{x_{i+1}}$.

REMARK 9.

1. Now that we have models, we can use modal operators and define satisfaction for them. The only modal operator of interest is the one having to do with delegation.

2. Note that in this model **a** can delegate to **b** to execute action α from state u. This is not general delegation but a very specific one.

DEFINITION 10 (Delegation modalities for switch double arrow). Let us add to the language two modalities $\Diamond$ and $\Diamond_\alpha$, defined as follows in the model $\mathfrak{M}$. $\Diamond$ corresponds to arbitrary legitimate paths. $\Diamond_\alpha$ corresponds to α delegation paths.

1. Let $\pi = a \to_{e_1} x_1 \to \ldots \to_{e_n} x_n$ be an arbitrary legitimate path. Let π' be an extension of π, namely

$$\pi' = a \to_{e_1} x_1 \to \ldots \to_{e_n} x_n \to_{e_{n+1}} y_1 \to \ldots \to_{e_{n+m}} y_m$$

   We say

   - $\pi \models \Diamond A$ iff for some extension π', $\pi' \models A$
   - $\pi \models A$ without modalities, iff $\mathbf{m}_{x_n} \models A$.

2. A sequence π is said to be α delegation path if for some t we have $x_i = t$ for all i and $e_i = (t, \mathbf{a}_i, \mathbf{b}_i \alpha)$. We can now define

   - $\pi \models \Diamond_\alpha A$ iff for some extension π' of π such that $x_n \to_{e_{n+1}} y_1 \to \ldots \to_{e_{n+m}} y_m$ is an α delegation path, we have $\pi' \models A$.

3. We similarly can define single modalities for each single connection of the form $t \twoheadrightarrow_{(\mathbf{a}, \alpha)} s$. The modality is possible $\Diamond_{(\mathbf{a}, \alpha)}$. Similarly we can define possible $\Diamond_{(\mathbf{a}, \mathbf{b}, \alpha)}$.

4 Comparison with modern literature

Let us start with the 2001 survey paper [Hagström et al., 2001] and the 2011 logical implementation paper [Aucher et al., 2011] based on it

Paper [Hagström et al., 2001] addresses an ownership-based framework for access control. Delegation here takes the form of granting access and administrative rights to other agents thus forming chains of granted accesses. The paper is a comprehensive study of the problem of revoking such rights, and of the impact different revocation schemes may have on the chains. Three main revocation characteristics are identified: the extent of the revocation to other grantees (propagation), the effect on other grants to the same grantee (dominance), and the permanence of the negation of rights (resilience). A classification is devised using these three dimensions. The different schemes thus obtained are described, and compared to other models from the literature of the time (up to 2001).

We begin by comparing with Delegation in the Talmud. The Talmud deals with a person delegating an action to another, to do the action on his behalf. The situation where a delegate can get instructions from two different people to do the same action as a delegate for each of them cannot arise.Think for example of delegating selling a house, only the owner of the house can delegate.If the ownership is shared , one can delegate selling his share only. We cannot have a situation where two different people delegate the same action to a third party. You may ask, how does the Talmud view, e.g. delegation of access control? The Talmud does not consider this as delegation. According to the Talmud, I cannot delegate an action which does not make a change in some legally recognised state of affairs. Selling propery, buying, divorcing, getting engaged, are candidates for delegation in the Talmud. Breaking a window, jumping over a fence are not. So revocation becomes rather simple here. In the long arm view of Talmudic delegation, the original master can revoke the last in the chain and reinstate him at will.The other members of the chain are not part of the long arm anyway.In the power of attorney view each element in the chain can revoke the next one only.So if the master delegates to an agent and the agent delegates to a subagent , the master can revoke the delegation to the agent only, and may be too late if the subagent was already delegated by the agent. There is one exception, according to the view of some Talmudic scholars. One can delegate murdering a person to an assassin agent. In the case of murdering somebody, you can have an assassin agent being delegated to murder by several people. In this case revocation policy is simple. If one of the people revokes the other delegations still stand. Note that according to the long arm view all masters who delegated to the assassin are individually each murdering the victim (because the assassin is their long arm) while with the power of attorney view, only the agent assassin is doing the murdering.

The second paper, [Aucher et al., 2011], gives a logical model to the first paper, [Hagström et al., 2001]. The implementation is according to Option 1, where the delegation chains only are modelled as an update system. Recently in paper [Barker et al., to appear] we offer a reactive model for access control, which is another way of modelling [Hagström et al., 2001] and more.

Other algorithmic papers dealing with chains of delegations and revocation systems are [Bertino et al., 1996; Barka and Sandhu, 2000; Firozabadi et al., 2001; Jaeger et al., 2002; Firozabadi and Sergot, 2002; Herzberg et al., 2000; Bandmann et al., 2002; Firozabadi et al., in press; Aucher et al., 2011].

5 Conclusion and discussion

In this paper we gave a preliminary study of delegation in the Talmud and compared it with modern delegation theory. In the Talmud the emphasis is more theoretical ; the Talmud is concerned more with the nature of delegation and circumstances for its cancellation, death or madness of the people involved and there is not so much emphasis on revocation. Talmudic delegation is more personal ,private persons delegate for the purpose of some legal action, divorce, buying and selling, and so revocation is a simple person to person act. Modern delegation is mainly for access control or the endowment of privileges usually involving large institutions and systems and so revocation protocols and the handling of delegation chains is more central. The emphasis is less conceptual and more algorithmic.

Acknowledgements

We are grateful to Xavier Parent and to Leon van der Torre from the University of Luxembourg, for penetrating comments and for visiting our Talmudic Project at Bar Ilan University. We also thank the participants of the Tel Aviv University computer science seminar for comments and suggestions on the paper.

BIBLIOGRAPHY

[Aucher et al., 2011] G. Aucher, S. Barker, G. Boella, V. Genovese and L. van der Torre. Dynamics in Delegation and Revocation Schemes: A Logical Approach. In *Conference on Data and Applications Security and Privacy (DBSec'11)*, Richmond, Virginia USA. July 11-13, 2011

[Bandmann et al., 2002] O. Bandmann, M. Dam, and B. S. Firozabadi. Constrained Delegation. In *IEEE Symbposum on Security and Privacy*, pp. 131–140, 2002.

[Barka and Sandhu, 2000] E. Barka and R. Sandhu. Framework for Role-Based Delegation Models,. In *Proceedings of the 16th Annual Computer Security Applications Conference*, May 2000, pp. 168-176.

[Barker et al., to appear] S. Barker, G. Boella, D. Gabbay and V. Genovese. Reactive Kripke Models and Answer Set Programming Applications to Delegation and Revocation Schemes. To appear in *Journal of Logic and Computation*.

[Bertino et al., 1996] E. Bertino, S. Jajodia, and P. Samarati. Non-timestamped Authorization Model for Data Management Systems. In *3rd ACM Conference on Computer and Communications Security*, March 1996, pp. 169-178.

[Final Amendment, 1999] Final Proposed Draft Amendment on Certificate Extensions(v6). Generated from Collaborative ITU and ISO/IEC meeting on the Directory, April 1999. Orlando, Florida, USA.

[Firozabadi et al., 2001] B. S. Firozabadi, M. Sergot, and O. Bandmann. Using Authority Certificates to Create Management Structures. In *Proceedings of Security Protocols, 9th International Workshop*, Cambridge, UK, pages 134–145. Springer Verlag, 2001.

[Firozabadi and Sergot, 2002] B. S. Firozabadi and M. Sergot. Revocation Schemes for Delegated Authorities. In *Proceedings of Policy 2002: IEEE 3rd International Workshop on Policies for Distributed Systems and Networks*. IEEE, June 2002.

[Firozabadi et al., in press] B. S. Firozabadi, M. Sergot, and O. Bandmann. Using Authority Certificates to Create Management Structures. In Proceeding of Security Protocols. In *9th International Workshop*, Cambridge, UK, April 2001. Springer Verlag. In press.

[Gabbay, 2008] D. Gabbay. Reactive Kripke Semantics and Arc Accessibility. In *Pillars of Computer Science: Essays Dedicated to Boris (Boaz) Trakhtenbrot on the Occasion of His 85th Birthday*, Arnon Avron, Nachum Dershowitz, and Alexander Rabinovich, editors, Lecture Notes in Computer Science, vol. 4800, Springer-Verlag, Berlin, 2008 pp 292-341.

[Grossi and Jones, to appear] D. Grossi and A. J. I. Jones Constitutive Norms and Countsas Conditionals, Handbook Chapter to appear.

[Hagström et al., 2001] Å. Hagström, S. Jajodia, F. Parisi.Persicce, and D. Wijesekera. Revocation — a Classification. In *Proceeding of the 14th Computer Security Foundation Workshop*. IEEE press, 2001.

[Herzberg et al., 2000] A. Herzberg, Y. Mass, J. Mihaeli, D. Naor, and Y. Ravid. Access control meets public key infrastructure, or: Assigning roles to strangers. In *IEEE Symposium on Security and Privacy*, pages 2–14, 2000.

[Jaeger et al., 2002] T. Jaeger, A. Edwards, and Xiaolan Zhang. Managing access control policies using access control spaces. In *Proceedings of the seventh ACM symposium on Access control models and technologies*, pages 3–12. ACM Press, 2002.

[Jones and Sergot, 1996] A. J. I. Jones and M. Sergot. A formal characterization of institutionalised power. *Journal of the IGPL*, 3:427–443, 1996.

[Rissanen et al., 2004] E. Rissanen, B. S. Firozabadi, and M. Sergot. Towards a mechanism for discretionary overriding of access control, position paper. Presented at Security Protocols, *12th International Workshop*, Cambridge, UK, 2004.

[Rissanen et al., 2005] E. Rissanen, B. S. Firozabadi and M. Sergot. Discretionary Overriding of Access Control in the Privilege Calculus. *IFIP International Federation for Information Processing*, 2005, Volume 173, 219-232, 2005. DOI: 10.1007/0-387-24098-5_16

[Searle, 1969] J. Searle. *Speech Acts. An Essay in the Philosophy of Language*. Cambridge University Press, Cambridge, 1969.

[Searle, 1995] J. Searle. *The Construction of Social Reality*. Free Press, 1995.

M. Abraham, R. Belfer, and U. Schild
Department of Computer Science
Bar Ilan University, Israel

D. Gabbay
Department of Informatics, King's College London, UK;
Department of Computer Science, Bar Ilan University, Israel; and
University of Luxembourg, Luxembourg.
E-mail: dov.gabbay@kcl.ac.uk

A Fibonacci estimation for the height of Schütte-like cut-free proofs

EDWARD HERMANN HAEUSLER AND LUIZ CARLOS PEREIRA

ABSTRACT. This note shows a Fibonacci based measure on the heights of derivations in the propositional $\{\bot, \to\}$-fragment of Schütte's system (SCP) for classical propositional logic. An observation on how this measure is related to a well-known measure for cut-free proofs in Sequent Calculus concludes this article.

1 Introduction

Considering proofs or derivations as trees and the usual definition of height on trees, the aim of this paper is to prove the following result on the estimation of the height of derivations in the $\{\bot, \to\}$-fragment of a Schütte-style [Schütte, 1977] calculus (SCP):

(*) If $\vdash^n \mathcal{P}[\mathcal{A}]$, $\vdash^m A \to B$ and $deg(A) = k$, then $\vdash^{f^k(n,m)} \mathcal{P}[\mathcal{B}]$.

Where:

- $\vdash^n C$ means that C has a derivation of height less than n in SCP;

- $f^k(n, m) = F_{k+1} \times m + F_k \times n$, where F_k is the k-th Fibonacci number;

- $deg(A)$ is the usual degree of A (the number of occurrences of $\to$ in A).

The main result (*) is obtained as a direct consequence of $F(k_1, m, F(k_2, n, m)) \leq F(k_1 + k_2 + 1, n, m)$ and the proof of cut-elimination for SCP (see [Schütte, 1977], theorem 4.6, pp24-25). A detailed estimation is defined along the construction of Schütte's original proof of the cut-elimination proof for SCP.

The definition of the Calculus SCP is based on Schütte's concepts of negative and positive parts/forms. Those forms are formulas with holes (indicated by $\star$'s, indexed or not), that indicate a position that determines the truth of the whole formula whenever a true (false) formula fills it. A form behaves at the same time as positive form and a negative form is called a NP-form, as for example $\star \to \star$ (see [Schütte, 1977]). A detailed definition of positive and negative forms, P-forms and N-forms respectively, follows. $\mathcal{F}[\star_1, \ldots, \star_n]$ is a formula with holes named by indexed $\star$'s, and $\mathcal{F}[A_1, \ldots, A_n]$ denotes the substitution of the symbols $\star_1, \ldots, \star_n$ by $A_1, \ldots, A_n$, respectively.

$\star$ is a P-form; If P is a P-form and A is a formula then $P[\star_1 \lor A]$, $P[A \lor \star_1]$ and $P[A \to \star_1]$ are P-forms and $P[\neg \star_1]$ and $P[\star_1 \to A]$ are N-forms; If N is a N-form and A is a formula then $N[\neg \star_1]$ and $N[\star_1 \to \bot]$ are P-forms and $N[A \land \star_1]$ and $N[\star_1 \land A]$ are N-forms.

In the final part of the article we compare our measure with the well-known estimatives for Classical Sequent Calculus that one can find in [Troelstra and Schwichtenberg, 1996].

Before proving the theorem we state the following auxiliary lemma concerning structural consequences and weak inferences. We say that B is a structural consequence of A ($A \vdash_S B$) iff every minimal positive part of A is also a positive part of B. An inference $A \vdash B$ is weak, iff, A is derivable with height n then B is derived with height less than or equal to n.

2 Estimating the height of Normal Proofs

The definition of the system SCP is as follows:

The axioms of SCP are:

AxI If $\mathcal{D}$ is a NP-form then $\mathcal{D}[v, v]$ is an axiom.

AxII If $\mathcal{N}$ is an N-form then $\mathcal{N}[\bot]$ is an axiom.

The rules of inference of SCP are:

I-→ $\mathcal{N}[(A \to \bot)], \mathcal{N}[B] \vdash \mathcal{N}[(A \to B)]$.

cut $\mathcal{P}[A], A \to B \vdash \mathcal{P}[B]$

Our main result is based on the admissibility of the cut-rule. The proof is standard: we consider derivations with just one application of the cut-rule that occurs as the last rule in the derivation and we show that this application can be eliminated. The proof then proceeds by induction on the length of the derivation. Let Π be a derivation of the form:

$$\frac{\begin{array}{cc}\Pi_1 & \Pi_2 \\ P[A] & A \to B\end{array}}{P[B]}$$

We will define a function $f^k(m, n)$ that will give an upper bound for the height of the cut-free (normal) proof obtained from Π, where:

- n is the height of Π_1;
- m is the height of Π_2;
- k is the degree of A.

The definition of f itself will proceed simultaneously with the proof that it is an upper-bound for cut-free proofs. In the sequel we use $T(\Pi)$ to denote the height of the cut-free proof associated to Π and Π^T to denote the cut-free proof itself. We will use $h(\Pi)$ to denote the height of Π

LEMMA 1. $F(k_1, m, F(k_2, n, m)) \leq F(k_1 + k_2 + 1, n, m)$

The inequality is proved as follows:
First, by induction on j and i, one shows that $F_i \times F_j + F_{i-1} \times F_{j-1} = F_{i+j}$ for $1 \leq i, j$.
Then, given the definition of F, one can clearly see that $F(k_1, m, F(k_2, n, m)) =$

$F(k_1, m, F_{k_2+1} \times m + F_{k_2} \times n) = F_{k_1+1} \times (F_{k_2+1} \times m + F_{k_2} \times n) + F_{k_1} \times m = (F_{k_1+1} \times F_{k_2+1} + F_{k_1}) \times m + F_{k_1+1} \times F_{k_2} \times n \leq F_{k_1+k_2+2} \times m + F_{k_1+k_2+1} \times n = F(k_1 + k_2 + 1, n, m)$.

Theorem 1. Let Π be a proof such that (1) Π has only one application of the cut, and (2) this application is the last rule in Π. Then, Π can be transformed into a cut-free proof Π^T having the same conclusion of Π, such that $h(\Pi^T) \leq f^k(m, n)$, where f is defined as below in the proof of this statement.

LEMMA 2. *If $A \vdash_S B$ then $A \vdash B$ is a weak inference.*

Proof of **Theorem 1.** By induction on k with subsidiary inductions on n and m.

basis ($k = 0$) In this case A is atomic and we define $f^0(n, m) = n + m$. We show that $h(\Pi^T) \leq n + m$ by induction on n.

- $n = 0$, then $P[A]$ is an axiom:

  1. $P[A] = N[\bot]$, then $P[B]$ is an axiom, $h(\Pi^T) = 0 \leq m = f^0(n, m)$.

  2. $P[A] = D[A, *_1]A$. In this case $P[B] = D[A, B]$ that follows structurally from $A \to B$. Thus, by 2 above we have that $h(\Pi^T) \leq n + m$.

- $n > 0$ In this case:

$$\Pi_1 = \frac{\begin{array}{cc} \Pi_{1,1} & \Pi_{1,2} \\ P_1[A] & P_2[A] \end{array}}{P[A]}$$

Thus,

$$\frac{P_1[B] \quad P_2[B]}{P[B]}$$

is a valid instance of an inference rule. So, let

$$\Pi' = \frac{\begin{array}{cc} \Pi'_1 & \Pi'_2 \\ P_1[B] & P_2[B] \end{array}}{P[B]}$$

where,

$$\Pi'_i - \frac{\begin{array}{cc} \Pi_{1,i} & \Pi_2 \\ P_1[A] & A \to B \end{array}}{P_i[B]}$$

By the inductive hypothesis we have that $h(\Pi'^T_i) \leq (n-1) + m$ and hence $h(\Pi^T) \leq (n-1) + m + 1 = n + m$.

Inductive Step $k > 0$. In this case A (the cut formula) is of the form $A_1 \to A_2$. Let k_1 and k_2 be the degrees of A_1 and A_2 respectively. Note that $k_1 + k_2 = k - 1$. We firstly observe that we can build:

$$\begin{array}{ccc} \Pi_{2,1} & & \Pi_{2,2} \\ (A_1 \to \bot) \to B & \text{and} & A_2 \to B \end{array}$$

from Π_2 and $h(\Pi_{2,i}) \leq m$. We set $P[A_1 \to A_2] \equiv P_1[A_2]$, then let

$$\Pi^* \equiv \frac{\begin{array}{cc}\Pi_1 & \Pi_{2,2} \\ P_1[A_2] & A_2 \to B\end{array}}{P_1[B]}$$

and hence by the inductive hypothesis we have that $h((\Pi^*)^T) \leq f^{k_2}(n,m)$. We can also observe that $P_1[B] \equiv P[A_1 \to B] \vdash_S A_1 \to P[B]$ and we can build $\Pi^{*\prime}$ with height at most $f^{k_2}(n,m)$, such that :

$$\begin{array}{c}\Pi_1^{*\prime} \\ A_1 \to P[B]\end{array}$$

We use $P_2[A_1]$ in order to denote $(A_1 \to \bot) \to B$. Thus, let

$$\Pi_2^* \equiv \frac{\begin{array}{cc}\Pi_{2,1} & \Pi_1^{\prime *} \\ P_2[A_1] & A_1 \to P[B]\end{array}}{P_2[P[B]]}$$

and hence it should be the case that $h((\Pi_2^*)^T) \leq f^{k_1}(m, f^{k_2}(n,m))$. Again by an structural deduction we obtain $P_2[P[B]] \equiv (P[B] \to \bot) \to B \vdash_S P[B]$ and we can build $\Pi_2^{*\prime}$ as a derivation of $P[B]$ with height at most $f^{k_1}(m, f^{k_2}(n,m))$. As the height of the cut-free deduction depends on k_1 and k_2, that is, it depends on the form of the formula A. Thus, based on the estimations done above, we define:

$$f^k(n,m) = max(\{f^{k1}(m, f^{k2}(n,m)) : K1 + K2 = K - 1\}))$$

with this definition, we have that $h(\Pi_2^{*\prime}) \leq f^k(n,m)$

□

The proof of the cut-elimination theorem is obtained by the iteration of the process described above. Take, for example, a proof of the following form:

$$\frac{\begin{array}{cc}\Pi_1 & \frac{\begin{array}{cc}\Pi_2 & \Pi_3 \\ P[A]_0 & A \to B\end{array}}{\begin{array}{c}P[B]_0 \\ \Pi_4 \\ \alpha \to \beta\end{array}} \\ P[\alpha]_1 & \end{array}}{P[B]_1},$$

where the heights of $\Pi_1, \Pi_2, \Pi_3, \Pi_4$ are $h(\Pi_1), h(\Pi_2), h(\Pi_3), h(\Pi_4)$ respectively, in Π_4 there is no cut rule, but the shown one, in Π_1 there is no cut rule either, the degree of A is k_1 and the degree of α is k_2. Then the cut-free proof obtained for the upper cut has height limited by $F_{k_1+1} \times h(\Pi_3) + F_{k_1} \times h(\Pi_2)$ resulting in a cut-free proof obtained for the lower cut limited by $F_{k_2+1} \times (F_{k_1+1} \times h(\Pi_3) + F_{k_1} \times h(\Pi_2)) + F_{k_2} \times h(\Pi_1)$, considering the worst case, which is $\Pi_4 = P[B]_0$ and the sub-derivation

containing the upper cut contributing for the height of the whole derivation[1]
From the fact that $F_i \times F_j \leq F_{i+j}$, and considering
$max(h(\Pi_3), h(\Pi_2), h(\Pi_4)) \leq h(\Pi'_4)$, where Π'_4 is the derivation of $\alpha \to \beta$, we
can conclude that the upper bound is $F_{k_1+k_2+2} \times h(\Pi'_4) + F_{k_2} \times h(\Pi_1)$.
Considering the presence of other cuts in Π_1, we have the following
proposition. Its proof is obtained by iterating the reasoning just explained.

Theorem 2. Let Π be the following proof in SCP:

$$\frac{\begin{array}{cc}\Pi_1 & \Pi_2 \\ \mathcal{P}[A] & A \to B\end{array}}{\mathcal{P}[B]}$$

Then, an upper bound for the $h(T(\Pi))$ is given by the following expression:

$$F_{\left(N_2+1+\sum_{\delta \in Cuts(\Pi_2) \cup \{A\}} d(\delta)\right)} \times h(\Pi_2) + F_{\left(N_1+\sum_{\delta \in Cuts(\Pi_1)} d(\delta)\right)} \times h(\Pi_1)$$

, where N_i is the number of cuts in Π_i, and $Cuts(\Pi_i)$ is the multiset of
cut-formulas in Π_i.

3 Conclusion: Comparing Schütte's system and Sequent Calculus

We compare our Fibonacci bound with the Schwichtenberg-Troelstra
numerical bound for cut-elimination in propositional Sequent Calculus (see
chapter 5 section 1 of [Troelstra and Schwichtenberg, 1996]). It is well-known
that the propositional fragment have exponential bounds. The bounds in
Natural Deduction are hyperexponential even for Propositional Case, for this
reason we restrict ourselves to the comparison to Sequent Calculus. Let Π be
the following derivation:

$$\frac{\begin{array}{cc}\Sigma_1 & \Sigma_2 \\ \Delta \Rightarrow \Gamma, A & \alpha, \Delta \Rightarrow \Gamma\end{array}}{\Delta \Rightarrow \Gamma}$$

The height of a cut-free sequent proof $Seq(\Pi)$ obtained from Π is
$max(h(\Pi_1), h(\Pi_2) + 1) \times 2^{cl(\Pi)}$, where $cl(\Pi)$ is

$$max_{\sigma \in \Pi} \left(\Sigma\{s(\varphi)/\varphi \text{ is a cut formula in } \sigma\}\right)$$

and $\sigma \in \Pi$ means that σ is a branch of Π, $s(\varphi)$ is the size (number of symbols
in φ). We know that $F_n \approx \phi^n/\sqrt{5}$, where ϕ is 1.61803398875..., the golden
ratio number. The SCP proof Π that is going to be used in the comparison
ends with an application of the cut-rule (the derivations of this last

---

[1] If the upper cut does not contribute for this height, we are in a case quite similar to one
cut only. Otherwise, the height of the derivation of $\alpha \to \beta$ is $n_4 + max(n_2, n_3) + 1$, and,
in such case the lower is the n_4 the higher will be the cut-free derivation obtained from the
upper cut.

application of the cut-rule may contain other cuts. The height of the cut-free Π' associated to Π is bounded by:

$$\frac{1}{\sqrt{5}} \times \phi^{\left(N_2+1+\sum_{\delta \in Cuts(\Pi_2) \cup \{\alpha\}} d(\delta)\right)} \times h(\Pi_2) + \frac{1}{\sqrt{5}} \times \phi^{\left(N_1+\sum_{\delta \in Cuts(\Pi_1)} d(\delta)\right)} \times h(\Pi_1)$$

where Π_i, $i = 1, 2$, are the derivations of the premises of the last cut-rule in Π. Using the fact that:

1. $d(\delta) \leq s(\delta)$;

2. $\sum_{\delta \in Cuts(\Pi_i)} s(\delta) \leq N_i \times cl(\Pi_i)$;

We have $\sum_{\delta \in Cuts(\Pi_i)} d(\delta) \leq \sum_{\delta \in Cuts(\Pi_i)} s(\delta) \leq N_i \times cl(\Pi_i)$, and hence the height of Π' is bounded by:

$$\frac{1}{\sqrt{5}} \times \phi^{(N_2+1+N_2 \times cl(\Pi_2))} \times h(\Pi_2) + \frac{1}{\sqrt{5}} \times \phi^{(N_1+N1 \times cl(\Pi_1))} \times h(\Pi_1)$$

that is bounded by:

$$\frac{1}{\sqrt{5}} \times max(h(\Pi_1), h(\Pi_2)) \times \phi^{(N_2 \times (1+cl(\Pi_2))+N_1 \times (1+cl(\Pi_1))+1)}$$

Finally we can observe that the above expression is in fact

$$\frac{1}{\sqrt{5}} \times (h(\Pi) - 1) \times \phi^{(N \times cl(\Pi)+1)}$$

, concluding that $h(SCP(\Pi)) \leq h(Seq(\Pi)) \times 2^{N+1}$, where N is the number of cuts in Π. So, SCP is barely more space-economic than Sequent Calculus, since $\frac{N}{cl(\Pi)} < 1$.

The results that we obtained can be easily extended to the fragment $\{\wedge, \vee, \rightarrow, \bot\}$, since the rules for $\wedge$ and $\vee$ are weak inferences in SCP.

BIBLIOGRAPHY

[Schütte, 1977] Kurt Schütte. *Proof Theory*. Springer-Verlag, 1977.
[Troelstra and Schwichtenberg, 1996] A. S. Troelstra and H. Schwichtenberg. *Basic Proof Theory*, volume 43 of *Cambridge tracts in theoretical computer science*. Cambridge University Press, 1996.

Edward Hermann Haeusler
Departamento de Informática
Pontifícia Universidade Católica do Rio de Janeiro (PUC-Rio)
Rio de Janeiro, Brazil
E-mail: hermann@inf.puc-rio.br

Luiz Carlos Pereira
Departamento de Filosofia
Universidade Federal de Rio de Janeiro
Departamento de Filosofia
Pontifícia Universidade Católica do Rio de Janeiro (PUC-Rio)
E-mail: luiz@inf.puc-rio.br

Valid Inferences
RICHARD L. EPSTEIN

This essay is dedicated to Walter Carnielli on the occasion of his 60th birthday: A good friend, a good colleague.

ABSTRACT. How we understand possibilities and valid inferences is crucial to our idea of one claim following from one or more other claims. Formal logics are one way to reduce possibilities to simpler notions when appropriate semantic and syntactic assumptions apply. In that context recent debates about logical consequence become clearer.[1]

1 Claims and inferences

The goal of this paper is to make clearer when we are justified in saying that the conclusion follows from the premises in an inference.

Claims and inferences A *claim* is a written or uttered piece of language that we agree to view as being either true or false but not both.

An *inference* is a collection of claims, one of which is designated the *conclusion* and the others the *premises*, that is intended by the person who sets it out either as showing that the conclusion follows from the premises or investigating whether that is the case.[2]

We use inferences in reasoning in at least five distinct ways: arguments, explanations, mathematical proofs, conditional inferences, and causal inferences. The criteria for what qualifies as a good inference depends on which of those ways we are considering. The following defintions, however, are crucial in the analysis of each of those.

Valid and strong inferences An inference is *valid* if it is impossible for the premises to be true and conclusion false at the same time and in the same way.[3]

---

[1] I am grateful to Fred Kroon and William S. Robinson for their comments on an earlier draft.

[2] Some say that what is true or false are abstract objects, propositions, and that claims as defined here are only representatives of those. Inferences, too, they take to be abstract. They would say that this paper is not about logic and possibility but how we can or should use logic and possibility in our reasoning.

[3] Some authors state the condition for validity as: a valid inference is one such that if the premises are true then the conclusion must be true; see, for example, J. L. Bell and M. Machover, A Course in Mathematical Logic, p. 5; Alfred Tarski, "On the concept of logical

Invalid inferences are classified on a *scale* from strong to weak: An inference is *strong* if it is very unlikely for the premises to be true and conclusion false at the same time; it is weak if it is not valid or strong.

With a valid inference if the premises are true, the conclusion is true; with a strong inference if the premises are true, then the conclusion most probably is true.

EXAMPLE 1. Felix teaches at a university.
Therefore, Felix is a professor.
Analysis Is this inference valid? Strong? Could it be that Felix teaches at a university but isn't a professor? Yes, he could be a temporary instructor or a teaching assistant. The inference isn't valid. It isn't even strong, since we cannot rule out those possibilities as unlikely.

As the example illustrates, an inference is either valid or it isn't; there are no degrees to it. A single example of a way the premises could be true and conclusion false suffices to show that the inference is not valid. In contrast, inferences are more or less strong, and that depends on our intersubjective evaluation about how likely certain possibilities are. Attempts to substitute objective criteria in terms of probabilities have been unsuccessful , and it is a contentious issue whether we can or should rely on any strong argument as showing that the conclusion follows. Here I will consider the nature of inferences only in terms of validity.

2 Possibilities

To evaluate an inference we talk about ways the world could be.

For the purposes of reasoning well, however, all we ever use are descriptions. This is what we did in Example 1, where we gave a description of how the world could be in which the premise is true and conclusion false: Felix is a teacher and Felix is a temporary instructor.

EXAMPLE 2. This is a rock. Therefore, it is solid.
Analysis Professor Zzzyzzx says this is not valid. He knows this, he says, by an intuition, a sense of the nature of the world. He says that he cannot communicate this great insight, but he knows that the inference is not valid.

Perhaps Professor Zzzyzzx is right. But we do not accept his evaluation, nor can we reason with him. Inchoate insights or mystical perceptions cannot be accepted if we wish to reason together, for reasoning together requires communication, and what is inchoate cannot, by its nature, be communicated. Of if

---

consequence", p. 411; John Etchemendy, The Concept of Logical Consequence. This appears to be a sloppy way of making the definition given in this paper, for if it were taken literally then only a necessary claim could be the conclusion of a valid inference.

There is no reason to think that the notion of validity presented here is a common one or is abstracted from one that is more or less universally held by most people: compare the research in "Truth" as Conceived by Those Who Are Not Professional Philosophers by Arne Naess. Logic is not descriptive of how people do reason, but is either descriptive of relations among abstract objects such as propositions, or else is, as I take it here, prescriptive of how to reason well.

it can be communicated, as the Zen master or the Christian mystic hopes, it cannot be done so in a way that is observable to all.

To invoke a way the world could be, a possibility, we have no choice but to use a description when we wish to reason together. A description of the world is a collection of claims: we suppose that this, and that, and this are true. We do not require that we give a complete description of the world, for no one is capable of presenting such a description or of understanding one if presented. By using collections of claims to stand in for or as descriptions of possibilities, we need not commit ourselves to a possibility being something real, such as a world in which I am not bald. A determinist can say that invoking possibilities is just a way to factor into our reasoning our ignorance of how the world is.

But what qualifies some collections of claims as describing a possibility and others not? What do we mean by saying that a dog giving birth to a donkey is a possibility, but a square circle is not? If you say that a dog giving birth to a donkey is not a possibility, how are we to decide if you are right?

Perhaps we have different ideas about what is possible. You might consider that it is physically impossible for a dog to give birth to a donkey, knowing all we now know about the biology of these animals. I might say that it is possible, we just don't know how yet.

Or I might say that it is not possible for a dog to give birth to a donkey by any ethical means, whereas you might say that it would be perfectly acceptable morally to interfere with the biology of dogs and donkeys to bring that about.

There are many different notions of possibility: physical possibility, moral possibility, possibility given what has happened up to this time, possibility given what we know up to this time, ... Regardless of which of these notions we are employing, we always seem to agree that a description of a way the world could be must at least be consistent. That is, it cannot have or entail a contradiction. It must be *logically possible*.

This seems to be the ground from which we all start in our reasoning. What is possible must be consistent. So there is no way the world could be in which there is a square circle. But there seems to be no contradiction inherent in postulating that a dog could give birth to a donkey: It is logically possible.[4]

It might seem we have made some progress in analyzing valid inferences. But the progress is illusory. What is logically possible is what contains or entails no contradiction. But that requires knowing what it means for a collection of claims to entail another claim, which is what we are trying to understand.

What is logically possible depends on how we understand valid inferences. But what is a valid inference depends on how we understand possibilities, particularly logical possibilities. We seem to be in a circle with no way out.

---

[4]Some logicians have attempted to formulate how to reason when the information we have is or might be inconsistent. A few of those have argued that contradictions, such as there being square circles, are possible. But such a bizarre assumption is not needed for reasoning around contradictions, as I show in "Paraconsistent logics with simple semantics", and it would leave us with no semantic basis from which to start our analysis of possibilities.

3 Logic and logical possibilities

One way we could extricate ourselves from the circle is to agree that we will use an informal notion of logic, our intuitive reasoning, to determine what is logically possible. But that is to deny the entire project of trying to understand valid inferences, for we would be trying to make explicit a concept that depends on our not making explicit our most fundamental assumptions.

What we can do is investigate parts of our reasoning, picking out just this or that kind of reasoning relative to restricted semantic assumptions that allow for clarity of analysis, and call that "a logic". Then we can have a clearer notion of possibility and of valid inference for that kind of reasoning. As we extend our investigations to allow for more kinds of reasoning, we will have fuller analysis of logical possibilities and valid inferences. But unless we should ever formalize all of reasoning, which seems very unlikely, we shall never have a complete analysis of logical possibilities and of valid inferences. We shall have analyzed at best only valid inferences relative to the specific semantic assumptions we have made, valid inferences for this logic.

Still, this is a useful goal.

4 Semantic reductions

Can we find some properties of claims that might be simpler and in terms of which we could understand possibilities?

We have one property of claims that we have accepted as clear or at least as so basic we cannot do without it: whether a claim is (or is to be considered) true or false. What further simple or clear properties of claims may be of use? One good candidate is their linguistic structure.

Some consider linguistic structure enough. They set out formal systems of reasoning—logics—solely in terms of linguistic form. That is, they give syntactic forms of claims that are deemed true for every possibility and syntactic forms of inferences that are deemed valid. These are then taken as the norms of reasoning. This, they say, is a way to extricate ourselves from the possibility-inference circle.

Though it may allow for a way out of the circle, it promises no insight or further clarity about inferences or possibility. It leaves the circle only to embrace intuitive justifications for why we should accept claims of those particular forms as always true and inferences of those forms of as valid. We are back at accepting an informal understanding of valid inference or an informal notion of possibility as the basis of our reasoning. This is not progress in understanding.

If, then, we look at the structure of claims it is to reduce possibilities and inferences to simpler semantic notions.

5 Propositional logics

We begin by looking at very gross structures of claims: how we can build some of them from other claims.

It is traditional to consider four ways to form new claims from other claims using the words, or as they are called *connectives*, "and", "or", "not", and "if ... then ...". For example, from "Ralph is a dog" and "George is a duck" we can form "Ralph is a dog and George is a duck", "Ralph is a dog or George is a duck", "Ralph is not a dog", "If Ralph is a dog, then George is a duck".

These connectives allow for a fairly rich analysis of reasoning. Once we have set out the semantic assumptions we will employ, we can ask whether other words that we apparently use to form new claims from old, such as "but", "despite", "nonetheless", "therefore", "in as much as", "because", and others can be assimilated to one or a composition of these original four. Doing that we will have a better idea of the scope of our analysis of inference and possibility.

Since we intend to state explicitly the semantic assumptions about how these connectives work, we typically replace the ordinary words with symbols such as "$\wedge$" for "and", "$\vee$" for "or", "$\neg$" for "not" as in "it is not the case that", and "$\to$" for "if ... then ...". These are the *formal connectives*. We can then be clear about what we mean when we say that we can make up new claims from others using these by defining a formal language that will give us the framework, the skeleton of all claims that we can consider in this analysis of reasoning.

We take the subscripted letters p_0, p_1, ... to stand for claims, that is, as *propositional variables*. These are meant to stand for claims that we consider to be, relative to our semantic assumptions, unanalyzable wholes: *atomic propositions*. We also need to be able to refer to these variables and the formulas we build up from them so we employ *metavariables* $A, B, C, \ldots$. Then we define a formal language $L(\neg, \to, \wedge, \vee, p_0, p_1, \ldots)$. The analogue of a sentence in English is a *well-formed formula* or *wff*:

(a) (p_i) is a wff for each $i = 0, 1, 2, \ldots$, an *atomic wff*.

(b) If A, B are wffs, then so are $(\neg A)$, $(A \to B)$, $(A \wedge B)$, and $(A \vee B)$.

(c) Only such concatenations of symbols as arise from repeated applications of (a) and (b) are wffs.[5]

Now we can begin our semantic analysis of claims of these forms. The simplest analysis of the connectives is to ignore all other semantic aspects of claims except the one that we began with: each has a truth-value. That plus the form of the claim then determines the semantic value of the whole. In this case, the analysis is:

$A \wedge B$ is true if and only if both are true.
$A \vee B$ is true if and only if one or the other or both is true.
$\neg A$ is true if and only if A is false.
$A \to B$ is true if and only if A is false or B is true.

These evaluations are usually set out in tabular form and are called the *classical truth-tables* for the connectives.
Schematically we have:

---

[5]This definition is usually replaced with an inductive one building the formal language in stages rather than relying on clause (c); see *Predicate Logic*.

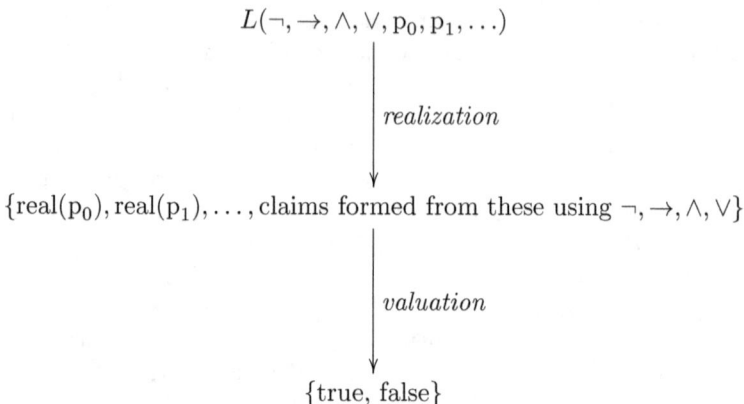

The first step, assigning claims to variables or (reversing the arrow) variables to claims and building formulas from those, is called a *realization* or *semi-formal language* where some of the atomic wffs are realized as particular claims and we have complex claims formed from them by the rules for defining wffs. The second step, called the *valuation*, is to assign truth-values to the atomic claims, from which the truth-value of every semi-formal wff follows by the classical truth-tables. The way in which we assign truth-values to the atomic claims cannot be taken into account in our analysis, for it is not for the logician to say what is actually true or false. The whole is called a *model* of *classical propositional logic*.

EXAMPLE 3. Ralph is barking and Juney is barking, but Ralph is not a dog.
  Howie is a cat and Howie is afraid of Juney.
  If Howie is afraid of Juney and Ralph is not a dog, then Juney is a dog.
  So Juney is a dog.

Analysis To use classical propositional logic to evaluate this inference, we first need to agree that "Ralph is barking", "Juney is barking", "Ralph is a dog", "Howie is a cat', "Howie is afraid of Juney", and "Juney is a dog" are claims, indeed atomic claims, and that it is appropriate to formalize both "and" and "but" here as $\wedge$, to formalize "not" as $\neg$, and to formalize "if ... then ..." as $\rightarrow$. In this example that seems O.K., but this first step is not always so straightforward.

So let's agree that the example is suitable to formalize as:

Premises
    (Ralph is barking $\wedge$ Juney is barking) $\wedge$ $\neg$(Ralph is a dog).
    Howie is a cat $\wedge$ Howie is afraid of Juney.
    (Howie is afraid of Juney $\wedge$ $\neg$(Ralph is a dog)) $\rightarrow$ Juney is a dog.
Conclusion
    Juney is a dog.

Under one choice of assigning claims to propositional variables, this is a realization of:

Premises
$(p_0 \wedge p_1) \wedge \neg(p_2)$
$p_{47} \wedge p_{312}$
$(p_{312} \wedge \neg(p_2)) \to p_{4318}$
Conclusion
p_{4318}

We need not realize every propositional variable: we only consider in our realization the claims in the reasoning we are analyzing.

We don't know which, if any, of these atomic claims are true. But we can analyze whether the inference is valid by looking at all the ways some of the premises could be true and some false. For example, we can suppose that the atomic claims "Ralph is barking", "Howie is a cat", "Howie is afraid of Juney", and "Ralph is a dog" are false, and that "Juney is barking" is true. In that case, as you can show, the semantic evaluation of the compound claims via our tables require that if all the premises are true, so is "Juney is a dog". The same happens for any other choice of truth-values for the atomic claims in the premises. So in any model if the premises of this inference are true, the conclusion is true, too. Thus, the inference is valid.

With this formalization of reasoning a possibility is reduced to a way to assign truth-values to atomic claims in a model for this logic. To say that an inference is valid relative to classical propositional logic is to say that there is no model in which the premises are (taken to be) true and the conclusion false. That is, there is no way to assign truth-values to the atomic claims in the semi-formal language that will make the premises true and the conclusion false. We have a reduction of the notion of possibility to simpler notions.

But this analysis still invokes our informal notion of possibility, for we talk about any possible assignment of truth-values. This, though is a restricted and much simpler notion than the idea of possibility in general. In an inference we have a finite number of premises plus a conclusion, and so only a finite number of propositional claims that appear in them, say n. There are 2^n different ways to assign truth-values to those, which we can inspect in a tabular form when n is small enough. When n is very large or where we choose to work with infinitely many premises, as we sometimes do in mathematics, we must rely on the method of specification of those premises to guide us in how we can assign truth-values to the atomic claims. The reduction of the notion of validity to simpler notions will then be as clear and convincing as our understanding of large numbers or an infinite number of premises.

We can expand our analysis to take into consideration further aspects of claims, such as how we may come to know them, or their subject matter, or how likely we deem them to be true. We can do that by making certain structural decisions about the nature of the additional aspect of claims and then factoring those into only slightly more complicated truth-tables. For each such analysis, that is, for each such logic, there is a corresponding notion of a model, which formalizes the notion of logical possibility, and hence a notion of valid inference. Depending on the semantic notions involved, the reduction of the notions of possibility and valid inference will be more or less clear and helpful.[6]

We can determine the forms of inferences that are valid for all such propositional logics. Perhaps these constitute the general form of valid inferences *relative to these connectives* that any reasoning creature would have to employ.[7] But it is not obvious that this constitutes an analysis of all reasoning even with claims as wholes, for there is no reason to think that for every connective of claims there is such a logic that aptly formalizes it.

Many logicians go further in abstracting from ordinary reasoning. They eliminate the realization part of models and say that a model is simply a valuation. In many textbooks a model for classical propositional logic is said to be an assignment of truth-values to the propositional variables. That is clearly wrong: a variable cannot be true or false. But the abstraction is a harmless and useful way to investigate valid inferences because given any assignment of truth-values there are certainly claims that could be assigned to the propositional variables to which we could assign those truth-values. Thus, the forms of inferences that are valid, that is the formal language correlates of inferences any of whose instances (realizations) will be valid, can be determined more easily by considering such abstracted versions of models.

We have now an analysis of possibility and valid inference for reasoning in the restricted context of paying attention to only claims as wholes and ways to build claims from them in a regular way using certain connectives, relative to assuming what for each logic are claimed to be simpler semantic notions that include the truth or falsity of claims that are considered structureless in this analysis. But the structure of claims we are allowed to consider is so restricted that the analysis of possibility and valid inference is far from what we need. Consider:

All men are mortal. Socrates is a man. Socrates is not mortal.

All dogs are brown. Some dogs are green.

Snow is white. Anything that's white isn't colored. Snow is colored.

Juney is barking loudly. Juney is not barking.

We know that not one of these collections of claims is a description of a way the world could be because we know—informally—that each is or leads to a contradiction. But we cannot show that any of them entails a contradiction if we are restricted to considering the forms of the claims relative to only propositional connectives. We need a way to understand possibilities that takes into account the internal structure of claims.

---

[6] See my *Propositional Logics* for a technical presentation of this analysis.
[7] See the last chapter of my Propositional Logics.

6 Predicate logic

There are several ways people have chosen to parse the internal structure of claims in analyses of reasoning. Each depends on a particular view of the world. We'll look here at the way that has been dominant in logical studies during the last one hundred years.

Consider the claim "Ralph is a dog". It is atomic relative to the propositional connectives. We can parse it, though, as we learned in school, into a name and predicate: "Ralph" and "is a dog". Extending the notion of predicate beyond our grammar studies in school, we can say that "Ralph loves Dick" can be parsed as two names, "Ralph" and "Dick", and a predicate "loves".

Modern predicate logic takes this way of parsing claims as fundamental. A *predicate* is a claim with the names (and pronouns) deleted, with clear marking for where those came from and where those or other names could be re-inserted.[8] Thus, "Ralph is standing between Fred and Dick" yields the predicate "— is standing between — and —".

This way of parsing atomic claims is based on the view that the world is made up of things. Names and pronouns stand for things, and predicates are what are said to apply to things. Things are said to satisfy a predicate if when names of those things are put into the blanks, a true claim results. Claims are about things.

This method of parsing claims was designed to facilitate modeling of reasoning with assertions about all or some things satisfying a predicate. The symbols "$\forall$" and "$\exists$" are used as formal counterparts to "all' and "exists" (or "some"). For example, "Everything loves Ralph" is formalized as "$\forall x(x$ loves Ralph)", and "There is something that is bigger than Ralph" is formalized as "$\exists x(x$ is bigger than Ralph)", where a variable is used to allow us to refer to things in a general way.

As before we delimit the scope of the syntax of claims we can consider in this logic by defining a formal language. The formal language has *predicate symbols* $P_0^1, P_0^2, P_0^3, \ldots, P_1^1, P_1^2, P_1^3, \ldots$ that stand for predicates where the superscript indicates how many blanks are in the predicate that have to be filled with names or variables. The formal language also has *name symbols* $c_0, c_1, \ldots$ that stand for names and *individual variables* $x_0, x_1, \ldots$ that can stand for things; together these are all called *terms*. The formal language also has the *quantifiers* "$\forall$" and "$\exists$", parentheses, and the connectives from our propositional logic analysis. The definition of *well-formed-formula* or *wff* is:

(a) For every $i \geq 0$, $n \geq 1$, and terms t_j, $1 \leq j \leq n$, $(P_i^n(t_1, \ldots, t_n))$ is an atomic wff.

(b) If A, B are wffs, then so are $(\neg A)$, $(A \to B)$, $(A \wedge B)$, and $(A \vee B)$.

(c) If A is a wff, then for any $i \geq 0$ so are $(\forall x_i A)$ and $(\exists x_i A)$.

(d) Only such configurations of symbols as arise by repeated applications of (a), (b), and (c) are wffs.

---

[8]Some say that predicates are abstract and that predicates as defined here only stand for or represent those.

In this formal language not every wff can correspond to a claim. For example, $P_0^2(x_0, x_1)$ cannot correspond to a claim because replacing the predicate symbol with, say, "is bigger than" we get "x_0 is bigger than x_1", which is not a claim until we say what x_0 and x_1 stand for. Only formulas in which every variable in the formula is quantified correspond to claims. Thus, $\forall x_0 \exists x_1 P_0^2(x_0, x_1)$ can correspond to the claim "For everything there is something that is bigger than it".

For the semantics of predicate logic, the use of variables is understood, for example, by treating not only "Juney is bigger than Ralph" but also "x_1 is bigger than Ralph" as atomic when we supply a reference for "x_1", that is, when we use "x_1" as a temporary name. When we supply references for variables an atomic wff is a claim, a *predication*. The references that are available for the names and for the variables used as temporary names are meant to come from some collection of things we are discussing, the *universe*. For the generality of our logical analyses we would prefer to have the universe be all things, but there are good reasons why that is not done.[9]

For predicate logic built on the analysis of the propositional connectives of classical propositional logic, the semantic analysis adds to the semantic notions used for classical propositional logic—truth-values of atomic claims—only what is needed to use variables: naming. Then, as with propositional logic, we have a method to extend the semantic analysis to compound propositions by using the tables for the connectives and a method of evaluation of the quantifiers. That evaluation is an inductive procedure but relies semantically only on the use of variables to name objects of the universe.

This, then, is *classical predicate logic*. Schematically, we have a *model* of classical predicate logic:

$$L(\neg, \rightarrow, \wedge, \vee, \forall, \exists, x_0, x_1, \ldots, P_0^1, P_0^2, P_0^3, \ldots, P_1^1, P_1^2, P_1^3, \ldots, c_0, c_1, \ldots)$$

| realization

$$L(\neg, \rightarrow, \wedge, \vee, \forall, \exists, x_0, x_1, \ldots; \text{realizations of predicate and name symbols})$$

universe: specified in some manner

assignments of references as needed; assignments of truth-values to atomic predications; truth-tables for $\neg, \rightarrow, \wedge, \vee$; evaluation of the quantifiers

{true, false}

---

[9] We have no way of reasoning about the collection of all things without contradictions, and we have no story of naming that is general enough to cover all things, as I discuss in *Predicate Logic*.

Here the realization is an assignment of linguistic predicates to some of the predicate variables and names to some of the name symbols, with the resulting complex claims formed from those on the pattern of the formal language. It is a *semi-formal language*, a formalized fragment of English (or some other ordinary language). The semantics, which I have described here only in very general terms, takes us from the semi-formal language to truth-values for formulas that correspond to claims.

Thus, we might have

$$L(\neg, \to, \wedge, \vee, \forall, \exists, x_0, x_1, \ldots, P_0^1, P_0^2, P_0^3, \ldots, P_1^1, P_1^2, P_1^3, \ldots, c_0, c_1, \ldots)$$

$$\downarrow$$

$L(\neg, \to, \wedge, \vee, \forall, \exists, x_0, x_1, \ldots;$ — is a dog, — is a cat, — eats grass, — is a wombat, — is the father of—; Ralph, Dusty, Howie, Juney)
universe: all animals, living or toy

$$\downarrow$$

{true, false}

In the case of predicate logics built on other propositional logics, additional aspects of predicates and names are incorporated into the semantics, such as the ways we can come to know references or the subject matter of predicates and names. These are sometimes called the *content*, or *connotation*, or *sense* of the names and predicates. To simplify this discussion I'll discuss only classical predicate logic.[10]

EXAMPLE 4. Juney is a dog.
Juney likes Richard L. Epstein.
So all dogs like Richard L. Epstein.
Analysis Informally, we recognize that this inference is not valid. It might be that Juney is a dog, Juney likes Richard L. Epstein, but there is some other dog, some unnamed mongrel, that doesn't like Richard L. Epstein.

For our analysis in classical predicate logic we first need to assume that each of the sentences in the inference is a claim, and we need to agree that "— is a dog" and "— likes —" are atomic predicates, and that "Juney" and "Richard L. Epstein" are names, and that "all" can be understood as "$\forall$". Further, we have to have some way to formalize the conclusion, since "dogs" is not a name and there isn't any variable there. It's not obvious and requires some motivation to see that with the semantic assumptions of this logic it's reasonable to formalize the conclusion as "$\forall x(x$ is a dog $\to$ x likes Richard L. Epstein). Then we can formalize the inference as:

Premises Juney is a dog
          Juney likes Richard L. Epstein
Conclusion $\forall x_1(x_1$ is a dog $\to$ x_1 likes Richard L. Epstein)

---

[10] See *Predicate Logic* for the general framework for such predicate logics, and Stanisław Krajewski's and my "Relatedness predicate logic" for a development of one such logic.

Under one choice of assignment of predicates to predicate symbols and names to name symbols, this is a realization of:

Premises $\quad P_0^1(c_0)$
$\quad\quad\quad\quad\quad P_{14}^2(c_0, c_1)$
Conclusion $\quad \forall x_1 (P_0^1(x_1) \to P_{14}^2(x_1, c_1))$

We need not realize every predicate symbol and every name symbol; we only consider in our realization the parts of the inference we are analyzing.

Now our analysis proceeds as the informal one. To show that the inference is invalid, consider a model in which "Juney" names some object, and the predications "Juney is a dog" and "Juney likes Richard L. Epstein" are true, and there is an object in the universe such that when x_1 is taken to refer to it the predication "x_1 is a dog" is true and the predication "x_1 likes Richard L. Epstein" is false. In that case, the premises are true and conclusion false. So our formal logic also classifies this inference as invalid.

A possibility, in this analysis, just as for classical propositional logic, is a way to assign truth-values to the atomic claims in a realization. But the assignment of truth-values to atomic claims here requires that we understand what we mean by naming, that is, providing reference for a variable in the universe of the realization, for atomic claims here include wffs such as "x_1 is a dog" whenever a reference is supplied for "x_1". Thus, the assignment of truth-values to atomic claims, naming, and the evaluation of the connectives and the quantifiers are all that we need to understand possibilities, once we have agreed on the formalization of any particular inference.

As the last example shows, this understanding of possibilities correlates well with how we informally analyze inferences that can be formalized in this logic. Here, as for propositional logic, any choice of assignments of truth-values to atomic predications must be allowed, for the atomic claims are meant to be composed of names and predicates that are unanalyzed, bearing no relation to each other that we can recognize with this logic. We must allow a model in which "Marilyn Monroe was a man" and "Napoleon lived in California" are deemed true. This is not to say they are true; rather, a possibility is a model, a description (in terms of the linguistic elements we recognize) of a way the world could be, and that is reduced to allowing any assignment of truth-values to atomic predications. We need only consider the predicates and names in the inferences: to see that the semi-formal version of Example 4 is invalid, we don't need to consider whether "Marilyn Monroe is a man" is true or false. Descriptions are only as full as needed in the analysis of the inference at hand.[11]

EXAMPLE 5. All men are mortal.
Socrates is a man.
Therefore, Socrates is mortal.

Analysis We informally recognize this as valid. But if asked to say why, we are at a loss to say more than that's just what the words mean—how could all men be mortal and Socrates not be mortal if he is a man?

---

[11] This is justified by the Partial Interpretation Theorem in *Classical Mathematical Logic*.

Classical predicate logic can make this clearer. We can formalize the example:

Premises $\forall x_1(x_1$ is a man $\to x_1$ is mortal)
                 Socrates is a man.
Conclusion   Socrates is mortal.

Any object in the universe to which "—is a man" applies must also satisfy "—is mortal". There must be an object in the model to which "Socrates" refers, and "—is a man" applies to it because of the second premise. So that object satisfies "—is mortal". So there is no way for the premises to be true and conclusion false.

Does classical predicate logic really give us a reduction of the notion of valid inference? To show that an inference is valid we have to consider all possible models of a particular semi-formal language. Often, as in the last example, this can be done by considering general properties of the models. However, we often reason without specifying each and every object in the universe, as when we reason about all the pigs in Denmark. We rely on our notion of naming: given that pig, we could name her "x_1", and then .... The reduction of possibility and valid inference here can be claimed to be simpler and clearer than the full informal notion in any particular case only to the extent that we believe our notion of naming is simpler and clearer. Still, we have made some progress: in terms of the semantic and syntactic assumptions of this logic, what is logically possible and what is a valid inference have been reduced to just the truth-values of atomic claims and to our notion of naming, along with our evaluation of the connectives and quantifiers.

But it is only progress and hardly the whole story of logical possibilities.

EXAMPLE 6. Ralph is a bachelor. So Ralph is married.
Analysis We know that this can't describe a way the world could be because we know that "bachelor" means in part "is not married". But "—is a bachelor" and "—is married" are distinct predicates that have no structure, so we could have a predicate logic model in which there is an object named by "Ralph' and both predicates "—is a bachelor" and "—is married" apply to it. This analysis shows that we should not take both "—is a bachelor" and "—is married" as atomic predicates in a model if we wish to respect our informal notion of possibility.

We can, however, incorporate our assumption about the meaning of those predicates into our analysis by requiring that the semi-formal claim "$\forall x_1(x_1$ is a bachelor $\to \neg(x_1$ is married)) $\land \forall x_1(x_1$ is married $\to \neg(x_1$ is a bachelor))" be true in any model in which we use these predicates. Such a definition is not part of an inference, for a definition is not true or false, but only apt or inept, good or bad. It is an assumption that underlies the choice of how we will formalize.

EXAMPLE 7. All dogs are brown and some dogs are green.
Analysis We know that it's not possible for all dogs to be brown and some

dogs to be green, understanding this as "almost entirely brown" and "almost entirely green". But "—is brown" and "—is green" have no internal structure, so they can only be taken as atomic predicates in this logic. Hence we could have a model in which "—is a dog" and "—is green" and "—is brown" apply to every object.

Here we do not want to say that we cannot use both "—is green" and "—is brown" together in our semi-formal languages, for we cannot derive "—is green" from "—is brown". Nor are we lacking a definition. Rather, it is our understanding of the use of these words relative to our experience of the world that convinces us the example is not a possibility. We can formalize part of that experience with two claims:

$\forall x_1 (x_1 \text{ is brown} \to \neg(x_1 \text{ is green}))$
$\forall x_1 (x_1 \text{ is green} \to \neg(x_1 \text{ is brown}))$

We can then restrict the analysis of inferences in which these predicates appear to those models in which these claims are true. Then the example will not be classified as a possibility.

These two examples show that we can adapt our formal logic to analyze possibilities and validity in cases where the semantic and syntactic resources of predicate logic are apparently too limited. The next example, however, shows a real limitation of predicate logic in analyzing possibilities.

EXAMPLE 8. Snow is white.
Anything that's white isn't colored.
Snow is colored.

Analysis We cannot use classical predicate logic to show that this is a contradictory description of the world. Snow, as used in the first claim, is not a thing; it cannot be a thing in a universe of predicate logic; it cannot serve as reference for a variable. It is a substance, and the assumption that the world is made up of things does not take into account the world being made up in part of substances that are not things. A different logic, Aristotelian logic, can be used to show that this example is not possible, for that logic is based on the view that the world is made up in part of substances. But no way of melding that logic and any predicate logic has been devised.[12]

Once we consider the internal structure of claims, our analyses of reasoning will depend on some view of the nature of the world, some metaphysics. So our formalization of reasoning will accomodate only reasoning compatible with that metaphysics. At best we can hope to combine various formalizations that depend on different but compatible metaphysics into one logic, such as a logic that is based on seeing the world as made up of things, and of substances, and of processes. But that has not been done yet.

---

[12] See *Predicate Logic* for a discussion of why reasoning about substances cannot be formalized in any predicate logic, or consult Francis Jeffrey Pelletier and Lenhart K. Schubert, "Mass expressions". A fuller analysis appears in my *The Internal Structure of Predicates and Names with an Analysis of Reasoning about Process*.

EXAMPLE 9. Birta is a dog.
Bon Bon is a donkey.
Therefore, Birta did not give birth to Bon Bon.

Analysis To formalize this inference in classical predicate logic we replace "not" with the connective "¬", taking "Birta" and "Bon Bon" as names and "—is a dog", "—is a donkey". "—gives birth to—" as atomic predicates. Since these predicates are atomic, we can have a model in which any truth-value can be assigned to any predication involving them. In particular, we could have a model in which the premises of this example are true and conclusion false, so the inference is not valid.

That's just to say our logic reconizes that a dog giving birth to a donkey is a possibility. Is that right? There seems to be nothing in the meaning of these predicates that determines whether it is possible for a dog to give birth to a donkey. The predicates are atomic, and anyone who wishes to show that a dog giving birth to a donkey is not a logical possibility must appeal to some other standard of reasoning to show that.

7 Conclusion

We have made some progress in analyzing valid inferences. We look at an inference and try to agree on what semantic assumptions we use in informally analyzing it. Then we ask whether we have or can devise a formal logic based on those semantic assumptions. If yes, then we try to formalize the inference in that logic relative to those semantic assumptions. If we are successful, we have clarified in this instance our claim that we do or do not have a logical possibility and whether we do or do not have a valid inference.

In some cases it is quite easy to formalize an inference. In other cases it is quite difficult, requiring us to question our understanding of language and the world. And in other cases there is no formal logic we can use; we have only our informal notions on which to rely.

To judge an inference as valid or invalid requires us to be explicit about the semantic assumptions we are making so that we can be clear enough to reason well together. When we can agree, we have made some progress.

Appendix 1

Two examples of the use of logical possibilities

Twin earth
Hilary Putnam describes what he calls "twin earth" in an analysis of the nature of necessity. Twin earth is very much like earth except that there is a substance there that has the usual properties we ascribe to water but isn't H_2O. It is, he says, a possibility which has to be considered in theories of necessity.[13]

---

[13]Putnam, *Mind, Language, and Reality*, p. 223.

Every time I read Putnam's description I find it too bizarre. I can't conceive of such a world. How could Putnam convince me otherwise?[14]

Conceivability isn't the issue. Perhaps you can't conceive of a dog giving birth to a donkey, but it is a possibility we have to take into account in some of our reasoning if we can show that no contradiction follows from it, which we have done relative to the assumptions of classical predicate logic.[15]

Similarly, when challenged, Putnam needs to show that no contradiction follows from his description of twin earth. But his job is much harder, for he cannot use the resources of predicate logic. "Water" and "H_2O" are mass terms; they stand for substances, and predicate logic is not suitable to formalize reasoning using those. Aristotelian logic is suitable, but much of the rest of his description requires the resources of predicate logic. And we have no accepted formal logic that combines Aristotelian logic with predicate logic.

To the challenge that his description is not a logical possibility, Putnam can reply only by showing informally that it leads to no contradiction. That is not easy, and he has not done it. It could very well require more assumptions about inferences and possibility that are not compatible with his views in that article. We do not know.[16]

Nothing

Jared Diamond relates how as a student at Harvard his professor Paul Tillich stymied him and his classmates with the question "Why is there something, when there could have been nothing?", using that question to lead to belief in the existence of a creator of all.[17]

But the question supposes what needs to be shown: there could have been nothing. All of our experience seems to deny that. Yes, there might not have

---

[14] Paul Needham in "Microessentialism: What is the argument?" says:

What could Putnam's twin earth fantasy show beyond the triviality that sufficiently ignorant people might not be able to distinguish similar substances? If it assumes that two substances are distinct at the microlevel and yet share all their macroproperties, so that they can't be distinguished in terms of macroproperties, then the scenario is wildly implausible. Assuming that it is in some sense possible doesn't show it to be possible. p. 11.

[15] Some logicians do think that what is possible is what is conceivable:

It is thus in principle inconceivable—that is, it is literally impossible—that the same events should occur at two distinct moments of time.
   Nicholas Rescher and Alasdair Urquhart, *Temporal Logic*, p. 151.

[16] One colleague commented to me:

The Chinese recognized a kind of stone, namely, jade. After chemical analysis was developed, it turned out that some jade was nephrite and some was a different mineral, jadeite. So, we know what it is like to regard something as a natural kind that turns out to have samples with different microstructures. The property of being water is the property of being H_2O. But the property of being watery is the property of being transparent, tasteless, having low viscosity, etc. Once these surface properties are distinguished from the microstructural properties, we can entertain the possibility that not everything watery is water, just as we can understand that not all jade is nephrite.

But using an argument by analogy to show that a description is consistent is a not adequate, for it would need to show that the differences between the two sides of the comparison do not matter.

[17] Jared Diamond, "The religious success story".

been Socrates, or atoms, or anything we currently know. But nothing?

A colleague of mine took me to task for saying this. After all, he said, it is logically possible that there could have been nothing, and to deny that is to offer one answer to the question.

But how do we know that's logically possible? It's not a matter of taking predicates we use in our daily lives and assigning different truth-values to predications using them. Predicate logic is no guide to reasoning about nothing.[18] No formal logic, no reduction of the notion of possibility and inference, is available to guide us in reasoning about nothing. And informal methods for doing that seem as suspect as assuming that there could be nothing.[19]

The burden of proof in establishing possibilities
In part, the issue here is the burden of proof in establishing that a collection of claims describes a way the world could be. If we can invoke a formal logic to analyze a collection of claims, we can justify that a particular collection of claims is consistent. If we have no such logic, then the burden of proof is on the person claiming that the collection is consistent to show that it is. Someone might say that putting the burden of proof on the person who asserts the existence of a possibility would seriously impede the use of thought experiments. But if it's bizarre, we are owed a justification. If you say that you saw a dog fly out of the sky last night and hover three feet above you for two minutes, I'd ask for some evidence. If you say that Putnam's description of Twin Earth is consistent, I'd ask for some evidence.

D. H. Mellor in "Theoretically structured time" also believes that the burden of proof is on the person who claims that a description is possible:

---

[18]There are two uses of "all" in English: with or without existential import. If we take "∀" to have no existential import, as is usually the case in formalizing, then every claim beginning with "∀" is true about nothing, and every claim beginning with "∃" is false. If we take "∀" to have existential import, then in reasoning about nothing every claim beginning with a quantifier is false. In either case the resulting logic is trivial and no help in reasoning about nothing. This is just an indication that the use of variables does not make sense when applied to a model with an empty universe, for there is no story of naming and predication for that. (Andrzej Mostowski, "On the rules of proof in the pure functional calculus of the first-order" axiomatizes the first reading invoking set-theoretic interpretations of predicates, but that does not explain how he understands the use of variables compatible with how they are used in models with non-empty universes.)

The platonist must interpret the original question as asking why there are concrete things, since he or she believes that abstract objects such as numbers necessarily exist. In that case, the issue is not how to reason about nothing, but why we should believe that there are abstract things.

[19]For an example of the confusions that arise in trying to reason about nothing see Robert Nozick, "Why is there something rather than nothing?"

Jared's question also assumes that the world is made up of things, which is incoherent on the view of the world as process, a view for which a formal logic can be devised; see my *The Internal Structure of Predicates and Names with an Analysis of Reasoning about Process*.

I have in any case a general objection to this fashionable kind of argument by fantasy. It presumes to show something possible by describing an imaginary world in which we should apparently be inclined to believe the possibility actual. That may serve to settle verificationist qualms about the sense of supposing (e.g.) that the Universe has doubled in size overnight. But to show possibility as well as sense, one has also to show the imaginary world possible, which a merely plausible sketch of it does not do. Impossibilities are all to easy to make plausible, as mathematics shows: "$10^{23}+1$ is prime", for example, is a plausible enough statement; but it may be false nonetheless, and if it is, it is impossible. But even if Newton-Smith's fantastic worlds were possible, the plausibility in them of multiple—or closed—time hypotheses would not show that they in turn might be true. For only if they too are really possible would their *prima facie* plausibility in those worlds give reason to think them true. In the face of contrary arguments *a priori*, such as the one I have alluded to against causal loops, fantasies supply no evidence for possibility at all.

Nor can fantasy be propped up in this respect by physics. Certainly the fashion for it is reinforced by that for taking current physics to be at once the arbiter of truth and subject to a neurotic post-positivist conviction that it will all end up falsified and replaced by something else. This daft combination of attitudes undoubtedly encourages unrestrained fantasizing about conceivable future physics, whose cognitive authority is then invoked to establish the present possibility of its consequences, however farfetched. But these are quite obviously worthless trains of thought. Unless our acceptance of current physics commits us to its truth, it has no cognitive authority over us at all, and a merely conceivable future physics cannot have more authority over us than our own current physics has. In any case, even acceptance of physics as true does not preclude a mistaken acceptance of the impossible. No doubt we *ought* only to accept the possible as true; but then rational acceptance will have to wait on proof of possibility, not supply it. pp. 66-67.

But W. H. Newton-Smith, in "Reply to Dr. Mellor", disagrees:

To show possibility, he says, "one has to show the imaginary world possible, which a merely plausible sketch of it does not do". Mellor has misunderstood the structure and point of such arguments. One deploys them in relation to a contentious idea (e.g. bringing about the past, branching time) which has not been shown by a knock-down argument to be incoherent. The first stage involves giving a description which it is uncontentious to claim characterizes a possible world. ... The second stage is consider what further explanatory description might be made of the possible world. ...

The third stage is to argue that in the possible world the evidence in fact supports invoking the contentious idea. ... *If* anyone had a proof of the incoherence of the contentious idea it could be deployed here to defeat the argument. The point of such arguments is twofold. First, they still doubts of a verificationist sort concerning the contentious idea. Secondly, they are intended to *persuade* one who feels that there is some absurdity in the contentious idea that this is not so. As Dummett points out in the prelude to his own argument by fantasy in the paper favourably cited by Mellor ["Bringing about the past"], if the idea of bringing about the past is logically absurd, that absurdity should show up no matter how things turn out. A description of how things might turn out as in my stories or Dummett's enables us to see that an idea which in the abstract seems absurd could in fact have fruitful application and is not absurd.

Deploying fantasy physics in this context does not involve assuming the worthless train of thought outlined by Mellor. The structure of the argument is as characterized above. One describes a way things could turn out to be ... and one shows that in such a context it would be reasonable to posit the applicability of a contentious idea ... Certainly Mellor is correct in claiming that even acceptance of present-day physics as true does not preclude a mistaken acceptance of the impossible. However, it remains true that the success of modern physics gives us a reason to think that it is not based on impossible notions. pp. 69-70.

And Jonathan Bennett in A Study of Spinoza's Ethics, says:

I would go further and argue postively that it is unreasonable to believe that there must be a maximally excellent being. Plantinga notes that if his thesis is possible, then it is true [Alvin Platinga, *The Nature of Necessity*, p. 219f] (In the most widely used systems of modal logic, any proposition of the form 'Necessarily *P*' is true if it is possible.) And this contributes to his sense of entitlement to believe it. Why, after all, should he give up his view that it *could* be true that there must be a maximally excellent being? Leibniz made a similar point about the premiss which he added to Descartes' ontological argument, namely, that there could be a being which fits Descartes' definition: 'One is entitled to presume the possibility of any Being ... until someone proves the contrary'.

That principle about the onus of proof looks right. But properly used it counts against Leibniz and Plantinga, not for them. Someone who says it is possible that there must be an F being is basically asserting that there must be an F being, and is thus asserting an infinity of denials of possibility: of every world description, which excludes the existence of an F being, he is saying that it is impossible for there to be such a world as that. So if F is maximal excellence, he will say that there could not possibly be a world in which the only concrete objects were time and space and portions of matter. I reply that I am entitled to presume the possibility of such a world until someone proves the contrary, and I add that it is not reasonable to believe in the impossibility of such a world without having positive reasons for doing so. pp. 71-72.[a]

---

[a]Bennett's parenthetical comment about systems of modal logic is wrong: it only holds for S5 if we take it as: $\Diamond \Box A \supset \Box A$. It is correct for all extensions of T (reflexive frames) if we take it as $\Diamond \Box A \vdash \Box A$, but that is a strange reading.

Bennett, Platinga, Leibniz all hold that for a collection of claims to be a description of a way the world could be it must be consistent. They all take it that the burden of proof is to show that the collection is not consistent. On the face of it, they say, anything we postulate is possible, though Bennett restricts that in his paper to "possibilities that do not themselves have modal concepts nested within them".

Appendix 2 Logical consequence

I'll try to relate some of the recent debates about what's called "logical consequence" to the analysis of valid inferences I've given.[20]

Form as a guide to valid inferences
William Hanson writes:

> Other texts try to avoid modal notions and hence make use of truth and falsity simpliciter, rather than truth and falsity at possible worlds. They say that the conclusion of an argument is a logical consequence of the premises just in case no argument with the same logical form has true premises and a false conclusion.
> William Hanson, "The concept of logical consequence", p. 366.

To decide what is the logical form of a claim we need to have in hand the logical constants. But, as Hanson and others point out, there is no unique collection of those that can be considered exhaustive. So Hanson offers his criteria:

> In order to satisfy the requirements of generality and apriority, the semantic element is subject to two constraints on its selection of constants: (a) it must designate some terms as logical constants, and it will include among them some that appear in discourse on a wide variety of subjects (that is, what I earlier called ubiquitous terms); and (b) taken together, the terms it chooses must allow us to distinguish arguments exhibiting the resulting relation of logical consequence from those that do not exhibit it in a strictly a priori manner, to the extent that we can make such distinctions

---

[20]See, for example, John Etchemendy, *The Concept of Logical Consequence*; William Hanson, "The concept of logical consequence"; Timothy Bays, "On Tarski on models"; José M. Sagüillo, "Logical consequence revisited"; Greg Ray, "Logical conseqeuence: A defense of Tarski"; G.V. Sher, "Did Tarski commit 'Tarski's Fallacy'?"; Stephen Read, "Formal and material consequence."

at all. The choice of logical constants is thus pragmatic in the sense that it is influenced by our goals that logic should exhibit the kinds of generality and apriority that I have discussed.

But there are no logical constants as he describes. English words, or words of any other ordinary language, cannot be logical constants in this sense because they are not used in a manner that is univocal. Consider the inference:

Bring me an ice cream cone and I'll be happy. I'm not happy. Therefore, you didn't bring me an ice cream cone.

We recognize this as valid. And doing so we recognize that "and" here cannot be formalized as the propositional logic "∧", for that would yield an invalid inference. The word "and" should be formalized with "→".

We must abstract from our usual understanding of ordinary words if we are to do anything like a formal analysis. Then it is the formal logical constants that determine the form of a claim. But to formalize claims relative to these constants, we must observe the most fundamental constraint on formalization, which the example above illustrates:

An informally valid inference must be formalized as a valid inference.

It is our understandings of possibilities and valid inference that determine whether any particular formalization is good. The reliance on forms of claims in this approach already assumes some informal notion of possibility.

In all the papers cited logical form is understood as some predicate logic form. But that's just to assume the hard work that needs to be done: why should we accept the metaphysical assumptions needed for predicate logic in order to formalize a particular inference?[21]

The actual world suffices in place of possibilities
Some people say we do not need to rely on possibilities in analyses of inferences. We need consider only the world as it is and vary our interpretations of the non-logical words in inferences.

Consider, for example:

Marilyn Monroe is a man.
Every man has two feet.
Therefore, Marilyn Monroe has a right foot.

The usual reasoning is that this is invalid: both "Marilyn Monroe is a man" and "Every man has two feet" could be true, while Marilyn Monroe could

---

[21]For example, Hanson in "The concept of logical consequence" considers at length the following inference:

$(\exists x) Water(x)$
Therefore, $(\exists x) H_2 O(x)$

But water, like snow, is a mass, a substance, and not a thing that we can reason about in predicate logic. Nor can we treat it as a predicate, as I discuss in *Predicate Logic*. We would have to accept "Water(Ralph)" as a claim, which would correspond to the nonsensical "Ralph is water".

have two left feet, and that is easy to formalize in classical predicate logic. But rather than using possibilities, it's said we can see that this example is invalid by understanding "Marilyn Monroe" as referring to me, "has two feet" as meaning "has two dogs", and "has a right foot" as meaning "has a black dog"; the premises are true and the conclusion false in the actual world we live in, so the inference is not valid. We need not invoke possibilities.

We need to know the logical form of the inference before we can apply this method, for we need to know for which terms we can vary the meaning. As we saw above, that requires the use of some prior notion of possibility or valid inference. In any case, this method is so far from any way that we do or can analyze inferences in our daily lives as to be useless.

Situations and possibilities
William Hanson considers the following example:

$(\exists x)(\exists y)(x \neq y)$
Therefore, $(\exists x)(\exists y)(\exists z)(x \neq y) \& (y \neq z) \& (x \neq z)$

To use classical predicate logic to show that this is invalid we need that there is a model with just two things. Hanson worries that the platonist says that mathematical objects necessarily exist, so for a platonist there couldn't be just two things.

To overcome this he says we could argue:

> In determining logical truth and logical consequence we are concerned not only with truth values at all possible worlds but also with truth values at all subworlds of situations that appear as possible worlds. . . . For the sake of promoting generality, one is willing to accept as counterexamples to arguments not only full-fledged ways things might have been but also fragments of such ways. I believe that logicians often think of logical models in this way, and that doing so does not commit them to the view that each such model is itself a full-fledged possible world (that is, a way things might have been). For the mathematical platonist, doing so can be seen as introducing further generality into logic, a generality that comes from taking logic to be applicable even to the tiniest and most bizarrely gerrymandered fragments of possible worlds.
>
> Hanson, "The concept of logical consequence", p. 388.

In all the examples above in describing possibilities we considered only the words that appeared in the inferences. No one gives a description of a way the world could be that is fully complete, and no one could. One assumption we always make is that some of the words in the inference are clear enough and distinct enough in meaning from the other words in the inference to be taken as basic, allowing us to describe a possibility—or a part of a possibility if you like—by considering only claims in which those words appear.[22] In using

---

[22]See the theory of formalizing presented in *Predicate Logic* and the Partial Interpretation

predicate logic to analyze an inference we need consider only models in which we put just enough objects in the universe for every name in the inference to have a reference and whatever other objects are needed to do our analysis, as the examples above illustrate. If that does not satisfy the platonist, then the platonist justification of the inference as valid would be for any mathematician a *reductio ad absurdum* of platonism.

The same misconception of the use of possibilities in inference analysis appears in a view described by J. C. Beall and Greg Restall:

> The world is made up of situations. They are simply *parts* of the world. Claims are true of not only the world as a whole, but some claims at least are true of situations. We will not spend time on the theory of situations and their individuation here: we will simply illustrate it. In the situation involving Greg's household as he writes this, it is *true* that Christine is reading the paper. It is also *true* that the stereo is playing. It is *false* that the television is on. It follows from this, and the fact that the television is in fact an inhabitant of the situation, that it is *true*, in this situation, that the television is off.
>
> Situations 'make' claims true and they 'make' others false. However, some situations, by virtue of being *restricted* parts of the world, may leave some claims undetermined. It is not true in this situation that JC is reading. It is also not false in this situation that JC is reading—that is, it is not true in this situation that JC is *not* reading. JC does not feature in this situation at all.
>
> It follows that the classical account of negation fails *for situations*.
>
> J. C. Beall and Greg Restall, "Logical pluralism", pp. 475-493.

A description of the world is never complete. We always look at some collection of claims that amount to our description: it's what we're paying attention to in our reasoning. We're reasoning about these claims. If someone says we should also consider some other claims, we'll do so, and then those, too, will be part of our description. A possibility is then just our description, so long as it is consistent (and satisfies whatever other semantic conditions our logic imposes). Classical negation does not fail for claims just because they weren't originally under consideration.

Set theory as a reduction of the notion of possibility
Hilary Putnam says,

> The notion of possibility does not have to be taken as a *primitive* notion in science. We can, of course, define a structure to be *possible* (mathematically speaking) just in case a model exists for

---

Theorem in *Classical Mathematical Logic*.

a certain theory, where the notion of a model is the standard set theoretic one. That is to say, we *can* take the existence of sets as basic and treat possibility as a derived notion.
  Putnam, "What is mathematical truth?", p. 71.

This is wrong, even for mathematical theories. It assumes that the mathematical possibilities that are of interest are for a theory that has been completely formalized, and very few mathematical theories have been. It also assumes that there is a uniquely correct logic for which set theoretic models are taken, presumably, given the era in which Putnam was writing and his other writings suggest, first-order classical mathematical logic. And it also assumes that there is a uniquely correct set theory. All these assumptions are wildly at odds with mathematical practice.[23]

We could proceed as Putnam suggests to get some grasp of possibilities relative to a set theory and logic, much as I have described for propositional logic and for predicate logic. But it seems misguided to use set theory as a foundation for the notion of possibility. The notion of the collection of all subsets of a set seems no clearer than the notion of possibility.

BIBLIOGRAPHY

[Bays, 2001] T. Bays. On Tarski on models. *The Journal of Symbolic Logic*, 66:1701–1726, 2001.
[Beall and Restall, 2000] J. C. Beall and G. Restall. Logical pluralism. *Australasian Journal of Philosophy*, 78:475–493, 2000.
[Bell and Machover, 1977] J. L. Bell and M. Machover. *A Course in Mathematical Logic*. North-Holland, 1977.
[Bennett, 1984] J. Bennett. *A Study of Spinoza's Ethics*. Hackett, 1984.
[Diamond, 2002] J. Diamond. The religious success story. *New York Review of Books*, 49(17), 2002.
[Epstein and Krakewski, 2004] R. L. Epstein and S. Krakewski. Relatedness predicate logic. *Bulletin of Advanced Reasoning and Knowledge*, 2:19–38, 2004.
[Epstein, 1990] R. L. Epstein. *Propositional Logics (The Semantic Foundations of Logic)*. Kluwer, 1st edition, 1990.
[Epstein, 1994] R. L. Epstein. *Predicate Logic (The Semantic Foundations of Logic)*. Oxford University Press, 1st edition, 1994.
[Epstein, 2006] R. L. Epstein. *Classical Mathematical Logic (The Semantic Foundations of Logic)*. Princeton University Press, 1st edition, 2006.
[Epstein, 2010] R. L. Epstein. The internal structure of predicates and names with an analysis of reasoning about process. Typescript, available at www.AdvancedReasoningForum.org, 2010.
[Epstein, 2011] R. L. Epstein. Mathematics as the art of abstraction. In Reasoning in Science and Mathematics, Advanced Reasoning Forum, 2011.
[Etchemendy, 1990] J. Etchemendy. *The Concept of Logical Consequence*. Harvard University Press, 1990.
[Hanson, 1997] W. Hanson. The concept of logical consequence. *The Philosophical Review*, 106:365–409, 1997.
[Mellor, 1982] D. H. Mellor. Theoretically structured time: A review of w. h. newton-smith *the structure of time*. *Philosophical Books*, 23:65–69, 1982.
[Mostowski, 1951] A. Mostowski. On the rules of proof in the pure functional calculus of the first-order. *Journal of Symbolic Logic*, 16:107–111, 1951.
[Naess, 1938] A. Naess. *"Truth" as Conceived by Those Who Are Not Professional Philosophers*. I Kommisjon hos Jacob Dybwad, Oslo, 1938.
[Needham, 2011] P. Needham. Micoressentialism: What is the argument? *Noûs*, 45:1–21, 2011.

---

[23]See my *Classical Mathematical Logic* and my article "Mathematics as the art of abstraction".

[Newton-Smith, 1982] W. H. Newton-Smith. Reply to Dr. Mellor. *Philosophical Books*, 23:69–71, 1982.

[Nozick, 1981] R. Nozick. *Philosophical Explanations*, chapter Why is there something rather than nothing?, pages 115–166. Harvard University Press, 1981.

[Putnam, 1975a] H. Putnam. *Mathematics, Matter, and Method: Philosophical Papers, vol. 1*, chapter What is mathematical truth?, pages 60–78. Cambridge University Press, 2nd edition, 1975.

[Putnam, 1975b] H. Putnam. *Mind, Language, and Reality: Philosophical Papers, vol. 2*. Cambridge University Press, 1975.

[Ray, 1996] G. Ray. Logical consequence: A defense of Tarski. *Journal of Philosophical Logic*, 25:617–677, 1996.

[Read, 1994] S. Read. Formal and material consequence. *Journal of Philosophical Logic*, 23:247–265, 1994.

[Rescher and Urquhart, 1971] N. Rescher and A. Urquhart. *Temporal Logic*. Springer-Verlag, 1971.

[Sagüillo, 1997] J. M. Sagüillo. Logical consequence revisited. *The Bulletin of Symbolic Logic*, 3:216–241, 1997.

[Sher, 1996] G. Sher. Did Tarski commit "Tarski's Fallacy"? *The Journal of Symbolic Logic*, 61:653–686, 1996.

[Tarski, 1956] A. Tarski. *Logic, Semantics, Metamathematics: Papers from 1923 to 1938*, chapter On the concept of logical consequence. Oxford University Press, 1956.

Richard L. Epstein
Advanced Reasoning Forum
Socorro, New Mexico, USA
Email: rle@advancedreasoningforum.org

Rules of the Game
TARCÍSIO PEQUENO AND JEAN-YVES BÉZIAU

ABSTRACT. In a first part we discuss what can be considered as a game or not, examining some typical games, giving a formula characterizing games, differentiating games from science, art, religion, sport and business. In a second part we discuss a central game notion, the notion of rule. We distinguish four categories of rules for a game: framework rules, deontic rules, strategic rules and teleological rules. Our discussion is illustrated by many examples ranging from chess to soccer via frescobol and doll's tea party.

Dedicated to Walter Carnielli.
60: the game is just starting!

We are, above all, rule-making and rule-following creatures. It follows immediately both that the philosophical clarification of the concept of a rule is a suitable large topic to engage philosophers and that any radically new insight about rules will have substantial repercussion on a wide range of philosophical reflections. ([Baker and Hacker, 1985], pp.56–67)

0. Start

The notion of game has been increasingly popular. But what is a game? In this paper we intend to deal with the notion of game taken in a wide and wild sense, not limited to game theory or the fashionable game semantics, trying to characterize it through deep and essential features.

Most logicians or/and mathematicians who are using the word "game" are using it only metaphorically, using abstract symbols such as "players" or "winning strategy". They are not developing a philosophical analysis of what a game is, or presenting a theory of what a game is or should be. Game theory is not a theory of games, it is a theory roughly inspired by games. Game semantics also is not the semantics of games but a semantics inspired by games.

The relation between game and logic is generally only a one-way road: logical tools inspired by games. In this paper we will work in the other direction, applying deontic logic to games. This is very illuminating to understand for example a difference between chess and football besides the physical/intellectual duality: as we will explain, chess is a allowance game by contract to football, a prohibition game.

There are books analyzing a specific game such as chess (e.g. [Lasker, 1950]), books about the sociology/ethnology of games (e.g. [Wendling, 2002]), but no books about the general nature of a game. The heterogeneity and many dimensions of games may explain this situation. The only exception is the excellent book by Roger Caillois (1913-1978), *Les jeux et les hommes* [1957]. Although this book has been translated in English (*Man, play and games*) it is not very well known, especially among philosophers. Our paper can be seen as an improvement of Caillois's analysis, but it does not presuppose the knowledge of this book (we advise nevertheless anybody interested in "game theory" or "ludology" to read this book). A central difference between our approach and Caillois's is the way that we are considering rules. The main subject of our paper is in fact beyond the notion of game, it is the notion of rule, but what we are dealing with are not rules in the air, they are rules of the game. Our paper is neither argumentative, nor an overview of previous works, philosophy does not reduce to this dichotomy, there is also conceptual analysis. An important objective of philosophy is clarification and understanding. Many philosophers have worked in this direction, recently Ludwig Wittgenstein. The central notion of our paper, the notion of rule, has been much discussed by Wittgenstein, but although we have been inspired by him, the present paper is not a comment about his work.

1 Game: what it is and what it is not

Maybe the world and/or the universe is a game (cf. [Israel and Gasca, 2009]). But before considering such a brainy issue, let us rather consider the universe of games. There are plenty of games, very different from each other. We can start with an eagle eye, a panoramic view of what are things called games and then try to classify and structure the universe of games.

A purely descriptive approach to games, as to many notions, would be chaotic, and classification is only a first step for developing a better understanding. The idea is to go further on by building a normative concept of game excluding on the one hand things that are usually considered as games and on the other hand including things that are not usually called games. But a too normative concept of game would be absurd as well as a trivial concept of game, according to which everything is a game. The notion of game can be illuminating but we must be careful not to be dazzled.

1.1 Four basic games

To start let us fix our ideas with four kinds of game we have carefully chosen to be representative of the different features of a game: *football, chess, bilboquet* and *monopoly*. The reason why we have chosen these games will clearly appear to the reader when we will analyze the main characteristic of what a game is. Moreover these games are more or less universal. Our study has been conducted by consulting encyclopedic books about games (e.g. [Encyclopedia of Games, 1998], visiting museums such as the beautiful Swiss Museum of Games (located in a Castle at La Tour de Peliz by Lake Geneva) and last but not least practicing games.

Despite the world cup, football is not very popular in the USA and Americans are using a different name for it ("soccer"). Football is from British origin, but the most popular footballing country is since many years Brazil, in the same way we can say that Russia is the kingdom of chess, monopoly typically American in spirit and bilboquet a frenchy entertainment, *intraduisible!* Most of the games can be classified into these four categories of game, or considered as a mixture of these categories. Let us briefly describe these games.

Football is a game based on physical activity opposing 11 against 11 players on a field, but also with additional actors of the game inside and outside such as referees and coaches, 90 minutes is the end of the game. A similar game is tennis, with only 2 players on a physical space called court, it is potentially unlimited in time, the end of the game is a score. A variant of it is called *frescobol* in Brazil (much popular in Copacabana), and *matkot* in Israel, where it is considered as the country's national sport. It is played on the beach with two players, with no loser or winner, the end of the game is when the ball disappears in the sand or on the sea.

Chess is a game on a board confronting 16 pieces against 16 pieces directed by 2 intelligent players (natural or artificial), checkmate is the end of the game, chance and physical skills are not part of the game.

Monopoly is a racing board games son of the *game of the goose*, the race does not reduce to chance, financial skill is essential, end of the game is bankruptcy. A variation of it is *hopscotch*, where ability is physical. Another variation is *game of life*, a Bradley game, updated version of Milton Bradley's original game: you start the game deciding whether jump ahead on the game board or pay your way into the college. Going to the college gets you the advantage of 3 shots at a good salary instead of 1 but will set you back financially to start. Along the road of life, you will get married, choose what kind of house to live in, buy insurance, and ride the ups and downs on the road of life. This Bradley game should not to be confused with Conways game of life, name related here to the biological sense of life. This is a zero player game: once the "pieces" are placed in the starting position, the rules determine everything that happens later. Nevertheless, Life is full of surprises! In most cases, it is impossible to look at a starting position (or pattern) and see what will happen in the future. The only way to find out is to follow the rules of the game.

Bilboquet was promoted by the king of France Henri III in the 16th century, he liked to play bilboquet to impress his mignons. There is a handle, a string and a ball with a hole. There is only one player winning when getting the (hole of the) ball on the handle. A variation of this game is much popular in Japan under the name *Kendama*. *Pinball* can be also seen as a variant of bilboquet, as well as the ancestor of pinball, *billiards*, and also *golf*. In all these games, there is a pin, a ball and hole. The British have a female version of bilboquet called *cup and balls*.

1.2 The chemistry of game

To understand what a game is, it is important to be able to argue with good criteria what is not a game. To do so we can distinguish 5 features of a game: competition (*Co*), fiction (*Fi*), chance (*Ch*), skill (*Sk*) and fun (*Fu*). Our idea is that an activity is a game if and only if it obeys the following formula:

$$Co + Fi + (Ch\ \%\ Sk) + Fu.$$

Translating the chemistry of symbolism[1], this means that a game must be a competition, a fiction, that it involves either or both chance and skill and last

---

[1] We are using chemical metaphors in honour of C.S.Peirce, great chemist and logician

but not least, a game is fun. It is quite easy to see that the games we have been discussing obey this formula, although the *Co* feature is not always explicit. Brazilians and Israelis like to say that frescobol/matkot favors cooperation rather that rivalry, but nevertheless the winning strategy is to keep the ball in the air and competitions is won by rival duos, champion is the duo who was able to keep the ball for the longer time. Football and other team games favor also cooperation, the difference is that football cannot really be played with only one team.

Based on this chemistry of game it is not difficult to eliminate from the universe of games activities such as business, cooking, praying. astronomy and playing piano. Business is maybe fun, it involves chance and skill, this can easily be seen as a competition, but when Pan American World Airways (Pan Am) reached bankruptcy leaving on the ground thousands of employees, this was no fiction. Praying may also be fun, it involves a certain skill, but it is not clearly a competition (race to heaven?) and not completely fictional (Jesus Christ is not essentially a movie character). What we can say about food and music is that they can be turned into games: the winner being for example the one producing the biggest sausage or the one conducting Beethoven 9th symphony in the fastest way. These activities are not fundamentally games, but like many activities, such as sexuality (erotic games), or even civil rights (American gambling of the green card), they can be degenerated into games. It would also be childish to consider astronomy, or any other science, as a game. Childish because science is not basically a competition to won the Nobel prize. Childish also because our world is not a fiction in a galaxy of possible worlds. It is true however than children are playing games to develop their cognitive faculties. But at some point (the skill of) the games turn into reality.

Talking about children games, we may wonder if our chemical formula is not eliminating too much: can we say for example that dolls tea party obeys the above magic game formula? What is missing is the *Co* factor. To solve this difficulty we may want to eliminate the *Co* factor from our formula, but then open the door of the house of games to many activities connected with acting. And in fact actors are said to play a role. A way then to exclude actors, like many sport men or women, would be to say they are not in a complete fictional state, since they are making a living of their fictional activities. A game that is essentially a work is no game. But this is not the only reason. The fact that most arts and sports are not necessarily fun is a connected reason to reject them of the game universe. Zeus will forgive us to consider than many Olympic games are not games according to our formula: is it fun to run 42,195 km?

But let us come back to children, another way to avoid the exclusion of children games is to understand the Co factor in a broad sense, the competition consisting in learning something, in most of cases, by imitation. In this sense of competition, doll's tea party is a competition with one's self, like bilboquet, this competition can be cooperative like frescobol, in the case of a many players doll's tea party.

The *Co* factor is one of the basic differences between our chemistry of game and the one of the most important game theoretist, Roger Caillois. For Caillois an activity is a game with the following features:

1. fun: the activity is chosen for its light-hearted character
2. separate: it is circumscribed in time and place
3. uncertain: the outcome of the activity is unforeseeable
4. non-productive: participation does not accomplish anything useful
5. fictitious: it is accompanied by the awareness of a different reality
6. governed by rules: the activity has rules that are different from everyday life

Skill is also pathologically missing from Caillois's definition. On the other hand in our definition we didn't put uncertainty, we rather implicitly included it through "Chance or Skill". In our formula we also don't have something corresponding to (6), the reason is not that we are considering games without rules but that rules are so important that we are no putting them at the same level as the five elements of our chemical formula.

Up to now, we have considered only human activities. What about animals, clouds and all the rest? Can we not say that the cat is playing with the mouse, that by chance a stone fell into the head of a monkey, that the world is a big game?

If we consider that fiction is an essential feature of a game and that fiction is crucially human, then there are no non human games. But some animals are quite similar to us it seems they can play roles. And more generally we may think that we are part of nature and that even our most sophisticated behaviors have some similarities with what is going in the world. It is no absurd to think, without much anthropomorphism, of a non-human activity involving fiction, fun, competition, competition, skill, at least an animal activity. If we are talking about stones, this is maybe another story, isn't it?

A way to clarify this issue, the question of the (rational) animal exclusivity with respect to games, is to examine one the most fundamental aspects of games we have not yet talked about, the notion of rules. Human games are based on rules created by human beings and to play a game is to take in account these rules. Can we say that a cow, a stone or a cloud creates rules and behave according to these rules?

2 Rules

A game is regulated by rules but to say that any activity regulated by rules is a game would quite naturally lead to the truism *everything is a game* ; Kant opens his book of logic [1800] by saying that "Everything in nature, in the inanimate as well as the animate world, happens according to rules, although we do not always know these rules".

Can we say that rules of human games are similar to rules of other human activities and also with rules of animals, vegetables, minerals, atomic activities?

2.1 Rules and laws

In Greece, 2700 years ago, human beings created some rules, the basis of a state: a democracy, an oligarchy, a monarchy, These are called *laws*, and we have the legislators, those creating the laws, and the legislation, a set of laws. It is also usual to speak about laws of nature and the Greeks were the first to describe such *laws of nature*, it was the beginning of science.

Why using the same word, what is the similarity between laws of a state and laws of nature? One could argue that a God (or some Gods) created the laws of nature, we have a divine legislator and a divine legislation. But does the stone obey the laws of God in the same way that citizen K pays her taxes? Kelly may not pay her taxes, but can the stone not obey the laws of God or Nature? It seems that when we are speaking of laws of nature there is an absolute necessity.

A different perspective to establish the bridge would be to argue that laws of nature are like civil laws, product of human minds. But the laws of nature are not completely arbitrary or/and product of our free will. They are describing natural phenomena, they are considered faithful, in particular they permit to make prediction, however this does not mean that they are absolutely necessary. If a stone does not follow the laws of nature, we will not put it in jail, we may on the contrary put it in a museum, as an extraordinary phenomenon. We may also change the laws to keep the reluctant stone in our net of scientific laws. According modern physics, the notion of chance is so important, it is not clear that there are absolute laws of nature, i.e. absolute regularity (see e.g. [Bohm, 1961]).

In case of doubt and to avoid confusion, we may decide to call *rules*, the legislative laws regulating human societies, by contrast to the laws of nature. But can we say that these rules, the rules of a society, are similar to rules of a game and that a society is a game? We will examine in the next section how rules of a game work, but it seems than since the start we can see an important difference. In a game, one *must* obey the rules, if he does not obey the rule he is *out of the game*. One may try to sustain the similarity saying that on the one hand cheating is part of the game, and that on the other hand one who does not obey the rules of society goes out of the society game, this outside being jail. Nevertheless it seems that in a game we are more like a rolling stone, mechanically and diligently following the rules. Rules in this sense look more imperative than laws of society. It is interesting to note that rules of chess and rules of football are officially qualified as laws, maybe to reinforce their imperative character (see Fide [2008] and Fifa [2010]).

Boole is famous for his book about the *laws* of thought [1854]. At first sight such laws look more like laws of nature. Our thoughts can be considered as part of nature, a position that could be reinforced by a neuroscientist establishing a relation between thought and brain activity. The wife of Boole, Mary Everest Boole, a gifted children teacher wrote many books, among them *Philosophy & Fun of Algebra*, where she says:

> Arithmetic means dealing logically with facts which we know (about questions of number). 'Logically'; that is to say, in accordance with the 'Logos' or hidden wisdom, i.e. the laws of normal action of the human mind And this law of the Logos is made not by

any King or Parliament, but by whoever or whatever created the human mind governments have grown wiser by experience and found out that, as far as arithmetic goes, there is no use in ordering people to go contrary to the laws of the Logos, because the Logos has the whip hand, and knows its own business, and is master of the situation. ([Boole, 1891], p.1).

But can we not decide, independently of any Parliament or God, according to which rules we will reason, in the same way that we decide according to which rules we will play chess? Is mathematics, often considered as the highest form of reasoning, not a game for which we can set the rules? Our analysis of rules of a game will hopefully clarify these questions about laws of nature, laws of society, laws of thought[2].

2.2 Four kinds of rules

For any kind of game, we will distinguish four kinds of rules, presented in the following table:

Framework rules	setting the game	Where, when, with whom and with what
Deontic rules	defining the act of playing	What can be done or not, what have to be done or not
Strategic rules	qualifying the playing	How to play
Teleological rules	objective of the game	Why to play

TABLE 1 : FOUR KINDS OF RULE

In our characterization of rules, the less obvious category is the category of strategic rules, we will explain why they are fundamental and what is the relation between the other categories that we will also be described, showing how all these rules harmoniously work together[3].

Framework rules

There is a tendency to directly connect rules with activities. But it is not absurd to consider rules in a wider sense, as conditions for activity. We can call rules the size of a football field, the duration of the game, the number of players, the sex of the players. We will qualify these rules as framework rules. Someone may think that this kind of rules are really different from a rule such as *one must not touch the ball with his hands*, because a football player, say Zidane, can obey or disobey (intentionally or not) this rule, but can he disobey a framework rules such as the size of the field? If he goes out of the field and goes on playing, we may say he is disobeying such rule. But there is a more radical way to disobey this framework rule, it is to play football within a field

---

[2] A special issue of *Logica Universalis* (vol.4, n2, 2010) on the theme "Is logic universal?" has recently been published discussing the question whether logic is eternal or not, including in particular a paper by Jaroslav Peregrin, "Logic and natural selection".

[3] Up to now, nobody has presented such analysis and classification of rules, for example Hintikka just presents a rough distinction between definitory and strategic rules (see e.g. [Hintikka, 2007], a dichotomy which confuses framework rules with deontic rules and which does not take into consideration the distinctive teleological rules, which are independent of deontic rules

of a size different from the official size. In this case the one who disobeys the rules (intentionally or not) is not an inside player like Zidane, but it can be someone in charge with the team business, say Bernard Tapie.

So with framework rules appear framework players. This is not really surprising, it is natural to say that the coach is part of the game, even if he is in some sense out of the game - he is outside of the field. In football there is also a strange inside player, this is the referee. In football the referee is only metaphorically inside the game, in fact he is outside the game, in the sense that he is not a competitor (contrarily to the coach). For him to be a competitor in the sense of playing in favor of a team is prohibited. This is the rule. The referee should be neutral. Also if the referee does not whistle the end of the match (intentionally or nor) at the right time, he is not obeying the rule (intentionally or not). So the referee also can obey or disobey rules, in particular framework rules.

Deontic rules

Deontic rules can be roughly characterized as what can be and cannot be done, have to be done and have not to be done, it is related with action once the framework is set, ruling the activity of the player within the game, although some framework rules can also be described by deontic modalities. We use the expression "deontic rules" because these rules can be described with the help of deontic modalities.

Deontic modalities can be explained with the following hexagon, which is an improvement of the square of deontic modalities, based on Blanchs ideas (see [Blanché, 1966], [Beziau and Payette, 2008], [Beziau and Payette, 2011], [Moretti, 2009]):

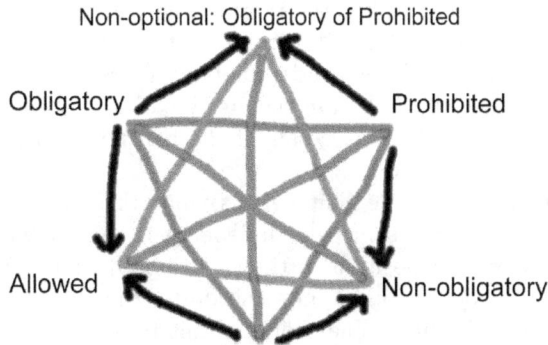

In standard systems of deontic modalities, the six modalities can be defined starting with only one and defining all the other ones with classical negation, conjunction and disjunction. But our present example of deontic rules show how poorer this reductionist viewpoint is.

Football is a game dominated by prohibition rules, by contrast with chess dominated by allowance rules. In a reductionist viewpoint this would be the same in the sense that prohibited = not allowed and vice versa allowed = not prohibited. These identities do not reflect a strong difference of behaviors between acting applying rules of allowance and acting doing whatever is not prohibited.

The deontic rules of chess are rules of allowance: the move of the pieces is clearly defined, they explain what moves can be done. What is prohibited is not stated explicitly it is by definition the classical negation of all what is allowed. Moreover these rules are really rules of allowance, not excluding obligatory moves, sometimes there are no options left.

Football rules are on the contrary prohibition rules, they indicate what cannot be done. Most important and explicit prohibition rules are: not touching the ball with the hand (except for the goal-keeper), and the off-side (law 11 of Fifa, rightly called in French *hors jeu*). What can be done cannot be precisely defined: we cannot give rules for the move of football players, in principle all what is not prohibited is allowed but this is not exactly the case for example it is not prohibited for a player to lay down on the ground if he is not injured, but generally a player will not do so.

A game like monopoly is a obligatory-optional game, oscillating between rules of obligation and optional rules. There are rules specifying what has to be done: to pay taxes. But other rules stipulated things which are optional (allowed and not obligatory): to buy houses.

Teleological rules

We can say that teleological rules are connected with the question: why playing? There is an interesting ambiguity here. The simplest answer is: play to win. But then we may ask: why winning the game? This is related to: why to play? We will not in this paper directly answer this abyssal question.

In general a game has a beginning and an end. The end is connected with winning and/or losing. Winning is defined by actions which are neither prohibited, nor obligatory, according to the rules of the game, *optional actions* that the players are trying to perform (Frescobol, not as competition between dual team, can be seen in a dual perspective, nobody is losing or winning, but the players are trying to keep the ball in the air. We can describe the game by: it is obligatory to keep the ball in the air, if the ball falls in the sand or in the sea, it is the end of the game.)

A teleological rule is a rule that defines what these optional actions are. Teleological rules are at the same time independent and related to other rules. Check-mate in chess is a position in the chess board; it is a position ending the game. We can choose another end position for the game, not changing the framework and deontic rules. The choice of this teleological rule what is the winning position of the game is therefore independent of other rules. But at the same time the teleological rules is related with other rules: the winning action is something that should be neither too easy nor too difficult to perform following the framework and deontic rules.

The winning action should in any case be possible, for example we cannot choose as an end position in chess, a position of the chess board which is not reachable from the starting position of the game. Herman Weyl [1927] has compared such non reachable positions with independent propositions in a mathematical system. Motivated by this similarity can we say that mathematics is a game? Someone is trying to prove a theorem in arithmetic. Maybe it is neither possible to prove this theorem, nor to find a counter-example of it, i.e. to disprove it by showing that its negation is a consequence of the axioms of arithmetic. The theorem is maybe independent, but it is possible to prove the independence of a theorem, such as an independence of the axiom of choice relatively to the axioms of ZF. An independence proof can be seen as end of the game.*En passant* let us note that Tsuji, da Costa and Doria [1998] have proven some interesting results of independence in theory of games by However a reason, among others, to say that mathematics is not a game would be to say that we can use neither chance nor a mechanism to win the game, i.e. to prove something, since chance is not considered as a mean of proof and there is not a general algorithm that can be apply to find a proof[4].

But can say that something which is purely algorithmic/mechanical is a game? After the birth of deep blue, can we still say that chess is a game? We can say that what is interesting is not to play algorithms but to find algorithms, and this can be considered as a game or as part of the game. This is connected with strategic rules.

Strategic rules

A strategy is a plan to develop an action. For example, in the famous movie *Rules of the game* by Jean Renoir, the following classical strategy of seduction is expounded: "Quand une femme rigole, elle est désarmée, vous en faites ce que vous voulez" (when a woman laugh, she is disarmed, you can do what you want of her).

---

[4]However theorems can be proved by algorithms. This was the case of a favorite conjecture of Tarski about Boolean algebra. Boolean algebra is an undecidable theory, but an algorithm to find a proof of this conjecture was run by a computer and after some time, it gave a positive solution with a proof. This is considered as the first true computer proof performed by a computer alone (see [Cipra, 1999]).

In football, the coach can consider that four players are attacking, in chess one may perform a Réti opening, in monopoly one may decide to buy everything he can buy. These ways of playing are not rules specified in the official regulations of the games. They correspond to optional actions, we can call them rules because they are regulating the game in the sense that the player will conduct his behavior according to them. Strategic rules depend not only on deontic rules, but also they depend on teleological rules: they are applied to win. Even in a pure gambling game there are rules of strategy: someone may always play his date of birth, luckily enough he may win before his 100th birthday.

To play a game is not only to apply allowance rules or developing hazardly or skillfully actions which are not prohibited, to play is also to create rules. That is maybe the most interesting aspect of a game and it is related with other human and non-human activities. François Le Lionnais and Raymond Queneau, the founders of the Oulipo, have argued that what is interesting in literature is not to write according to existing rules (such as alexandrine in poetry, the climax in drama - happening at the minute 69 in any good Hollywood movie) or reject rules (Surrealism), but to create new rules and write according to them, more generally: the seeking of new structures and patterns which may be used by writers in any way they enjoy [Oulipo, 1973]. One of the most famous Oulipian product is the book by Georges Pérec (1978), *La vie mode d'emploi* (*Life a user's manual*).

Most of the oulipians are mathematicians (such as Claude Berge) or have a mathematical background (like Le Lionnais en Queneau themselves), this does not mean that they are just applying mathematical rules to literature, but rather that they are inspired in the game of literature by mathematics where creation of rules is also a fundamental feature (about philosophy of mathematics, we recommend the masterpiece edited by Le Lionnais, *Les grands courants de la pensée mathématique* [1948], with papers by Bourbaki, Le Corbusier, etc). We can see for example Cantor's diagonal argument as a new rule. In this proof Cantor is using some rules of reasoning such as the excluded middle, but the form of the proof itself is new and this new idea can be generalized and applied to develop new ways of reasoning, in this sense it is a rule, or a new way to use the usual rules. As pointed out recently by Brady and Rush [2008], if we change the rules of the logic game, then we cannot perform anymore Cantors argument.

Someone may argue that strategy is at the meta-level: it is how to use rules. And for this reason, one may be radical saying that strategic rules are not rules but meta-rules. Here we consider that strategic rules can be either rules about rules or rules tout court. Even in logic the distinction between rules and meta-rules is not absolute, from a certain point view the deduction theorem can be seen as a meta-rule, from another perspective, it can be seen as a rule (see [Beziau, 1999] about this discussion). For us a meta-rule has some common features with a rule and in particular we agree with the possibility of formally describing strategic rules, as developed by Batens and Provijn [2001, 2009].

2.3 Rules and the four Aristotelian causes

It is possible that nature is acting according to strategic rules, and that new rules are being created all the time. The main distinction between nature (the universe) and a game is that the notion of teleological rules is not clear, unless we consider a God with a secret objective or the universe as a frescobol game: the goal being to stay in the air ...

To finish we present a connection between Aristotle's four causes and the four kinds of rules we have described. We will not here enter in details or justifications but leave it as a source of inspiration for future work.. Here is a table:

Framework rules	Material cause
Deontic rules	Formal cause
Strategic rules	Efficient cause
Teleological rules	Final cause

TABLE 2 : THE FOUR KINDS OF RULE
AND THE FOUR ARISTOTELIAN CAUSES

The value of this table can be checked by using it on the one hand to apply Aristotle's theory to some specific games and on the other hand by establishing a connection between rules of the games and activities other than games.

If we consider that the Aristotelian theory of four causes explains everything, then we may argue that everything can be explained by these four categories of rules via this concordance table. But can we then jump to the claim that a tsunami, the solar system or whatever going around in the world is a game?

Not necessarily, because a game is not only characterized by rules, we have also characterized games with a chemical formula. And it is not obvious that such phenomena obey this chemical formula.

Acknowledgments

Both authors were members of the CombLog project (CNPq ; Brazilian research council) when preparing this paper. Thank you to Catherine Chantilly for discussions, to Arthur Ronald de Vallauris Buchsbaum for his help in bibliographical supplies and to two anonymous referees whose remarks were useful to improve the paper.

BIBLIOGRAPHY

[Baker and Hacker, 1985] G.P.Baker and P.M.S.Hacker, *Scepticism, rules and language*, Basil Blackwell, London, 1985.
[Batans and Provijn, 2001] D.Batens and D.Provijn, "Pushing the search paths in the proofs. A Study in proof heuristics." *Logique et Analyse*, **173-174-175**, 2001, 113–134.
[Beziau, 1999] J.-Y.Beziau, "Rules, derived rules, permissible rules and the various types of systems of deduction", in *in Proof, types and categories*, E.H.Hauesler and L.C.Pereira (eds), PUC, Rio de Janeiro,1999, pp.159–184.
[Beziau, 1999] J.-Y.Beziau (ed), *Is logic universal?*, Special issue of *Logica Universalis*, Issue 2, Volume 4, 2010.
[Beziau and Payette, 2008] J.-Y.Beziau and G.Payette (ed), Special issue of *Logica Universalis* on the square of opposition, Issue 1, Volume 2, 2008
[Beziau and Payette, 2011] J.-Y.Beziau and G.Payette (ed), *New perspectives on the square of opposition*, Peter Lang, Bern, 2011.
[Blanché, 1966] R.Blanch'e, *Structures Intellectuelles - Essai sur l'organisation systématique des concepts*, Vrin, Paris, 1966.

[Bohm, 1961] D.Bohm, *Causality and chance in modern physics*, Harper, New York, 1961.
[Boole, 1854] G.Boole, *An Investigation of the laws of thought, on which are founded the mathematical theories of logic and probabilities*, Macmillan & Co., London, and Cambridge, 1854.
[Boole, 1891] M.E.Boole, *Philosophy & fun of algebra*, C.W. Daniel, London, 1891.
[Brady and Rush, 2008] R.Brady and P.Rush, "What is wrong with Cantors diagonal argument?", *Logique & Analyse*, **202**, 2008, 185–219.
[Caillois, 1957] R.Caillois, *Des jeux et des hommes*, Gallimard, Paris, 1957.
[Chaitin, 2011] G.Chaitin, "Life as evolving software", *Journal of Applied Non-Classical Logic*, to appear.
[Cipra, 1999] B.Cipra, "As easy as EQP", in *What is happening in mathematical science 1998-1999*, American Mathematical Society, Providence, 1999
[Encyclopedia of Games, 1998] *Encyclopedia of Games: Rules and Strategies for More Than 250 Indoor and Outdoor Games*, Aurum Press, London, 1998.
[Fide, 2008] Fide, *Laws of Chess*, World Chess Federation, Dredsen, 2008.
[Oulipo, 2010] Fifa, *Laws of the game 2009-2010*, Fédération Internationale de Football Association, Zürich, 2010.
[Hintikka, 2007] J.Hintikka, *Socratic Epistemology: Explorations of Knowledge-Seeking by Questioning*, Cambridge University Press, Cambridge, 2007.
[Israel and Gasca, 2009] G.Israel and A.M.Gasca, *The world as a mathematical game- John von Neumann and Twentieth century science*, Birkäuser, Basel, 2009.
[Kant, 1800] I.Kant, *Logik*, 1800.
[Le Lionnais, 1948] F.Le Lionnais (ed), *Les grand courants de la pense mathmatiques*, Cahiers du Sud, 1948. English translation: *Great currents of mathematical thought*, Dover, New-York, 1971.
[Lasker, 1950] E.Lasker, *Modern chess strategy*, G Bell & Sons, London, 1950.
[Moretti, 2009] A.Moretti, "The geometry of standard deontic logic", *Logica Universalis*, **3**, 2009, 19–57.
[Oulipo, 1973] Oulipo, *La littérature potentielle*, Gallimard, Paris, 1973.
[Pequeno, 2006] T.Pequeno, "Pedras podem seguir regras?", in *Colóquio Wittgenstein*, G.Imaguire, M.A.Montenegro and T.Pequeno (eds), UFC Edições, Fortaleza, 2006, pp.137–154.
[Provijn, 2009] D.Provijn, "Strategies. Whats in a name?" in Walter Carnielli, Marcello E. Coniglio & Itala M. Loffredo D'Ottaviano, (eds.), *The Many Sides of Logic*, College Publications, London, 2009, pp.287–306.
[Renoir, 1939] J.Renoir, *La règle du jeu*, film, screenplay by Jean Renoir and Carl Koch, 1939.
[Tsuji *et al.*, 1998] Tsuji, M., N.C.A. da Costa and F.A,Doria, ' "The incompleteness of theories of games", *Journal of Philosophical Logic*, **27**, 1998, 553–568.
[Wendling, 2002] T.Wendling, *Ethnologie des joueurs d'échecs*, Presses universitaires de France, Paris, 2002.
[Weyl, 1927] H.Weyl, *Philosophie der Mathematik und Naturwissenschaft*, Oldenbourg, Berlin, 1927.

Tarcisio Pequeno
University of Fortaleza - UNIFOR
Fortaleza, Brazil
E-mail: tarcisio@lia.ufc.br

Jean-Yves Béziau
Federal University of Rio de Janiero - UFRJ
Rio de Janeiro, Brazil
E-mail: jyb.logician@gmail.com

Proving Completeness for Nested Sequent Calculi[1]

MELVIN FITTING

ABSTRACT. Proving the completeness of classical propositional logic by using maximal consistent sets is perhaps the most common method there is, going back to Lindenbaum (though not actually published by him). It has been extended to a variety of logical formalisms, sometimes combined with the addition of Henkin constants to handle quantifiers. Recently a deep-reasoning formalism called *nested sequents* has been introduced by Kai Brünnler, able to handle a larger variety of modal logics than are possible with standard Gentzen-type sequent calculi. In this paper we sketch how yet another variation on the maximality method of Lindenbaum allows one to prove completeness for nested sequent calculi. It is certainly not the only method available, but it should be entered into the record as one more useful tool available to the logician.

1 Introduction

Recently Kai Brünnler introduced a version of deep reasoning called *nested sequents*, [Brünnler, 2009], with systems for a number of common modal logics including some that lack cut-free sequent calculi in the ordinary sense. Proving completeness for such systems is most commonly done by what might be called a *systematic backward proof search*. One starts with the desired goal and works backward until either a proof is found or enough information is produced to generate a counter-model. In [Fitting, 2011] it was shown that there are close connections between nested sequents and prefixed tableaus, and this allows one to transfer completeness results from prefixed tableau calculi to nested sequent systems. In addition, syntactic proof of cut-elimination can be given, see [Brünnler, 2009], and this allows transferal of completeness results from axiomatic formulations as well.

In this paper we sketch another approach to proving completeness for nested sequent systems of modal logic. It makes use of a suitably generalized maximal consistency construction, combined with a version of the introduction of Henkin witnesses as in first-order completeness arguments. This is not really a new approach. It has been applied to standard sequent calculi and to tableau systems, including prefixed ones. For nested sequents the form the construction takes is somewhat peculiar. A maximality requirement must be met by every nested subsequent, and Henkin witnesses are themselves nested sequents. We

---

[1]This paper is dedicated to Walter Carnielli on the occasion of his sixtieth birthday. Thank you Walter, and I wish you many more years and theorems.

believe the construction, while fairly simple, has interest so we are presenting it to the logic community.

We begin with a discussion about nested sequent calculi for the logic K, with some minor modifications to the original formulation. Then we give our completeness proof for it. Finally we make some brief remarks about how the work can be extended to other modal logics.

2 Formulas and Uniform Notation

Formuas are built up from atomic formulas, propositional letters, P, Q, ... They are built up using propositional connectives $\land$, $\lor$, $\neg$, $\supset$, and modal operators $\Box$ and $\Diamond$, in the usual way.

We use the common grouping of formulas into classes that behave alike—α, β, ν, π. It is often referred to as *uniform notation*. Compound formulas and their negations are grouped into those that behave conjunctively, α formulas, and those that behave disjunctively, β formulas. For each, *components* are defined, α_1 and α_2 for α formulas, and β_1 and β_2 for β formulas. These are given in the following tables.

α	α_1	α_2
$X \land Y$	X	Y
$\neg(X \lor Y)$	$\neg X$	$\neg Y$
$\neg(X \supset Y)$	X	$\neg Y$

β	β_1	β_2
$X \lor Y$	X	Y
$\neg(X \land Y)$	$\neg X$	$\neg Y$
$X \supset Y$	$\neg X$	Y

In a similar way there are the necessary formulas, ν, and possible formulas, π, and their components. These are given in the following tables.

ν	ν_0
$\Box X$	X
$\neg \Diamond X$	$\neg X$

π	π_0
$\Diamond X$	X
$\neg \Box X$	$\neg X$

3 Nested Sequents

We sketch a modal nested sequent system for the logic K. There are some changes from the original formulation in [Brünnler, 2009], making our work somewhat easier. The changes are as follows. We do not assume formulas are in negation normal form. The paper [Brünnler, 2009] works with multisets, assuming contraction and weakening rules. We use sets in place of multisets, and drop all structural rules. We do not allow the empty sequent. And finally, we make use of uniform notation, α, β, ν, and π. The definition of nested sequent that we use is a recursive one.

DEFINITION 1. A *nested sequent* is a non-empty finite set of formulas and nested sequents.

Nested sequents generalize *Tait* or *one-sided* sequents. All formulas have been moved to the right sides of arrows, and the arrows deleted. In effect, they are disjunctions. Nested sequents for modal logic iterate this idea; nesting corresponds to necessitation. More formally, let $\Gamma = \{X_1, \ldots, X_n, \Delta_1, \ldots, \Delta_k\}$ be a nested sequent, where each X_i is a formula and each Δ_j is a nested sequent. Then one defines a translation to ordinary formulas; think of $\Gamma^\dagger$ as the 'meaning' of Γ.

Axioms	$\Gamma(A, \neg A)$,
	A a propositional letter
Double Negation Rule	$\dfrac{\Gamma(X)}{\Gamma(\neg\neg X)}$
α **Rule**	$\dfrac{\Gamma(\alpha_1) \quad \Gamma(\alpha_2)}{\Gamma(\alpha)}$
β **Rule**	$\dfrac{\Gamma(\beta_1, \beta_2)}{\Gamma(\beta)}$
ν **Rule**	$\dfrac{\Gamma([\nu_0])}{\Gamma(\nu)}$
π **Rule**	$\dfrac{\Gamma(\pi, [\pi_0, \ldots])}{\Gamma(\pi, [\ldots])}$

Figure 1. Nested Sequent Rules for K

$$\Gamma^\dagger = X_1 \vee \ldots \vee X_n \vee \Box\Delta_1^\dagger \vee \ldots \vee \Box\Delta_k^\dagger$$

A proof of the soundness of nested sequent systems can be based on this translation. In this present brief paper we omit discussion of soundness issues.

There are standard notational conventions for nested sequents. Enclosing outer set brackets are often omitted. A nested sequent that is a member of another nested sequent has its members listed in square brackets, and is called a *boxed sequent*. For example, $A, B, [C, [D, E], [F, G]]$, is the conventional way of writing $\{A, B, \{C, \{D, E\}, \{F, G\}\}\}$. For this, the 'meaning' defined above is $A \vee B \vee \Box(C \vee \Box(D \vee E) \vee \Box(F \vee G))$. We systematically use $\Gamma, \Delta, \ldots$ for nested sequents, boxed or top level.

DEFINITION 2. *Subsequents* are defined as follows.

1. Γ is a subsequent of Γ.

2. If $\Delta \in \Gamma$, any subsequent of Δ is a subsequent of Γ.

Suppose Γ is a nested sequent in which propositional letter P occurs once— we write $\Gamma(P)$ for this. Subsequently, $\Gamma(X)$ is the result of replacing P in Γ with X. Similarly for $\Gamma(X, Y)$, $\Gamma(\Delta)$, and so on. Using this convention, Figure 1 displays the nested sequent rules for K, from [Brünnler, 2009] (extended to allow arbitrary formulas and not just those in negation normal form). Assume $\Gamma(P)$ is some nested sequent with one occurrence of propositional letter P, implicit behind the formulation of the rules displayed. Also we use $[\ldots]$ to stand for a *non-empty* nested sequent, and $[Z, \ldots]$ is the same sequent but with Z added.

Sequent proofs start with axioms and end with the nested sequent being proved. Proof of a formula is a derivative notion: a proof of the nested sequent consisting of just the formula X is taken to be a proof of X itself. Figure 2 contains an example of a nested sequent K proof.

$$
\cfrac{\cfrac{\cfrac{\lozenge(P\vee R),\big[\neg P,P,R\big],\big[\neg Q\big] \qquad \lozenge(Q\vee S),\big[\neg P\big],\big[\neg Q,Q,S\big]}{\lozenge(P\vee R),\big[\neg P,P\vee R\big],\big[\neg Q\big] \qquad \lozenge(Q\vee S),\big[\neg P\big],\big[\neg Q,Q\vee S\big]}}{\cfrac{\lozenge(P\vee R),\big[\neg P\big],\big[\neg Q\big] \qquad \lozenge(Q\vee S),\big[\neg P\big],\big[\neg Q\big]}{\cfrac{\lozenge(P\vee R)\wedge\lozenge(Q\vee S),\big[\neg P\big],\big[\neg Q\big]}{\cfrac{\lozenge(P\vee R)\wedge\lozenge(Q\vee S),\neg\lozenge Q,\big[\neg P\big]}{\cfrac{\lozenge(P\vee R)\wedge\lozenge(Q\vee S),\neg\lozenge P,\neg\lozenge Q}{\cfrac{\neg(\lozenge P\wedge\lozenge Q),\lozenge(P\vee R)\wedge\lozenge(Q\vee S)}{(\lozenge P\wedge\lozenge Q)\supset(\lozenge(P\vee R)\wedge\lozenge(Q\vee S))}}}}}}
$$

Figure 2. K Proof Example

4 Dual-Consistency

When proving propositional completeness axiomatically, a set $\{X_1,\ldots,X_n\}$ is called consistent if it is not the case that $\neg X_1 \wedge \ldots \wedge \neg X_n$ is provable. One shows that a consistent set extends to a maximal consistent one. Then if one calls a member of a maximal consistent set true, and a non-member false, it is verified that this is a standard truth-functional assignment. As it happens, the introduction of negation is complex to deal with when nested sequents are being used. We avoid the problem by dualizing everything, both at the beginning of the argument and at the end. The end game will be discussed in Section 6 but anticipating, we will call members (never mind members of what for now) false instead of true. Here at the beginning we make use of the following, which does not bring negation into the picture.

DEFINITION 3. Call a nested sequent Γ *dual-consistent* if Γ is not provable in the sequent calculus for K given in Figure 1 of Section 3.

In the usual treatments of logic, subsets of consistent sets are consistent. Here is the analog of this for the present setting.

PROPOSITION 4. *Suppose Δ is a subsequent of $\Gamma(\Delta)$, sequent Δ^* is such that $\Delta \subseteq \Delta^*$, and $\Gamma(\Delta^*)$ is like $\Gamma(\Delta)$ but with Δ replaced by Δ^*. If $\Gamma(\Delta^*)$ is dual-consistent, so is $\Gamma(\Delta)$.*

Proof. We just sketch the basic idea. Suppose $\Gamma(\Delta)$ is not dual-consistent, that is, $\Gamma(\Delta)$ is provable. In the final line of the proof of $\Gamma(\Delta)$, enlarge Δ to Δ^* getting $\Gamma(\Delta^*)$, and then 'propagate upward' throughout the proof the addition of formulas and nested sequents that effected this enlargement. Rule applications remain rule applications, and axioms remain axioms. This converts the proof of $\Gamma(\Delta)$ into one for $\Gamma(\Delta^*)$, so $\Gamma(\Delta^*)$ is not dual-consistent. ■

We also need a version of maximality, and we need it relativized it to a specified set of formulas. What is maximal now is a subsequent, in a sequent. Note that formulas *and their negations* are taken into account. The following terminology addresses this.

DEFINITION 5. For a set S of formulas, we say a formula X is S *determinate* if X belongs to S or is the negation of a formula belonging to S.

LEMMA 6. *Suppose S is a set of formulas that is closed under subformulas. If α is S determinate so are α_1 and α_2; if β is S determinate so are β_1 and β_2; if $\neg\neg X$ is S determinate so is X.*

Proof. We show the result for α formulas—the argument is similar in the β case and simpler for double negations.

Suppose first that $\alpha \in S$. There are three cases to consider. (1) If $\alpha = X \wedge Y$ then $X \in S$ and $Y \in S$ because of closure under subformulas, that is, $\alpha_1 \in S$ and $\alpha_2 \in S$, so these are S determinate. (2) If $\alpha = \neg(X \vee Y)$ then $X \in S$ and $Y \in S$, again because of subformula closure. Then $\alpha_1 = \neg X$ is S determinate because it is the negation of a member of S, and similarly for α_2. (3) $\alpha = \neg(X \supset Y)$. This case is a mixture of (1) and (2).

Next suppose that α is the negation of a member of S. Then α cannot be $X \wedge Y$, so there are only two cases to consider. (1) $\alpha = \neg(X \vee Y)$ where $X \vee Y \in S$. Then both X and Y are in S by subformula closure, so $\alpha_1 = \neg X$ and $\alpha_2 = \neg Y$ are S determinate since they are negations of members of S. (2) $\alpha = \neg(X \supset Y)$. This is similar to case (1). ∎

Now we have the central notion of maximal dual-consistency.

DEFINITION 7. Let $\Gamma(\Delta)$ be a nested sequent that is dual-consistent, with Δ as a subsequent. Also let S be a set of formulas that is closed under subformulas. We say Δ is *maximal* in $\Gamma(\Delta)$ with respect to S provided, for each X that is S determinate, if $X \notin \Delta$, the addition of X to Δ makes $\Gamma(\Delta)$ dual-inconsistent.

THE LINDENBAUM CONSTRUCTION: Let $\Gamma(\Delta)$ be a dual-consistent sequent with a subsequent Δ, and let S be a *finite* subformula closed set. It is straightforward to extend Δ to Δ^* so that Δ^* is maximal in $\Gamma(\Delta^*)$ with respect to S. A standard Lindenbaum-style construction will do—we omit the details. We restrict S to being finite because otherwise we would be forced to deal with infinite nested sequents, and we would rather not do so.

Maximality has something like the usual properties one expects from axiomatic completeness proofs, but dualized. For example in the α case below one might think the conclusion should involve "and" since α formulas are conjunctive, but in fact the case involves "or."

PROPOSITION 8. *Suppose S is a finite set of formulas that is closed under subformulas. Let $\Gamma(\Delta)$ be dual-consistent, with a subsequent Δ that is maximal with respect to the set S.*

1. *If A is atomic, not both A and $\neg A$ are in Δ.*
2. *If $\neg\neg X$ is S determinate and $\neg\neg X \in \Delta$ then $X \in \Delta$.*
3. *If α is S determinate and $\alpha \in \Delta$ then $\alpha_1 \in \Delta$ or $\alpha_2 \in \Delta$.*
4. *If β is S determinate and $\beta \in \Delta$ then $\beta_1 \in \Delta$ and $\beta_2 \in \Delta$.*

Proof. The cases are as follows.

1. If both A and $\neg A$ are in Δ, then $\Gamma(\Delta)$ is an axiom and hence provable.

2. Assume that $\neg\neg X$ is S determinate, and hence so is X by Lemma 6. Now we proceed contrapositively. Suppose $X \notin \Delta$. Using maximality, $\Gamma(\Delta \cup \{X\})$ is not dual-consistent, and hence is provable. It follows by the Double Negation Rule that $\Gamma(\Delta \cup \{\neg\neg X\})$ is also provable, and thus is not dual-consistent. Since $\Gamma(\Delta)$ is dual-consistent, it follows that $\neg\neg X \notin \Delta$.

3. Assume that α is S determinate, and hence both α_1 and α_2 also are by Lemma 6. Again the argument is contrapositive. Suppose $\alpha_1 \notin \Delta$. Maximality implies $\Gamma(\Delta \cup \{\alpha_1\})$ is not dual-consistent, and hence is provable. Similarly suppose $\alpha_2 \notin \Delta$; then $\Gamma(\Delta \cup \{\alpha_2\})$ is provable. Using the α Rule, $\Gamma(\Delta \cup \{\alpha\})$ is provable and so not dual-consistent. It follows that $\alpha \notin \Delta$.

4. Similar to the preceding case.

■

5 Henkin Witnesses

Dual-consistency and maximality will allow us to take care of propositional connectives. We still need machinery for the modal operators. What is presented in this section is an analog of a familiar step in standard axiomatic completeness proofs in modal logic.

HENKIN-LINDENBAUM EXPANSION: Let $\Gamma(\Delta)$ be a dual-consistent nested sequent with Δ as a subsequent. We define a *Henkin-Lindenbaum expansion* of Δ in $\Gamma(\Delta)$ to be a nested sequent $\Gamma(\Delta^*)$, where Δ^* is constructed in the following way.

1. First some definitions. Suppose $\nu \in \Delta$. The *Henkin witness for ν in Δ* is the nested sequent $[\nu_0, \pi_0^1, \ldots, \pi_0^n]$ where $\pi^1, \ldots, \pi^n$ are all the π formulas in Δ. A *Henkin witness* in Δ is a Henkin witness for ν in Δ, for some ν. We say $[\nu_0, \pi_0^1, \ldots, \pi_0^n]$ is *new to* Δ if it is not a subset of any nested sequent that is a member of Δ. We call the set S of all subformulas of $\nu_0, \pi_0^1, \ldots, \pi_0^n$ the *foundation set* of $[\nu_0, \pi_0^1, \ldots, \pi_0^n]$.

   Let Δ' be the result of adding to Δ all Henkin witnesses that are new to Δ. $\Gamma(\Delta')$ is also dual-consistent (shown below in Lemma 10).

2. For each Henkin witness $[\nu_0, \pi_0^1, \ldots, \pi_0^n]$ that was added to Δ to get Δ', enlarge it to a subsequent that is maximal in Γ with respect to its foundation set, using the Lindenbaum construction as sketched in Section 4. Let Δ^* be the result of thus expanding every Henkin witness in Δ'.

3. The outcome is a dual-consistent nested sequent, $\Gamma(\Delta^*)$ where $\Delta \subseteq \Delta^*$, Δ and Δ^* contain the same formulas, but Δ^* also contains, for each Henkin witness that is new to Δ, a nested sequent extending that Henkin witness, maximally dual-consistent in Γ with respect to the foundation set of that Henkin witness.

EXAMPLE 9. Suppose $\Delta = [\Box B, \Box C, \Diamond D, \Diamond E, F]$ and $\Gamma(\Delta) = A, [\Box B, \Box C, \Diamond D, \Diamond E, F]$. Then Δ has two Henkin witnesses, $[B, D, E]$ and $[C, D, E]$, in Γ. The outcome of step 1 above is the nested sequent $A, [\Box B, \Box C, \Diamond D, \Diamond E, F, [B, D, E], [C, D, E]]$. Then the outcome of step 2 is the nested sequent $A, [\Box B, \Box C, \Diamond D, \Diamond E, F, [B, D, E]^*, [C, D, E]^*]$ where $[B, D, E]^*$ is some extension of $[B, D, E]$ that is maximally dual-consistent in Γ with respect to the set of all subformulas of B, D, E, and similarly for $[C, D, E]^*$.

In step 1 of the Henkin-Lindenbaum Expansion process we said certain nested sequents would be dual-consistent. We now show this.

LEMMA 10. *Suppose $\Gamma(\Delta)$ is dual-consistent and Δ' is the result of adding to Δ all Henkin witnesses that are new to Δ. Then $\Gamma(\Delta')$ is also dual-consistent.*

Proof. We show the result of adding a single Henkin witness preserves dual-consistency. The Proposition then follows by iterating this.

Assume that $\nu \in \Delta$, and $\pi^1, \ldots, \pi^n$ are all the π formulas in Δ. Suppose $\Gamma(\Delta \cup \{[\nu_0, \pi_0^1, \ldots, \pi_0^n]\})$ were not dual-consistent. Then it would be provable. By repeated application of the π Rule, $\Gamma(\Delta \cup \{[\nu_0]\})$ would be provable. Then by the ν Rule, $\Gamma(\Delta)$ would be provable, and hence not dual-consistent. It follows that if $\Gamma(\Delta)$ is dual-consistent, so is the result of adding one Henkin witness to Δ. ∎

6 Completeness

With the basic work out of the way, we can now prove completeness itself. Let X be a formula that is fixed for the rest of this section, and suppose X is not provable in the nested sequent system for K. Using the unprovability of X we describe a process for generating a sequence of more and more elaborate nested sequents $\Gamma_1, \Gamma_2, \ldots$. Every subsequent of each Γ_i will have an associated foundation set, as specified in the Henkin-Lindenbaum Expansion of Section 5, and will be maximally dual-consistent in Γ_i with respect to its foundation set. As a bookkeeping device, certain subsequents of each Γ_i will be marked as *finished*.

Since X is not provable, the nested sequent $\{X\}$ is dual-consistent. This sequent is treated a little differently from later ones in the process since it is not a subsequent of anything else. Take the set of all subformulas of X as the foundation set for $\{X\}$, enlarge $\{X\}$ to a set that is maximally dual-consistent, with respect to its foundation set, and call the result Γ_1.

Next suppose Γ_n has been defined, every subsequent of it has an associated foundation set and is maximally dual-consistent with respect to it. If every subsequent of Γ_n is marked as finished, the construction stops. Otherwise choose a subsequent, Δ, that is not marked as finished, and consider $\Gamma_n(\Delta)$. Let $\Gamma_n(\Delta^*)$ be a Henkin-Lindenbaum expansion of Δ in $\Gamma_n(\Delta)$, as specified in Section 5, and set $\Gamma_{n+1} = \Gamma_n(\Delta^*)$. Mark Δ^* itself as finished in Γ_{n+1}, and also mark as finished any subsequents that were carried over unchanged from Γ_n and were marked as finished there. Note that if $\Gamma_n = \Gamma_1$, it is its only unfinished subsequent. The process just described still applies, but Γ_1 is, in effect, in an empty context.

This process must stop after a finite number of steps, for the following reason. The modal degree of each member of a Henkin witness in Δ must be less (by 1) than the modal degree of some member of Δ. Conseqently, if $\Gamma_{n+1} = \Gamma_n(\Delta^*)$, the maximal modal degrees of the foundation sets of members of Δ^* must be less than the maximal modal degree of the foundation set of Δ^* itself. When modal degrees reach 0 the process stops. Let us say Γ_∞ is the final member of the sequence just described.

Now we construct a a Kripke model $\mathcal{M} = \langle \mathcal{G}, \mathcal{R}, \Vdash \rangle$ from Γ_∞, as follows.

Let the set of possible worlds, $\mathcal{G}$, be the collection of all subsequents of Γ_∞.

Next we specify the accessibility relation, $\mathcal{R}$. For $\Delta, \Omega \in \mathcal{G}$, let $\Delta \mathcal{R} \Omega$ provided $\Omega \in \Delta$.

Finally we have the truth-at-a-world relation. For an atomic formula A, set $\mathcal{M}, \Delta \Vdash A$ just when $A \notin \Delta$. Note that this condition is dual to the one usually seen in completeness proofs.

We now have a model $\mathcal{M} = \langle \mathcal{G}, \mathcal{R}, \Vdash \rangle$. For it we have a kind of *dual truth lemma*, Proposition 12. In proving this we make use of structural induction, but it is not quite the usual version based on complexity of formulas as measured by degree. Instead the version we use is the following—it is reasonably intuitive, and a formal proof that it works can be found in [Fitting, 1996], as Theorem 2.6.3.

PROPOSITION 11. *Every formula has property* **Q** *provided:*

1. *every atomic formula and its negation has property* **Q**;

2. *if X has property* **Q** *so does $\neg\neg X$;*

3. *if α_1 and α_2 have property* **Q** *so does α;*

4. *if β_1 and β_2 have property* **Q** *so does β;*

5. *if ν_0 has property* **Q** *so does ν;*

6. *if π_0 has property* **Q** *so does π.*

Now here is our *dual truth lemma*, proved using Proposition 11.

PROPOSITION 12. *Let $\mathcal{M} = \langle \mathcal{G}, \mathcal{R}, \Vdash \rangle$ be the model constructed above, let $\Delta \in \mathcal{G}$, and let S be the foundation set of Δ. For each formula Z that is S determinate: $Z \in \Delta \Longrightarrow \mathcal{M}, \Delta \not\Vdash Z$;*

Proof. By induction on complexity we show that, if Z is S determinate where S is the foundation set of Δ, and $Z \in \Delta$, then $\mathcal{M}, \Delta \not\Vdash Z$. There are several cases.

Atomic Case: Suppose A is atomic and $A \in \Delta$. Then $\mathcal{M}, \Delta \not\Vdash A$ by definition of the model.

Negated Atomic Case: Suppose A is atomic and $\neg A \in \Delta$. Then $A \notin \Delta$ by Proposition 8 part 1, so $\mathcal{M}, \Delta \Vdash A$, again by definition of the model, and hence $\mathcal{M}, \Delta \not\Vdash \neg A$.

Double Negation Case: Suppose the result is known for Z, $\neg\neg Z$ is S determinate where S is the foundation set of Δ, and $\neg\neg Z \in \Delta$. By Lemma 6, Z is S determinate, and by Proposition 8 part 2, $Z \in \Delta$. Then by the induction hypothesis, $\mathcal{M}, \Delta \not\Vdash Z$. It follows that $\mathcal{M}, \Delta \not\Vdash \neg\neg Z$.

α *Case:* Suppose α is S determinate where S is the foundation set of Δ, the result is known for α_1 and α_2, and $\alpha \in \Delta$. By Lemma 6 both α_1 and α_2 are S determinate, and $\alpha_1 \in \Delta$ or $\alpha_2 \in \Delta$, by Proposition 8 part 3. By the induction hypothesis, $\mathcal{M}, \Delta \not\Vdash \alpha_1$ or $\mathcal{M}, \Delta \not\Vdash \alpha_2$. In either case, $\mathcal{M}, \Delta \not\Vdash \alpha$.

β *Case:* Similar to the previous case.

ν *Case:* Suppose ν is S determinate where S is the foundation set of Δ, the result is known for ν_0, and $\nu \in \Delta$. In the process of constructing Γ_∞, at some point a Henkin witness in Δ, $[\nu_0, \pi_0^1, \ldots, \pi_0^n]$ has been expanded to produce a member, Ω of Δ. Since ν_0 must be in the foundation set of Ω, by the induction hypothesis, $\mathcal{M}, \Omega \not\Vdash \nu_0$. And since $\Omega \in \Delta$, $\Delta \mathcal{R} \Omega$. Then $\mathcal{M}, \Delta \not\Vdash \nu$.

π *Case:* Suppose π is S determinate where S is the foundation set of Δ, the result is known for π_0, and $\pi \in \Delta$. In this case π_0 must be in every Henkin witness in Δ. It follows that π_0 must belong to every subsequent that is a member of Δ, and must be a member of its foundation set. Then by the induction hypothesis, $\mathcal{M}, \Omega \not\Vdash \pi_0$ for every $\Omega \in \mathcal{G}$ with $\Delta \mathcal{R} \Omega$, and so $\mathcal{M}, \Delta \not\Vdash \pi$.

∎

Since the construction of Γ_∞ begins with a sequent containing X, then X must be a member of Γ_∞ itself. By the Proposition above, $\mathcal{M}, \Gamma_\infty \not\Vdash X$, and so X is not K-valid.

7 Other Modal Logics

There are extensions of the nested sequent system described in Section 3 for many standard modal logics, see [Brünnler, 2009; Fitting, 2011]. We do not state the rules here. Completeness for K is easiest to establish, with that for T and D a close second. Things become harder when transitivity is involved because the construction process described in Section 6, when appropriately adapted to these logics, need not terminate. There are two solutions, at least, for this problem.

First, we can simply accept the fact that the construction process goes on forever. Then we need to define an appropriate notion of limit for the sequence $\Gamma_1, \Gamma_2, \ldots$, and this is not difficult. Conceptually the limit would be a nested sequent that allowed infinitely deep nesting. This is not 'legal' given the way we have defined nested sequents, as sets, because it violates well-foundedness. It is, however, an intuitively plausible thing, and the notion of direct limit, from category theory, is a formal substitute. We then carry out the construction of a model using this limit, instead of using the last term of the sequence as we did above.

Second, since all formulas are subformulas of the formula we are trying to prove, if a construction goes on forever there must be repetition. One can terminate work on a subsequent when it duplicates one of its 'ancestors.' In this way work can be forced to halt, as it did for K, after a finite number of steps. Unfortunately, this method won't extend to admit quantifiers, though the one with limits, described above, can be made to work.

We do not go into details of these more complex constructions here. In this paper we merely wanted to show how the maximal consistent set construction, familiar from classical propositional logic, could be extended to nested sequent calculi, and enough has now been said to give the general idea.

BIBLIOGRAPHY

[Brünnler, 2009] K. Brünnler. Deep sequent systems for modal logic. *Archive for Mathematical Logic*, 48(6):551–577, 2009.

[Fitting, 1996] M. C. Fitting. *First-Order Logic and Automated Theorem Proving*. Springer-Verlag, first edition edition, 1996. Errata at http://comet.lehman.cuny.edu/fitting/errata/errata.html.

[Fitting, 2011] M. C. Fitting. Prefixed tableaus and nested sequents. *Annals of Pure and Applied Logic*, 2011. To appear.

Melvin Fitting
Department of Mathematics and Computer Science
Lehman College (CUNY)
New York, USA
Email: melvin.fitting@lehman.cuny.edu

Real Semigroups and Rings

MAX DICKMANN AND FRANCISCO MIRAGLIA

ABSTRACT. We show, among other results, that the real semigroup (RS) associated to any preordered ring, $\langle A, T \rangle$, is naturally isomorphic to the RS of reduced bounded inversion ring, canonically and functorially associated to $\langle A, T \rangle$.

Our goal here is to show that the real semigroup (RS) of any preordered ring, $\langle A, T \rangle$, is naturally isomorphic to the RS of a reduced bounded inversion ring (BIR), canonically and functorially associated to $\langle A, T \rangle$. This will be accomplished in section 3. Section 1 contains basic material concerning real semigroups, in particular, those associated to preordered rings, while section 2 describes the relations between the real semigroup associated to a p-ring and to its ring of fractions by a multiplicative set. We also take the opportunity to present interesting examples of RS-congruences in real semigroups associated to rings.

1 Preliminaries

For the theory of real semigroups, the reader is referred to [Dickmann and Petrovich, 2004] and to the more comprehensive [Dickmann and Petrovich, 2011]. For lack of a convenient reference, we give a succinct account of the natural functors from the category of preordered rings to that of real semigroups and from the latter into the category pre-special groups.

1.1. Notation and Basic Definitions. In all that follows, "ring" means commutative unitary ring. Let R be a ring.

a) If $D \subseteq R$:

(1) $R^\times = \{u \in R : a \text{ is a unit in } R\}$ is the (multiplicative) group of units in R;

(2) $D^\times = D \cap R^\times$;   (3) $D^2 = \{d^2 \in R : d \in D\}$;

(3) $\Sigma D^2 = \{\sum_{i=1}^n d_i^2 \in R : n \geq 1 \text{ is an integer and } \{d_1, \ldots, d_n\} \subseteq D\}$.

b) A **preorder** on R is a subset P of R, closed under sums and products, containing R^2 and such that $-1 \notin P$. If $-1 \notin \Sigma R^2$, R is said to be **semi-real**; in this case, ΣR^2 is the least preorder on R.

c) A **preordered ring (p-ring)** is a pair $\langle A, T \rangle$, where A is a ring and T is a preorder on A.

d) A p-ring $\langle A, T \rangle$ is a **bounded inversion ring (BIR)** if $1 + T \subseteq A^\times$.

e) If $\langle A, T \rangle$, $\langle R, P \rangle$ are p-rings, a map $f : \langle A, T \rangle \longrightarrow \langle R, P \rangle$ is a **p-ring morphism** if it is a morphism of unitary rings, satisfying $f[T] \subseteq P$. ∎

Forthwith, all rings will be assumed to be semi-real.

1.2. T-convex and T-radical Ideals. (cf. Chapter 4 of [Bochnak et al., 1998]).

a) An ideal I in a p-ring $\langle A, T \rangle$ is

- **T-convex** if for all $s, t \in T$, $s + t \in I \Rightarrow s, t \in I$;
- **T-radical** if for all $a \in A$ and $t \in T$, $a^2 + t \in I \Rightarrow a \in I$.

A ΣA^2-radical ideal is called **real**.

By Proposition 4.2.5 in [Bochnak et al., 1998] an ideal of A is T-radical iff it is T-convex and radical. In particular, a prime ideal is T-radical iff it is T-convex. ∎

Note that if $T \subseteq \alpha \in \mathrm{Sper}(A)$, then the prime ideal $\mathrm{supp}(\alpha)$ is T-convex. Conversely,

PROPOSITION 1. ([Bochnak et al., 1998], Prop. 4.3.8, p. 90) *If I is a proper prime ideal, T-convex for a given preorder T of A, then there is $\alpha \in \mathrm{Sper}(A, T)$ such that $\mathrm{supp}(\alpha) = \alpha \cap -\alpha = I$.* ∎

PROPOSITION 2. a) ([Bochnak et al., 1998], Prop. 4.2.7, p. 87) *A preorder T on a ring A is proper iff A has a proper T-convex ideal.*

b) *If $\langle A, T \rangle$ is a p-ring, any ideal of A, maximal for the property of being T-convex, is prime.* ∎

PROPOSITION 3. ([Bochnak et al., 1998], Prop. 4.2.6, p. 87) *Given a preorder T of A, every ideal I of A is contained in a smallest T-radical ideal (possibly improper), namely:*

$$\sqrt[T]{I} = \{a \in A : \exists\, m \in \mathbb{N} \text{ and } t \in T \text{ such that } a^{2m} + t \in I\},$$

*called the **T-radical of I**, the intersection of all T-convex prime ideals containing I.* ∎

REMARK 4. With notation as in 3:

a) If $a \in A$, write $\sqrt[T]{a}$ for the T-radical of the principal ideal (a). In particular, $\sqrt[T]{0}$ is the T-radical of the zero ideal. By 3, an ideal I is T-radical iff $\sqrt[T]{I} = I$.

b) If $T = \Sigma A^2$ and I is an ideal in A, we write $\sqrt[re]{I}$ for $\sqrt[T]{I}$, the **real radical** of I, equal to the intersection of all real primes of A containing I.

c) Recall that a ring A is **reduced** if it has no non-zero nilpotent elements, i.e., the intersection of all prime ideals in A is the zero ideal; the analog of this notion in the case of preordered rings appears in the next definition. ∎

DEFINITION 5. A p-ring $\langle A, T \rangle$ is **T-reduced** if $\sqrt[T]{0} = (0)$. If $T = \Sigma A^2$, i.e., $\sqrt[re]{0} = (0)$, we say A is a **real ring**. Clearly, a T-reduced ring is reduced and semi-real [1].

Every p-ring has a **BIR hull**, as follows:

PROPOSITION 6. (Proposition 6.5.(a), p. 72ff of [Dickmann and Miraglia, 2011]) *If $\langle A, T \rangle$ is a p-ring, then $S = 1 + T$ is a proper multiplicative subset of A. Moreover, if $\nu : A \longrightarrow A^* = AS^{-1}$ is the ring of fractions of A by S and*

$$T^* = \{t/s^2 \in A^* : t \in T \text{ and } s \in S\},$$

then

(1) T^* is a proper preorder of A^* and $\langle A^*, T^* \rangle$ is a BIR.

(2) ν is a p-ring morphism; moreover, if A is T-reduced (cf. 5), then ν is injective.

(3) If $f : \langle A, T \rangle \longrightarrow \langle R, P \rangle$ is a p-ring morphism and $\langle R, P \rangle$ is a BIR, there is a unique p-ring morphism, $g : \langle A^*, T^* \rangle \longrightarrow \langle R, P \rangle$, such that $g \circ \nu = f$.

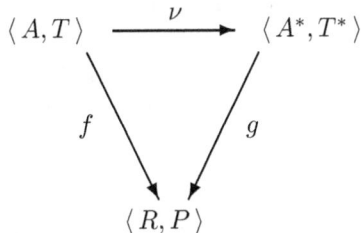

∎

1.3. Ternary Semigroups ([Dickmann and Petrovich, 2004], [Dickmann and Petrovich, 2011]). A structure $\langle S, \cdot, 1, 0, -1 \rangle$ is a **ternary semigroup (TS)** if

[TS 1] $\langle S, \cdot, 1 \rangle$ is an Abelian semigroup (monoid) with identity 1;

[TS 2] $x^3 = x$, for all $x \in S$;   [TS 3] $-1 \neq 1$ and $(-1)(-1) = 1$;

[TS 4] $x \cdot 0 = 0$, for all $x \in S$;   [TS 5] For all $x \in S$, $x = -1 \cdot x \Rightarrow x = 0$.

If $x \in S$, write $-x$ for $-1 \cdot x$.

b) If S is a TS, $R \subseteq S^3$ is a ternary relation on S and $a, b, c \in S$, write $a \in R(b, c)$ in place of $R(a, b, c)$. Define the **transversal** of R, R^t, by

[t-rep] $a \in R^t(b, c) \Leftrightarrow a \in R(b, c) \wedge -b \in R(c, -a) \wedge -c \in R(b, -a)$.

b) If S, S' are TSs, a map, $f : S \longrightarrow S'$ is **TS-morphism** if it preserves product, 0, 1 and -1. ∎

DEFINITION 7. A set-theoretic map, $f : D \longrightarrow E$, induces a map

$$f \times f : D \times D \longrightarrow E \times E, \text{ given by } \langle a, b \rangle \longmapsto \langle f(a), f(b) \rangle.$$

Define [2],

---

[1] Since $\sqrt[T]{0} = (0)$, A has a proper real prime ideal, and 2.(a) guarantees that ΣA^2 is a proper preorder of A. Moreover, our definition of real ring coincides with the usual one, i.e. (0) is a real ideal (cf. 4.(a)).

[2] Sometimes called the *fibered product of A over f*.

$$\ker f \;=\; (f \times f)^{-1}[\Delta_E] \;=\; \{\langle a,b\rangle \in D \times D : f(a) = f(b)\},$$

called the **kernel of f**, where Δ_E is the diagonal of $E \times E$. If $D \xrightarrow{f} E \xrightarrow{g} F$ are set-theoretic maps, we clearly have

[ker-comp] $\qquad\qquad \ker(g \circ f) \;=\; (f \times f)^{-1}[\ker g].$

1.4. TS-Congruences. For more detailed information on this topic, the reader is referred to section 1 of Chapter I in [Dickmann and Petrovich, 2011] (cf. Definition I.1.9ff).

a) Let S be TS; an equivalence relation, θ, on S is a **TS-congruence** if it is a congruence with respect to the product in S, i.e, $a\,\theta\,a'$ and $b\,\theta\,b'$ implies $(ab)\,\theta\,(a'b')$.

Let $S/\theta = \{a/\theta : a \in S\}$ be the set of equivalence classes of elements of S by θ and let $\pi_\theta : S \longrightarrow S/\theta$, $a \longmapsto a/\theta$ be the canonical quotient map. Notice that $\ker \pi_\theta = \theta$. With the operation induced by the product in S, S/θ has a natural structure of ternary semigroup, wherein 1, 0, and -1 are the classes of these constants modulo θ. Moreover, π_θ is a TS-morphism and the diagram $S \xrightarrow{\pi_\theta} S/\theta$ has the following property:

> If $f : S \longrightarrow S'$ is a TS-morphism, such that $\theta \subseteq \ker f$, there is a *unique* TS-morphism, $\widehat{f} : S/\theta \longrightarrow S'$, making the diagram below left commutative:

[TS-UFP]

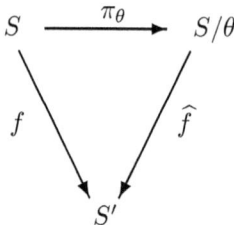

Indeed, it is straightforward that $\widehat{f}(a/\theta) = f(a)$, $a \in S$, has the required properties.

b) If $f : S_1 \longrightarrow S_2$ is a TS-morphism, then $\ker f$ is a TS-congruence on S_1 and there is a *unique* TS-morphism, $\widehat{f} : S_1/\theta \longrightarrow S_2$ such that $\widehat{f} \circ \pi_{\ker f} = f$. Moreover, it is straightforward to show:

(1) $\widehat{f}$ is injective $\Leftrightarrow$ $\ker f = \theta$;

(2) $\widehat{f}$ is a TS-isomorphism $\Leftrightarrow$ $\ker f = \theta$ and f is surjective. ∎

1.5. Real Semigroups ([Dickmann and Petrovich, 2004], [Dickmann and Petrovich, 2011]). a) A **real semigroup (RS)** is:

- A TS, $\mathscr{G}$, together with a ternary relation, $\mathscr{D}_\mathscr{G} = \mathscr{D}$, **representation by binary forms**, satisfying, for all $a, b, c, d, e \in \mathscr{G}$, (where $\mathscr{D}^t$ is the transversal of $\mathscr{D}$):

[RS 0] : $c \in \mathscr{D}(a,b) \Leftrightarrow c \in \mathscr{D}(b,a)$; $\qquad$ [RS 1] : $a \in \mathscr{D}(a,b)$;

[RS 2] : $a \in \mathscr{D}(b,c) \Rightarrow ad \in \mathscr{D}(bd, cd)$;

[RS 3] (Strong associativity) :
$a \in \mathscr{D}^t(b,c)$ and $c \in \mathscr{D}^t(d,e)$ $\Rightarrow$ $\exists\, x \in \mathscr{D}^t(b,d)$ so that $a \in \mathscr{D}^t(x,e)$.
[RS 4] : $e \in \mathscr{D}(c^2a, c^2b)$ $\Rightarrow$ $e \in \mathscr{D}(a,b)$;
[RS 5] : $ad = bd$, $ae = be$ and $c \in \mathscr{D}(d,e)$ $\Rightarrow$ $ac = bc$;
[RS 6] : $c \in \mathscr{D}(a,b)$ $\Rightarrow$ $c \in \mathscr{D}^t(c^2a, c^2b)$;
[RS 7] (Reduction) : $\mathscr{D}^t(a,-b) \cap \mathscr{D}^t(-a,b) \neq \emptyset$ $\Rightarrow$ $a = b$;
[RS 8] : $a \in \mathscr{D}(b,c)$ $\Rightarrow$ $a^2 \in \mathscr{D}(b^2, c^2)$.

b) $L_{RS} = \langle \cdot, \mathscr{D}, 1, 0, -1 \rangle$ is the language of RSs.

c) If $\mathscr{G}$ is a RS, write $\mathscr{G}^{\times} = \{x \in \mathscr{G} : x^2 = 1\}$ for the group of **units in** $\mathscr{G}$.

d) If $\mathscr{G}_1$, $\mathscr{G}_2$ are RSs, a map, $f : \mathscr{G}_1 \longrightarrow \mathscr{G}_2$, is a **RS-morphism** if it preserves 1, 0, -1, product and representation. ∎

REMARK 8. If $\mathscr{G}$ is a RS, the representation and transversal representation are interdefinable as follows. for a, b, $c \in \mathscr{G}$:

- By [t-rep] in 1.3.(b),

$a \in \mathscr{D}^t(b,c)$ $\Leftrightarrow$ $a \in \mathscr{D}(b,c) \wedge -b \in \mathscr{D}(b,-a) \wedge -c \in \mathscr{D}(b,-a)$;

- The axioms for RSs in 1.5.(a) entail $a \in \mathscr{D}(b,c)$ $\Leftrightarrow$ $a \in \mathscr{D}^t(a^2 b, a^2 c)$.

Hence, if $f : \mathscr{G}_1 \longrightarrow \mathscr{G}_2$ is a TS-morphism and $\mathscr{G}_i$, $i = 1, 2$, are RSs, the following are equivalent:

(1) f is a RS-morphism (i.e., it preserves representation);

(2) f preserves transversal representation, i.e, for all a, b, $c \in \mathscr{G}_1$,

$a \in \mathscr{D}^t_{\mathscr{G}_1}(b,c)$ $\Rightarrow$ $f(a) \in \mathscr{D}^t_{\mathscr{G}_2}(f(b), f(c))$. ∎

1.6. RS-Congruences. For an extensive discussion of the theme, the reader is referred to Chapter II of [Dickmann and Petrovich, 2011]. We shall here mildly change the presentation, in order to emphasize the importance of the unique factorization property contained in Definition II.2.1 of [Dickmann and Petrovich, 2011]. To keep matters straight, if $\mathscr{G}$ is a RS, we write $|\mathscr{G}|$ *for the ternary semigroup underlying* $\mathscr{G}$.

DEFINITION 9. A RS-morphism, $f : \mathscr{G} \longrightarrow \mathscr{G}'$ has the **RS-unique factorization property (RS-UFP)** if for all RS-morphisms, $g : \mathscr{G} \longrightarrow \mathscr{H}$ such that $\ker f \subseteq \ker g$, there is a *unique* RS-morphism, $h : \mathscr{G}' \longrightarrow \mathscr{H}$ making the following diagram commute:

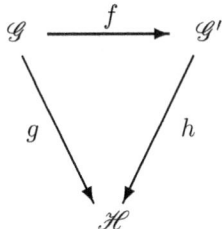

Notation as in 1.4, we have

DEFINITION 10. (Essentially Def. II.2.1, [Dickmann and Petrovich, 2011]) An equivalence relation θ on a RS $\mathscr{G}$ is a **RS-congruence** if

[RS-cong 1] : θ is a congruence of ternary semigroups (1.4);

[RS-cong 2] : There is a ternary relation $\mathscr{D}_\theta$ in the quotient TS, $|\mathscr{G}|/\theta$, such that $\mathscr{G}/\theta := \langle |\mathscr{G}|/\theta, \cdot, \mathscr{D}_\theta, -1, 0, 1 \rangle$ is a RS and the canonical projection, $\pi_\theta : \mathscr{G} \longrightarrow \mathscr{G}/\theta$, is a RS-morphism;

[RS-cong 3] : The map $\pi_\theta : \mathscr{G} \longrightarrow \mathscr{G}/\theta$ has the RS-UFP.

Write $Con_{RS}(\mathscr{G})$ for the set of RS-congruences in $\mathscr{G}$.

LEMMA 11. *a) Let $f_i : \mathscr{G} \longrightarrow \mathscr{G}_i$, $i = 1, 2$, be RS-morphisms with the RS-UFP. If $\ker f_1 = \ker f_2$, there is a unique RS-isomorphism, $h : \mathscr{G}_1 \longrightarrow \mathscr{G}_2$, making the triangle below left commutative.*

b) Let $f : \mathscr{G} \longrightarrow \mathscr{G}'$ be a surjective RS-morphism. If f has the RS-UFP, then $\ker f$ is a RS-congruence in $\mathscr{G}$.

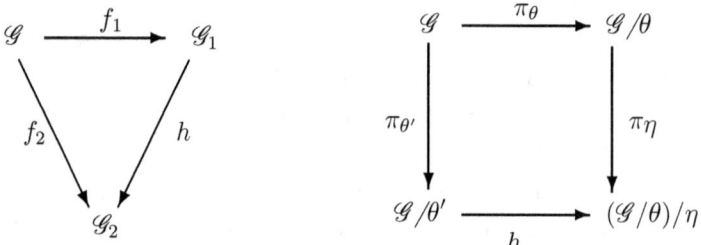

c) (Double quotient) Let $\mathscr{G}$ be a RS and let θ be a RS-congruence in $\mathscr{G}$. If η is a RS-congruence on $\mathscr{G}/\theta$, then, with notation as in 10, $\theta' := (\pi_\theta \times \pi_\theta)^{-1}[\eta]$ is a RS-congruence on $\mathscr{G}$. Moreover, there is a unique RS-isomorphism, $h : \mathscr{G}/\theta' \longrightarrow (\mathscr{G}/\theta)/\eta$ making the square above right commutative.

PROOF. Item (a) is clear. For (b), by 1.4.(b), $\theta := \ker f$ is a TS-congruence on $\mathscr{G}$. Since f is surjective, the [TS-UFP] in 1.4.(a), together with (2) in 1.4.(b), yield a unique *TS-isomorphism*, $\widehat{f} : |\mathscr{G}|/\theta \longrightarrow |\mathscr{G}'|$, making the following diagram commute:

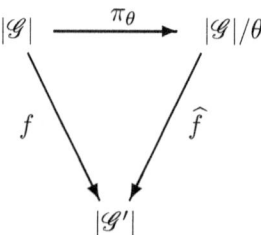

Since $\mathscr{G}'$ is a RS, $\widehat{f}$ may be made into a *RS-isomorphism*, by which $|\mathscr{G}|/\theta$ becomes a RS, $\mathscr{G}/\theta$, and $\pi_\theta : \mathscr{G} \longrightarrow \mathscr{G}/\theta$ becomes a RS-morphism. Since, $\theta = \ker f = \ker \pi_\theta$ and f has the RS-UFP, the same will be true of $\pi_\theta : \mathscr{G} \longrightarrow$

$\mathscr{G}/\theta$, and θ is a RS-congruence, as claimed.

c) Let $g = \pi_\eta \circ \pi_\theta$; g is clearly surjective and $\ker g = \theta' = (\pi_\theta \times \pi_\theta)^{-1}[\eta]$. Note that

(I)   (i) $\theta \subseteq \theta'$;   (ii) $(\pi_\theta \times \pi_\theta)[\theta'] = \eta = \ker \pi_\eta$.

Indeed, (i) follows from $\Delta_{\mathscr{G}/\theta} \subseteq \eta$ (inverse image is increasing), while (ii) from the surjectivity of π_θ. We claim that g has the RS-UFP. To see this, let $f : \mathscr{G} \longrightarrow \mathscr{H}$ be a RS-morphism, such that $\theta' \subseteq \ker f$. Because θ is a RS-congruence and $\theta \subseteq \ker f$ (by (I.(i)) above), there is a unique RS-morphism, $f_\theta : \mathscr{G}/\theta \longrightarrow \mathscr{H}$, making the upper triangle $(*)$ in the diagram below commutative.

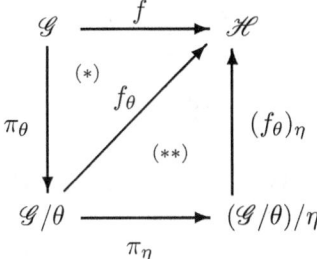

Since $f = \pi_\theta \circ f_\theta$, we have $\theta' \subseteq \ker f = (\pi_\theta \times \pi_\theta)^{-1}[\ker f_\theta]$, and the surjectivity of π_θ together with (I.(ii)) above, entail $\eta \subseteq \ker f_\theta$. Now, the fact that η is a RS-congruence yields a unique RS-morphism, $(f_\theta)_\eta : (\mathscr{G}/\theta)/\eta \longrightarrow \mathscr{H}$, making the lower triangle $(**)$ in the above square commutative, establishing the RS-UFP for $g = \pi_\eta \circ \pi_\theta$. Now it follows immediately from (b) that θ' is a RS-congruence on $\mathscr{G}$. Moreover, since both $\pi_{\theta'} : \mathscr{G} \longrightarrow \mathscr{G}/\theta'$ and $g : \mathscr{G} \longrightarrow (\mathscr{G}/\theta)/\eta$ have the the same kernel and the RS-UFP, item (a) yields the unique RS-isomorphism making the displayed square in the statement commute, ending the proof. ∎

In what follows, we shall see applications of the above results to RSs arising from p-rings.

1.7. The Real Semigroup of a p-Ring. Let $\langle A, T \rangle$ be a p-ring. For details on the constructions about to be presented, the reader is referred to [Dickmann and Petrovich, 2004], [Dickmann and Petrovich, 2011] and [Marshall, 1996]

a) Let $\mathrm{Sper}(A)$ be the real spectrum of A (cf. Chapter 7 and Chapter 4 in [Bochnak et al., 1998]) and set

$$\mathrm{Sper}(A,T) = \{\alpha \in \mathrm{Sper}(A) : T \subseteq \alpha\},$$

called the **real spectrum** of $\langle A, T \rangle$.

Each $a \in A$ gives rise to map, $\bar{a}_T : \mathrm{Sper}(A, T) \longrightarrow 3 = \{-1, 0, 1\}$, given by

$$\bar{a}_T(\alpha) = \begin{cases} 1 & \text{if } a \in \alpha \setminus -\alpha; \\ 0 & \text{if } a \in \mathrm{supp}(\alpha) = \alpha \cap -\alpha; \\ -1 & \text{if } a \in -\alpha \setminus \alpha. \end{cases}$$

If T is clear from context, we write $\bar{a}$ for $\bar{a}_T$.

b) Write $\mathscr{G}_{A,T} = \{\bar{a} : a \in A\}$. With the product induced by A, $\mathscr{G}_{A,T}$ is a ternary semigroup with identity 1 (the constant function 1) and distinguished elements 0 and -1 (the corresponding constant valued maps).

Define a **representation relation** on $\mathscr{G}_{A,T}$, as follows: for $a, b, c \in A$,

($\mathscr{D}$) $\quad \overline{a} \in \mathscr{D}_{\mathscr{G}_{A,T}}(\overline{b}, \overline{c}) \Leftrightarrow \exists\, t, t_1, t_2 \in T$, s.t. $\overline{at} = \overline{a}$ and $ta = t_1 b + t_2 c$.

The corresponding **transversal representation relation** is given by

($\mathscr{D}^t$) $\qquad \overline{a} \in \mathscr{D}^t_{\mathscr{G}_{A,T}}(\overline{b}, \overline{c}) \Leftrightarrow \begin{cases} \exists\, a', b', c' \in A \text{ so that } \overline{a} = \overline{a'}, \\ \overline{b} = \overline{b'},\ \overline{c} = \overline{c'} \text{ and } a' = b' + c'. \end{cases}$

With these representation relations, $\mathscr{G}_{A,T}$ **is a real semigroup** in the sense of [Dickmann and Petrovich, 2004] and [Dickmann and Petrovich, 2011]. As above (1.5.(c)), $\mathscr{G}^\times_{A,T} = \{\overline{a_T} \in \mathscr{G}_{A,T} : \overline{a_T}^2 = 1\}$ is the group of units in $\mathscr{G}_{A,T}$. ∎

In the present setting, and with notation as in 1.7.(a), the following result is important:

THEOREM 12. (Thm. 5.4.2, Cor. 5.4.3 (p. 93ff) in [Marshall, 1996]) *Let $\langle A, T \rangle$ be a p-ring. For $a, b \in A$:*

a) $\overline{a_T} = 0$ *iff there is $k \geq 0$ such that $-a^{2k} \in T$.*
b) $\overline{a_T} = 1$ *iff there are $s, t \in T$ such that $(1+s)a = 1+t$.*
c) $\overline{a_T} \geq 0$ *iff there are $s, t \in T$ and $k \geq 0$ so that $(a^{2k} + s)a = a^{2k} + t$.*
d) $\overline{a_T} = \overline{b_T}$ *iff there are $s, t \in T$ and $k \geq 0$ so that $sab = (a^2 + b^2)^k + t$.* ∎

1.8. The functor $\mathscr{G}$ from p-Rings to RS. Notation as above, let **p-Rings** and **RS** be the categories of p-rings and RSs, respectively. If $f : \langle A, T \rangle \longrightarrow \langle A', T' \rangle$ is a p-ring morphism, define

$$\mathscr{G}(f) : \mathscr{G}_{A,T} \longrightarrow \mathscr{G}_{A',T'}, \text{ given by } \overline{a_T} \longmapsto \overline{f(a)_{T'}}.$$

To see that $\mathscr{G}(f)$ is well-defined, let $a, c \in A$ verify $\overline{a_T} = \overline{c_T}$; by Theorem 12.(d), there are $s, t \in T$ and an integer $k \geq 0$ such that

(I) $\qquad\qquad\qquad (a^2 + c^2)^k + t = sac.$

Applying f to both sides of (I) and recalling the inclusion $f[T] \subseteq T'$, obtains

$$(f(a)^2 + f(c)^2)^k + f(t) = f(s)f(a)f(c),$$

with $f(t), f(s) \in T'$; whence, another application of 12.(d) yields $\overline{f(a)_{T'}} = \overline{f(c)_{T'}}$, as needed. It is straightforward that $\mathscr{G}(f)$ is a semigroup morphism, preserving 1, 0 and -1. For $\mathscr{G}(f)$ to be a RS-morphism, it suffices to prove

(II) $\qquad \overline{a_T} \in \mathscr{D}_{\mathscr{G}_{A,T}}(\overline{b_T}, \overline{c_T}) \Rightarrow \overline{f(a)_{T'}} \in \mathscr{D}_{\mathscr{G}_{A',T'}}(\overline{f(b)_{T'}}, \overline{f(c)_{T'}}),$

with $a, b, c \in A$. By ($\mathscr{D}$) in 1.7.(b), the hypothesis in (II) is equivalent to the existence of $t, t_1, t_2 \in T$ such that

(III) $\qquad\qquad ta = t_1 b + t_2 c$ and $\overline{ta} = \overline{t} \cdot \overline{a} = \overline{a}.$

Applying f to the first equation in (III) yields $f(t)f(a) = f(t_1)f(b) + f(t_2)f(c)$, with $f(t), f(t_1), f(t_2) \in T'$; moreover, since $\mathscr{G}(f)$ is a semigroup morphism, the second equation in (III) entails $\overline{f(a)_{T'}} = \mathscr{G}(f)(\overline{(at)_T}) = \mathscr{G}(f)(\overline{a_T}) = \overline{f(a)_{T'}}$, and the conclusion in (II) is immediately forthcoming from ($\mathscr{D}$) in 1.7.(b).

Clearly, the maps
$$\begin{cases} \langle A, T \rangle & \longmapsto & \mathscr{G}_{A,T} \\ \langle A, T \rangle \xrightarrow{f} \langle A', T' \rangle & \longmapsto & \mathscr{G}_{A,T} \xrightarrow{\mathscr{G}(f)} \mathscr{G}_{A',T'}, \end{cases}$$
yield we have a covariant functor from **p-Rings** to **RS**. ∎

EXAMPLE 13. Let $\langle A, T \rangle$ be a p-ring. The identity map, $Id_T : \langle A, \Sigma A^2 \rangle \longrightarrow \langle A, T \rangle$ is a p-ring morphism; let $\rho_T = \mathscr{G}(Id_T) : \mathscr{G}_A \longrightarrow \mathscr{G}_{A,T}$ be the induced RS-morphism, as in 1.8. We have
$$\rho_T(\overline{a}) = \overline{a_T} = \overline{a} \restriction \mathrm{Sper}(A, T).$$
Hence, $\ker \rho_T = \{\langle \overline{a}, \overline{b} \rangle \in \mathscr{G}_A \times \mathscr{G}_A : \overline{a_T} = \overline{b_T}\}$ and ρ_T is clearly surjective. Note that the description of $\ker \rho_T$ in $\langle A, T \rangle$ is given by 12.(d).

We claim that ρ_T has the RS-UFP. Indeed, let $\mathscr{G}$ be a RS and let $f : \mathscr{G}_A \longrightarrow \mathscr{G}$ be a RS-morphism, such that $\ker \rho_T \subseteq \ker f$. Since ρ_T is onto, the uniqueness of the factor RS-morphism — if it exists at all —, is clear. For $a \in A$, define
$$\widehat{f} : \mathscr{G}_{A,T} \longrightarrow \mathscr{G} \text{ by } \widehat{f}(\overline{a_T}) = f(a).$$
Since $\ker \rho_T \subseteq \ker f$, $\widehat{f}$ is well defined; moreover, it is straightforward that $\widehat{f}$ is a TS-morphism, verifying $\widehat{f} \circ \rho_T = f$. It remains to check that $\widehat{f}$ is a RS-morphism. taking into account the definition of $\widehat{f}$, this amount to showing for $a, b, c \in A$,

(I) $\qquad \overline{a_T} \in \mathscr{D}^t_{\mathscr{G}_{A,T}}(\overline{b_t}, \overline{c_T}) \quad \Rightarrow \quad f(a) \in \mathscr{D}^t_{\mathscr{G}}(f(b), f(c)).$

and we conclude by the equivalence between (1) and (2) in 8. By ($\mathscr{D}^t$) in 1.7, the antecedent in (I) is equivalent to the existence of $a', b', c' \in A$ so that

(II) (i) $\overline{a'_T} = \overline{a_T}$, $\overline{b'_T} = \overline{a_T}$, $\overline{c'_T} = \overline{c_T}$ and (ii) $a' = b' + c'$.

From (II.(i)) and the hypothesis that $\ker \rho_T \subseteq \ker f$ we obtain $f(a) = f(a')$, $f(b) = f(b')$ and $f(c) = f(c')$, while (II.(ii)) and ($\mathscr{D}^t$) in 1.7.(b) entail $\overline{a'} \in \mathscr{D}^t_{\mathscr{G}_A}(\overline{b'}, \overline{c'})$. Since f is a RS-morphism, the latter relation implies $f(a') \in \mathscr{D}^t_{\mathscr{G}}(f(b'), f(c'))$, which in turn yields $f(a) \in \mathscr{D}^t_{\mathscr{G}}(f(b), f(c))$, establishing (I), as desired.

By items (a) and (b) of Lemma 11, $\ker \rho_T$ **is a RS-congruence** on $\mathscr{G}_A$ and the diagram $\rho_T : \mathscr{G}_A \longrightarrow \mathscr{G}_{A,T}$ is naturally isomorphic to the projection of $\mathscr{G}_A$ onto $\mathscr{G}_A / \ker \rho_T$. ∎

1.9. The functor $\mathscr{U}$ from RS to pRSG. Recall that **pRSG** is the category of reduced *pre-special* groups (pRSG) and SG-morphisms (cf. Definition 1.2, p. 2 and Definition 1.11, p. 10 in [Dickmann and Miraglia, 2000]).

If $\mathscr{G}$ is a RS, then (cf. I.2.10, p. 23 of [Dickmann and Petrovich, 2011]), $\mathscr{G}^\times = \{x \in \mathscr{G} : x^2 = 1\}$, the *group of units* of $\mathscr{G}$, with the induced representation relation [3], is a p-RSG. The canonical embedding, $u_{\mathscr{G}} : \mathscr{G}^\times \longrightarrow \mathscr{G}$, is a semigroup morphism, preserving 1 and −1. In fact, the passage from $\mathscr{G}$ to $\mathscr{G}^\times$ constitutes

---
[3] In $\mathscr{G}^\times$, $\mathscr{D}$ and $\mathscr{D}^t$ coincide.

a *functor*, as follows. If $f : \mathscr{G}_1 \longrightarrow \mathscr{G}_2$ is a RS-morphism, clearly we have $f[\mathscr{G}_1^\times] \subseteq \mathscr{G}_2^\times$. Hence, $f\restriction \mathscr{G}_1^\times := f^\times$ is a map from $\mathscr{G}_1^\times$ into $\mathscr{G}_2^\times$. Moreover,

- $f^\times$ takes 1 to 1 and -1 to -1;
- $\forall\ a,b,c \in \mathscr{G}_1^\times,\ a \in \mathscr{D}_{\mathscr{G}_1}(b,c) \Rightarrow f(a) \in \mathscr{D}_{\mathscr{G}_2}(f(b),f(c))$,

i.e, $f^\times$ is a pRSG-morphism and the following diagram commutes:

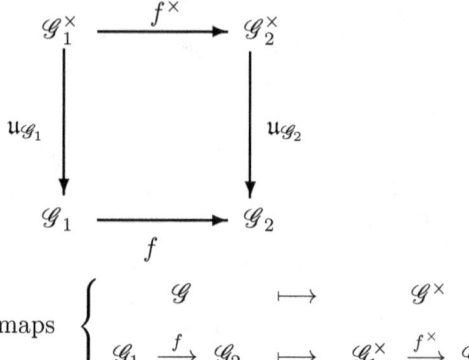

It is easily seen that the maps $\begin{cases} \mathscr{G} \longmapsto \mathscr{G}^\times \\ \mathscr{G}_1 \xrightarrow{f} \mathscr{G}_2 \longmapsto \mathscr{G}_1^\times \xrightarrow{f^\times} \mathscr{G}_2^\times, \end{cases}$
constitute a covariant functor, $\mathscr{U} : \mathbf{RS} \longrightarrow \mathbf{pRSG}$. ∎

2 The Real Semigroup of a Ring of Fractions

We here describe the basic relations between the real semigroups associated to a p-ring and to its ring of fractions by a multiplicative set. Firstly, we register the following (well-known) result:

LEMMA 14. *Let $\langle A, T \rangle$ be a p-ring and let S be a multiplicative subset of A and let $R := AS^{-1}$ be the ring of fractions of A by S and let $\iota_A : A \longrightarrow R$ be canonical ring morphism.*

a) The following are equivalent:

(1) $P = \left\{ \dfrac{t}{s^2} \in R : t \in T \text{ and } s \in S \right\}$ *is a proper preorder of R;*

(2) $S \cap (T \cap -T) = \emptyset$.

If these equivalent conditions are met, then S is a proper multiplicative subset of A and ι_A is a morphism of unitary p-rings.

b) If $S \cap \sqrt[T]{0} = \emptyset$, then the set P in (1) of item (a) is a proper preorder and ι_A is a p-ring morphism.

PROOF. Since $(T \cap -T) \subseteq \sqrt[T]{0} = \bigcap \{\mathrm{supp}(\alpha) : \alpha \in \mathrm{Sper}(A, T)\}$, (b) is follows immediately from (a). For (a), it is clear that ι_A is a p-ring morphism (whether $\langle R, P \rangle$ is proper or not). To prove (1) $\Rightarrow$ (2), assume that (2) fails; hence, there is $s \in S$ so that $s, -s \in T$. Thus, $-s^2 \in S$ and $-1 = \dfrac{-s^2}{s^2} \in P$, contradicting (1). To show that (2) $\Rightarrow$ (1), if $-1 = \dfrac{t}{s^2} \in P$, there is $w \in S$ such that $-ws^2 = w\,t$; multiplying through by w obtains $-w^2 s^2 = w^2 t$ and so $w^2 s^2$ is in $(T \cap -T) \cap S$, contradicting (2). ∎

THEOREM 15. *Let $\langle A, T \rangle$ be a p-ring and let S be a multiplicative subset of A, such that $S \cap (T \cap -T) = \emptyset$. Let $R := AS^{-1}$ be the ring of fractions of A by S and let $\iota_A : A \longrightarrow R$ be the canonical ring morphism.*

a) Let $h := \mathscr{G}(\iota_A) : \mathscr{G}_{A,T} \longrightarrow \mathscr{G}_{R,P}$ be the induced RS-morphism. Then, for all $a \in A$ and $s \in S$, $\overline{\left(\dfrac{a}{s}\right)} = \overline{\left(\dfrac{as}{1}\right)}$. In particular, h is surjective.

b) For $a, b \in A$, the following are equivalent:

(1) $\overline{\left(\dfrac{a}{1}\right)} = \overline{\left(\dfrac{b}{1}\right)}$ *in* $\mathscr{G}_{R,P}$;

(2) *There is $s \in S$ so that $\overline{as} = \overline{bs}$ in $\mathscr{G}_{A,T}$;*

(3) *There is $s \in S$ such that $\overline{as^2} = \overline{bs^2}$ in $\mathscr{G}_{A,T}$.*

c) h is injective $\Leftrightarrow \{\overline{s} \in \mathscr{G}_{A,T} : s \in S\} \subseteq \mathscr{G}_{A,T}^{\times}$.

d) For $a, b \in A$, the following are equivalent:

(1) $\overline{\left(\dfrac{a}{1}\right)} \in \mathscr{D}_{\mathscr{G}_{R,P}}\left(\overline{\left(\dfrac{b}{1}\right)}, \overline{\left(\dfrac{c}{1}\right)}\right)$;

(2) *There are $s_1, s_2, s_3 \in S$ such that, with $a' = s_1^2 a$, $b' = s_2^2 c$ and $c' = s_3^2 c$, we have $\overline{a'} \in \mathscr{D}_{\mathscr{G}_{A,T}}(\overline{b'}, \overline{c'})$.*

e) Suppose $\mathscr{G}$ is a RS, $f : \mathscr{G}_{A,T} \longrightarrow \mathscr{G}$ is a RS-morphism and f verifies the following condition

[ker] *For all $a, b \in A$, if there is $s \in S$ such that $\overline{as} = \overline{bs}$, then $f(\overline{a}) = f(\overline{b})$.*

Then, there is a unique RS-morphism, $\widehat{f} : \mathscr{G}_{R,P} \longrightarrow \mathscr{G}$, making the following diagram commutative:

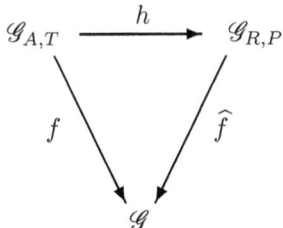

f) $\ker h$ is a RS-congruence on $\mathscr{G}_{A,T}$ and the diagram $h : \mathscr{G}_{A,T} \longrightarrow \mathscr{G}_{R,P}$ is canonically RS-isomorphic to the quotient $\pi_{\ker h} : \mathscr{G}_{A,T} \longrightarrow \mathscr{G}_{A,T}/\ker h$.

g) With notation as in 13, the kernel of the composition $\mathscr{G}_A \xrightarrow{\rho_T} \mathscr{G}_{A,T} \xrightarrow{h} \mathscr{G}_{R,P}$ is a RS-congruence, θ, on $\mathscr{G}_A$, canonically RS-isomorphic to the quotient $\pi_\theta : \mathscr{G}_A \longrightarrow \mathscr{G}_A/\theta$.

PROOF. By Lemma 14.(a), $\langle R, P \rangle$ is a proper p-ring and ι_A is a morphism of unitary p-rings.

a) For $a \in A$ and $s \in S$, we have

$$\frac{a^2 s^2}{1} + \frac{a^2}{s^2} = \frac{a^2}{1}\left(\frac{s^2}{1} + \frac{1}{s^2}\right) = \frac{a^2}{1}\frac{1+s^4}{s^2} = \frac{as}{1}\frac{a}{s}\frac{1+s^4}{s^2},$$

with $\dfrac{1+s^4}{s^2} \in P$ and Theorem 12.(d) yields the desired conclusion.

b) Since for all $\xi \in \mathscr{G}_{A,T}$, $\xi^3 = \xi$, (2) and (3) are clearly equivalent. For (2) $\Rightarrow$ (1), suppose $\overline{as} = \overline{bs}$ in $\mathscr{G}_{A,T}$, with a, $b \in A$ and $s \in S$. Then,

$$\overline{\left(\dfrac{a}{1}\right)} = \overline{\left(\dfrac{as}{s}\right)} = \overline{\left(\dfrac{bs}{s}\right)} = \overline{\left(\dfrac{b}{1}\right)},$$

as needed. It remains to establish (1) $\Rightarrow$ (2). If $\overline{\left(\dfrac{a}{1}\right)} = \overline{\left(\dfrac{b}{1}\right)}$ in $\mathscr{G}_{R,P}$, by Theorem 12.(d) there are t, $t_1 \in T$, u, $v \in S$ and an integer $k \geq 0$ such that, in $R = AS^{-1}$,

(II) $\qquad \left(\dfrac{a^2}{1} + \dfrac{b^2}{1}\right)^k + \dfrac{t}{u^2} = \dfrac{t_1}{v^2}\dfrac{a}{1}\dfrac{b}{1}.$

Since $\dfrac{a^2}{1} + \dfrac{b^2}{1} \in P$, we may assume that $k \geq 2$ (and in fact, to be any prescribed positive integer greater than to the original k). The definition of ring of fractions yields $w \in S$ so that, after clearing denominators, we obtain [4]

(III) $\qquad u^2 v^2 w^2 (a^2 + b^2)^k + tv^2 w^2 = t_1 u^2 w^2 ab.$

Multiplying (III) by $(uvw)^{2k-2}$, obtains, with $t' = tv^2 w^2 (uvw)^{2k-2} \in T$,

$[(auvw)^2 + (buvw)^2]^k + t' = t_1 u^2 w^2 (uvw)^{2k-2} ab = t_1 \, (uw)^{2k-2} \, v^{2k-4} \, (auvw)(buvw)$

$\qquad\qquad\qquad\qquad = t''\,(auvw)(buvw),$

with $t'' \in T$; setting $s := uvw \in S$, the immediately preceding equality and Theorem 12.(d) entail $\overline{as} = \overline{bs}$ in $\mathscr{G}_{A,T}$, as needed.

c) Suppose h is injective and $s \in S$. Then, (a) yields

$$\overline{1} = \overline{\left(\dfrac{1}{s}\right)}\overline{\left(\dfrac{s}{1}\right)} = \overline{\left(\dfrac{s}{1}\right)}\overline{\left(\dfrac{s}{1}\right)} = \overline{\left(\dfrac{s^2}{1}\right)},$$

and the injectivity of h entails $\overline{1} = \overline{s^2}$ in $\mathscr{G}_{A,T}$, i.e, $\overline{s} \in \mathscr{G}_{A,T}^{\times}$. The converse is an immediate consequence of the equivalence in (b).

d) (1) $\Rightarrow$ (2). By ($\mathscr{D}$) in 1.7.(b), there are t, t_1, $t_2 \in T$ and x, y, $z \in S$ such that, in $R = AS^{-1}$,

(IV) $\begin{cases}(i) \quad \dfrac{t}{x^2}\dfrac{a}{1} = \dfrac{t_1}{y^2}\dfrac{b}{1} + \dfrac{t_2}{z^2}\dfrac{c}{1}, \text{ and} \\[2mm] (ii) \quad \overline{\left(\dfrac{t}{x^2}\right)}\overline{\left(\dfrac{a}{1}\right)} = \overline{\left(\dfrac{ta}{x^2}\right)} = \overline{\left(\dfrac{a}{1}\right)}.\end{cases}$

By (a) above, $\overline{\left(\dfrac{ta}{x^2}\right)} = \overline{\left(\dfrac{tax^2}{1}\right)}$ and so IV.(ii) entails $\overline{\left(\dfrac{tax^2}{1}\right)} = \overline{\left(\dfrac{a}{1}\right)}$, whence, by (c), there is $s \in S$ such that

(V) $\qquad \overline{tax^2 s^2} = \overline{as^2}.$

---

[4] Recall: $a/s = a'/s'$ in R iff there is $w \in S$ so that $was' = wsa'$; multiplying by w, yields $w^2 as' = w^2 sa'$.

By IV.(i), there is $w \in S$, so that, after clearing denominators, we get
$$w^2 y^2 z^2 \, ta \;=\; t_1 \, (w^2 x^2 z^2 b) \;+\; t_2 \, (w^2 x^2 y^2 c).$$
Multiplying this equality by $x^2 s^2$ yields
(VI) $\qquad (tx^2)(w^2 y^2 z^2 s^2 a) \;=\; t_1 \, (w^2 x^4 z^2 s^2 b) \;+\; t_2 \, (w^2 x^4 y^2 s^2 c).$

Set $s_1 = wyzs$, $s_2 = wx^2 zs$ and $s_3 = wx^2 ys$; clearly $s_i \in S$, $i = 1, 2, 3$. Moreover, if $a' = s_1^2 a$, $b' = s_2^2 b$ and $c' = s_3^2 c$, (VI) takes the form
(VII) $\qquad\qquad\qquad tx^2 \, a' \;=\; t_1 b' + t_2 c'.$

Now note that (V) yields, multiplying by $\overline{w^2 y^2 z^2}$,
$$\overline{tx^2 a'} \;=\; \overline{tx^2 a s^2 (w^2 y^2 z^2)} \;=\; \overline{\quad} \;=\; \overline{s^2 a (w^2 y^2 z^2)} \;=\; \overline{a'},$$
which, together with (VII), entails $\overline{a'} \in \mathscr{D}_{\mathscr{G}_{A,T}}(\overline{b'}, \overline{c'})$, as desired.

(2) $\Rightarrow$ (1). Since h is a RS-morphism and $h(\overline{a}) = h(\overline{a'})$, $h(\overline{b}) = h(\overline{b'})$ and $h(\overline{c}) = h(\overline{c'})$ (by (b)), (2) entails, because $h(\overline{c}) = \overline{\left(\dfrac{c}{1}\right)}$ ($c \in A$), $h(\overline{a}) \in \mathscr{D}_{\mathscr{G}_{R,P}}(h(\overline{b}), h(\overline{c}))$, as needed.

e) The uniqueness of a map (if it exists) making the diagram commutative is clear. Define $\widehat{f} : \mathscr{G}_{R,P} \longrightarrow \mathscr{G}$ by
$$\widehat{f}\left(\overline{\left(\dfrac{a}{x}\right)}\right) \;=\; f(\overline{ax}).$$

To see $\widehat{f}$ is well-defined, assume $\overline{\left(\dfrac{a}{x}\right)} = \overline{\left(\dfrac{b}{y}\right)}$; by (a) we have $\overline{\left(\dfrac{ax}{1}\right)} = \overline{\left(\dfrac{by}{1}\right)}$ and so (b) yields $s \in S$ such that $\overline{axs} = \overline{bys}$. Since f verifies [ker] in the statement, we obtain $f(\overline{ax}) = f(\overline{by})$, as needed. It is straightforward that $f = \widehat{f} \circ h$ and $\widehat{f}$ preserves product, as well as the constants 1, 0 and -1. It remains to check that $\widehat{f}$ preserves representation. Since h is surjective, by the definition of $\widehat{f}$ it suffices to prove, for $a, b, c \in A$:

(VIII) $\qquad \overline{\left(\dfrac{a}{1}\right)} \in \mathscr{D}_{\mathscr{G}_{R,P}}\left(\overline{\left(\dfrac{b}{1}\right)}, \overline{\left(\dfrac{c}{1}\right)}\right) \;\Rightarrow\; f(\overline{a}) \in \mathscr{D}_{\mathscr{G}}(f(\overline{b}), f(\overline{c})).$

By item (d), the antecedent in (VIII) implies the existence of $s_i \in S$, $i = 1, 2, 3$, such that
(IX) $\qquad\qquad\qquad \overline{as_1^2} \in \mathscr{D}_{\mathscr{G}_{A,T}}(\overline{bs_2^2}, \overline{cs_3^2}).$

Now, note that for $u \in A$ and $x \in S$, $\overline{ux^2} = \overline{u}\,\overline{x^2}$, and so condition [ker] implies $f(\overline{ux^2}) = f(\overline{u})$; this observation and (IX) entail, because f is a RS-morphism, the conclusion in (VIII), as needed.

f) By item (b),
$$\ker h \;=\; \{\langle \overline{a}, \overline{b}\rangle \in \mathscr{G}_{A,T} \times \mathscr{G}_{A,T} : \exists \, s \in S \text{ so that } \overline{as} = \overline{bs}\}.$$
Hence, condition [ker] in (e) is equivalent to $\ker h \subseteq \ker f$ and the conclusion in (e) shows that h is a surjective RS-morphism with the RS-UFP. The conclusion in (f) then an immediate consequence of Lemma 11.(b), while item (g) follows from (f) and Lemma 11.(c), ending the proof. ∎

3 A Representation Theorem. Applications

Our first step in representing the RS of any p-ring by that of a BIR, is to show that the RS of a p-ring is (naturally) isomorphic to the RS of a *reduced* p-ring.

PROPOSITION 16. *Let $\langle A, T \rangle$ be a p-ring and let $I = \sqrt[T]{0}$ be the T-radical of the zero ideal. Set $R = A/I$. Then,*

a) $\langle R, T/I \rangle$ *is a reduced p-ring and the canonical projection, $\pi_I : \langle A, T \rangle \longrightarrow \langle R, T/I \rangle$ is a p-ring morphism.*

b) The RS-morphism $\mathscr{G}(\pi_I) : \mathscr{G}_{A,T} \longrightarrow \mathscr{G}_{R,T/I}$ is an isomorphism of real semigroups.

PROOF. a) Clearly, T/I is closed under sums, products and contains the squares in R; if $\mathfrak{p}$ is a proper T-convex prime ideal in A, then for all $t \in T$, $1 + t \notin \mathfrak{p}$ (otherwise, $1 \in \mathfrak{p}$). By Proposition 3, for all $t \in T$, $1 + t \notin \sqrt[T]{0}$; whence $-1/I \notin T/I$, and T/I is a proper preorder of R. It is clear that π_I is a p-ring morphism. The verification that R is T/I-reduced is the same as that of item (3) in the proof of part (b.3) of Proposition 6.5 (p. 73ff) in [Dickmann and Miraglia, 2011].

b) To ease notation, we shall write $\bar{a}$ for the elements of $\mathscr{G}_{A,T}$ and $\mathscr{G}_{R,T/I}$, omitting the subscripts T and T/I, respectively. By 1.8, $h := \mathscr{G}(\pi_I)$ is a RS-morphism. To show it to be an isomorphism we must verify:

(1) h is bijective;

(2)[5] For all $a, b, c \in A$, $h(\bar{a}) \in \mathscr{D}^t_{\mathscr{G}_{R,T/I}}(h(\bar{b}), h(\bar{c})) \;\Rightarrow\; \bar{a} \in \mathscr{D}^t_{\mathscr{G}_{A,T}}(\bar{b}, \bar{c})$.

Note that, because h is a RS-morphism, the converse of (2) also holds.

Proof of (1). It is clear that h is surjective. To show h is injective, let $x, y \in A$ and suppose
$$h(\bar{x}) = \overline{x/I} = \overline{y/I} = h(\bar{y}).$$
By Theorem 12.(d), there are $t, s \in T$ and an integer $k \geq 0$ such that
$$((x/I)^2 + (y/I)^2)^k + t/I = (s/I)(x/I)(y/I),$$
that is, setting $u := (x^2 + y^2)^k + t$, we have $u - sxy \in I = \sqrt[T]{0}$. Note that $u \in T$. By Proposition 3, there are $m \geq 0$ and $t' \in T$ such that

(I) $\qquad\qquad\qquad (u - sxy)^{2m} + t' = 0.$

But we have $(u - sxy)^{2m} = E - F$, where:

$E = \sum_{k\,even} \binom{n}{k} u^{2m-k}(sxy)^k \quad$ and $\quad F = \sum_{k\,odd} \binom{n}{k} u^{2m-k}(sxy)^k.$

Then:

• $E = \sum_{k\,even} \binom{n}{k} u^{2m-k}(sxy)^k = u^{2m} + \sum_{k\,even \geq 2} \binom{n}{k} u^{2m-k}(sxy)^k;$

Since $u \in T$ and k is even, $\sum_{k\,even \geq 2} \binom{n}{k} u^{2m-k}(sxy)^k$ is in T and we may write

---

[5] Because representation and transversal representation are interdefinable in a RS, $\mathscr{G}$, with $\mathscr{D}^t_\mathscr{G} \subseteq \mathscr{D}_\mathscr{G}$; see (†), [RS4] and [RS6] in Def. 2.1, p. 106 of [Dickmann and Petrovich, 2004] or [t-rep], [RS4] and [RS6] in Def. I.2.1, p. 19 of [Dickmann and Petrovich, 2011]

(II) $$E = u^{2m} + t^*, \text{ with } t^* \in T;$$

• $F = \sum_{k \text{ odd}} \binom{n}{k} u^{2m-k}(sxy)^k = sxy \sum_{k \text{ odd}} u^{2m-k}(sxy)^{k-1}$; just as above, $\sum_{k \text{ odd}} u^{2m-k}(sxy)^{k-1}$ is in T and we obtain

(III) $$F = s^*xy, \text{ with } s^* \in T.$$

Substituting (II) and (III) into (I) yields

(IV) $$u^{2m} + t^* + t' = s^*xy.$$

Now observe that $u^{2m} = [(x^2 + y^2)^k + t]^{2m} = (x^2 + y^2)^{2mk} + s'$, with $s' \in T$. Hence, (IV) entails
$$(x^2 + y^2)^{2mk} + s' + t^* + t' = s^*xy,$$
whence, by Theorem 12.(d), $\overline{x} = \overline{y}$, establishing the injectivity of $h = \mathscr{G}(\pi_I)$.

Proof of (2). By $(\mathscr{D}^t)$ in 1.7.(b), there are $a', b', c' \in A$, such that

(V) $$\begin{cases} (i) \ \overline{a/I} = \overline{a'/I}, \ \overline{b/I} = \overline{b'/I}, \ \overline{c/I} = \overline{c'/I} \\ \text{and} \\ (ii) \ a'/I = b'/I + c'/I. \end{cases}$$

By (V).(ii), there is $r \in I = \sqrt[T]{0}$ such that

(VI) $$a' = b' + c' + r = b' + (c' + r).$$

We now register the following

FACT 17. Let S be a semi-real ring, let $u, v \in S$ and let $\beta \in \text{Sper}(S)$. If $v - u \in \text{supp}(\beta)$, then $\overline{u}(\beta) = \overline{v}(\beta)$.

Proof. Clearly, $u \in \text{supp}(\beta)$ iff $v \in \text{supp}(\beta)$. Next, if $u \in \beta \setminus (-\beta)$, then $v = u + (v - u)$ and so $v \in \beta$. If $-v \in \beta$, then $-u = -v + (v - u)$ implies $-u \in \beta$, which is impossible. Hence, $v \in \beta \setminus (-\beta)$. Because $\text{supp}(\beta)$ is an ideal, the argument is symmetric in u and v and so $u \in \beta \setminus (-\beta)$ iff $v \in \beta \setminus (-\beta)$. Since $(-u - (-v)) = (v - u) \in \text{supp}(\beta)$, the reasoning above applies to yield $-u \in \beta \setminus (-\beta)$ iff $-v \in \beta \setminus (-\beta)$. Hence, $\overline{u}(\beta) = \overline{v}(\beta)$, as desired. □

Since r is in the intersection of all T-convex ideals in A, Fact 17 yields, with $c^* := c' + r$,

(VII) $$\overline{c' + r} = \overline{c^*} = \overline{c'}.$$

Now the injectivity of h, the equalities in (V).(i), (VI) and (VII) entail
$$\overline{a} = \overline{a'}, \ \overline{b} = \overline{b'}, \ \overline{c^*} = \overline{c'} \text{ and } a' = b' + c^*,$$
which, by $(\mathscr{D}^t)$ in 1.7.(b), guarantee $a \in \mathscr{D}^t_{\mathscr{G}_{A,T}}(b, c)$, establishing (2) and ending the proof. ∎

The next step in our construction is the following

THEOREM 18. Let $\langle A, T \rangle$ be a T-reduced p-ring. Let
$$\mathcal{U} = A \setminus \bigcup_{\alpha \in \text{Spec}_R(A,T)} \text{supp}(\alpha)$$
be the complement of the union of all T-convex prime ideals in A. Let S be a multiplicative set contained in $\mathcal{U}$, let $A_S = AS^{-1}$ be the ring of fractions of A

by S and set $T_S = \left\{ \dfrac{t}{s^2} \in A_S : t \in T \text{ and } s \in S \right\}$. Then,

a) $\mathcal{U}$ is a proper saturated [6] multiplicative set in A, whose elements are all non-zero divisors.

b) $\langle A_S, T_S \rangle$ is a proper p-ring and the canonical morphism, $\iota_S : A \longrightarrow A_S$, is a p-ring embedding.

c) A_S is T_S-reduced.

d) The map $\mathscr{G}(\iota_S) : \mathscr{G}_{A,T} \longrightarrow \mathscr{G}_{A_S, T_S}$ is a RS-isomorphism.

e) Consider the following conditions:

   (1) For all $s \in S$ and $t \in T$, $s^2 + t \in S$;   (2) $\langle A_S, T_S \rangle$ has bounded inversion.

Then, (1) $\Rightarrow$ (2); if S is saturated, these conditions are equivalent.

REMARK 19. a) If $\langle A, T \rangle$ is a p-ring, the set $\mathcal{U}$ in the statement of Theorem 18 consists of the elements $a \in A$ satisfying $\bar{a}^2 = 1$, i.e., $\mathcal{U} = \{a \in A : \bar{a} \in \mathscr{G}_{A,T}^\times\}$, where $\mathscr{G}_{A,T}^\times$ is the group of units of the RS $\mathscr{G}_{A,T}$.

b) If $\langle A, T \rangle$ is a BIR, then $\mathcal{U} = A^\times$, the group of units in A. Indeed, by Proposition 6.3 (p. 71) of [Dickmann and Miraglia, 2011], every maximal ideal in A is T-convex and so the set of elements outside every T-convex prime ideal in A is $A^\times$. Hence, with notation as in 18, if $\langle A, T \rangle$ is a BIR, $\langle A, T \rangle$ and $\langle A_S, T_S \rangle$ are naturally isomorphic. ∎

Proof of Theorem 18. a) It is well-known that the complement of a union of prime ideals is a proper saturated multiplicative set in A. For $x \in \mathcal{S}$, suppose $xu = 0$, for some $u \in A$. Since x is outside all T-convex primes in A, we get $u \in \bigcap_{\alpha \in Spec_R(A,T)} \operatorname{supp}(\alpha) = \sqrt[T]{0}$ and so $u = 0$ because A is T-reduced. It now clear that no element of $\mathcal{S}$ is a zero-divisor.

b) Since $\mathcal{U} \cap \sqrt[T]{0} = \emptyset$, we also have $S \cap \sqrt[T]{0} = \emptyset$ and so, by 14.(b), $\langle A_S, T_S \rangle$ is a proper p-ring and ι_S a p-ring morphism. Moreover, since no element of S is a zero-divisor, it is well-known that ι_S is an embedding.

c) For $a \in A$ and $x \in S$, suppose $\dfrac{a}{x}$ is in the T_S-radical of 0 in A_S. By Proposition 3, there are $t \in T$, $y \in S$ and an integer $m \geq 0$ such that $\dfrac{a^{2m}}{x^{2m}} + \dfrac{t}{y^2} = 0$. Hence, in A we obtain

(I) $\qquad\qquad y^2 a^{2m} + x^{2m} t = 0.$

Multiplying the equation in (I) by y^{2m-2} yields

$$(ay)^{2m} + y^{2m-2} x^{2m} t = 0,$$

and another application of 3 entails $ay \in \sqrt[T]{0}$ in A. Since $y \in S$ is outside all T-convex primes in A, we get $a \in \sqrt[T]{0}$; whence, the T-reducibility of A implies $a = 0$ and, in turn, the T_S-reducibility of A_S.

---

[6] $xy \in \mathcal{U} \Rightarrow x, y \in \mathcal{U}$.

d) Write $\mathscr{G}$ for $\mathscr{G}_{A,T}$ and $\mathscr{G}_S$ for $\mathscr{G}_{A_S,T_S}$. By 1.8, $h := \mathscr{G}(\iota_S)$ is a RS-morphism. To show it is an isomorphism, it suffices to prove:

(♮.1) h is surjective; (♮.2) h is injective;

(♮♮) For all $a, b, c \in A$,

$$h(a) = \overline{\left(\frac{a}{1}\right)} \in \mathscr{D}_{\mathscr{G}_S}\left(\overline{\left(\frac{b}{1}\right)}, \overline{\left(\frac{c}{1}\right)}\right) \;\Rightarrow\; \overline{a} \in \mathscr{D}_{\mathscr{G}}(\overline{b}, \overline{c}).$$

Since h is a RS-morphism, the converse of (♮♮) is also true. Property (♮.1) follows from item (a) in Theorem 15, while (♮.2) is a consequence of 15.(c) and 19.(a). By item (d) in Theorem 15, the hypothesis in (♮♮) yields $s_i \in S$, $i = 1, 2, 3$, such that $\overline{as_1^2} \in \mathscr{D}_{\mathscr{G}}(\overline{s_2^2 b}, \overline{s_3^2 c})$, which is equivalent to $\overline{a} \in \mathscr{D}_{\mathscr{G}}(\overline{b}, \overline{c})$, because $\overline{s_i} \in \mathscr{G}^\times$, $i = 1, 2, 3$, as needed.

e) For $s \in S$ and $t \in T$, $1 + \dfrac{t}{s^2} = \dfrac{s^2 + t}{s^2} \in A_S^\times$ iff there is $u \in S$ such that $u(s^2 + t) \in S$ (recall that ι_S is an embedding). It is now clear that (1) $\Rightarrow$ (2) (with $u = 1$), while, if S is saturated, the converse is immediately forthcoming. ∎

3.1. Notation. Let $\langle A, T \rangle$ be p-ring and let $\mathcal{U}$ be the complement of the T-convex primes in A, as in 18. Write $\mathscr{G}^\natural$ for $\mathscr{G}_{A^\natural, T^\natural}$, where

- $A^\natural = A\mathcal{U}^{-1}$ for the ring of fractions of A by $\mathcal{U}$;
- $T^\natural = \left\{ \dfrac{t}{s^2} \in A\mathcal{U}^{-1} : t \in T \text{ and } s \in \mathcal{U} \right\}$;
- $\iota^\natural : A \longrightarrow A^\natural$ for the canonical p-ring embedding. ∎

We now have

COROLLARY 20. *Notation as above, let $\langle A, T \rangle$ be a p-ring. Let*
- $\nu : \langle A, T \rangle \longrightarrow \langle A^*, T^* \rangle$ *be its BIR hull (cf. Proposition 6);*
- $\iota^\natural : \langle A, T \rangle \longrightarrow \langle A^\natural, T^\natural \rangle$ *be as in 3.1.*

Write $\mathscr{G}$ for $\mathscr{G}_{A,T}$ and $\mathscr{G}^*$ for $\mathscr{G}_{A^*,T^*}$. Then,

a) $\langle A^\natural, T^\natural \rangle$ *is a reduced BIR and $\iota : \langle A^*, T^* \rangle \longrightarrow \langle A^\natural, T^\natural \rangle$ is the unique p-ring embedding making the diagram (D) commutative.*

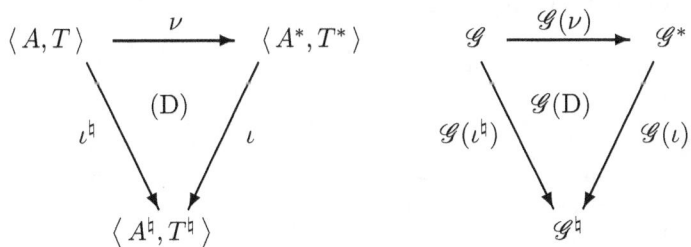

b) *Diagram $\mathscr{G}(D)$ is commutative and its arrows are RS-isomorphisms.*

PROOF. a) By items (a), (c) in 18, $\langle A^\natural, T^\natural \rangle$ is a $T^\natural$-reduced proper p-ring and $\iota^\natural$ is an injective p-ring morphism. Note that $s \in \mathcal{U}$ and $t \in T$ implies $s^2 + t$

$\in \mathcal{U}$; otherwise, $s^2 + t \in \mathfrak{p}$, for some T convex prime ideal in A, and so $t, s^2 \in \mathfrak{p}$, which entails $s \in \mathfrak{p}$, an impossibility. Hence, by 18.(e), $\langle A^\natural, T^\natural \rangle$ is a BIR. By the universal property of the BIR hull in 6.(3), there is a unique p-ring morphism, $\langle A^*, T^* \rangle \xrightarrow{\iota} \langle A^\natural, T^\natural \rangle$, making diagram (D) commute. Since, in fact, $1 + T \subseteq \mathcal{U}$ [7], ι is an embedding.

b) Diagram $\mathscr{G}$(D) arises by applying the functor $\mathscr{G}$ to diagram (D), whence it is commutative. As noted above, the multiplicative set $1 + T \subseteq \mathcal{U}$ and so, by 18.(d), $\mathscr{G}(\nu)$ and $\mathscr{G}(\iota^\natural)$ are both RS-isomorphisms; consequently, the same must be true of $\mathscr{G}(\iota)$, ending the proof. ∎

REMARK 21. a) If $\langle A, T \rangle$ is a proper p-ring, all constructions employed to go from $\langle A, T \rangle$ to $\langle A^\natural, T^\natural \rangle$ – quotients and rings of fractions –, are functorial. In fact, it is well-known that these constructions commute with each other.

b) The isomorphism $\mathscr{G}(\pi_I)$ of 16 yields a conclusion analogous to that of 20 for *all* proper p-rings. More precisely, given $\langle A, T \rangle$, let $I = \sqrt[T]{0}$ and let $\langle R, P \rangle$ be the BIR hull of $\langle A/I, T/I \rangle$; then, $\mathscr{G}_{A,T}$ is RS-isomorphic to $\mathscr{G}_{R,P}$. ∎

BIBLIOGRAPHY

[Bochnak et al., 1998] J. Bochnak, M. Coste, and M-F. Roy. *Real Algebraic Geometry*, volume 36 of *Ergeb. Math.* Springer-Verlag, Berlin, 1998.

[Dickmann and Miraglia, 2000] M. Dickmann and F. Miraglia. *Special Groups : Boolean-Theoretic Methods in the Theory of Quadratic Forms*, volume 689 of *Memoirs Amer. Math. Soc.* AMS, Providence, R.I., 2000.

[Dickmann and Miraglia, 2011] M. Dickmann and F. Miraglia. Faithfully quadratic rings. 161pp. To appear, 2011.

[Dickmann and Petrovich, 2004] M. Dickmann and A. Petrovich. Real Semigroups and Abstract Real Spectra, I. *Contemporary Math.*, 344:99–119, 2004. AMS.

[Dickmann and Petrovich, 2011] M. Dickmann and A. Petrovich. Real semigroups and abstract real spectra. 260pp. To appear, 2011.

[Marshall, 1996] M. Marshall. *Spaces of Orderings and Abstract Real Spectra*, volume 1636 of *Lecture Notes in Mathematics*. Springer-Verlag, Berlin, 1996.

Max Dickmann
Équipe de Logique Mathématique
Université de Paris VII
Projet Topologie et Géométrie Algébriques
Institut de Mathématiques de Jussieu
Paris, France
Email: dickmann@logique.jussieu.fr

Francisco Miraglia
Departamento de Matemática
Instituto de Matemática e Estatística
Universidade de São Paulo
São Paulo, Brazil
Email: miraglia@ime.usp.br

---

[7] $\overline{(1+t)} = \overline{1}$, for all $t \in T$.

Once More on the Unexpected Examination

CHRIS MORTENSEN AND JENNIE LOUISE

ABSTRACT. We describe the so-called paradox of the unexpected examination. We aim to show that it is not especially paradoxical, and indeed that the matter is rather straightforward. We argue that the most important ingredient is to understand the one day case: the professor asserts that there will be an unexpected exam today. Along the way, we show that several variants of the paradox are essentially the same, including the version that utilises the Godel incompleteness theorem

1 Introduction

The literature on the so-called "paradox' of the unexpected examination is extensive. We do not propose to survey it, but we believe that it remains in an unsettled state even today. One of our aims will be to show that the so-called paradox, when properly understood, is not really paradoxical at all, and indeed rather straightforward, turning on a relatively benign feature of language. We show in passing that various versions of the paradox are essentially the same. There are several useful surveys of the literature, such as [Chow, 1998]. Chow distinguishes between two broad approaches to formulating the problem. One approach, for the more mathematically-minded impressed by the parallels with the Gödel incompleteness theorems, takes "unexpected" to mean something like "unprovable". As Chow notes, this involves dealing with self-reference (and its mathematical analog, diagonalization). As will be seen in our discussion, we think that this introduces unnecessary complications which do not get to the core of it. For examples of this approach, early and late, see [Nerlich, 1961], also [Kritchman and Raz, 2010]. Another of our aims will be to issue a corrective to Kritchman and Raz. We will argue that the problem is not best understood with the Gödel incompleteness theorems, but by means of epistemic logic, which does not need the expressive power adequate for Gödel. This leads us to the other approach, which takes "unexpected" to mean something like "unknown ahead of time", see *e.g.* [Sorensen, 2006]. We adopt this latter approach, applying epistemic logic to the concept of "know". This means in effect that we are addressing what [Jackson, 1987] called the "hard paradox". We will argue that the so-called hard paradox is no more than a gratuitous contradiction, and as such does not threaten a true contradiction and therefore is not paradoxical. We also indicate how what we have to say bears on variants of the paradox employing belief and reasonable belief rather than knowledge. Finally we show how the analysis bears on Gödel's result.

2 The Unexpected Examination

The Unexpected Examination can be described as follows. The professor tells the students that there will be an examination one day next week, and it will be unexpected in the sense that on the day it is on, the students will not know this. The students then reason as follows. The unexpected exam cannot be on Friday. This is because if it is on Friday, then it isn't on Monday to Thursday. Hence on Friday morning we would know this fact, and so be able to deduce (know) that it must be on Friday, which contradicts the professor's original assertion that it will be unknown on that day.

Then, they reason, since it cannot be on Friday, it cannot be on Thursday either. On Wednesday evening we will know that it hasn't been on Monday to Wednesday, and we have just deduced (and so know) it cannot be on Friday. Hence, reasoning as before, it would be known to be Thursday, contradicting the information that it will be unknown on the day. By similar reasoning, each day is ruled out. There cannot be an unexpected examination. However, the professor's assertion does not seem obviously problematic: unexpected examinations can and do occur. So it seems there must be an error in the students' reasoning. But what?

3 Knowledge: The One-day Case.

What does it take to "solve" a "paradox"? In this discussion, we will take a paradox to be *an argument where there are plausible reasons to believe both of a pair of incompatible propositions*. Two strategies for solving paradoxes then suggest themselves. One strategy is to show that, after all, there are *not* plausible reasons to believe both of an incompatible pair; that is, one or both of the plausible reasons breaks down under scrutiny. This is the kind of solution that we will be defending. Indeed, we are prepared to stick our necks out and say that it is rather simple. We are not alone in thinking that there is no real paradox here, just about everyone writing on the topic has thought so too. Even so, saying just why it is not paradoxical is not so simple, as we see.

The other strategy for dealing with paradox, a more radical strategy, is to show that the plausible reasons for incompatible conclusions are very good indeed, good enough to provide reasons to believe, even a proof, of a true contradiction. We do not think that the unexpected examination is like this, but we do think that there might be defensible examples in paradoxes such as The Liar, Grelling's, Russell's and The Sorites.

Now, an important element in understanding of the puzzle is found in the one-day case. The professor simply tells the students that there will be an unexpected examination *today*. Something has obviously gone wrong; this seems like a trivial error.

A central point to understand is that *some propositions can be true but cannot be known to be true*. This was shown by an argument due to [Fitch, 1963]. Let E be the proposition that there is an examination on today. Then the proposition "E is true but unknown", that is $E \,\&\, \sim K(E)$, cannot consistently be known, even if $E \,\&\, \sim K(E)$ by itself is consistent. That is, the proposition $K(E \,\&\, \sim KE)$ is contradictory. This is straightforwardly proved in propositional epistemic logic, as follows:

Beginning with the supposition to be proved contradictory:

(1) $K(E \mathbin{\&} \sim K(E))$ Then:

(2) $E \mathbin{\&} \sim K(E)$

This follows from (1) by the principle $K(X) \to X$, which is definitional of knowledge. Note that (2) expresses the content of the professor's assertion. But now, from (1), distributing K over $\&$, that is $K(X \& Y) \longleftrightarrow (K(X) \& K(Y))$:

(3) $K(E) \mathbin{\&} K \sim K(E)$

But the first conjunct of (3) contradicts the second conjunct of (2). Hence, discharging the assumption (1):

(4) $\sim K(E \mathbin{\&} \sim K(E))$

The contradiction arises from supposing that E, the first conjunct of (2), is known. Notice that the professor's assertion $E \mathbin{\&} \sim K(E)$, that is line (2), is not by itself contradictory: the contradiction follows when it is supposed to be known. The point we want to make, however, that the contradiction between the second conjunct of (2) and the first conjunct of (3) is a *mere* contradiction: there is no paradox here. Indeed, we can say something even stronger: *there is surely no paradox if it fails to be known that E*. If E is not known (for example because the professor does not speak), then of course E will be unexpected. So, *if the problem is to be at all interesting, then it must be known that there is an examination on*.

Of course, there are thus many examples where the first conjunct of (3), $K(E)$, holds (so that the second conjunct of (2), $\sim K(E)$, fails); indeed any known E will do. But also, we can *make E* known, or *make known E*. One way to ensure this, for example, is to provide a good, sound proof of E (assuming E the kind of thing amenable of proof, such as a mathematical theorem). But there are many other ways to make it known beyond a reasonable doubt. Even an announcement of an exam carries a public commitment, and could be backed by the legal system or an independent and reputable custodian, etc. The point here is: pick your standards of knowledge, and if any of the foregoing fails as a guarantee of knowledge, then there is no paradox, because there is no knowledge. Thus, if E is made known, so that $K(E)$ is true, then $\sim K(E)$ is false. There is no alternative. This is no big loss however; the professor's assertion that E is unknown appears as an unsubstantiated stipulation, and without further support makes no contribution to forcing a paradox on us.

On the other hand, in different circumstances, but faced with a similar choice between $K(E)$ and $\sim K(E)$, the audience might have a good reason to think that they do not know E. The question then arises: how could someone fail to know what is openly said to them? This might seem faintly mysterious, but there are plenty of examples of circumstances in which an assertion might reasonably be doubted, including honest mistakes, misplaced confidence and dishonest lies.

The deductive resources used in the argument above are minimal. But we should also notice that Fitch's argument provides a template for other versions of the problem. For example, instead of $K = $ knowledge, take $K = $ belief (or reasonable belief), so that unexpected = not believed. Truth is not definitional of belief as it is of knowledge, so the move from (1) to (2) fails; but (2) can simply be included as an extra premiss. The argument then proceeds as above, appealing only to the distribution of belief over conjunctions. (This might be questioned, but we surely would not want to solve the paradox by disallowing the students the ability to believe the consequences of their beliefs.) The moral of this version of the paradox is thus that *some propositions can be believed, and can be true, but cannot be truly believed.* We also discuss in the final section a version of the problem where K = provability, but it is apparent that the analysis given so far applies to that also.

There is no paradox in the one-day case; examples abound. Such as:

(a) The Arras. There can be an unknown dagger hidden behind the arras; but were the presence of the dagger to be made known, that would *make false the other part of what was said*, namely that it is unknown.

(b) The Surprise Party. I blurt out to you that your friends are having a surprise birthday party for you today on your birthday. Thanks, that ruined the surprise.

In sum, the lesson to be learned from the one-day case is that there is no paradox, only an act of assertion on the part of the professor which is infelicitous in that some part of it cannot be known by the target audience. Making one part known makes the other part false. Sorensen (see eg. 2006) calls these "blindspots", and we will say more on that later. Moreover, we can tell often enough which part of the professor's assertion is unknown, for example the claim to ignorance. If on the other hand ignorance is assured for whatever reason, then the claim to knowledge of what is asserted is defeated.

4 Multiple Days.

We now proceed to the case of multiple days. We see that there are additional complications, but that no further paradox emerges. We concentrate on the two-day case, where the main points can be seen. Later we briefly mention the three-day case.

We let F be the proposition "An exam is on Friday" and let T be the proposition "An exam is on Thursday". We also need to distinguish two occasions of knowing, K_t (known on Thursday morning, or now) and K_f (known on Friday morning, or after Thursday). They are related by the principle that whatever is known on Thursday morning, is known on Friday morning (and not forgotten). This is a reasonable assumption in the present context, and one that the students' reasoning requires.

There is an obvious issue of how to formulate the professor's problematic assertion. There are a number of options. We adopt one here, and consider an alternative later. The important thing is that there be *a modelling of the students argument that makes it comprehensible why they would argue in the way they do,* and so affords a natural reconstruction thereof. It seems to have been a widespread assumption in the literature that the students argue fallaciously (see also Section 6), but we prefer to emphasise what is right in their argumentation, as much as the errors.

The simplest formulation would seem to be:

(5) $(T \leftrightarrow \sim F) \mathbin{\&} \sim K_t(T) \mathbin{\&} \sim K_f(F)$

This reads that there is an examination on exactly one day, and it is not known on Thursday morning that it is on Thursday, and it is not known on Friday morning that it is on Friday. We now proceed to a formal reconstruction, which will aid us in later analysis.

As in the one-day case, the assertion (5) can be true, as long as it is not made known. Strictly, *unless it is made known from the start that there is an exam on one of the days*, then the exam will certainly be unexpected, even if it occurs on Friday. Hence, for there to be any chance of a paradox, we must have:

(6) $K_t(T \leftrightarrow \sim F)$

As in the one-day case, there are many imaginable cases in which (6) can be made true (Ayer 1973). For example, the professor publicly throws dice with evens determining Thursday, odds determining Friday, and seals the outcome until it is acted on. But it can also be false, in which case a Friday surprise is possible. [Quine, 1953] argued that since the truth of the assertion that it is on one of the days is incompatible with its being known, it must be concluded that it isn't known, whence surprise is inevitable. But this isn't justified, if we accept that it can be made known that an exam in on. Still, Quine was right to this extent, that knowledge of the assertion that an exam is on, is an essential ingredient, without which no paradox follows.

Proceeding with the students' argument:

(7) F Premiss for conditional proof
(8) $\sim T$ From (5) and (7)
(9) $K_f(\sim T)$ From (8), if T doesn't happen, this will be obvious on Friday
(10) $K_f(T \leftrightarrow \sim F)$ From (6), making the assumption that knowledge is preserved over time (to be considered later)
(11) $K_f(F)$ From (9) and (10), assuming minimal powers of deduction
(12) $F \to K_f(F)$ From (7)-(11), conditional proof
(13) $\sim F$ From (12) and the last conjunct of (5). Thus on these premisses, Friday is ruled out.
(14) $K_t(F \to K_f(F))$ From (12), assuming the ability to argue
(15) $K_t(\sim K_f(F) \to \sim F)$ From (14), substituting logical equivalences inside K context.
(16) $K_t \sim K_f(F) \to K_t(\sim F)$ From (15), distributing K over implication
(17) $K_t \sim K_f(F)$ Premiss (to be considered later)
(18) $K_t(\sim F)$ From (16) and (17) by modus ponens
(19) $K_t(T)$ From (6) and (15)

But (19) contradicts the middle conjunct of (5).

Now there are a number of steps in this argument that bear examination.

First, line (10). A common theme in the literature is that it is the, or one of the, main fallacious moves made by the students. Frank Jackson argues that one can know at the beginning of the week that an exam will be on one day (line (6)), while acknowledging at the beginning that if it were to get to Friday without an exam, then it would no longer be reasonable to believe that an exam will be on. That is, (10) doesn't follow from (6), and moreover the students are not entitled to (10) when reasoning at the beginning of the week.

We acknowledge this point in general terms. Nonetheless, there are epistemic concepts for which (10) does seem to be reasonable (Jackson suggests $K = $ certainty). We certainly think that it is possible make things known at the beginning sufficient to ensure knowledge on the last day that it will be on that day. Indeed, that is what Ayer was arguing. For example, let a coin be tossed to determine the day and let the result of the toss be concealed until the day. For the present, our point is that without (10) there is no chance of a paradox developing, as Friday cannot then be ruled out. Hence, (10) is essential for there to be any chance of a paradox.

A second point to take note of, is that *if it is not on Thursday, then this will be known by Friday morning.* This is obvious, so obvious that it is fair to say that it is known throughout the week. But it needs (10) to be able to work it into a usable form such as from (14), so we note the dependency on (10) at this point,

Third, there is line (17). The students will readily realise at any time that a Thursday exam can hardly be known (or believed) on Friday to be on Friday: $K_t(T \to \sim K_f(F))$. Distributing the K gives $K_t(T) \to K_t \sim K_f(F)$, and then contraposing gives that *if it is unknown on Thursday that it is unknown in Friday, then a Thursday exam will surely be unknown.* In sum, for there to be any chance of a paradox, it must be *known on Thursday that it is unknown on Friday.* This is (17).

There are ways for (17) to be false, such as when the asserter is known to be unreliable. Then a Thursday surprise is possible. That (17) is an essential but dubious ingredient was pointed out by [Wright and Sudbury, 1977], among others. Jackson makes a similar point. He argues that even though it might be reasonable for the students to believe at the beginning of the week in their ignorance on Thursday, they should acknowledge then that this might not be so reasonable later in the week. But the falsity of (17) allows a Thursday exam to be unknown, so assuming (17) is fallacious. However, as in the one-day case, there are ways to make (17) true, by making known on Thursday that $\sim K_f(F)$. For example, after tossing a coin, the professor (or an independent arbiter) inspects it and informs the students that they won't know on Friday that the exam is on that day.

With these various premises, then, as in the one day case, the argument goes through and we have a contradiction. But, as in the one day case, it is a gratuitous contradiction, and so not paradoxical. It is similar to the diagnosis of The Barber (non-)paradox: the Barber is a contradictory story, but so what? We have no reason to believe in the contradiction of the Barber's existence, indeed every reason to believe in its falsehood. In passing, this does not diminish Russell's Paradox, which is of the same form as The Barber, but quite different in import. This is because there is independent reason to believe in the existence of the Russell Set, in the form of the simplicity of unrestricted Comprehension. The latter is the mark of a real paradox.

Thus, given the ancillary provisos, the incompatibility between (5) and (6) is just a falsehood. If it is made known that it is on one of the two days, then one of the other premises must fail. If instead $K = $ belief, the same argument can be made, as long as both (5) and (6) are asserted as independent premises, as in the one-day case.

With these complications, the message is essentially the same as the one-day case. The professor's announcement can be true but can't be known in any sense which postulates the retention of knowledge across the week. An unexpected exam can occur on either day, but the attempt to make known the key parts of the professor's assertion, eventually produces a contradiction, and so forces us to abandon the attempt.

5 Alternatives

An alternative formulation of the professor's assertion is open to a similar analysis. We take the professor's assertion to be that **on the day that it is on**, it is unexpected. Noting as before that to be interesting it must be made known that it is on one of the days, we have:

(20) $K_t(T \leftrightarrow \sim F)$ & $((T\ \&\sim K_t(T)) \vee (F\ \&\sim K_f(F)))$.

Clearly, the first conjunct of (20) is (6) and can be made to hold, Of course, there can be situations where it is not known, despite being asserted: perhaps the professor is known to be unreliable or a trickster. Then, if the exam is on Thursday, the students do not know this, and if the exam is on Friday, the students do not know this on either Thursday or Friday. The professor's assertion (5) is true, but because it is unknown that there is an exam on just one day, it is unexpected. It is apparent that this parallels our discussion of the one-day case.

The second conjunct of (20) holds if its first disjunct holds, and that can hold if (17) fails. The second disjunct can hold if (10) fails. So, again, we can demonstrate to be consistent, and hence not paradoxical, this version of what the professor asserts or makes known.

We can readily extend the argument to the case of three days or more. With three days, the assertion that there is an exam on one of those days is $((W \leftrightarrow \sim T) \leftrightarrow \sim F)$. But obviously, the same point applies as in the two-day case: if it is not known ahead of time that it is unknown on Friday, then it can be on, but unexpected, on Thursday. Interestingly, that then allows that an unexpected exam can be on Wednesday or Thursday.

6 Loose Ends

We note a useful distinction made by Jackson, between the "easy" and "hard" paradoxes. The easy paradox involves vindicating the strong intuition that (5) and (6) can be true, and in addition that the latter two conjuncts of (5) can be reasonably believed at the beginning of the week. The hard paradox involves showing that (5) and (6) can be true, and in addition that the latter two conjuncts of (5) can be known, in a sense of knowledge which does not erode across the week. Jackson claims only to solve the easy paradox. We do not dispute his diagnosis, nor that it is only the easy paradox that he has diagnosed. Indeed, it is clear that our own discussion also supplies the conclusion that the easy paradox is no paradox. But, if the professor's announcement is made in terms of unexpectedness stipulated as persisting across the week, then it should be addressed. But that amounts to the hard paradox.

We also think that the premises of the hard paradox, particularly (10) and (17), can severally be made true. But not collectively made true. The hard paradox is really just the observation that the premises of the argument above are jointly contradictory. One thing or another has to give, they can't all be true. A vigorous assertion by the professor doesn't by itself make them all true, let alone plausible, and thus doesn't make for a paradox.

One minor complication arises if the professor attempts to re-instate the contradiction as a paradox by resorting to non-Boolean logic in their assertion. Thus, the professor might assert that *even if* there is an exam on Friday, it will *still* be unknown on that day. Now, in the students' argument above, logic is taken truth-functionally, so that this conditional comes out true just because its antecedent is false. But there are non-truth-functional conditionals that are not made true just by the falsity of the antecedent, and this is one of them. Semantics for such conditionals exist, including the Lewis-Stalnaker semantics and the Routley-Meyer relevant semantics. We can sidestep the details by treating the conditional informally. What one should say in this case is clear enough: the professor has spoken falsely, and indeed the strong conditional is the culprit. If the professor has made known that there is an exam on just one day, then it *cannot* be that if it really gets to Friday then it will still be unknown on that day. If it gets to Friday, then it is expected.

To return to questions of methodology, we noted above that many authors have seen their job to be identifying the fallacy in the students' argument, correlatively with vindicating appropriate intuitions, such as that the exam can be on and unexpected even though announced. We do not reject this, but our view has been broader, in that we have taken seriously the possibility of a genuine paradox. We have, however, found no reason to think so. Even so, there are interesting lessons to be learned from the prospect of a contradiction, which the one-day case is particularly useful to illustrate.

To take another example: as we noted earlier, [Sorensen, 2006] uses "blindspots" for propositions which can be true but cannot be known or truly believed. As we have seen, blindspots are common enough, and no semantic big deal. Blindspots are blindspots *for* some target audience, such as the students. Hence, one can make an announcement which the audience can deduce is one of their blindspots. However, we think that the name is a little misleading, in that it suggests some sort of irremediable deficiency. All that really happens as a consequence of the public announcement is that $E \, \& \sim K(E)$ remains not known for the very good reason it becomes false since its second conjunct becomes false when its first conjunct becomes known. Even so, the world changes from unknown E to known E, which is a gain in knowledge.

7 The Gödel Incompleteness Theorems

We can now apply it all to Gödel. As noted by [Kritchman and Raz, 2010], there is clearly a *structural similarity* with the Gödel incompleteness theorems. But the paradox does not *reduce* to the Gödel result. If anything, things are the other way around.

The construction of the Gödel sentence G, informally "This sentence is unprovable", requires *self-reference*. There is nothing especially suspect about self-reference, it is routine in natural languages. But to construct G in arithmetic requires the representation of self-reference in *diagonalization*, and diagonalization requires syntax and proof theory to be arithmetized.

Once the Diagonal Lemma for the incompleteness theorems is proved, we have that $\vdash G \leftrightarrow \sim Prov(\ulcorner G \urcorner)$. Thus G is equivalent to $G \& \sim Prov(\ulcorner G \urcorner)$. Clearly, therefore, we can substitute K = provability into our analysis above. By a similar argument to the one-day case, $G \& \sim Prov(\ulcorner G \urcorner)$ cannot be proved, unless arithmetic is inconsistent. The argument is exactly the argument of the one-day case above. It therefore can be carried out in epistemic logic. Thus, our analysis is simpler in that it does not depend on self-reference: propositional epistemic logic falls well short of arithmetic in expressive power.

To be completely analogous, the argument would need the assumption that what is provable is also true. The principle of the truth of knowledge is definitional of knowledge, but the truth of what is provable (in some formalized arithmetical theory) is definitely not definitional. Notoriously, the truth of provability is assumed by people trying to prove informally from the Gödel theorems that we are not Turing machines: assuming that arithmetic is consistent, we have the truth of G. But the second Gödel theorem shows us that we cannot prove that consistency if we are mere Turing machines.

However, there is no real problem. We can add the truth of G as an extra assumption, as we did for belief, and our points above apply. This yields the familiar result that G cannot be both true and provable: if it is true then it is unprovable.

8 Conclusion

We take it, then, that we have shown that one important lesson to be learned in all these cases is encapsulated in the one-day case. One reaction we sometimes hear voiced is that after deducing the contradiction then we know that something is false somewhere, so we should just give up and conclude that none of the premisses are known. However, then there can be an unexpected exam, in apparent conflict with the fact that the contradictory conclusion cannot be true. But this is to misunderstand. A contradiction cannot be deduced from the simple announcement of an unexpected exam, that much is agreed by all. The contradiction only follows from the announcement plus additional premisses about knowledge which might or might be true, but they cannot all be true. Making appropriate premisses known leads, for example, to the correct conclusion that Friday cannot be a surprise, and there is no way to make it a surprise short of refusing to think, which is no news.

BIBLIOGRAPHY

[Ayer, 1973] Ayer, A.J. (1973). On a Supposed Antinomy. *Mind* 82.
[Chow, 1998] Chow, T. (1998), The Surprise Examination or Unexpected Hanging Paradox, *American Mathematical Monthly*, 105, 41–51.
[Fitch, 1963] Fitch, F. (1963), A Logical Analysis of Some Value Concepts, *Journal of Symbolic Logic*, 28/2, 135–142.
[Jackson, 1987] Jackson, F. (1987), The Easy Examination Paradox, in *Conditionals*, Oxford, Blackwell, 115–126.
[Kritchman and Raz, 2010] Kritchman, S. and Raz, R. (2010), The Surprise Examination Paradox and the Second Incompleteness Theorem, *Notices of the American Mathematical Society*, Vol 57 No 11, 1454–1458.
[Nerlich, 1961] Nerlich, G. (1961), Unexpected Examinations and Unprovable Statements, *Mind* 70, 503–514.
[Quine, 1953] Quine, W.V.O. (1953), On a So-called Paradox, *Mind* 62, 65–66.
[Sorensen, 2006] Sorensen, R. (2006), Epistemic Paradoxes, *Stanford Encyclopedia of Philosophy*.

[Wright and Sudbury, 1977] Wright, C. and Sudbury, A, (1977), The Paradox of the Unexpected Examination, *Australasian Journal of Philosophy*, 55, 41–58.

Chris Mortensen and Jennie Louise
Department of Philosophy
The University of Adelaide
Adelaide, Australia
E-mails: {Chris.Mortensen, Jennie.Louise}@adelaide.edu.au

Revisiting 'Generally' and 'Rarely'

PAULO A. S. VELOSO AND SHEILA R. M. VELOSO

ABSTRACT. Assertions and arguments involving some vague notions (like 'generally' and 'rarely') occur often in ordinary language and in some branches of science. A precise treatment of such ideas has been a basic motivation for logics of qualitative reasoning.

1 Introduction

Here we revisit and discuss some issues underlying 'generally' and 'rarely'.

Assertions and arguments involving some vague notions occur often, not only in ordinary language, but also in some branches of science. For instance, one often encounters assertions such as "Eagles generally fly" and "Penguins rarely fly" as well as "Bodies generally expand when heated" and "Metals rarely are liquid under ordinary conditions". The vagueness is given by "modifiers", such as 'generally' and 'rarely'. A precise treatment of such ideas has been a basic motivation underlying logics for qualitative reasoning [11].[1]

Several aspects of logics concerning 'generally' and 'rarely' have been developed [11]. Some issues, however, may be left somewhat unclear, leading to misunderstandings. Also, some objections against the approach have been raised, such as lack of intuitive justification.[2]

Here, we will revisit and discuss some issues concerning logics for 'generally' and 'rarely', with the aim of clarifying the approach, its underlying motivations, intuitions and justification. We will also briefly comment on some recent developments.

The structure of this paper is as follows.[3] In Section 2, we examine some intuitions and accounts for notions of 'generally' and 'rarely'. In Section 3, we indicate how one can arrive at families of sets as capturing some notions of 'generally' and 'rarely'. In Section 4, we examine some logical aspects of 'generally' and 'rarely', like oppositions, expressive and deductive powers, as well as axioms; we also briefly consider natural deduction systems and sequent calculi. Section 5 presents some concluding comments.

2 Notions of 'Generally' and 'Rarely'

We will now examine some intuitions behind notions of 'generally' and 'rarely'.

We wish to express assertions involving notions, such as 'generally' and 'rarely', and reason about them in a precise manner. For this purpose, one needs a clear understanding of these notions, which appear to be quite vague.

---
[1]Helpful discussions with Walter A. Carnielli are gratefully acknowledged.
[2]The authors are grateful to Richard Epstein for his remarks and criticisms.
[3]Most footnotes contain details that may be disregarded at a first reading.

Various possible interpretations seem to be associated with the somewhat vague notions of 'generally' and 'rarely'. We shall now consider a few reasonable ones and examine some intuitions underlying them. Consider assertions of the form "objects generally have P" or "objects rarely have P", where P is a given property. How is one to understand these assertions? What would be the possible grounds for accepting them? We shall now examine some answers to these questions stemming from possible accounts for 'generally' and 'rarely'.

We will consider extensional accounts for 'generally' and 'rarely', as illustrated in the next example.

EXAMPLE 1. The Cariocas are the natives of Rio. If one accepts the assertion "Brazilians generally like the Cariocas", then one also accepts the assertion "Brazilians generally like the natives of Rio". Similarly, if one accepts the assertion "People rarely like the natives of Rio", then one also accepts the assertion "People rarely like the Cariocas".

2.1 Numerical Accounts

We will now examine some numerical accounts for 'generally' and 'rarely'.

EXAMPLE 2. Consider the assertion "Brazilians generally like soccer". Imagine that among the Brazilians, 180 millions like soccer. Then, the Brazilians that like soccer form a sizable set of the universe of Brazilians (which has about 190 millions people).

EXAMPLE 3. Consider the assertion "Viennese enjoy music". If 75% of the Viennese like music, then the set E of Viennese enjoying music form a likely set of the universe of Viennese. We can visualize this situation as follows:

$\overline{E}$	E
25%	75%

Similarly, one may understand "Birds generally fly" as The flying birds form a sizable or likely set (of birds). These two accounts of generally' may be termed "metric", as they try to reduce it to a measurable aspect, so to speak. One explains "people generally have property P" as "the people having this property P form a likely (or sizable) set", i. e. a set having "high" relative frequency (or cardinality), where 'high' is understood as above a given threshold.[4]

These metric accounts, however, differ in one important aspect, as can be seen by considering the relation of having the same size. On the one hand, the size accounts – cardinality above a given threshold – clearly fail to distinguish sets with the same cardinality: they are all either above or below the threshold. We may say that we have a non-local notion. In contrast, sets with the same size may very well have distinct probabilities.[5] So, likely sets are not invariant under permutations of the universe.

---

[4] Such a threshold may depend on the situation or the person.

[5] For instance, one may find unlikely to find a prime number, even though the sets of prime and composite numbers have the same cardinality as that of naturals.

These metrical accounts support some inferences.

EXAMPLE 4. Imagine that everyone that likes sports watches channel SportTV. If one accepts the assertion "Boys generally like sports", then one would also accept the assertion "Boys generally watch SportTV". These metrical accounts explain this as follows: the set S, of boys that like sports, is a subset of the set T, of boys that watch SportTV; so, if S is above the threshold, then so is T.

Some inferences, however are not supported by these metrical accounts.

EXAMPLE 5. Consider the universe of Brazilians and imagine that one accepts the assertions: "Brazilians generally have their beards shaved" and "Brazilians generally shave their legs". Then, one does not have to accept the assertion "Brazilians generally have their beards shaved and shave their legs". Consider the sets B, of Brazilians that have their beards shaved, and L, of Brazilians that shave their legs. They are both above the threshold, but their intersection may very well be below it. (In this case, one would expect $B \cap L$ to be nearly empty.[6])

2.2 Relaxed Accounts

We will now examine some relaxed accounts for 'generally' and 'rarely'.

One may understand "Birds generally fly" as "Birds rarely fail to fly", in the sense that the exceptions are very few, i. e. the non-flying birds form a rather small set. Thus, the intended meaning of "objects generally have P" can also be given by means of the set of exceptions, namely those objects failing to have P.

EXAMPLE 6. Consider the assertion "Natural numbers rarely divide twelve". One may explain it by saying that the divisors of twelve form a small set (of naturals), where 'small' is understood as finite.

Similarly, one would understand the assertion "Real numbers rarely are rational" as the rationals form a small set of reals, with 'small' now taken as (at most) denumerable.[7]

These relaxed accounts sanction some inferences.

EXAMPLE 7. Consider the universe of natural numbers with 'small' being finite. Then, one accepts the assertions: "Naturals rarely divide twelve" and "Naturals rarely are below seven". Indeed, both sets T, of divisors of twelve, and S, of naturals below seven, are finite. Thus, so are their intersection $T \cap S$ and their union $T \cup S$. Hence, one should also accept both assertions "Naturals rarely divide twelve and are below seven' "Naturals rarely divide twelve or are below seven".

These relaxed accounts still rely on a threshold, but it is less arbitrary.

2.3 Qualitative Accounts

We will now examine some qualitative accounts for 'generally' and 'rarely'.

---

[6]The Brazilians that have their beards shaved are generally males, whereas the Brazilians that shave their legs are generally females. So, the Brazilians that have their beards shaved and shave their legs form a rather small fraction of the population.

[7]An example with finite universe will be given later: Example 11 (in 2.4).

The accounts mentioned in 2.1 and 2.2 may be termed "quantitative". These accounts may suffice for various cases, but they do not seem to cover some examples, where these notions appear to have a qualitative character.

EXAMPLE 8. Consider the assertion "Real numbers generally are rational". How is one to understand this assertion? What would be the possible grounds for accepting it? The rationals do not seem to form a likely, sizable or large set of reals in a quantitative sense: there are too few of them.[8] Yet, there seems to be a sense in which one may accept that "Real numbers generally are rational". Indeed, every open interval with a rational also has irrationals; so one may say that the rationals are "almost everywhere" within the reals: they are ubiquitous within the reals.[9] The irrationals are also ubiquitous within the reals. So, in this account, one should also accept that "Real numbers generally are irrational".

This example illustrates a local qualitative notion of 'generally'. One explicates "objects generally have a given property" by saying that "the set of objects having this property is a dense set" in a given topology.

Example 8 relies on the usual topology of the reals: given by open intervals. In other topologies for the reals, the situation may change.

EXAMPLE 9. One may consider other topologies for the reals.

1. Consider as closed sets those including or included in $\mathbb{Q}$. Then, the rationals are not dense, but the irrationals are dense.

2. Consider as closed sets those including or included in $\overline{\mathbb{Q}}$. Then, the rationals are dense, but the irrationals are not dense.

3. Consider every set of reals as closed. Then, neither the rationals nor the irrationals are dense.

So, we have diverse accounts for 'generally' and 'rarely'.

Example	Account
Brazilians and soccer (2)	generally by sizable
Viennese and music (3)	generally by likely
Naturals (6)	rarely by finite
Reals (8)	generally by dense

These accounts differ with respect to their reliance on a threshold and to invariance (local or non-local notions). We may summarise these features as follows:

Account	Reading	Threshold	Invariance
Size (2.1)	sizeable	+	Y
Frequency (2.1)	likely	+	N
Relaxed (2.2)	small	±	Y
Qualitative (2.3)	ubiquitous	−	N

---

[8] It is the assertion "Real numbers generally are irrational" that appears to be more reasonable, as explained above (in 2.2).

[9] More precisely, the rational reals form a dense set of reals. [5]

2.4 Abstract Accounts

We will now examine some abstract accounts for 'generally' and 'rarely'.

We have various distinct notions of 'generally' and 'rarely'. We would like to give them a unified treatment. As more neutral names encompassing these notions, we will prefer to use 'important' in lieu of 'sizable', 'likely' or 'large' (corresponding to 'generally'), and, accordingly 'negligible' for 'non-sizable', 'unlikely' or 'small' (corresponding to 'rarely').

The preceding examples illustrate some cases of 'important' and 'negligible'.

Example	Notion
Brazilians and soccer (2)	important = sizable
Viennese and music (3)	important = likely
Naturals (6)	negligible = finite
Reals (8)	important = dense

We can see that the earlier versions can be subsumed under the more flexible abstract notions. The previous terms are somewhat vague, the more so with the new ones. Nevertheless, they present some advantages. First, the reliance on a – somewhat arbitrary – threshold is less stringent. Also, they have a wider range of applications, stemming from the liberal interpretations of 'important' and 'negligible'. For instance, when saying "Meetings attended only by junior staff can be discarded", one seems to be considering sets including only junior staff members as 'negligible'.

The following examples give some versions of 'important' and 'negligible'.

EXAMPLE 10. Imagine that a socialite visiting Hollywood and eager to attend interesting parties receives the following pieces of advice.

1. "Important parties are those attended by the celebrities".

2. "Important parties are those attended by Madonna".

Then, 'important' sets of guests are those including the celebrities, for the former advisor, and those where Madonna is, for the latter advisor.

EXAMPLE 11. Consider a driver having to transport birds, dogs and elephants. The elephants are hard to transport; so sets having no elephants may be left to his assistants: the negligible sets are those having no elephants. In this case, the union of 2 negligible sets is still negligible.[10]

As these examples indicate, the notions of 'important' and 'negligible' are relative to the situation or person. Further examples can illustrate this point.

1. First, consider two sets with the same size: one consisting of a horse and an ox, and another one consisting of a horse and a dog. These two sets may be just as important to a conservationist. But, the former may be more important to a farmer, whereas the latter might be preferred by an English gentleman, keen on fox hunting.

---

[10] This is similar to the situation illustrated by Example 7 (in 2.2).

2. Now, consider two sets with distinct sizes: one consisting of thirty birds, and another one consisting of a couple of elephants. The Zoo director is likely to consider them equally important. But, an ornithologist might rank the former as more important, whereas a circus manager (or a driver in charge of transporting them) would probably give more attention to the latter. So, a smaller set may be more important than a larger set, or just as important.

We have seen that one has various distinct notions of 'generally' and 'rarely', which may be explicated in terms of important and negligible sets, respectively. Under the light of the preceding considerations, one has explanations as follows.

- One explains "objects generally have property P" as "the objects having property P form an important set (of objects)".
- One explains "objects rarely have property P" as "the objects with property P form a negligible set (of objects)".

The preceding examples illustrate how some inferences can be sanctioned, or not, by properties of 'important' and 'negligible' sets.

Example	Property
Boys and sports (4)	superset of important is important
Brazilians (5)	intersection of important sets may be non-important.
Naturals (7)	intersection of negligible sets is negligible
Elephants (11)	union of negligible sets is negligible

3 Families for 'Generally' and 'Rarely'

We now indicate how one can arrive at families of sets, with certain properties, as capturing some notions of 'generally' and 'rarely'.

We have seen that one has various distinct notions of 'important' (corresponding to 'generally') and negligible' (corresponding to 'rarely'), which are relative to a situation. Inferences are sanctioned, or not, by properties of 'important' and 'negligible' sets. Even though these properties may vary according to the situation, one may expect them to share some properties.

The quantitative accounts seen before may suggest comparing sets by a relation $\approx$ of 'having about the same size'. One may be tempted to consider that one has an equivalence relation. Indeed, reflexivity and symmetry seem reasonable. But, what about transitivity: are we prepared to accept that the extremes X_0 and X_n of a long chain $X_0 \approx X_1 \approx \ldots \approx X_n$ are still sets with about the same size?

Actually, a notion like 'having about the same size' does not seem to be such a good starting point. This is so because one is naturally led to think that sets with the same size should have about the same size. In other words, this is a non-local notion. The notion we are seeking should be, in contrast, a local one: we wish to be able to connect it to important and negligible sets.

In view of these considerations, we will prefer to use a relation of dominance as our basic comparison between subsets of a given universe, which we will understand as being practically included, i. e. included except for a negligible part.

So, given a universe V, we have:

- two families of subsets: $\mathcal{N}$, of negligible ones, and $\mathcal{W}$, of important ones;

- a binary relation of dominance $\sqsubseteq$ on V.

Also, instead of assuming at the outset some properties (like transitivity), we shall put forward some – hopefully palatable – postulates. This enterprise is somewhat reminiscent of reverse mathematics, with a difference: we will suggest some postulates and try to justify them on intuitive grounds. [11]

3.1 Basic Properties

We now examine some basic properties of 'important' and 'negligible'.

Intuitively, 'important' can be understood as "carrying considerable weight or importance". Dictionaries explain 'negligible' as follows.

- Something negligible is "fit to be neglected or discarded". [12].

- "Something that is negligible is so small or unimportant that is not worth considering or worrying about". [2].

Thus, the following considerations seem reasonable.

A set X is to be considered important when its exceptions form a negligible set; the exceptions being the elements outside X, forming the complement $\overline{X}$. So, it seems reasonable to say that set X is important when its complement $\overline{X}$ is negligible. This situation can be visualized as follows:

V

We thus have the following basic postulate connecting important and negligible sets under complementation.

$[\overline{\mathcal{W}} = \mathcal{N}]$ $X \in \mathcal{W}$ iff $\overline{X} \in \mathcal{N}$

Also, it seems reasonable to consider a set practically within another one when the possible counterexamples to the inclusion form a negligible set. The set of exceptions to $X \subseteq Y$ is their difference $X \setminus Y$.[12] So, it seems reasonable to say that X is dominated by Y when their difference $X \setminus Y$ is negligible. This

---

[11]"The fundamental question in reverse mathematics is to determine which set of existence axioms are required to prove particular theorems of mathematics". [7] (p. 45). Here, instead of locating familiar axioms, we will be suggesting some new postulates, whence the need for justifying their acceptance on intuitive grounds.

[12]Note that $X \subseteq Y$ iff $X \setminus Y = \emptyset$.

situation can be visualized as follows:

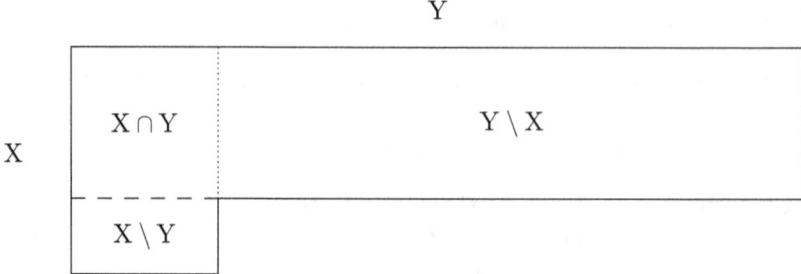

We thus have the following basic postulate connecting dominance with negligible sets via difference.

$[\sqsubseteq \setminus \mathcal{N}]$ $X \sqsubseteq Y$ iff $X \setminus Y \in \mathcal{N}$

These two basic postulates have some immediate consequences.

COROLLARY 3.1 (Consequences of the basic postulates).

$[\sqsubseteq]$ Behavior of dominance under complementation: X is dominated by Y iff $\overline{Y}$ is dominated by $\overline{X}$: $X \sqsubseteq Y$ iff $\overline{Y} \sqsubseteq \overline{X}$.

$[\mathcal{N} \sqsubseteq \emptyset]$ Characterization of the negligible sets as those dominated by the empty set: $X \in \mathcal{N}$ iff $X \sqsubseteq \emptyset$.

$[\mathcal{W} \sqsubseteq V]$ Characterization of the important sets as those dominating the universe: $Y \in \mathcal{W}$ iff $V \sqsubseteq Y$.

Proof. Consequences $[\sqsubseteq]$ and $[\mathcal{N} \sqsubseteq \emptyset]$ follow from postulate $[\sqsubseteq \setminus \mathcal{N}]$.[13] $[\mathcal{W} \sqsubseteq V]$ follows from postulate $[\overline{\mathcal{W}} = \mathcal{N}]$ and consequences $[\sqsubseteq]$ and $[\mathcal{N} \sqsubseteq \emptyset]$.[14]

3.2 Common Properties

We now examine some common properties of 'important' and 'negligible'.

Our relation of dominance might be trivial in two respects: no dominance or dominance for any pair of subsets. This is not exactly what one might expect.

In the quantitative accounts, the empty set is rather small and the universe is rather large. These ideas seem to apply to 'important' and 'negligible' as well. Intuitively, 'important' means carrying considerable weight or importance and, dually, 'negligible' means carrying almost no weight or importance. In fact, in Hasse diagram for inclusions, one would tend to regard the important sets as

---

[13] Indeed, $X \setminus Y = \overline{Y} \setminus \overline{X}$ and $X \setminus \emptyset = X$.
[14] Indeed, $Y \in \mathcal{W}$ iff $\overline{Y} \in \mathcal{N}$ iff $\overline{Y} \sqsubseteq \emptyset$ iff $\overline{\emptyset} \sqsubseteq \overline{\overline{Y}}$, i. e. $V \sqsubseteq Y$.

being high (near the top) and negligible sets as being low (near the bottom).

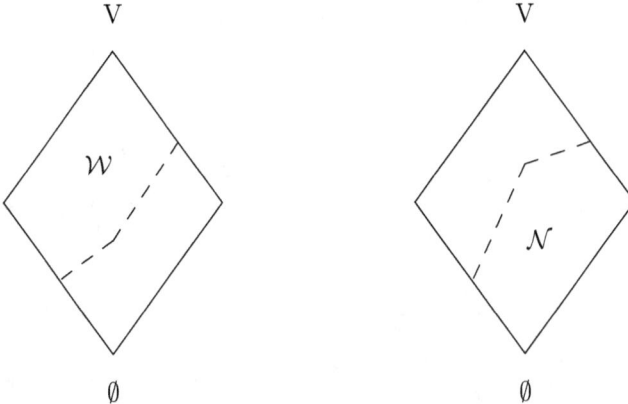

So, one would probably regard the empty set as (rather) negligible and the universe as (rather) important.

Let us see what is involved in considering the empty set as negligible.

One would probably accept that the empty set is practically within itself. Actually, it seems reasonable to say that every set is practically within itself. Also, a subset of a set seems to be practically within it.

LEMMA 3.1. *The following assertions are equivalent.*

|⊆⊑| A subset of Y is dominated by Y: if $X \subseteq Y$, then $X \sqsubseteq Y$.

|=⊑| Every set X is dominated by itself: $X \sqsubseteq X$.

|∅⊑| The empty set ∅ is dominated by itself: $\emptyset \sqsubseteq \emptyset$.

|∅𝒩| The empty set ∅ is negligible: $\emptyset \in \mathcal{N}$.

|∅V| The empty set is dominated by the universe: $\emptyset \sqsubseteq V$.

|V𝒲| The universe V is important: $V \in \mathcal{W}$.

Proof. These equivalences follow from the basic postulates.[15]

By accepting any one of the assertions in Lemma 3.1, one has bounds for dominance: every set dominates the empty set and is dominated by the universe.

|⊥⊤| For every subset $X \subseteq V$: $\emptyset \sqsubseteq X \sqsubseteq V$

Now, our relation of dominance might collapse too many subsets. Our intuitive ideas about dominance suggest that the (nonempty) universe is not practically within the empty set: $V \not\sqsubseteq \emptyset$.

In the metric accounts (in 2.1) the universe (being nonempty) is not below threshold 0; so it is not sizable. These ideas seem to apply to 'negligible' as

---

[15] Clearly |⊆⊑| ⇒ |=⊑| and |∅⊑| is a special case of |=⊑|. Now, |∅⊑| ⇒ |∅𝒩| (by [𝒩 ⊑ ∅]) and |∅𝒩| ⇒ |⊆⊑|. By [$\overline{\mathcal{W}} = \mathcal{N}$], |∅V| ⇔ |∅𝒩| and |∅𝒩| ⇔ |V𝒲|.

well: one should consider the universe as non-negligible (V $\notin \mathcal{N}$) and, dually, the empty set as non-important ($\emptyset \notin \mathcal{W}$).

Lemma 3.1 yields the equivalence of these three assertions:

$$V \notin \mathcal{N} \quad \Leftrightarrow \quad V \not\subseteq \emptyset \quad \Leftrightarrow \quad \emptyset \notin \mathcal{W}$$

Now, let us call a family $\mathcal{K}$ of subsets of a universe V *proper* when it has the universe but not the empty set: $V \in \mathcal{K}$ and $\emptyset \notin \mathcal{K}$.

Thus, for a proper family of important sets, we have the properties in Table 1

	Important	Negligible
($\in$)	$V \in \mathcal{W}$ (V is important)	$\emptyset \in \mathcal{N}$ ($\emptyset$ is negligible)
($\notin$)	$\emptyset \notin \mathcal{W}$ ($\emptyset$ is not important)	$V \notin \mathcal{N}$ (V is not negligible)
($\emptyset$)	$\mathcal{W} \neq \emptyset$ (important sets exist)	$\mathcal{N} \neq \emptyset$ (negligible sets exist)
($\wp$)	$\mathcal{W} \neq \wp$ (V) (non-important sets exist)	$\mathcal{N} \neq \wp$ (V) (non-negligible sets exist)

Table 1. Properties of a proper family of important sets

3.3 Specific Properties

We now examine some specific properties of 'important' and 'negligible'.

First, let us examine the case of subsets. Example 4 (in 2.1) shows a case where a superset of an important set is important. Should one consider a subset of a negligible set still negligible? This seems reasonable in quantitative accounts, as illustrated in 2.1 and 2.2. Let us examine what is involved in accepting this idea. Should one imagine that the addition of a negligible set N produces negligible impact, in the sense $X \cup N$ is practically within X?

LEMMA 3.2. *The following assertions are equivalent.*

($\subseteq \mathcal{N}$) A subset of a negligible set is negligible: if $X \subseteq N \in \mathcal{N}$, then $X \in \mathcal{N}$.

($\cap \mathcal{N}$) Intersection with a negligible set is negligible: if $N \in \mathcal{N}$, then $Y \cap N \in \mathcal{N}$.

($\cup \mathcal{N}$) Addition of negligible set has negligible impact: $X \cup N \sqsubseteq X$, if $N \in \mathcal{N}$.

($\supseteq \mathcal{W}$) A superset of an important set is important: $Y \in \mathcal{W}$, if $Y \supseteq W \in \mathcal{W}$.

Proof. These equivalences follow from the basic postulates.[16]

Now, let us call a family $\mathcal{K}$ of subsets of a universe *superset* when it closed under supersets: $Y \in \mathcal{K}$ whenever $Y \supseteq X \in \mathcal{K}$.[17]

Thus, one should accept that the family $\mathcal{W}$ of important sets is superset exactly when one finds reasonable the assertions in Lemma 3.2.[18]

For a superset family of important sets, we have the properties in Table 2.

---

[16] Clearly, we have ($\subseteq \mathcal{N}$) $\Leftrightarrow$ ($\cap \mathcal{N}$) (since $Z \subseteq N$ iff $Z \cap N = Z$). Also, ($\cap \mathcal{N}$) $\Leftrightarrow$ ($\cup \mathcal{N}$), because $(Z \cup N) \setminus Z = \overline{Z} \cap N$ and ($\subseteq \mathcal{N}$) $\Leftrightarrow$ ($\supseteq \mathcal{W}$) by $[\overline{\mathcal{W}} = \mathcal{N}]$.

[17] Note that a nonempty superset family has the universe in it.

[18] Another equivalent assertion is: if $X \subseteq Y$ and $Y \sqsubseteq Z$, then $X \sqsubseteq Z$.

Important	Negligible
$Y \supseteq W \in \mathcal{W} \Rightarrow Y \in \mathcal{W}$ (supersets)	$X \subseteq N \in \mathcal{N} \Rightarrow X \in \mathcal{N}$ (subsets)
$W \in \mathcal{W} \Rightarrow X \cup W \in \mathcal{W}$ (union)	$N \in \mathcal{N} \Rightarrow Y \cap N \in \mathcal{N}$ (intersection)
$W \in \mathcal{W} \Rightarrow X \sqsubseteq X \cap W$ (∩-dominance)	$N \in \mathcal{N} \Rightarrow X \cup N \sqsubseteq X$ (∪-dominance)

Table 2. Properties of superset families of important sets

Now, let us consider the case of intersection of negligible sets, should it be negligible? This is the case of Example 7 (in 2.2). Is it reasonable that if two sets are practically within a set, so is their intersection?

LEMMA 3.3. The following assertions are equivalent.

($\mathcal{N}\cap$) Intersection of negligible sets is negligible: if $M, N \in \mathcal{N}$, then $M \cap N \in \mathcal{N}$.

($\mathcal{N} \cup \mathcal{W}$) Addition of a negligible set to an important set has negligible impact: if $W \in \mathcal{W}$ and $N \in \mathcal{N}$, then $W \cup N \sqsubseteq W$.

($\cap \sqsubseteq$) If both X and Y are practically within Z, then so is their intersection $X \cap Y$: if $X \sqsubseteq Z$ and $Y \sqsubseteq Z$, then $X \cap Y \sqsubseteq Z$.

($\mathcal{W}\cup$) Union of important sets is important: if $T, W \in \mathcal{W}$, then $T \cup W \in \mathcal{W}$.

Proof. These equivalences follow from the basic postulates.[19].

Now, let us call a family $\mathcal{K}$ of subsets of a universe *union-closed* when it closed under union: $X \cup Y \in \mathcal{K}$ whenever $X, Y \in \mathcal{K}$.[20]

So, one should accept that the family $\mathcal{W}$ of important sets is union-closed when one finds reasonable the assertions in Lemma 3.3.[21]

For a union-closed family of important sets, we have the properties in Table 3.

$T, W \in \mathcal{W} \Rightarrow T \cup W \in \mathcal{W}$ (union)	$M, N \in \mathcal{N} \Rightarrow M \cap N \in \mathcal{N}$ (intersection)
$T, W \in \mathcal{W} \Rightarrow T \cup \overline{W} \sqsubseteq T$ (∪-⊑)	$M, N \in \mathcal{N} \Rightarrow M \sqsubseteq M \setminus N$ (\-⊑)
$X \sqsubseteq Z \,\&\, Y \sqsubseteq Z \Rightarrow X \cap Y \sqsubseteq Z$ (intersection-dominance)	

Table 3. Properties of union-closed families of important sets

Next, let us consider the case of union. Should the union of negligible sets be negligible? This is the case in Examples 7 (in 2.2) and 11 (in 2.4). Is it reasonable that if two sets are practically within a set, so is their union?

LEMMA 3.4. The following assertions are equivalent.

---

[19] First, by $[\overline{\mathcal{W}} = \mathcal{N}]$, ($\mathcal{N}\cap$) ⇔ ($\mathcal{N} \cup \mathcal{W}$) (as $(W \cup N) \setminus W = \overline{W} \cap N$) and ($\mathcal{N}\cap$) ⇔ ($\mathcal{W}\cup$). Now, ($\mathcal{N}\cap$) ⇒ ($\cap \sqsubseteq$) as $(X \cap Y) \setminus Z = (X \setminus Z) \cap (Y \setminus Z)$ and ($\cap \sqsubseteq$) ⇒ ($\mathcal{N}\cap$) by $[\mathcal{N} \sqsubseteq \emptyset]$ (cf. Corollary 3.1)

[20] Note that a superset family is union-closed.

[21] We explain Example 7 (in 2.2) as follows: as $\mathcal{W}$ is union-closed, $T \in \mathcal{W}$ and $S \in \mathcal{W}$ yield $T \cup S \in \mathcal{W}$.

($\mathcal{N}\cup$) Union of negligible sets is negligible: if $M, N \in \mathcal{N}$, then $M \cup N \in \mathcal{N}$.

($\sqsubseteq \cap$) If both X and Y are practically within Z, then so is their union $X \cup Y$:
if $X \sqsubseteq Z$ and $Y \sqsubseteq Z$, then $X \cup Y \sqsubseteq Z$.

($\mathcal{W}\cap$) Important sets closed under intersection: $T \cap W \in \mathcal{W}$, if $T, W \in \mathcal{W}$.

Proof. These equivalences follow from the basic postulates.[22]

Now, let us call a family $\mathcal{K}$ of sets *intersection-closed* when it closed under intersection: $X \cap Y \in \mathcal{K}$ whenever $X, Y \in \mathcal{K}$.

So, one should accept that the family $\mathcal{W}$ of important sets is intersection-closed when one finds reasonable the assertions in Lemma 3.4. [23]

An intersection-closed family of important sets has the properties in Table 4.

$$T, W \in \mathcal{W} \Rightarrow T \cap W \in \mathcal{W} \text{ (intersection)} \qquad M, N \in \mathcal{N} \Rightarrow M \cup N \in \mathcal{N} \text{ (union)}$$

$$X \sqsubseteq Z \,\&\, Y \sqsubseteq Z \Rightarrow X \cup Y \sqsubseteq Z \text{ (union-dominance)}$$

Table 4. Properties of intersection-closed families of important sets

Finally, let us consider the case of complementation. Note that a set is negligible iff it is practically within its complement: $X \in \mathcal{N}$ iff $X \sqsubseteq \overline{X}$ (as $X \setminus \overline{X} = X$).

Should a set be both important and negligible? Should the complement of a negligible set be negligible? In 2.3, we have Example 8, where the rationals are both important and negligible, and Example 9, where this is not so. Are these situations reasonable?

LEMMA 3.5. The following assertions are equivalent.

($\cap$) No set is both negligible and important: $\mathcal{N} \cap \mathcal{W} = \emptyset$.

($\overline{\mathcal{N}}$) Complement of negligible is non-negligible: if $N \in \mathcal{N}$, then $\overline{N} \notin \mathcal{N}$.

($\overline{\mathcal{W}}$) Complement of important is non-important: if $W \in \mathcal{W}$, then $\overline{W} \notin \mathcal{W}$.

Proof. These equivalences follow from the basic postulate $[\overline{\mathcal{W}} = \mathcal{N}]$.[24]

Now, let us call a family $\mathcal{K}$ of subsets of a universe *repelling* when it never has the complement of its sets: $\overline{Z} \notin \mathcal{K}$ whenever $Z \in \mathcal{K}$.

Thus, one should accept that the family $\mathcal{W}$ of important sets is repelling exactly when one finds reasonable the assertions in Lemma 3.5. [25]

For a repelling family of important sets, we have the properties in Table 5.

---

[22]Indeed, ($\mathcal{N}\cup$) $\Rightarrow$ ($\sqsubseteq \cap$) as $(X \cup Y) \setminus Z = (X \setminus Z) \cup (Y \setminus Z)$ and ($\sqsubseteq \cap$) $\Rightarrow$ ($\mathcal{N}\cup$) by $[\mathcal{N} \sqsubseteq \emptyset]$ (cf. Corollary 3.1). Finally, ($\mathcal{N}\cup$) $\Leftrightarrow$ ($\mathcal{W}\cap$) by $[\overline{\mathcal{W}} = \mathcal{N}]$.

[23]We explain Example 11 (in 2.4) as follows: as $\mathcal{W}$ is intersection-closed, $T \in \mathcal{W}$ and $W \in \mathcal{W}$ yield $T \cap W \in \mathcal{W}$.

[24]Indeed, ($\cap$) $\Leftrightarrow$ ($\overline{\mathcal{N}}$) as $X \in \mathcal{N} \cap \mathcal{W}$ iff $X \in \mathcal{N}$ and $\overline{X} \in \mathcal{W}$ and ($\overline{\mathcal{N}}$) $\Leftrightarrow$ ($\overline{\mathcal{W}}$) by $[\overline{\mathcal{W}} = \mathcal{N}]$.

[25]We can explain Example 9 (in 2.3) as follows: since $\mathcal{W}$ is not repelling, $W \in \mathcal{W}$ does not yield $\overline{W} \notin \mathcal{W}$.

($\overline{\mathcal{W}}$) $W \in \mathcal{W} \Rightarrow \overline{W} \notin \mathcal{W}$ (complement of important is non-important)

($\overline{\mathcal{N}}$) $N \in \mathcal{N} \Rightarrow \overline{N} \notin \mathcal{N}$ (complement of negligible is non-negligible)

($\cap$) $\mathcal{N} \cap \mathcal{W} = \emptyset$ (no set is both negligible and important)

Table 5. Properties of repelling families of important sets

Should every set be negligible or important? Should the complement of a non-negligible set be negligible? This is the case of the Madonna interpretation in Example 10 (in 2.4).[26] Are such situations reasonable?

LEMMA 3.6. The following assertions are equivalent.

($\cup$) Every set is negligible or important: $\mathcal{N} \cup \mathcal{W} = \wp$ (V).

($\widetilde{\mathcal{N}}$) Complement of non-negligible is negligible: if $X \notin \mathcal{N}$, then $\overline{X} \in \mathcal{N}$.

($\widetilde{\mathcal{W}}$) Complement of non-important is important: if $Y \notin \mathcal{W}$, then $\overline{Y} \in \mathcal{W}$.

Proof. These equivalences follow from the basic postulate $[\overline{\mathcal{W}} = \mathcal{N}]$.[27]

Now, let us call a family $\mathcal{K}$ of subsets of a universe *attracting* when it has the complement of the sets outside it: $\overline{Z} \in \mathcal{K}$ whenever $Z \notin \mathcal{K}$.

Thus, one should accept that the family $\mathcal{W}$ of important sets is attracting exactly when one finds reasonable the assertions in Lemma 3.6.[28]

For an attracting family of important sets, we have the properties in Table 6.

($\widetilde{\mathcal{W}}$) $W \notin \mathcal{W} \Rightarrow \overline{W} \in \mathcal{W}$ (complement of non-important is important)

($\widetilde{\mathcal{N}}$) $X \notin \mathcal{N} \Rightarrow \overline{X} \in \mathcal{N}$ (complement of non-negligible is negligible)

($\cup$) $\mathcal{N} \cup \mathcal{W} = \wp$ (V) (every set is negligible or important)

Table 6. Properties of attracting families of important sets

The accounts presented in Section 2 have some of the properties seen above (in 3.2 and 3.3).

Account	Important sets
Numerical (2.1)	proper, superset, union-closed, repelling
Relaxed (2.2)	proper, superset, union-closed, intersection-closed
Qualitative (2.3)	proper, superset, union-closed

---

[26] If one calls important the sets with more examples than counterexamples and dually for negligible ($X \in \mathcal{W}$ iff $|X| < |\overline{X}|$ and $Y \in \mathcal{N}$ iff $|\overline{X}| < |X|$), then a set with half the elements of the universe is neither negligible nor important.

[27] Indeed, ($\cup$) $\Leftrightarrow$ ($\widetilde{\mathcal{N}}$) as $X \notin \mathcal{N} \cup \mathcal{W}$ iff $X \notin \mathcal{N}$ and $\overline{X} \notin \mathcal{W}$ and ($\widetilde{\mathcal{N}}$) $\Leftrightarrow$ ($\widetilde{\mathcal{W}}$) by $[\overline{\mathcal{W}} = \mathcal{N}]$.

[28] For the Madonna interpretation in Example 10 (in 2.4) we have: since $\mathcal{W}$ is attracting, $Z \notin \mathcal{W}$ yields $\overline{Z} \in \mathcal{W}$.

4 Logical Aspects of 'Generally' and 'Rarely'

We now examine some logical aspects of 'generally' and 'rarely'.

To express 'generally' and 'rarely', we add to the usual first-order syntax two new quantifiers: ∇ (for 'generally') and $\triangle$ (for 'rarely'). Thus, we now have new quantified formulas, namely general and rare formulas, in addition to the familiar universal and existential formulas.[29]

4.1 Logics for 'Generally' and 'Rarely'

We now examine logics for 'generally' and 'rarely'.

The intended meanings of general and rare formulas are as follows. Given a property P, consider its extension P, consisting of the elements with property P, and its complement $\overline{P}$, consisting of the elements not having property P.[30]

(*gnr*) Property P holds *generally* iff the its extension P belongs to the given family $\mathcal{W}$ of important sets: $\nabla v\, P$ holds iff $P \in \mathcal{W}$.

(*rrl*) Property P holds *rarely* iff the set $\overline{P}$ (of its counterexamples) belongs to the given family $\mathcal{W}$ of important sets: $\triangle v\, P$ holds iff $\overline{P} \in \mathcal{W}$.

What are those families of important sets? As seen in Section 2, we have diverse notions of 'generally' and 'rarely', giving rise to families with distinct properties (cf. Section 3).

We first examine oppositions. We will examine contrary, sub-contrary, contradictory and sub-alternate assertions.[31] We will use the following notation:

⟶ for sub-alternation   ∼ ∼ for contradictory

– – – – for contrariety   ·········· for subcontrariety

The classical square displays the oppositions between assertions, classified as affirmative or negative and universal or particular.[32]

Now, in addition to universal and particular assertions, we have general and rare assertions. We wish to examine oppositions between them.

---

[29] We actually have new formulas: see Proposition 4.1 (at the end of 4.1).

[30] One can give the meaning of the familiar quantifiers by means of families, e. g. property P holds universally iff the its extension P belongs to the family consisting of the universe.

[31] Contrary assertions cannot be both true, subcontrary assertions cannot be both false, and contradictory assertions cannot be both true nor false. Sub-alternation amounts to implication.

[32] The classical square of oppositions is as follows:

	Affirmative	Negative
Universal		
Particular		

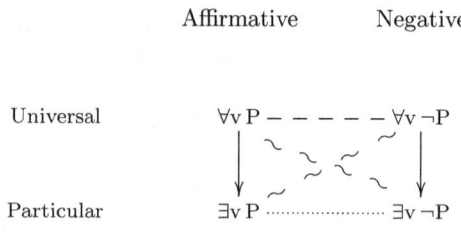

Let us first examine the general case: an arbitrary family $\mathcal{W}$ of important sets. Then, the basic postulate $[\overline{\mathcal{W}} = \mathcal{N}]$ (in 3.1) gives the equivalence:

$[\nabla \neg \triangle]$ $\nabla v\, P \leftrightarrow \triangle v\, \neg P$ (P holds generally iff ¬P holds rarely)

We thus have the equivalences general and rare assertions in Figure 1.

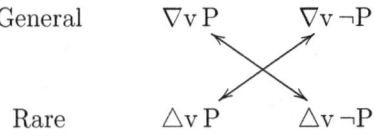

Figure 1. Equivalences between general and rare assertions

We now examine families of important sets that are repelling or attracting. We then have some implications.

- For repelling families, the properties in Table 5 (in 3.2) give:

  $(\nabla \neg \triangle)$ $\nabla v\, P \rightarrow \neg \triangle v\, P$ (if P holds generally, then P does not hold rarely), by property $(\cap)$, or $(\nabla \neg) : \nabla v\, \neg P \rightarrow \neg \nabla v\, P$ (cf. $(\overline{\mathcal{W}})$).

- For attracting families, the properties in Table 6 (in 3.2) give:

  $(\neg \nabla \triangle)$ $\neg \nabla v\, P \rightarrow \triangle v\, P$ (if P does not hold generally, then P holds rarely), by property $(\cup)$, or $(\neg \nabla) : \neg \nabla v\, P \rightarrow \nabla v\, \neg P$ (cf. $(\widetilde{\mathcal{W}})$).

Repelling and attracting families of important sets behave as in Figure 2.

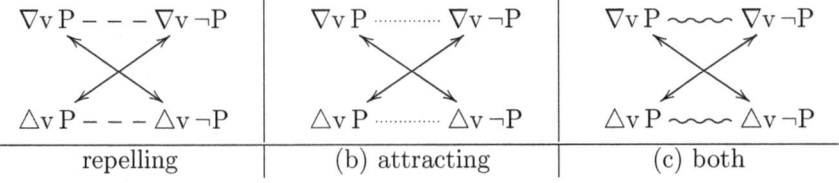

Figure 2. Oppositions for repelling and attracting families of important sets

Now, consider a proper family $\mathcal{W}$ of important sets. Then, properties $(\in)$ and and $(\not\ni)$ of Table 1 (in 3.3) yield the following implications:

$|\forall \nabla|$ $\forall v\, P \rightarrow \nabla v\, P$ (if P holds universally, then P holds generally);

$|\nabla \exists|$ $\nabla v\, P \rightarrow \exists v\, P$ (if P holds generally, then P has some instance).

In this proper case, we have the oppositions shown in Figure 3.

Let us now examine intersection-closed families of important sets. Then, property $(\mathcal{W}\cap)$ of Lemma 3.4 (in 3.3) gives the following implication:

$(\wedge \nabla)$ $(\nabla v\, P \wedge \nabla v\, Q) \rightarrow \nabla v\, (P \wedge Q)$ (if both P hold generally, then the conjunction $P \wedge Q$ also holds generally),

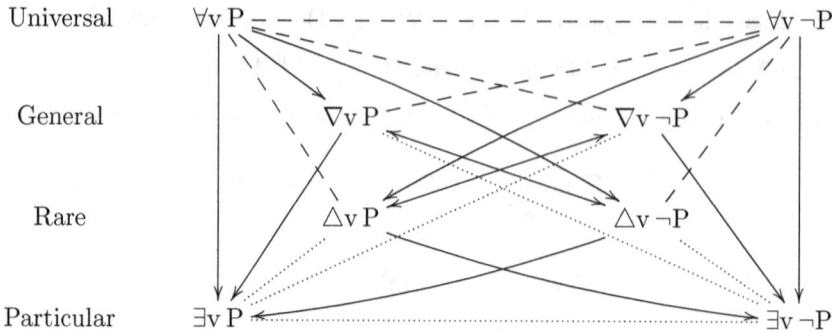

Figure 3. Oppositions for proper families of important sets

Now, a proper intersection-closed family $\mathcal{W}$ of important sets is repelling.[33] This is the case of a proper filter.[34] Thus (see Figures 2(a) and 3), proper intersection-closed families of important sets have the oppositions shown in Figure 4.

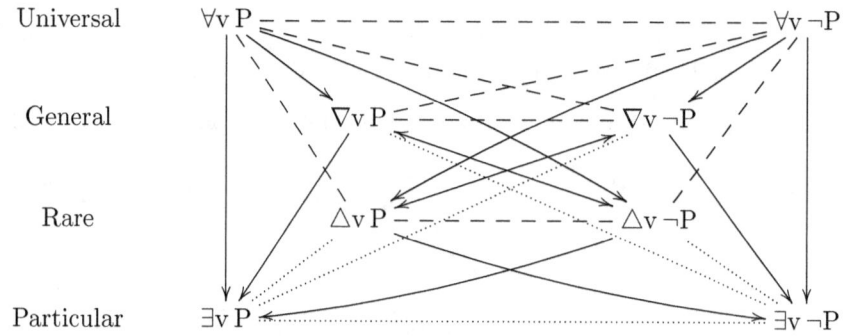

Figure 4. Oppositions for proper intersection-closed families of important sets

Finally, consider a proper intersection-closed family of important sets that is attracting. This is the case of an ultrafilter.[35] Thus (see Figures 2(c) and 3), proper attracting intersection-closed families of important sets have the oppositions shown in Figure 5.

The meanings of general and rare formulas given above – in (gnr) and (rrl) – suffice for some purpose, e. g. examining oppositions. For other purposes, it is more convenient to have more encompassing definitions.

The *general* and *rare* formulas are $\nabla v\, F$ and $\triangle v\, F$, where F is a formula and v is a variable v. Variable v is bound in $\nabla v\, F$ and $\triangle v\, F$.

---

[33] Indeed, if both $W \in \mathcal{W}$ and $\overline{W} \in \mathcal{W}$, then $\emptyset = W \cap \overline{W} \in \mathcal{W}$.

[34] A proper filter is a superset intersection-closed family without the universe. [1]

[35] An ultrafilter is a maximal proper, equivalently a proper filter that is attracting. [1]

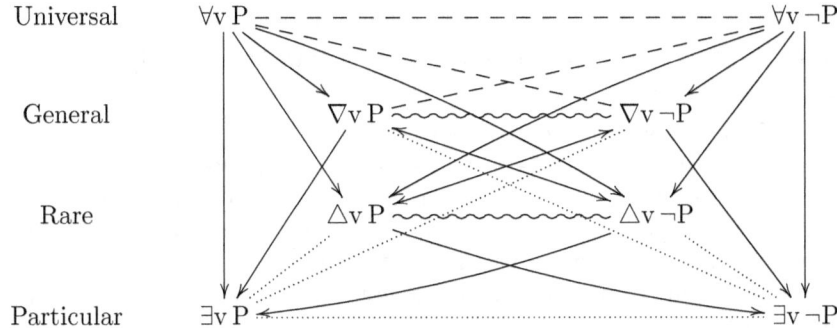

Figure 5. Oppositions for proper attracting intersection-closed important sets

A *modulated structure* $\mathfrak{A}^{\mathcal{W}}$ consists of a usual (first-order) structure $\mathfrak{A}$ together with a family $\mathcal{W}$ of subsets of its universe A. We now define *satisfaction* of a formula under an assignment **s** of values to variables, using updated assignments.[36]

(∇) For a general formula: $\mathfrak{A}^{\mathcal{W}} \models \nabla v\, F\, [\![s]\!]$ iff the set $\{\, a \in A \,/\, \mathfrak{A}^{\mathcal{W}} \models F\, [\![s[v/a]]\!] \,\}$ belongs to family $\mathcal{W}$.

($\triangle$) For a rare formula: $\mathfrak{A}^{\mathcal{W}} \models \triangle v\, F\, [\![s]\!]$ iff the set $\{\, a \in A \,/\, \mathfrak{A}^{\mathcal{W}} \not\models F\, [\![s[v/a]]\!] \,\}$ belongs to family $\mathcal{W}$.

We say that a formula G *holds* in a modulated structure when it is satisfied by every assignment: $\mathfrak{A}^{\mathcal{W}} \models G$ iff $\mathfrak{A}^{\mathcal{W}} \models G\,[\![s]\!]$, for every assignment into A.

We now wish to compare our new quantifiers with the familiar ones. We now wish to examine how adjacent classical and generalised quantifiers interact. For a proper family, we have ∇ between $\forall$ and $\exists$ (see $|\forall\nabla|$ and $|\nabla\exists|$ or Figure 3).

We first consider some cases where the behavior of the new quantifier is similar to that of the classical ones. In first-order logic, $\forall v\, F$ yields $\exists v\, F$ and $\exists u\forall z\, F$ yields $\forall z\exists u\, F$. Since ∇ is between $\forall$ and $\exists$, one might expect some similar transfer principles for ∇.[37] Now, analogues of $\exists u\forall z\, F \to \forall z\exists u\, F$ are the transfers of ∇: $\nabla u\forall z\, F \to \forall z\nabla u\, F$ (over $\forall$) and $\exists u\nabla z\, F \to \nabla z\exists u\, F$ (over $\exists$). These transfer principles are easily interpreted.[38]

For a superset family of important sets, property ($\supseteq \mathcal{W}$) of Lemma 3.2 (in 3.3) gives the following implication:

($\to \nabla$) $\forall v\,(F \to G) \to (\nabla v\, F \to \nabla v\, G)$.

---

[36] The updated assignment s[v/a] agrees with **s** on all variables, but for v (where it gives a).

[37] For a proper family of important sets, $\forall u\forall z\, F$ yields both and $\nabla u\forall z F$ and $\nabla u\forall z F$ (by $|\forall\nabla|$) and either one of them yields $\exists u\exists z\, F$. (by $|\nabla\exists|$). Also $\forall u\nabla z\, F$ yields $\exists u\nabla z\, F$, $\nabla u\exists z\, F$ and $\exists u\exists z\, F$ (by $|\nabla\exists|$).

[38] The behavior of ∇ is reminiscent of that of $\exists$ in the first transfer (over $\forall$), and of that of $\forall$ in the second transfer (over $\exists$).

For a superset family of important sets, we have both transfer principles: over $\forall$ and over $\exists$. The converse implications fail (see Lemma 4.1).

COROLLARY 4.1. Consider a modulated structure $\mathfrak{A}^{\mathcal{W}}$, where $\mathcal{W}$ is a superset family over A. Then, $\nabla u \forall z\, F \to \forall z \nabla u\, F$ and $\exists u \nabla z\, F \to \nabla z \exists u\, F$ hold on $\mathfrak{A}^{\mathcal{W}}$.

Proof. Both assertions follow from $(\to \nabla)$.

We now consider a case where the behavior of the new quantifiers contrasts with that of the classical ones. The familiar quantifiers commute: we have the equivalences $\forall u \forall z\, F \leftrightarrow \forall z \forall u\, F$ and $\exists u \exists z\, F \leftrightarrow \exists z \exists u\, F$. What about ∇: do we have the equivalence between $\nabla u \nabla z\, F$ and $\nabla z \nabla u\, F$?

We have a counter-example for this commutativity as well as for the converse transfer principles.

LEMMA 4.1. Over the universe of naturals assertion, consider the family $\mathcal{W}$ of important sets to have no finite set.[39] Consider the modulated structure $\mathfrak{N}^{\mathcal{W}}$, where $\mathfrak{N}$ is the structure of the naturals with the usual order $<$. Let F be the formula $u < z$. Then, on $\mathfrak{N}^{\mathcal{W}}$, none of the the formulas holds:

$$\forall u \nabla z\, F \to \nabla z \forall u\, F, \quad \nabla u \exists z\, F \to \exists z \nabla u\, F, \quad \nabla u \nabla z\, F \to \nabla z \nabla u\, F.$$

Proof. For each $n \in \mathbb{N}$, the set $\{ m \in \mathbb{N} \,/\, m < n \}$ is finite.

We now examine the expressive power of the new quantifiers. One feels that assertions such as "Eagles generally fly" and "Penguins rarely fly" cannot be expressed by purely first-order means. Now, consider assertion "Objects generally are equal to c". where c is a given object. It holds exactly when the singleton $\{c\}$ is important. Can one express this assertion by first-order means?

We now see that the new quantifiers actually increase the expressive power.

PROPOSITION 4.1. Consider the sentence $\nabla u \exists z\, u \doteq z$[40], where u and z are distinct variables. Then, there exists no first-order sentence F, such that the equivalence $\nabla u \exists z\, u \doteq z \leftrightarrow F$ holds in every modulated structure.

Proof. For an ultrafilter $\mathcal{W}$, $\nabla u \exists z\, u \doteq z$ holds in $\mathfrak{A}^{\mathcal{W}}$ iff $\mathcal{W}$ is principal.[41]

4.2 Logical Axioms for 'Generally' and 'Rarely'

We now examine axiomatizations for 'generally' and 'rarely'.[42]

We wish to examine consequence. Some examples in Section 2 illustrate how some inferences are sanctioned, or not, by properties of the family $\mathcal{W}$ of important sets. The next example extends Example 4 (in 2.1).

EXAMPLE 12. Consider the formulas $\forall v\, (F \leftrightarrow G)$, $\nabla v\, F$ and $\nabla v\, G$. Given a modulated structure $\mathfrak{A}^{\mathcal{W}}$ where both $\forall v\, (F \leftrightarrow G)$ and $\nabla v\, F$ hold, if $\mathcal{W}$ is a superset family, then $\nabla v\, G$ also holds in $\mathfrak{A}^{\mathcal{W}}$. Thus, $\nabla v\, G$ holds in every modulated structure with a superset family of important sets where both $\nabla v\, F$ and $\forall v\, (F \leftrightarrow G)$ hold.

---

[39] This is the so-called Fréchet filter, which is an ultrafilter.

[40] We use $\doteq$ for equality.

[41] A principal filter is generated by a singleton and there exist principal and non-principal ultrafilters over each infinite universe.

[42] Note that 'rarely' is definable from 'generally' in view of $[\nabla \neg \triangle]$ (in 4.1).

We wish to to examine such consequences relying on properties of the family of important sets, i. e. for some classes of families of important sets. In addition to the class $\mathbb{B}$ of unrestricted families of sets, examples of such classes are:

Class	Family $\mathcal{K}$ of sets	Properties
$\mathbb{P}$	proper	$V \in \mathcal{K}$ and $\emptyset \notin \mathcal{K}$
$\mathbb{S}$	superset	$Y \supset X \in \mathcal{K} \Rightarrow Y \in \mathcal{K}$
$\mathbb{I}$	intersection-closed	$X, Y \in \mathcal{K} \Rightarrow X \cap Y \in \mathcal{K}$
$\mathbb{R}$	repelling	$X \in \mathcal{K} \Rightarrow \overline{X} \notin \mathcal{K}$
$\mathbb{A}$	attracting	$X \notin \mathcal{K} \Rightarrow \overline{X} \in \mathcal{K}$

Other familiar such classes are the following classes:

1. $\mathbb{L}$ of proper lattices (proper families closed under union and intersection),

2. $\mathbb{F}$ of proper filters (superset intersection-closed proper families),

3. $\mathbb{U}$ of proper ultrafilters (attracting proper filters).

Consider a set Γ of formulas. We extend satisfaction to set Γ of formulas: $\mathfrak{A}^{\mathcal{W}} \models \Gamma \llbracket s \rrbracket$ iff $\mathfrak{A}^{\mathcal{W}} \models G \llbracket s \rrbracket$, for every $G \in \Gamma$.[43] Now, consider a class $\mathbb{S}$ of families of sets. We say that formula F is a $\mathbb{S}$ *consequence* of set Γ of formulas (noted $\Gamma \models^{\mathbb{K}} F$) iff $\mathfrak{A}^{\mathcal{W}} \models F \llbracket s \rrbracket$ whenever $\mathfrak{A}^{\mathcal{W}} \models \Gamma \llbracket s \rrbracket$ with $\mathcal{K} \in \mathbb{K}$.

We begin with axiomatic calculi for 'generally'.

Our semantics is extensional (cf. Section 2). So, a set is important or not regardless of how it is expressed. Thus, we have basic extensionality principles.

- We have an extensionality axiom (similar to $(\rightarrow \nabla)$ in 4.1):

  $[\leftrightarrow \nabla] \; \forall v \, (F \leftrightarrow G) \rightarrow (\nabla v \, F \rightarrow \nabla v \, G)$.

- We also have an extensionality axiom concerning variable substitution:

  $[\nabla^u] \; \nabla v \, F \rightarrow \nabla u \, F^{[v/u]}$, if variable u does not occur bound in F.[44]

These two axioms give rise to the corresponding axiom schemas $\{\leftrightarrow \nabla\}$ and $\{\nabla^\sigma\}$, which give the basic set $\mathsf{Bsc} := \{\leftrightarrow \nabla\} \cup \{\nabla^\sigma\}$.

The basic set Bsc gives a sound and complete axiomatization for unrestricted consequence: over the unrestricted class $\mathbb{B}$ of families of sets. Given a set $\Gamma \cup \{F\}$ of formulas, formula F is a $\mathbb{B}$ consequence of Γ ($\Gamma \models^{\mathbb{B}} F$) iff F is a first-order consequence of the set $\Gamma \cup \{\leftrightarrow \nabla\}$ ($\Gamma \cup \{\leftrightarrow \nabla\} \vdash F$).

We have seen other axioms corresponding to properties of families of important sets (in 4.1), namely those shown in Table 7.

These axioms give rise to axiom schemas that can be added to the basic set Bsc to axiomatize consequence over the corresponding classes of families. For unrestricted families, however, one can only have general consequences in the presence of some general hypothesis.

EXAMPLE 13. Consider consistent theories with information about which metals are solid under ordinary conditions.

---

[43] So, satisfaction of a set of formulas amounts to satisfying every one of its formulas.
[44] For the result of replacing v by u in F(v) we use $F(v)^{[v/u]}$ or simply $F(u)$.

$|\forall\nabla| : \forall v\, P \to \nabla v\, P$ $|\nabla\exists| : \nabla v\, P \to \exists v\, P$
$(\to \nabla) : \forall v\, (F \to G) \to (\nabla v\, F \to \nabla v\, G)$ $(\wedge\nabla) : (\nabla v\, P \wedge \nabla v\, Q) \to \nabla v\, (P \wedge$
$(\nabla\neg) : \nabla v\, \neg P \to \neg \nabla v\, P$ $(\neg\nabla) : \neg \nabla v\, P \to \nabla v\, \neg P$

Table 7. Axioms for families of important sets

1. First, consider a purely first-order theory Δ, with two axioms expressing "Mercury is not solid" and "Every metal, other than mercury, is solid". In this case, we cannot decide whether "metals generally are solid".

2. Now, consider a consistent theory G extending Δ with the genera "metals generally are distinct from mercury". Then, one concludes that "metals generally are solid".

The above modulated consequences are conservative extensions of first-order consequence: given a set $\Delta \cup \{G\}$ of first-order formulas, formula G is a **U** consequence of Δ ($\Delta \models^{\mathbf{U}} G$) iff G is a first-order consequence of Δ ($\Delta \vdash G$). For proper families, some general consquences are trivialized: given a set $\Delta \cup \{G\}$ of first-order formulas, formula $\nabla v\, G$ is a $\mathbb{P}$ consequence of Δ ($\Delta \models^{\mathbb{P}} \nabla v\, G$) iff $\forall v\, G$ is a first-order consequence of Δ ($\Delta \vdash \forall v\, G$).

4.3 Logical Calculi for 'Generally' and 'Rarely'

We now examine some logical calculi for 'generally' and 'rarely'.

The axioms of Table 7 (in 4.2) can be expressed as inference rules. It is, however, more convenient to use a slightly different approach via generic object: the generic object has a property P exactly when P holds generally.

We introduce a new symbol _ for the *generic object* and *marked formulas* as the result of replacing a variable by the generic object _. A marked formula has the form $\langle F(\_)\rangle$ and is meant to represent the formula $\nabla v\, F(v)$, with '$\langle$' and '$\rangle$' indicating that $F(v)$ is the scope of ∇.

We can now introduce natural deduction systems for 'generally'. [9] We will employ familiar terminology and notation. [4], [6], [8].

We start with a natural deduction system for first-order logic and we extend to it so as to cover new formulas: with quantifier ∇ and marked ones. The expected structure of a derivation is a first-order logic derivation with local manipulations of ∇ and marked formulas (see Examples 14 and 15).

We have *introduction* and *elimination rules* for ∇ (to leave and enter an environment for manipulating 'generally') as follows:

$$(\nabla E)\ \frac{\nabla v\, F(v)}{\langle F(\_)\rangle} \qquad (\nabla I)\ \frac{\langle F(\_)\rangle}{\nabla z\, F(z)}\ \text{for } z \text{ not occurring in } F(\_)$$

We also have a transformation rule for marked formulas, which corresponds to the extensionality axiom $[\leftrightarrow \nabla]$. The *equivalence rule* ($\updownarrow$) is as follows (where variable v does not occur free in $(\Gamma \cup \Delta)$:

$$(\updownarrow) \quad \cfrac{\begin{array}{c}\Gamma, [F(v)]^i \\ \vdots \\ G(v)\end{array} \quad \langle F(\_)\rangle \quad \begin{array}{c}\Delta, [G(v)]^i \\ \vdots \\ F(v)\end{array}}{\langle G(\_)\rangle} \, i$$

In this rule ($\updownarrow$), the marked formula $\langle F(\_)\rangle$ is called its *major premise* and the discharged unmarked formulas $F(v)$ and $G(v)$ are called its *minor premises*.[45]

One can reduce consecutive applications of ($\updownarrow$) to a single one.

The *basic system* has these three rules: (∇E), (∇I) and ($\updownarrow$).

EXAMPLE 14. In the basic system, we have a derivation of $\nabla x \nabla y\, G(x,y)$ from $\forall x \forall y (F(x,y) \leftrightarrow G(x,y))$ and $\nabla x \nabla y\, F(x,y)$. Using H for $\forall x \forall y (F(x,y) \leftrightarrow G(x,y))$, we can construct such a derivation as follows.

1. We have derivations Σ', from H and $F(x,y)$ to $G(x,y)$, and Σ'', from H and $G(x,y)$ to $F(x,y)$, both using ($\forall$E) and ($\leftrightarrow$E).

2. With these derivations Σ' and Σ'', we can construct a derivation Π', from H and $\nabla y\, F(x,y)$ to $\nabla y\, G(x,y)$, using (∇E), ($\updownarrow$) and (∇I), as follows:

$$(\nabla \text{I}) \cfrac{(\updownarrow) \cfrac{\begin{array}{c}\text{H}, [F(x,y)]^1 \\ \Sigma' \\ G(x,y)\end{array} \quad (\nabla\text{E})\cfrac{\nabla y\, F(x,y)}{F(x,\_)} \quad \begin{array}{c}\text{H}, [G(x,y)]^1 \\ \Sigma'' \\ F(x,y)\end{array}}{G(x,\_)} \, 1}{\nabla y\, G(x,y)}$$

Derivation Π' has 2 lateral tracks: from H to $G(x,y)$ and from H to $F(x,y)$.

3. Similarly, we can construct a derivation Π'', from H and $\nabla y\, G(x,y)$ to $\nabla y\, F(x,y)$, using the derivations Σ' and Σ''.

4. With these derivations Π' and Π'', we can construct a derivation from H and $\nabla x \nabla y\, F(x,y)$ to $\nabla x \nabla y\, G(x,y)$, using ($\nabla$E), ($\updownarrow$) and ($\nabla$I), as follows:

$$(\nabla \text{I}) \cfrac{(\updownarrow) \cfrac{\begin{array}{c}\text{H}, [\nabla y\, F(x,y)]^3 \\ \Pi' \\ \nabla y\, G(x,y)\end{array} \quad (\nabla\text{E})\cfrac{\nabla x \nabla y\, F(x,y)}{\nabla y\, F(\_,y)} \quad \begin{array}{c}\text{H}, [\nabla y\, G(x,y)]^3 \\ \Pi'' \\ \nabla y\, F(x,y)\end{array}}{\nabla y\, G(\_,y)} \, 3}{\nabla x \nabla y\, G(x,y)}$$

The main track of this derivation goes from $\nabla x \nabla y\, F(x,y)$ to $\nabla x \nabla y\, G(x,y)$.

A derivation in the basic system has tracks as follows:

1. applications of elimination rules, followed by

---

[45] This rule may be said to eliminate the marked premise and introduce the marked conclusion.

2. at most one application of the equivalence rule, followed by

3. applications of introduction rules.

So, each track has at most 2 minimal formulas. Thus, the basic system has the sub-formula property: every formula occurring in a derivation is a sub-formula of the hypotheses or of the conclusion.

We can set up systems for properties of families.[46]

Introduction	Elimination
$(\underline{\top}\text{I}) \; \dfrac{}{\langle \top \rangle}$	$(\underline{\bot}\text{E}) \; \dfrac{\langle \bot \rangle}{\bot}$
$(\underline{\wedge}\text{I}) \; \dfrac{\langle F(\_) \rangle \quad \langle G(\_) \rangle}{\langle F(\_) \wedge G(\_) \rangle}$	$(\underline{\wedge}\text{E}) \; \dfrac{\langle F_1(\_) \wedge F_2(\_) \rangle}{\langle F_i(\_) \rangle}$
$(\underline{\vee}\text{I}) \; \dfrac{\langle F(\_) \rangle \quad \langle G(\_) \rangle}{\langle F(\_) \vee G(\_) \rangle}$	$(\underline{\vee}\text{E}) \; \dfrac{\langle F(\_) \vee G(\_) \rangle \quad \begin{array}{c} \Gamma, [\langle F(\_) \rangle]^i \\ \vdots \\ H \end{array} \quad \begin{array}{c} \Delta, [\langle G(\_) \rangle]^i \\ \vdots \\ H \end{array}}{H} \; i$
$(\underline{\neg}\text{I}) \; \dfrac{\begin{array}{c} \Gamma, [\langle H(\_) \rangle]^i \\ \vdots \\ \bot \end{array}}{\langle \neg H(\_) \rangle} \; i$	$(\underline{\neg}\text{E}) \; \dfrac{\langle \neg H(\_) \rangle \quad \langle H(\_) \rangle}{\bot}$

These rules can be added to basic system to provide natural deduction systems for consequence over corresponding classes of families.

EXAMPLE 15. We have a derivation Π of $\langle Q(\_) \rangle$ from $\forall v\,(P(v) \to Q(v))$ and $\langle P(\_) \rangle$, using ($\updownarrow$) and ($\underline{\wedge}$E). This derivation can be constructed as follows (where derivation Σ is as expected):

$$(\underline{\wedge}\text{E}) \; \dfrac{(\updownarrow) \; \dfrac{\begin{array}{c} [P(v)]^1, \forall v(P(v) \to Q(v)) \\ \Sigma \\ P(v) \wedge Q(v) \end{array}}{\langle P(\_) \wedge Q(\_) \rangle} \quad \langle P(\_) \rangle \quad (\underline{\wedge}\text{E}) \; \dfrac{[P(v) \wedge Q(v)]^1}{P(v)} \; 1}{\langle Q(\_) \rangle}$$

The main track of derivation Π goes from $\langle P(\_) \rangle$ to $\langle Q(\_) \rangle$, using ($\updownarrow$) and ($\underline{\wedge}$E).

We thus have a derivation of $\nabla v\,Q(v)$ from $\forall v\,(P(v) \to Q(v))$ and $\nabla v\,P(v)$,

---
[46]We use $\top$ to abbreviate $\bot \to \bot$.

using (∇E), $(\updownarrow)$, $(\wedge E)$ and (∇I), as follows:

$$(\nabla I) \dfrac{\forall v(P(v) \to Q(v)) \qquad (\nabla E)\dfrac{\nabla v P(v)}{\langle P(\_)\rangle}}{\dfrac{\begin{array}{c}\Pi\\\langle Q(\_)\rangle\end{array}}{\nabla v Q(v)}}$$

For proper families, the system has properties similar to the basic one. Derivations in other systems, however, may have tracks with several (non consecutive) applications of $(\updownarrow)$, which cannot be reduced to a single one.

EXAMPLE 16. The usual way of establishing that a set is in a filter is by showing that it includes the intersection of sets in the filter. So, we wish to derive $\langle H(\_)\rangle$ from $\langle F(\_)\rangle$, $\langle G(\_)\rangle$ and $\forall v\,[(F(v) \wedge G(v)) \to H(v)]$. We can do this as follows.

1. First, we obtain derivation Π' from derivation Π in Example 15, by replacing $P(v)$ by $F(v) \wedge G(v)$ and $Q(v)$ by $H(v)$.

2. Then, we construct the following derivation:

$$\forall v\,[(F(v)\wedge G(v)) \to H(v)] \qquad (\wedge I)\dfrac{\langle F(\_)\rangle \quad \langle G(\_)\rangle}{\langle F(\_)\wedge G(\_)\rangle}$$
$$\dfrac{\Pi'}{\langle H(\_)\rangle}$$

This derivation has a single application of $(\updownarrow)$. It has main tracks from $\langle F(\_)\rangle$ and $\langle G(\_)\rangle$ to $\langle H(\_)\rangle$, both using the rules $(\wedge I)$, $(\updownarrow)$ and $(\wedge E)$ in this order.

Example 16 illustrates a feature of systems having none of the rules $(\vee E)$, $(\neg E)$ and $(\neg I)$. This is the case of systems for some families of important sets, like (proper) superset families, lattices and filters. In such systems, one can reduce every derivation to a special form, having at most one application of $(\updownarrow)$. The marked part of a track will then have one of the following forms.

- No application of $(\updownarrow)$ and marked eliminations followed by marked introductions: $(E)^*\,\langle H\rangle\,(I)^*$, with single minimal formula $\langle H\rangle$.

- A single application of $(\updownarrow)$ after marked eliminations and before marked introductions: $(E)^*\,\langle F\rangle\,(\updownarrow)\,\langle G\rangle\,(I)^*$, with 2 minimal formulas $\langle F\rangle$ and $\langle G\rangle$.

- A single application of $(\updownarrow)$ after marked introductions and before marked eliminations: $(I)^*\,\langle P\rangle\,(\updownarrow)\,\langle Q\rangle\,(E)^*$, with 2 maximal formulas $\langle P\rangle$ and $\langle Q\rangle$.

We can also introduce sequent calculi for 'generally'. [10]

We start with a sequent calculus for first-order logic and we extend it so as to cover new formulas: with quantifier ∇ and marked ones. Instead of introduction and elimination rules, we now have introduction on the right or on the left of a sequent.

The rules for ∇ are as follows (for a variable v not occurring in F(_)):

$$[\nabla \Rightarrow] \frac{\langle F(\_)\rangle, \Gamma \Rightarrow \Delta}{\nabla v\, F(v)\rangle, \Gamma \Rightarrow \Delta} \qquad [\Rightarrow \nabla] \frac{\Gamma \Rightarrow \Delta, \langle F(\_)\rangle}{\Gamma \Rightarrow \Delta, \nabla v\, F(v)}$$

The *equivalence rule* $[\updownarrow]$ is as follows (where v does not occur free $\Gamma \cup \Delta$):

$$[\updownarrow] \frac{F(v), \Gamma \Rightarrow \Delta, G(v) \qquad G(v), \Gamma \Rightarrow \Delta, F(v)}{\langle F(\_)\rangle, \Gamma \Rightarrow \Delta, \langle G(\_)\rangle}$$

One can reduce consecutive applications of $[\updownarrow]$ to a single one.

The *basic calculus* has these three rules: $[\nabla \Rightarrow]$, $[\Rightarrow \nabla]$ and $[\updownarrow]$.

EXAMPLE 17. We can derive the sequent $\nabla v\, F(v), \forall v(F(v) \leftrightarrow G(v)) \Rightarrow \nabla v\, G(v)$ in the basic calculus as follows (where Π' and Π'' are first-order derivations):

$$[\Rightarrow \nabla] \frac{[\nabla \Rightarrow] \frac{[\updownarrow] \frac{\Pi' \qquad \Pi'}{\langle F(\_)\rangle, \forall v(F(v) \leftrightarrow G(v)) \Rightarrow G(v) \qquad G(v), \forall v(F(v) \leftrightarrow G(v)) \Rightarrow F(v)}}{\nabla v\, F(v), \forall v(F(v) \leftrightarrow G(v)) \Rightarrow G(v) \Rightarrow \langle G(\_)\rangle}}{\nabla v\, F(v), \forall v(F(v) \leftrightarrow G(v)) \Rightarrow \nabla v\, G(v)}$$

One can eliminate the cut rule form derivations in the basic calculus. Thus, the basic calculus has the sub-formula property: every formula occurring in a derivation is a sub-formula of the formulas in the final sequent.

One can also set up sequent calculi for properties of families.[47]

Right		Left	
$[\Rightarrow \langle\,\rangle]$	$\dfrac{\Gamma \Rightarrow \Delta, H(v)}{\Gamma \Rightarrow \Delta, \langle H(\_)\rangle}$	$[\langle\,\rangle \Rightarrow]$	$\dfrac{H(v), \Gamma \Rightarrow \Delta}{\langle H(\_)\rangle, \Gamma \Rightarrow \Delta}$
$[\Rightarrow \wedge]$	$\dfrac{\Gamma \Rightarrow \Delta, \langle F(\_)\rangle \quad \Gamma \Rightarrow \Delta, \langle G(\_)\rangle}{\Gamma \Rightarrow \Delta, \langle F(\_) \wedge G(\_)\rangle}$	$[\wedge \Rightarrow]$	$\dfrac{\langle F_i(\_)\rangle\rangle, \Gamma \Rightarrow \Delta}{\langle F_1(\_) \wedge F_2(\_)\rangle\rangle, \Gamma \Rightarrow \Delta}$
$[\Rightarrow \vee]$	$\dfrac{\Gamma \Rightarrow \Delta, \langle F_i(\_)\rangle}{\Gamma \Rightarrow \Delta, \langle F_1(\_) \vee F_2(\_)\rangle}$	$[\vee \Rightarrow]$	$\dfrac{\langle F(\_)\rangle, \Gamma \Rightarrow \Delta \quad \langle G(\_)\rangle, \Gamma \Rightarrow \Delta}{\langle F(\_) \vee G(\_)\rangle, \Gamma \Rightarrow \Delta}$
$[\Rightarrow \neg]$	$\dfrac{\Gamma \Rightarrow \Delta, \langle H(\_)\rangle}{\langle \neg H(\_)\rangle, \Gamma \Rightarrow \Delta}$	$[\neg \Rightarrow]$	$\dfrac{\langle H(\_)\rangle, \Gamma \Rightarrow \Delta}{\Gamma \Rightarrow \Delta, \langle \neg H(\_)\rangle}$

5 Concluding Remarks

We now present some concluding comments.

Assertions and arguments involving some vague notions occur often, both in ordinary language and in some branches of science. The vagueness is given by "modifiers", such as 'generally' and 'rarely'. A precise treatment of such ideas has been a basic motivation underlying logics for qualitative reasoning [11],

---

[47] In rules $[\Rightarrow \langle\,\rangle]$ and $[\langle\,\rangle \Rightarrow]$, variable v should not occur free in Γ or Δ.

where several aspects have been developed. Some issues, however, may be left somewhat unclear, leading to misunderstandings. Also, some objections, such as lack of intuitive justification, have been raised.

We have revisited and discussed some issues concerning logics for 'generally' and 'rarely', with the aim of clarifying the approach, its underlying motivations, intuitions and justification.

In Section 2, we have examined some intuitions underlying notions of 'generally' and 'rarely', with the aim of indicating why some accounts are not sufficient and explaining how several accounts can be subsumed under an abstract one (in 2.4). In Section 3, we have indicated how one can arrive naturally at families of sets, with certain properties, as capturing some notions of 'generally' and 'rarely'. In Section 4, we have examined some logical aspects of 'generally' and 'rarely', like oppositions, expressive and deductive powers ((in 4.1), as well as axioms (in 4.2); we have also briefly commented on some recent developments concerning natural deduction systems and sequent calculi (in 4.3).

Thus, the abstract approach to 'generally' and 'rarely', via families of sets appears to be well motivated and justified. It leads to logical systems with convenient properties, like extending conservatively their first-order counterparts.

BIBLIOGRAPHY

[1] J. L. Bell and A. B. Slomson. *Models and Ultraproducts: an introduction.* North-Holland, 1971 (2nd rev. pr.).
[2] Collins Cobuild. *Collins Cobuild English Language Dictionary.* Collins, 1987.
[3] H. B. Enderton. *A Mathematical Introduction to Logic.* Academic Press, 1972.
[4] G. Gentzen. *Investigations into logical deduction.* In M. E. Szabo, editor, *The Collected Papers of Gerhard Gentzen.* North-Holland, 1969.
[5] J. L. Kelley. *General Topology.* D. van Nostrand, 1955.
[6] D. Prawitz. *Natural Deduction: a proof-theoretical study.* Almquist & Wiksdell, 1965.
[7] R. Solomon - Ordered groups: a case study in reverse mathematics. *Bulletin of Symbolic Logic* 5(1), pages 45–58, 1999.
[8] D. van Dalen. *Logic and Structure.* Springer-Verlag, 1989 (2. edn, 3. prt.).
[9] L. B. Vana, P. A. S. Veloso, Veloso and S. R. M. Veloso. Natural deduction for 'generally'. *Logic Journal of the IGPL* 15 (5), pages 775–800, 2007.
[10] L. B. Vana, P. A. S. Veloso, Veloso and S. R. M. Veloso. Sequent calculi for 'generally'. *Electronic Notes in Theoretical Computer Science* 205, pages 49–65, 2008.
[11] P. A. S. Veloso and W. A Carnielli. Logics for qualitative reasoning. In S. Rahman et al., editors, *Logic, Epistemology, and the Unity of Science*, pages 487–526, Kluwer Press, 2004.
[12] N. Webster. *Webster's Seventh New Collegiate Dictionary.* Merriam Co., 1970.

Paulo A. S. Veloso
Departamento de Ciência da Computação
Universidade Federal do Rio de Janeiro
Rio de Janeiro, Brazil
E-mail: srmv@bridge.com.br

Sheila R. M. Veloso
Departamento de Engenharia de Sistemas de Computação
Universidade do Estado do Rio de Janeiro
Rio de Janeiro, Brazil
E-mail: sheila@cos.ufrj.br

Rational measure of rational simplexes
DANIELE MUNDICI

dedicated to Walter A. Carnielli on his 60th birthday

ABSTRACT. The rational m-dimensional measure of a rational m-simplex T in $\mathbb{R}^n$ is proportional to the m-dimensional Hausdorff measure of T, where the constant of proportionality κ only depends on the affine hull of T. We give an explicit formula for κ. We survey recent applications of the rational measure to the theory of conditionals for the continuous events described by formulas in infinite-valued Łukasiewicz logic.

1 Introduction

For all unexplained notions and notation we refer to [Mundici, 2011b]. Let P be a *rational polyhedron* in $\mathbb{R}^n$, i.e., a finite union of rational simplexes in $\mathbb{R}^n$. For any triangulation $\mathcal{T}$ of P we denote by $\mathcal{T}^{\max}$ the set of maximal simplexes in $\mathcal{T}$. For all $i = 0, 1, 2, \ldots$ we let $P^{(i)} = \bigcup \{T \in \mathcal{T}^{\max} \mid \dim(T) = i\}$, and we say that $P^{(i)}$ is the *i-dimensional part of P*. $P^{(i)}$ is a (possibly empty) polyhedron and does not depend on the triangulation $\mathcal{T}$ of P. For every regular simplex $S \subseteq \mathbb{R}^n$ the *denominator* $\den(S)$ of S is defined by $\den(S) = \den(v_0) \cdots \den(v_m)$. As usual, $\den(v_i)$ denotes the least common denominator of the coordinates of v_i.

THEOREM 1. *For $i = 0, 1, \ldots$ and Δ a regular triangulation of a rational polyhedron $P \subseteq \mathbb{R}^n$, let $\lambda^{(i)}_\Delta(P) = \sum \left\{ \frac{1}{i!\,\den(S)} \mid S \in \Delta^{\max},\ \dim(S) = i \right\}$, where the sum equals zero if there are no maximal i-simplexes in Δ. Then $\lambda^{(i)}_\Delta(P) = \lambda^{(i)}_\nabla(P)$ for any regular triangulation ∇ of P.*

Proof. [Mundici, 2008, Theorem 2.1(i)] . ∎

Accordingly, we will write $\lambda^{(i)}(P)$ instead of $\lambda^{(i)}_\Delta(P)$, and call $\lambda^{(i)}(P)$ the *i-dimensional rational measure of P*.

In [Mundici, 2011b, §3] and [Mundici, 2011a] one can find an account of the deep relationship between rational polyhedra and finitely axiomatizable theories in Łukasiewicz logic. In the final section of this paper the rational measure will be shown to have a crucial role in defining invariant states and invariant conditionals in this logic.

2 A characterization of the rational measure

We assume familiarity with the basic properties of Hausdorff measure, and in particular, with the corollary of the isodiametric inequality [Evans and Gariepy,

1992, 2.2], [Federer, 1969, 2.10.33] stating that for any rational polyhedron P the d-dimensional Hausdorff measure of $P^{(d)}$ agrees with the Lebesgue d-dimensional measure of any congruent copy P' of $P^{(d)}$ lying on $\mathbb{R}^d$.

$\mathcal{L}^{(n)}$ and $\mathcal{H}^{(n)}$ respectively denote n-dimensional Lebesgue and Hausdorff measure

---

By a *flat* A in $\mathbb{R}^n$ we mean a rational affine subspace of $\mathbb{R}^n$, i.e., a finite intersection A of rational hyperplanes in $\mathbb{R}^n$. If the origin belongs to A we say that A is *homogeneous*.

We will write $\mathcal{P}^{(n)}$ for the set of all rational polyhedra in $\mathbb{R}^n$.

We let $\mathcal{G}_n = GL(n, \mathbb{Z}) \ltimes \mathbb{Z}^n$ denote the group of transformations of the form

$$x \mapsto Ax + t \quad \text{for} \quad x \in \mathbb{R}^n,$$

where $t \in \mathbb{Z}^n$ and A is an $n \times n$ matrix with integer elements and determinant ± 1. Any such transformation preserves the lattice $\mathbb{Z}^n$ of integer points in $\mathbb{R}^n$.

To appreciate the crucial role of $\mathcal{G}_n$ in Lukasiewicz logic, suffice to say that $\mathbb{Z}$-homeomorphism is a dual notion of isomorphism for finitely presented MV-algebras, [Mundici, 2011b, §3]. In particular, two rational polyhedra $P, Q \subseteq \mathbb{R}^n$ are $\mathbb{Z}$-homeomorphic iff there is a piecewise linear homeomorphism η of P onto Q such that every linear piece of both η and η^{-1} is extendible to a member of $\mathcal{G}_n$.

The rational measure has the following characterization:

THEOREM 2. *For each $n = 1, 2, \ldots$ and $d = 0, 1, \ldots$, the map $\lambda^{(d)} \colon \mathcal{P}^{(n)} \to \mathbb{R}_{\geq 0}$ has the following properties, for all $P, Q \in \mathcal{P}^{(n)}$:*

(i) (Invariance) *If $P = \gamma(Q)$ for some $\gamma \in \mathcal{G}_n$ then $\lambda^{(d)}(P) = \lambda^{(d)}(Q)$.*

(ii) (Valuation) *$\lambda^{(d)}(\emptyset) = 0$, $\lambda^{(d)}(P) = \lambda^{(d)}(P^{(d)})$, and the restriction of $\lambda^{(d)}$ to the set of all rational polyhedra P, Q in $\mathbb{R}^n$ having dimension at most d is a* valuation: *in other words,*

$$\lambda^{(d)}(P) + \lambda^{(d)}(Q) = \lambda^{(d)}(P \cup Q) + \lambda^{(d)}(P \cap Q). \tag{1}$$

(iii) (Conservativity) *For any $P \in \mathcal{P}^{(n)}$ let $(P, 0) = \{(x, 0) \in \mathbb{R}^{n+1} \mid x \in P\}$. Then $\lambda^{(d)}(P) = \lambda^{(d)}(P, 0)$.*

(iv) (Pyramid) *For $k = 1, \ldots, n$, if $\operatorname{conv}(v_0, \ldots, v_k)$ is a regular k-simplex in $\mathbb{R}^n$ with $v_0 \in \mathbb{Z}^n$ then*

$$\lambda^{(k)}(\operatorname{conv}(v_0, \ldots, v_k)) = \lambda^{(k-1)}(\operatorname{conv}(v_1, \ldots, v_k))/k. \tag{2}$$

(v) (Normalization) Let $j = 1, \ldots, n$. Suppose the set $B = \{w_1, \ldots, w_j\} \subseteq \mathbb{Z}^n$ is part of a basis of the free abelian group $\mathbb{Z}^n$. Let the closed parallelepiped $P_B \subseteq \mathbb{R}^n$ be defined by

$$P_B = \left\{ x \in \mathbb{R}^n \mid x = \sum_{i=1}^{j} \gamma_i w_i, \ 0 \leq \gamma_i \leq 1 \right\}. \tag{3}$$

Then $\lambda^{(j)}(P_B) = 1$.

(vi) (Proportionality) Let A be an m-dimensional rational affine subspace of $\mathbb{R}^n$ for some $m = 0, \ldots, n$. Then there is a constant $\kappa_A > 0$, only depending on A, such that $\lambda^{(m)}(Q) = \kappa_A \cdot \mathcal{H}^{(m)}(Q)$ for every rational m-simplex $Q \subseteq A$.

Proof. [Mundici, 2011, 4.2]. ∎

THEOREM 3. *Properties (i)-(vi) in Theorem 2 uniquely characterize the rational measures* $\lambda^{(0)}, \ldots, \lambda^{(n)}$, *for each* $n = 1, 2, \ldots$, *among all maps from* $\mathcal{P}^{(n)}$ *to* $\mathbb{R}_{\geq 0}$.

Proof. [Mundici, 2011, 8.2]. ∎

3 The value of κ_A

Fix $e = 0, \ldots, n$ and let F be an e-dimensional flat in $\mathbb{R}^n$. Let $u_0, \ldots, u_e$ be the vertices of a regular e-simplex T in F. Following [Mundici, 2011b, §2.1] and [Mundici, 2008, §2], for any $y = (y_1, \ldots, y_n) \in \mathbb{Q}^n$ let the vector $\tilde{y} \in \mathbb{Z}^{n+1}$ be defined by

$$\tilde{y} = \text{den}(y)(y, 1) = (\text{den}(y) \cdot y_1, \ldots, \text{den}(y) \cdot y_n, \text{den}(y)).$$

Finally, let the integer $d \geq 1$ be defined by

$$d = d_F = \min\{q \in \mathbb{Z} \mid q \text{ is the denominator of a rational point of } F\}. \tag{4}$$

DEFINITION 4. As usual, [Lekkerkerker, 1969], for any k-dimensional sublattice Λ of $\mathbb{Z}^n$, the *determinant* $\det(\Lambda)$ of Λ is the k-dimensional volume of a fundamental region for Λ in the k-dimensional rational linear subspace spanned by Λ.

Suppose we have an e-dimensional flat $E = E_0 + r$, where E_0 is a homogeneous flat and r is a rational translation in $\mathbb{R}^n$.

Let $d = d_E$ be the smallest denominator of a rational point of E. By [Mundici, 2011, 8.1] we have rational points $a_0, \ldots, a_e \in E$, all with the same denominator $d = d_E$, which are the vertices of a regular e-simplex $W = \text{conv}(a_0, \ldots, a_e) \subseteq E$. Thus the vectors $\tilde{a}_0, \ldots, \tilde{a}_e \in \mathbb{Z}^{n+1}$ are a basis of the lattice $L \cap \mathbb{Z}^{n+1}$ of integer points in the linear span L of the $\tilde{a}_i$ in $\mathbb{R}^{n+1}$. L is an $(e+1)$-dimensional (homogeneous) linear subspace of $\mathbb{R}^{n+1}$.

The rational measure of W is immediately shown to satisfy $\lambda^{(e)}(W) = \frac{1}{e! \, d^{e+1}}$ because $\dim(W) = e$ and W has $e+1$ vertices all of denominator d.

To compute the Hausdorff e-dimensional measure of W, let the e-simplex $W' = (W, 1) \subseteq \mathbb{R}^{n+1}$ be obtained by vertically lifting W to the hyperplane $x_{n+1} = 1$. Let dW' be the e-simplex in $\mathbb{R}^{n+1}$ obtained by multiplying by d each vector of W'. All vertices of dW' lie at the same height d in $\mathbb{R}^{n+1}$. Clearly, $dW' = \operatorname{conv}(\tilde{a}_0, ..., \tilde{a}_e)$. We have the identity $\mathcal{H}^{(e)}(W) = \mathcal{H}^{(e)}(dW')/d^e = \mathcal{H}^{(e)}(\operatorname{conv}(\tilde{a}_0, ..., \tilde{a}_e))/d^e$. Translating the e-simplex $\operatorname{conv}(\tilde{a}_0, ..., \tilde{a}_e) \subseteq \mathbb{R}^{n+1}$ by a shift of $-\tilde{a}_0$, since Hausdorff measure is translation invariant, we have $\mathcal{H}^{(e)}(W) = \mathcal{H}^{(e)}(\operatorname{conv}(0, \tilde{a}_1 - \tilde{a}_0, ..., \tilde{a}_e - \tilde{a}_0))/d^e$. On the hyperplane $x_{n+1} = 0$ of $\mathbb{R}^{n+1}$ we now have an e-simplex with integer vertices $0, \tilde{a}_1 - \tilde{a}_0, ..., \tilde{a}_e - \tilde{a}_0$. Let us write w_i for the vector in $\mathbb{R}^n$ obtained by forgetting the last (zero) coordinate of $\tilde{a}_e - \tilde{a}_0$. We then have points $0, w_1, ..., w_e \in \mathbb{Z}^n \cap E_0 \subseteq \mathbb{R}^n$. Evidently, $\mathcal{H}^{(e)}(W) = \mathcal{H}^{(e)}(\operatorname{conv}(0, w_1, ..., w_e))/d^e$. Let $\mathcal{P}(w_1, ..., w_e)$ be the parallelopiped spanned by the vectors $w_1, ..., w_e$ in $\mathbb{R}^n$. The well known relation between the volumes of parallelopiped and simplex yields now

$$\mathcal{H}^{(e)}(W) = \frac{\mathcal{H}^{(e)}(\mathcal{P}(w_1, ..., w_e))}{d^e \, e!}$$

Note that the flat E_0 coincides with E translated by $-a_0$. Further, $w_1, ..., w_e$ are a basis of the lattice $E_0 \cap \mathbb{Z}^n$. Therefore, by Definition 4,

$$\mathcal{H}^{(e)}(W) = \frac{\det(E_0 \cap \mathbb{Z}^n)}{d^e \, e!},$$

whence

$$k_E = \frac{\lambda^{(e)}(W)}{\mathcal{H}^{(e)}(W)} = \frac{\frac{1}{e! \, d^{e+1}}}{\frac{\det(E_0 \cap \mathbb{Z}^n)}{d^e \, e!}} = \frac{1}{d \times \det(E_0 \cap \mathbb{Z}^n)}.$$

We have proved

THEOREM 5. *Let $E = E_0 + t$ be an e-dimensional flat in $\mathbb{R}^n$, $e = 0, 1, \ldots, n$, with E_0 homogeneous and $t \in \mathbb{Q}^n$. Then the ratio k_E between the e-dimensional rational measure and the e-dimensional Hausdorff measure of any e-simplex T in E is equal to $(d_E \times \det(E_0 \cap \mathbb{Z}^n))^{-1}$.*

If the lattice $E_0 \cap \mathbb{Z}^n$ is equipped with a basis $b_1, \ldots, b_e$, where each b_i is an integer vector in $\mathbb{R}^n$, then letting A be the matrix with the coordinates of b_i in the ith row, we have the identity

$$\det(E_0 \cap \mathbb{Z}^n) = (AA')^{1/2},$$

where A' is the transpose of A. See, e.g., [Birkhoff and Mac Lane, 1953, Theorem 7, p.308].

4 Applications to conditionals in Łukasiewicz logic

By definition, a *state* σ of an MV-algebra A is a map $\sigma \colon A \to [0, 1]$ such that $\sigma(1) = 1$ and $\sigma(x \oplus y) = \sigma(x) + \sigma(y)$ whenever $x \odot y = 0$. Every state yields a procedure to compute the "average truth-value" of the formulas of Łukasiewicz propositional logic, in the precise sense of de Finetti's analysis of probability in terms of coherent bookmaking. See [Kühr and Mundici, 2007, 3.2], [Mundici, 2009] and [Mundici, 2011a, 4.1] for the extension of de Finetti's probability theory to events described by formulas in Łukasiewicz logic. See [Fedel et al., 2011] for a further extension to non-reversible bookmaking.

Valuations in Łukasiewicz logic coincide with extreme states, i.e., extreme (de Finetti) coherent probability assessments on the continuous-spectrum events described by formulas in Łukasiewicz logic, [Mundici, 2011a, §7 (iii)]. Using the above invariant measures and the categorical duality [Marra and Spada, to appear], [Mundici, 2011b] between rational polyhedra and finitely presented MV-algebras one can show ([Mundici, 2008, 4.1]) that *every finitely presented MV-algebra has an invariant state σ such that $\ker(\sigma) = \{0\}$*; This latter property is referred to as the *faithfulness* of σ.

Among all invariant measures of rational polyhedra, the measures $\lambda^{(i)}$ allow the introduction of MV-algebraic Lebesgue integration theory on finitely presented MV-algebras, [Manara et al., 2007], [Marra, to appear], [Marra, 2007], [Panti, 2009]. This is the main ingredient of the theory of invariant conditionals in Łukasiewicz logic.

Let F_m be the set of formulas in the variables $X_1, \ldots, X_m$. A *conditional* is a map $\mathsf{P} \colon \theta \mapsto \mathsf{P}_\theta$ such that, for every $m = 1, 2, \ldots$ and every satisfiable formula $\theta \in \mathsf{F}_m$, P_θ is a state of the Lindenbaum MV-algebra $\mathsf{L}(\{\theta\})$ of θ. We say that P is *invariant* if for any two satisfiable formulas $\phi \in \mathsf{F}_m$, $\psi \in \mathsf{F}_n$, and isomorphism η of $\mathsf{L}(\{\phi\})$ onto $\mathsf{L}(\{\psi\})$, we have $\mathsf{P}_\phi = \mathsf{P}_\psi \circ \eta$, where $\circ$ denotes composition. P is said to be *faithful* if so is every state P_θ.

For every formula $\psi \in \mathsf{F}_m$, $\psi/\equiv_{\{\psi\}}$ denotes the the image of ψ in $\mathsf{L}(\{\psi\})$. For simplicity we will write $\mathsf{P}_\theta(\psi)$ instead of $\mathsf{P}_\theta(\frac{\psi}{\equiv_{\{\theta\}}})$, and say that $\mathsf{P}_\theta(\psi)$ is *the probability of ψ given θ*.

A formula α is *P-independent of* (a satisfiable formula) θ if the probability of α given θ coincides with the unconditional probability of α. In other words, $\mathsf{P}_\theta(\alpha) = \mathsf{P}_{\theta \leftrightarrow \theta}(\alpha) = \mathsf{P}_{\alpha \leftrightarrow \alpha}(\alpha)$.

THEOREM 6. *Łukasiewicz logic L_∞ has a faithful invariant conditional P^* satisfying the following two conditions:*

(i) (Substitutivity) *For any formula ψ and satisfiable formula θ, and variable X not occurring in θ and ψ, $\mathsf{P}^*_\theta(\psi) = \mathsf{P}^*_{\theta \odot (\psi \leftrightarrow X)}(X)$.*

(ii) (Independence) *If α and θ are two formulas in disjoint sets of variables $\{Y_1, \ldots, Y_m\}$ and $\{Z_1, \ldots, Z_n\}$, and θ is satisfiable, then α is P^* independent of θ.*

Proof. [Mundici, 2006], [Mundici, 2009]. ∎

THEOREM 7. *Further, Łukasiewicz logic has an invariant conditional P with the following properties:*

(i) *For any formula ψ and satisfiable formula θ, and variable X not occurring in θ and ψ, $\mathsf{P}_\theta(\psi) = \mathsf{P}_{\theta \odot (\psi \leftrightarrow X)}(X)$.*

(ii) If α and θ are two formulas in disjoint sets of variables $\{Y_1,\ldots,Y_m\}$ and $\{Z_1,\ldots,Z_n\}$, and θ is satisfiable, then α is P-independent of θ.

(iii) (Rényi's multiplication law): $\mathsf{P}_\theta((\phi\odot\psi)^\infty) = \mathsf{P}_{\psi\odot\theta}(\phi^\infty)\cdot\mathsf{P}_\theta(\psi^\infty)$, whenever $\psi\odot\theta$ is satisfiable, where $\mathsf{P}_\beta(\alpha^\infty\mid\beta) = \lim_{t\to\infty}\mathsf{P}_\beta(\alpha^t\mid\beta)$ and $\alpha^t = \underbrace{\alpha\odot\ldots\odot\alpha}_{t\text{ occurrences of }\alpha}$

Proof. [Mundici, 2011b, §15]. ∎

BIBLIOGRAPHY

[Birkhoff and Mac Lane, 1953] G.Birkhoff, S. Mac Lane, *A survey of modern algebra*, The Macmillan Company, New York, Revised Edition, 1953.

[Evans and Gariepy, 1992] L. C. Evans, R. F. Gariepy, *Measure theory and fine properties of functions*, CRC Press, Boca Raton, FL, 1992.

[Fedel et al., 2011] M.Fedel, K.Keimel, F. Montagna, W. Roth, Imprecise probabilities, bets and functional analytic methods in Łukasiewicz logic, *Forum Mathematicum*, 2011 doi 10.1515/FORM.2011.123

[Federer, 1969] H. Federer, *Geometric measure theory*, Springer, New York, 1969.

[Kühr and Mundici, 2007] J. Kühr, D.Mundici, De Finetti theorem and Borel states in [0,1]-valued algebraic logic, *International Journal of Approximate Reasoning*, 46:605-616, 2007.

[Lekkerkerker, 1969] C.G. Lekkerkerker, *Geometry of numbers*, Wolters-Noordhoff, Groningen and North-Holland, Amsterdam, 1969.

[Manara et al., 2007] C.Manara, V. Marra, D.Mundici, Lattice-ordered abelian groups and Schauder bases of unimodular fans, *Transactions of the American Mathematical Society*, 359:1593-1604, 2007.

[Marra, to appear] V. Marra, Lattice-ordered abelian groups and Schauder bases of unimodular fans, II, *Transactions of the American Mathematical Society*, to appear.

[Marra, 2007] V. Marra, D.Mundici, The Lebesgue state of a unital abelian ℓ-group, *Journal of Group Theory*, 10:655-684, 2007.

[Marra and Spada, to appear] V. Marra, L. Spada, The dual adjunction between MV-algebras and Tychonoff spaces, *Studia Logica*, special issue in memoriam Leo Esakia. To appear.

[Mundici, 2008] D.Mundici, The Haar theorem for lattice-ordered abelian groups with order-unit, *Discrete and continuous dynamical systems*, 21:537-549, 2008.

[Mundici, 2006] D.Mundici, Faithful and invariant conditional probability in Łukasiewicz logic. In: D. Makinson, et al. (Eds.), Proceedings of the conference *Trends in Logic IV*, Torun, Poland, 2006, Trends in Logic, vol. 28. Springer, New York, (2008) 213-232.

[Mundici, 2009] D.Mundici, Conditionals and independence in many-valued logics, In: Proceedings 10th European Conference on Symbolic and Quantitative Approaches to Reasoning with Uncertainty ECSQARU09, Verona, Italy, 2009. C. Sossai and G. Chemello (Eds.), Lecture Notes in Artificial Intelligence, 5590:16-21, 2009.

[Mundici, 2009] D.Mundici, Interpretation of de Finetti coherence criterion in Łukasiewicz logic, *Annals of Pure and Applied Logic*, 161:235-245, 2009.

[Mundici, 2011a] D. Mundici, Finite axiomatiability in Łukasiewicz logic, *Annals of Pure and Applied Logic*, 162:1035-1047, 2011.

[Mundici, 2011b] D.Mundici, *Advanced Łukasiewicz calculus and MV-algebras*, Trends in Logic, vol. 35, Springer, Berlin, 2011.

[Mundici, 2011] D.Mundici, Measure theory in the geometry of $GL(n,\mathbb{Z})\ltimes\mathbb{Z}^n$, arxiv 1102.0897v1, 2011.

[Panti, 2009] G. Panti, Invariant measures in free MV-algebras, *Communications in Algebra*, 36: 2849-2861, 2009.

Daniele Mundici
Department of Mathematics "Ulisse Dini"
University of Florence
Florence, Italy
E-mail: mundici@math.unifi.it

Principal and Boolean congruences on θ-valued Łukasiewicz–Moisil algebras

ALDO V. FIGALLO, INÉS PASCUAL AND ALICIA ZILIANI

ABSTRACT. The first system of many–valued logic was introduced by J. Łukasiewicz, his motivation was of philosophical nature as he was looking for an interpretation of the concepts of possibility and necessity. Since then, plenty of research has been developed in this area. In 1968, when Gr.C. Moisil came across Zadeh's fuzzy set theory he found the motivation he had been looking for in order to legitimate the introduction and study of infinitely–valued Łukasiewicz algebras, so he defined θ–valued Łukasiewicz algebras (or LM_θ–algebras, for short) (without negation), where θ is the order type of a chain.

In this article, our main interest is to investigate the principal and Boolean congruences and θ–congruences on LM_θ–algebras. In order to do this we take into account a topological duality for these algebras obtained in (A.V. Figallo, I. Pascual, A. Ziliani, *A Duality for θ– Valued Łukasiewicz-Moisil Algebras and Applications*. J. Mult-Valued Logic & Soft Computing. Vol. 16, pp 303-322. (2010). Furthermore, we prove that the intersection of two principal θ–congruences is a principal one. On the other hand, we show that Boolean congruences are both principal congruences and θ–congruences. This allowed us to establish necessary and sufficient conditions so that a principal congruence is a Boolean one. Finally, bearing in mind the above results, we characterize the principal and Boolean congruences on n–valued Łukasiewicz algebras without negation when we consider them as LM_θ–algebras in the case that θ is an integer n, $n \geq 2$.

1 Introduction

In 1940, Gr.C. Moisil defined the 3–valued and the 4–valued Łukasiewicz algebras and in 1942, the n–valued Łukasiewicz algebras ($n \geq 2$). His goal was to study Łukasiewicz's logic from the algebraic point of view. It is well-known that these algebras are not the algebraic counterpart of n–valued Łukasiewicz propositional calculi for $n \geq 5$ ([Boicescu et al., 1991; Cignoli, 1970]). R. Cignoli ([Cignoli, 1982]) found algebraic counterparts for $n \geq 5$ and he called them proper n–valued Łukasiewicz algebras. On the other hand, in 1968, Gr.C. Moisil ([Moisil, 1972]) introduced θ–valued Łukasiewicz algebras (without negation), where θ is the order type of a chain. These structures were thought by Moisil as models of a logic with infinity nuances, but the need to find a strong motivation for them delayed the announcement. The motivation was found when Moisil came across Zadeth's fuzzy set theory, in which he saw a confirmation of his old ideas.

For a general account of the origins and the theory of Lukasiewicz many valued logics and Lukasiewicz algebras the reader is referred to [Boicescu et al., 1991; Cignoli et al., 2000; Iorgulescu, 1984].

This paper is organized as follows: in Section 2, we summarize the principal notions and results of θ−valued Lukasiewicz algebras ([Boicescu et al., 1991]), in particular the topological duality for these algebras obtained in [Figallo et al., 2010]. In Section 3, we characterize the open subsets of the dual space associated with an LM_θ−algebra which determine both the principal LM_θ and θLM_θ−congruences. This last result enables us to prove that the intersection of two principal θLM_θ−congruences is a principal one. Furthermore, whenever θ is an integer n, $n \geq 2$, we obtain the filters which determine principal congruences on n−valued Lukasiewicz algebras (or LM_n−algebras) and, we are also in a position to show that the intersection of two principal LM_n−congruences is a principal one. In Section 4, attention is focused on Boolean congruences. Firstly, we characterize them by means of certain closed and open subsets of the associated space. These results allow us to prove that the boolean LM_θ and θLM_θ−congruences coincide, and also that they are principal congruences associated with filters generated by Boolean elements of these algebras.

2 Preliminaries

In this paper, we take for granted the concepts and results on distributive lattices, universal algebra, Lukasiewicz algebras and Priestley duality. To obtain more information on these topics, we direct the reader to the bibliography indicated in [Balbes and Dwinger, 1974; Boicescu, 1984; Boicescu et al., 1991; Burris and Sankappanavar, 1981; Priestley, 1970; Priestley, 1972; Priestley, 1984]. However, in order to simplify the reading, we will summarize the main notions and results we needed throughout this paper.

In what follows, if X is a partially ordered set and $Y \subseteq X$ we will denote by $[Y)$ $((Y])$ the set of all $x \in X$ such that $y \leq x$ $(x \leq y)$ for some $y \in Y$, and we will say that Y is increasing (decreasing) if $Y = [Y)$ $(Y = (Y])$. In particular, we will write $[y)$ $((y])$ instead of $[\{y\})$ $((\{y\}])$. Furthermore, if $x, y \in X$ and $x \leq y$, a segment is the set $\{z \in X : x \leq z \leq y\}$ which will be denoted by $[x, y]$.

Let $\theta \geq 2$ be the order type of a totally ordered set J with least element 0 being $J = \{0\} + I$ (ordinal sum). Following V. Boisescu et al ([Boicescu et al., 1991]) recall that:

A θ−valued Lukasiewicz–Moisil algebra (or LM_θ−algebra) is an algebra $\langle A, \vee, \wedge, 0, 1, \{\phi_i\}_{i \in I}, \{\overline{\phi}_i\}_{i \in I}\rangle$ of type $(2, 2, 0, 0, \{1\}_{i \in I}, \{1\}_{i \in I})$ where $\langle A, \vee, \wedge, 0, 1\rangle$ is a bounded distributive lattice and for all $i \in I$, ϕ_i and $\overline{\phi}_i$ satisfy the following conditions:

(L1) ϕ_i is an endomorphism of bounded distributive lattices,

(L2) $\phi_i x \vee \overline{\phi}_i x = 1$, $\phi_i x \wedge \overline{\phi}_i x = 0$,

(L3) $\phi_i \phi_j x = \phi_j x$,

(L4) $i \leqslant j$ implies $\phi_i x \leqslant \phi_j x$,

(L5) $\phi_i x = \phi_i y$ for all $i \in I$ imply $x = y$.

It is well known that there are LM_θ–congruences (or congruences) on LM_θ–algebras such that the quotient algebra doesn't satisfy the determination principle (L5). That is the reason why a new notion was defined as follows: a θLM_θ–congruence (or θ–congruence) on an LM_θ–algebra is a bounded distributive lattice congruence ϑ such that $(x, y) \in \vartheta$ if and only if $(\phi_i x, \phi_i y) \in \vartheta$ for all $i \in I$. ([Boicescu, 1984; Iorgulescu, 1984]).

The following characterization of the Boolean elements of an LM_θ–algebra will be useful for the study of the Boolean congruences on these algebras:

(L6) Let A be an LM_θ–algebra and let $C(A)$ be the set of all Boolean elements of A. Then, for each $x \in A$, the following conditions are equivalent:

   (i) $x \in C(A)$,
   (ii) there are $y \in A$ and $i \in I$ such that $x = \phi_i y$ ($x = \overline{\phi}_i y$),
   (iii) there is $i_0 \in I$ such that $x = \phi_{i_0} x$ ($x = \overline{\phi}_{i_0} x$),
   (iv) for all $i \in I$, $x = \phi_i x$ ($x = \overline{\phi}_i x$).

In [Figallo et al., 2010], we extended Priestley duality to LM_θ–algebras considering θ–valued Łukasiewicz–Moisil spaces (or $l_\theta P$-spaces) and $l_\theta P$–functions. More precisely,

A θ–valued Łukasiewicz–Moisil space (or $l_\theta P$–space) is a pair $(X, \{f_i\}_{i \in I})$ provided the following conditions are satisfied:

(lP1) X is a Priestley space, ([Priestley, 1970; Priestley, 1972; Priestley, 1984])

(lP2) $f_i : X \to X$ is a continuous function,

(lP3) $x \leqslant y$ implies $f_i(x) = f_i(y)$,

(lP4) $i \leqslant j$ implies $f_i(x) \leqslant f_j(x)$,

(lP5) $f_i \circ f_j = f_i$.

(lP6) $\bigcup_{i \in I} f_i(X)$ is dense in X (i.e. $\overline{\bigcup_{i \in I} f_i(X)} = X$, where $\overline{Z}$ denotes the closure of Z).

An $l_\theta P$– function from an $l_\theta P$– space $(X, \{f_i\}_{i \in I})$ into another $(X', \{f'_i\}_{i \in I})$ is an increasing continuous function f from X into X' satisfying $f'_i \circ f = f \circ f_i$ for all $i \in I$.

It is worth mentioning that condition (lP6) is equivalent to the following one:

(lP7) If U and V are increasing closed and open subsets of X and $f_i^{-1}(U) = f_i^{-1}(V)$ for all $i \in I$, then $U = V$ ([Figallo et al., 2010]).

Besides, if $(X, \{f_i\}_{i \in I})$ is an $l_\theta P$-space, then for all $x \in X$, the following properties are satisfied ([Figallo et al., 2010]):

(lP8) $x \leqslant f_i(x)$ or $f_i(x) \leqslant x$ for all $i \in I$,

(lP9) $f_0(x) \leq x$ and $f_0(x)$ is the unique minimal element in X that precedes x,

(lP10) $x \leq f_1(x)$ and $f_1(x)$ is the unique maximal element in X that follows x.

Furthermore, the above properties allow us to assert that

(lP11) X is the cardinal sum of the sets $[\{f_i(x)\}_{i \in I}] \cup (\{f_i(x)\}_{i \in I}]$ for $x \in X$. If I has least element 0 and greatest element 1, then X is the cardinal sum of the sets $[f_0(x), f_1(x)]$ for $x \in X$.

Although in [Figallo et al., 2010] we developed a topological duality for LM_θ-algebras, next we will describe some results of it with the aim of fixing the notation we are about to use in this paper.

(A1) If $(X, \{f_i\}_{i \in I})$ is an $l_\theta P$-space and $D(X)$ is the lattice of all increasing closed and open subset of X, then $\mathbb{L}_\theta = \langle D(X), \cup, \cap, \emptyset, X, \{\phi_i^X\}_{i \in I}, \{\overline{\phi}_i^X\}_{i \in I} \rangle$ is an LM_θ-algebra, where the operations ϕ_i^X and $\overline{\phi}_i^X$ are defined by means of the formulas: $\phi_i^X(U) = f_i^{-1}(U)$ and $\overline{\phi}_i^X(U) = X \setminus f_i^{-1}(U)$ for all $U \in D(X)$ and for all $i \in I$.

(A2) If $\langle A, \vee, \wedge, 0, 1, \{\phi_i\}_{i \in I}, \{\overline{\phi}_i\}_{i \in I} \rangle$ is an LM_θ-algebra and $X(A)$ is the set of all prime filters of A, ordered by inclusion and with the topology having as a sub-basis the sets $\sigma_A(a) = \{P \in X(A) : a \in P\}$ and $X(A) \setminus \sigma_A(a)$ for each $a \in A$, then $L_\theta = (X(A), \{f_i^A\}_{i \in I})$ is the $l_\theta P$-space associated with A, where the functions $f_i^A : X(A) \longrightarrow X(A)$ are defined by the prescription: $f_i^A(P) = \phi_i^{-1}(P)$ for all $i \in I$ and for all $P \in X(A)$.

Then the category of $l_\theta P$-spaces and $l_\theta P$-functions is naturally equivalent to the dual of the category of LM_θ-algebras and their corresponding homomorphism, where the isomorphisms σ_A and ϵ_X are the corresponding natural equivalences.

In addition, this duality allowed us to to characterize the LM_θ and θLM_θ-congruences on these algebras for which we introduced these notions:

(A3) A subset Y of X is semimodal if $\bigcup\limits_{i \in I} f_i(Y) \subseteq Y$, and Y is a θ-subset of X if $\bigcup\limits_{i \in I} f_i(Y) \subseteq Y \subseteq \overline{\bigcup\limits_{i \in I} f_i(Y)}$. Hence, for each closed θ-subset Y of X we have that $Y = \overline{\bigcup\limits_{i \in I} f_i(Y)}$ ([Figallo et al., 2010]).

Then, we proved that

(A4) The lattice $\mathcal{C}_S(\mathbf{L}_\theta(A))$ of all closed and semimodal subsets of $\mathbf{L}_\theta(A)$ is isomorphic to the dual lattice $Con_{LM_\theta}(A)$ of all congruences on A, and the isomorphism is the function Θ_S defined by the prescription

$$\Theta_S(Y) = \{(a,b) \in A \times A : \sigma_A(a) \cap Y = \sigma_A(b) \cap Y\}$$

$$= \{(a,b) \in A \times A : \sigma_A(b) \triangle \sigma_A(a) \cap Y = \emptyset\} \ ([\text{Figallo et al., 2010,}$$
Theorem 2.1.1]).

(A5) The lattice $\mathcal{C}_\theta(\mathbf{L}_\theta(A))$ of all closed θ−subsets of $\mathbf{L}_\theta(A)$ is isomorphic to the dual lattice $Con_{\theta LM_\theta}(A)$ of all θ−congruences on A, and the isomorphism is the function Θ_θ defined as in (A4) ([Figallo et al., 2010, Theorem 2.1.2]).

Finally, we will emphasize the following properties of Priestley spaces and so, $l_\theta P$−spaces which will be quite useful in order to characterize the principal congruences on LM_θ−algebras.

(A6) $[A)$ and $(A]$ are closed subsets of X, for each closed subset A of X.

(A7) R is a closed, open and convex subset of X if and only if there are $U, V \in D(X)$ such that $V \subseteq U$ and $R = U \setminus V$.

3 Principal congruences

In this section our first objective is to characterize the principal LM_θ and θLM_θ−congruences on an LM_θ−algebra by means of certain open subsets of its associated $l_\theta P$-space.

THEOREM 1. *Let A be an LM_θ−algebra and let $L_\theta(A)$ be the $l_\theta P$-space associated with A. Then, it holds:*

(i) *the lattice $\mathcal{O}_{CS}(X(A))$ of all open subsets of $X(A)$ whose complements are semimodal is isomorphic to the lattice $Con_{LM_\theta}(A)$ of all congruences on A, and the isomorphism is the function Θ_{OS} defined by the prescription $\Theta_{OS}(G) = \{(a,b) \in A \times A : \sigma_A(b) \triangle \sigma_A(a) \subseteq G\}$.*

(ii) *The lattice $\mathcal{O}_{C\theta}(X(A))$ of all open subsets of $X(A)$ whose complements are θ−subsets of $X(A)$ is isomorphic to the lattice $Con_{\theta LM_\theta}(A)$ of all θ−congruences on A and the isomorphism is the function $\Theta_{O\theta}$ defined as in (i).*

Proof. It is a direct consequence of (A4) and (A5), bearing in mind that there is a one-to-one correspondence between the closed and open subsets of a topological space and that $\Theta_{OS}(G) = \Theta_S(X(A)\setminus G)$ and $\Theta_{O\theta}(G) = \Theta_\theta(X(A)\setminus G)$. ∎

From now on, we will denote by $\Theta(a,b)$ and $\Theta_\theta(a,b)$ the principal LM_θ and θLM_θ−congruence generated by (a,b), respectively.

PROPOSITION 2. *Let A be an LM_θ−algebra and let $L_\theta(A)$ be the $l_\theta P$-space associated with A. Then the following conditions are equivalent for all $a, b \in A$, $a \leqslant b$:*

(i) $\Theta(a,b) = \Theta_{OS}(G)$ for some $G \in \mathcal{O}_{CS}(X(A))$,

(ii) G is the least element of $\mathcal{O}_{CS}(X(A))$, ordered by inclusion, which contains $\sigma_A(b) \setminus \sigma_A(a)$,

(iii) $G = (\sigma_A(b) \setminus \sigma_A(a)) \cup \bigcup_{i \in I} f_i^{A^{-1}}(\sigma_A(b) \setminus \sigma_A(a))$,

(iv) $G = (\sigma_A(b) \setminus \sigma_A(a)) \cup \bigcup_{i \in I} \sigma_A(\phi_i b \wedge \overline{\phi}_i a)$.

Proof. (i) $\Rightarrow$ (ii): By Theorem 1 we have that $\sigma_A(b) \setminus \sigma_A(a) \subseteq G$. On the other hand, suppose that there is $H \in \mathcal{O}_{CS}(X(A))$ such that $\sigma_A(b) \setminus \sigma_A(a) \subseteq H$. Then, $(a,b) \in \Theta_{OS}(H)$ and so, $\Theta_{OS}(G) \subseteq \Theta_{OS}(H)$. This assertion and Theorem 1 imply that $G \subseteq H$.

(ii) $\Rightarrow$ (i): From the hypothesis and Theorem 1 we conclude that $(a,b) \in \Theta_{OS}(G)$. Furthermore, if $\varphi \in Con_{LM_\theta}(A)$ and $(a,b) \in \varphi$, then by Theorem 1 we have that $\varphi = \Theta_{OS}(H)$ for some $H \in \mathcal{O}_{CS}(X(A))$. From these last assertions and the fact that $a \leq b$ we infer that $\sigma_A(b) \setminus \sigma_A(a) \subseteq H$ and so, by the hypothesis we conclude that $G \subseteq H$. This means that $\Theta_{OS}(G) \subseteq \Theta_{OS}(H) = \varphi$, which allows us to assert that $\Theta_{OS}(G) = \Theta(a,b)$.

(ii) $\Rightarrow$ (iii): From the hypothesis we have that $(\sigma_A(b) \setminus \sigma_A(a)) \cup \bigcup_{i \in I} f_i^{A^{-1}}(\sigma_A(b) \setminus \sigma_A(a)) \subseteq G$. Besides, from (lP2) and (lP5) we infer that $(\sigma_A(b) \setminus \sigma_A(a)) \cup \bigcup_{i \in I} f_i^{A^{-1}}(\sigma_A(b) \setminus \sigma_A(a)) \in \mathcal{O}_{CS}(X(A))$ and hence, by (ii) we conclude that $G = (\sigma_A(b) \setminus \sigma_A(a)) \cup \bigcup_{i \in I} f_i^{A^{-1}}(\sigma_A(b) \setminus \sigma_A(a))$.

(iii) $\Rightarrow$ (ii): From (lP2) and (lP5) we have that $G \in \mathcal{O}_{CS}(X(A))$. Suppose now, that there is $H \in \mathcal{O}_{CS}(X(A))$ and that $\sigma_A(b) \setminus \sigma_A(a) \subseteq H$. Since $X(A) \setminus H$ is semimodal, we infer that $f_i^{A^{-1}}(\sigma_A(b) \setminus \sigma_A(a)) \subseteq H$ for all $i \in I$ from which it follows that $G \subseteq H$.

(iii) $\Leftrightarrow$ (iv): It is a direct consequence from the fact that σ_A is an isomorphism. ∎

COROLLARY 3. *Let A be an LM_θ–algebra and let $L_\theta(A)$ be the $l_\theta P$-space associated with A. Then the following conditions are equivalent:*

(i) *$G \in \mathcal{O}_{CS}(X(A))$ and $\Theta_{OS}(G)$ is a principal congruence on A,*

(ii) *there is a closed, open and convex subset R of $X(A)$ such that $G = R \cup \bigcup_{i \in I} f_i^{A^{-1}}(R)$.*

Proof. It follows immediately from (A7) and Proposition 2 taking into account that σ_A is an LM_θ–isomorphism. ∎

Next, bearing in mind the above results we will obtain different descriptions of the elements of $\mathcal{O}_{C\theta}(X(A))$ by means of the duality which will be useful later on.

PROPOSITION 4. Let A be an LM_θ-algebra and let $L_\theta(A)$ be the $l_\theta P$-space associated with A. Then the following conditions are equivalent for all $a, b \in A$, $a \leqslant b$:

(i) $H \in \mathcal{O}_{C\theta}(X(A))$ and $\Theta_{O\theta}(H) = \Theta_\theta(a,b)$,

(ii) $H = X(A) \setminus \overline{\bigcup_{i \in I} f_i^A(X(A) \setminus G)}$, where $G = (\sigma_A(b) \setminus \sigma_A(a)) \cup \bigcup_{i \in I} f_i^{A^{-1}}(\sigma_A(b) \setminus \sigma_A(a))$,

(iii) H is the least element of $\mathcal{O}_{C\theta}(X(A))$, ordered by inclusion, which contains $\sigma_A(b) \setminus \sigma_A(a)$,

(iv) $H = X(A) \setminus \overline{\bigcup_{i \in I} f_i^A(\bigcap_{i \in I} f_i^{A^{-1}}(X(A) \setminus (\sigma_A(b) \setminus \sigma_A(a))))}$,

(v) $H = X(A) \setminus \overline{\bigcup_{i \in I} f_i^A(\bigcap_{i \in I} \sigma_A(\phi_i a \vee \overline{\phi}_i b))}$.

Proof. (i) $\Rightarrow$ (ii): Let us observe that $\Theta(a,b) \subseteq \Theta_\theta(a,b)$. Hence, by Theorem 1 we have that there are $G \in \mathcal{O}_{CS}(X(A))$ and $H \in \mathcal{O}_{C\theta}(X(A)$ such that $\Theta(a,b) = \Theta_{OS}(G)$, $\Theta_\theta(a,b) = \Theta_{O\theta}(H)$ and $G \subseteq H$. Then, by Proposition 2 we infer that $G = (\sigma_A(b) \setminus \sigma_A(a)) \cup \bigcup_{i \in I} f_i^{A^{-1}}(\sigma_A(b) \setminus \sigma_A(a))$. Besides, it holds that $\Theta_\theta(a,b)$ is the least θ-congruence on A which contains $\Theta(a,b)$. Hence, from these statements and Theorem 1 we have that $X(A) \setminus H$ is the greatest closed θ-subset of $X(A) \setminus G$. On the other hand, taking into account that $X(A) \setminus G$ is semimodal, we conclude that $\overline{\bigcup_{i \in I} f_i^A(X(A) \setminus G)}$ is the greatest closed θ-subset of $X(A) \setminus G$. Hence, from the last assertions we conclude that $X(A) \setminus H = \overline{\bigcup_{i \in I} f_i^A(X(A) \setminus G)}$ and so, $H = X(A) \setminus \overline{\bigcup_{i \in I} f_i^A(X(A) \setminus G)}$.

(ii) $\Rightarrow$ (iii): It is easy to check that $H \in \mathcal{O}_{C\theta}(X(A))$. Furthermore, since $X(A) \setminus G$ is semimodal, we have that $\overline{\bigcup_{i \in I} f_i^A(X(A) \setminus G)} \subseteq X(A) \setminus G$ and so, $G \subseteq H$. This condition and the hypothesis imply that $\sigma_A(b) \setminus \sigma_A(a) \subseteq H$. On the other hand, if $W \in \mathcal{O}_{C\theta}(X(A)$ and $\sigma_A(b) \setminus \sigma_A(a) \subseteq W$, then $H \subseteq W$. Indeed, since $X(A) \setminus W$ is semimodal, from Proposition 2 we have that $G \subseteq W$ which implies that $\overline{\bigcup_{i \in I} f_i^A(X(A) \setminus W)} \subseteq \overline{\bigcup_{i \in I} f_i^A(X(A) \setminus G)}$. This last statement and the fact that $X(A) \setminus W$ is a closed θ-subset of $X(A)$ we infer that $X(A) \setminus W \subseteq \overline{\bigcup_{i \in I} f_i^A(X(A) \setminus G)}$ which completes the proof.

(iii) $\Rightarrow$ (i): This follows using an analogous reasoning to the proof of (ii) $\Rightarrow$ (i) in Proposition 2.

(ii) $\Leftrightarrow$ (iv): We only prove that $f_i^A((X(A) \setminus (\sigma_A(b) \setminus \sigma_A(a))) \cap \bigcap_{i \in I} f_i^{A^{-1}}$
$(X(A) \setminus (\sigma_A(b) \setminus \sigma_A(a)))) = f_i^A(\bigcap_{i \in I} f_i^{A^{-1}}(X(A) \setminus (\sigma_A(b) \setminus \sigma_A(a))))$ for all $i \in I$,
from which the proof follows immediately. Indeed, suppose that $y = f_j(x)$ for some $j \in I$ and that $f_i(x) \notin \sigma_A(b) \setminus \sigma_A(a)$ for all $i \in I$. Then, by (lP5) we infer that $f_i(y) = f_i(x)$ and $f_i(y) \notin \sigma_A(b) \setminus \sigma_A(a)$ for all $i \in I$. From these statements we have that $y \in (X(A) \setminus (\sigma_A(b) \setminus \sigma_A(a))) \cap \bigcap_{i \in I} f_i^{A^{-1}}(X(A) \setminus (\sigma_A(b) \setminus \sigma_A(a)))$
and so, $y \in f_j^A((X(A) \setminus (\sigma_A(b) \setminus \sigma_A(a))) \cap \bigcap_{i \in I} f_i^{A^{-1}}(X(A) \setminus (\sigma_A(b) \setminus \sigma_A(a))))$.
Therefore, we conclude that $f_j^A(\bigcap_{i \in I} f_i^{A^{-1}}(X(A) \setminus (\sigma_A(b) \setminus \sigma_A(a)))) \subseteq f_j^A((X(A) \setminus (\sigma_A(b) \setminus \sigma_A(a))) \cap \bigcap_{i \in I} f_i^{A^{-1}}(X(A) \setminus (\sigma_A(b) \setminus \sigma_A(a))))$ for all $j \in I$. The other inclusion is obvious.

(iv) $\Leftrightarrow$ (v): It is routine. ∎

COROLLARY 5. *Let A be an LM_θ–algebra and let $L_\theta(A)$ be the $l_\theta P$-space associated with A. Then the following conditions are equivalent:*

(i) *$\Theta_{0\theta}(H)$ is a principal θ–congruence on A,*

(ii) *there is a closed, open and convex subset R of $X(A)$ such that*

$$H = X(A) \setminus \overline{\bigcup_{i \in I} f_i^A(\bigcap_{i \in I} f_i^{A^{-1}}(X(A) \setminus R))}.$$

Proof. It is an immediate consequence of Proposition 4 and (A7). ∎

COROLLARY 6. *Let A be an LM_θ–algebra. Then the intersection of two principal θ–congruences is a principal one.*

Proof. Let ϑ_1 and ϑ_2 be principal θ–congruences on A. Then, by Corollary 5, there are closed, open and convex subsets R_1 and R_2 of $X(A)$ such that $\vartheta_1 = \Theta_{0\theta}(H_1)$ and $\vartheta_2 = \Theta_{0\theta}(H_2)$, where $H_1 = X(A) \setminus \overline{\bigcup_{i \in I} f_i^A(\bigcap_{i \in I} f_i^{A^{-1}}(X(A) \setminus R_1))}$
and $H_2 = X(A) \setminus \overline{\bigcup_{i \in I} f_i^A(\bigcap_{i \in I} f_i^{A^{-1}}(X(A) \setminus R_2))}$. Bearing in mind Theorem 1 we infer that $\vartheta_1 \cap \vartheta_2 = \Theta_{0\theta}(H_1 \cap H_2)$. On the other hand, we have that
$$\begin{aligned} H_1 \cap H_2 &= X(A) \setminus \overline{\bigcup_{i \in I} f_i^A(\bigcap_{j \in I} f_j^{A^{-1}}(X(A) \setminus R_1)} \cup \overline{\bigcap_{k \in I} f_k^{A^{-1}}(X(A) \setminus R_2))} \\ &= X(A) \setminus \overline{\bigcup_{i \in I} f_i^A(\bigcap_{i \in I} (f_i^{A^{-1}}(X(A) \setminus R_1) \cup f_i^{A^{-1}}(X(A) \setminus R_2)))} \\ &= X(A) \setminus \overline{\bigcup_{i \in I} f_i^A(\bigcap_{i \in I} (f_i^{A^{-1}}(X(A) \setminus (R_1 \cap R_2)))}. \end{aligned}$$
From these last equalities and the fact that $R_1 \cap R_2$ is a closed, open and convex subset of $X(A)$ we conclude, by Corollary 5, that $\vartheta_1 \cap \vartheta_2$ is a principal θ–congruence on A. ∎

In the sequel, we will determine sufficient conditions for the intersection of two principal congruences is not a principal one, in the particular case that the l_θP-space associated with an LM$_\theta$–algebra is the cardinal sum of an arbitrary but not finite set of segments or by [Figallo et al., 2010, Corollary 2.1.5, Theorem 2.2.2] when this is isomorphic to a subdirect product of an arbitrary but not finite set of subalgebras of the LM$_\theta$–algebra $B_2^{[I]}$.

PROPOSITION 7. *Let $\varphi_1 = \Theta_{OS}(G_1)$ and $\varphi_2 = \Theta_{OS}(G_2)$ principal congruences on A where $G_j = R_j \cup \bigcup_{i \in I} f_i^{A^{-1}}(R_j)$ and R_j is a closed, open and convex subset of $X(A)$, $1 \le j \le 2$. If $(R_1 \setminus (R_2 \cup \bigcup_{i \in I} f_i^{A^{-1}}(R_1))) \cap \bigcup_{i \in I} f_i^{A^{-1}}(R_2)$ is a proper and dense subset of $R_1 \setminus (R_2 \cup \bigcup_{i \in I} f_i^{A^{-1}}(R_1))$, then $\varphi_1 \cap \varphi_2$ is not a principal congruence on A.*

Proof. By virtue of Theorem 1 we have that (1) $\varphi_1 \cap \varphi_2 = \Theta_{OS}(G_1 \cap G_2)$. Besides, from the hypothesis we infer that $G_1 \cap G_2$ is partitioned into mutually disjoint sets as follows (2) $G_1 \cap G_2 = (R_1 \cap R_2) \cup (((R_1 \setminus R_2) \setminus \bigcup_{i \in I} f_i^{A^{-1}}(R_1)) \cap \bigcup_{i \in I} f_i^{A^{-1}}(R_2)) \cup (((R_2 \setminus R_1) \setminus \bigcup_{i \in I} f_i^{A^{-1}}(R_2)) \cap \bigcup_{i \in I} f_i^{A^{-1}}(R_1)) \cup (\bigcup_{i \in I} f_i^{A^{-1}}(R_1) \cap \bigcup_{i \in I} f_i^{A^{-1}}(R_2))$. Suppose now, that $\varphi_1 \cap \varphi_2$ is a principal congruence on A. Then, from Corollary 3 there is a closed, open and convex subset R of $X(A)$ such that (3) $G_1 \cap G_2 = R \cup \bigcup_{i \in I} f_i^{A^{-1}}(R)$. By (2) and (3) we conclude that $((R_1 \setminus R_2) \setminus \bigcup_{i \in I} f_i^{A^{-1}}(R_1)) \cap \bigcup_{i \in I} f_i^{A^{-1}}(R_2) \subseteq R \cup \bigcup_{i \in I} f_i^{A^{-1}}(R)$ and $((R_1 \setminus R_2) \setminus \bigcup_{i \in I} f_i^{A^{-1}}(R_1)) \cap \bigcup_{i \in I} f_i^{A^{-1}}(R_2) \cap \bigcup_{i \in I} f_i^{A^{-1}}(R)) = \emptyset$, from which we get that (4) $((R_1 \setminus R_2) \setminus \bigcup_{i \in I} f_i^{A^{-1}}(R_1)) \cap \bigcup_{i \in I} f_i^{A^{-1}}(R_2) \subseteq R$. Therefore, we have that (4) $((R_1 \setminus R_2) \setminus \bigcup_{i \in I} f_i^{A^{-1}}(R_1)) \cap \bigcup_{i \in I} f_i^{A^{-1}}(R_2) \subseteq R \cap ((R_1 \setminus R_2) \setminus \bigcup_{i \in I} f_i^{A^{-1}}(R_1))$. From the hypothesis and the fact that $R \cap (R_1 \setminus R_2) \setminus \bigcup_{i \in I} f_i^{A^{-1}}(R_1))$ is a closed subset of $(R_1 \setminus R_2) \setminus \bigcup_{i \in I} f_i^{A^{-1}}(R_1)$ we conclude that $R \cap ((R_1 \setminus R_2) \setminus \bigcup_{i \in I} f_i^{A^{-1}}(R_1)) = (R_1 \setminus R_2) \setminus \bigcup_{i \in I} f_i^{A^{-1}}(R_1)$ and hence, (5) $(R_1 \setminus R_2) \setminus \bigcup_{i \in I} f_i^{A^{-1}}(R_1) \subseteq R$. Furthermore, from the hypothesis there is $x \in X(A)$ such that (6) $x \in (R_1 \setminus R_2) \setminus \bigcup_{i \in I} f_i^{A^{-1}}(R_1))$ and (7) $x \notin \bigcup_{i \in I} f_i^{A^{-1}}(R_2)$. Then, (5) and (6) imply that $x \in R$ and so, by (3) it follows that (8) $x \in G_1 \cap G_2$. On the other hand, by (6) we infer that $x \notin R_1 \cap R_2$, $x \notin \bigcup_{i \in I} f_i^{A^{-1}}(R_1) \cap \bigcup_{i \in I} f_i^{A^{-1}}(R_2)$ and $x \notin ((R_2 \setminus R_1) \setminus \bigcup_{i \in I} f^A{}_i^{-1}(R_2)) \cap \bigcup_{i \in I} f_i^{A^{-1}}(R_1)$. Besides, by (7) we have that $x \notin ((R_1 \setminus R_2) \setminus \bigcup_{i \in I} f_i^{A^{-1}}(R_1)) \cap \bigcup_{i \in I} f_i^{A^{-1}}(R_2)$. These las assertions and (2) allow us to conclude that $x \notin G_1 \cap G_2$, which contradicts (8). Therefore, $\varphi_1 \cap \varphi_2$ is not a principal congruence on A. ∎

Bearing in mind the above results, our next task is to characterize the principal congruences on n-valued Lukasiewicz algebras without negation (or LM_n-algebras) when we consider them as LM_θ-algebras in the case that θ is an integer n, $n \geq 2$. It is well-known that each congruence on an LM_n-algebra is a θ-congruence on this algebra. For this aim, first we will determine the following properties of the $l_n P$-spaces.

PROPOSITION 8. Let $(X, \{f_1, \ldots, f_{n-1}\})$ be an $l_n P$-space. Then condition (LP6) is equivalent to any of these conditions:

(l_nP6) $X = \bigcup_{i=1}^{n-1} f_i(X)$,

(l_nP7) if Y, Z are subsets of X and $f_i^{-1}(Y) = f_i^{-1}(Z)$ for all i, $1 \leq i \leq n-1$, then $Y = Z$,

(l_nP8) for each $x \in X$, there is i_0, $1 \leq i_0 \leq n-1$, such that $x = f_{i_0}(x)$.

Proof. (lP6) $\Leftrightarrow$ (l_nP6): By (lP1), X is a Hausdorff and compact space, from which it follows, by (lP2), that $f_i : X \longrightarrow X$ is a closed function for all i, $1 \leq i \leq n-1$. Therefore, $\bigcup_{i=1}^{n-1} f_i(X)$ is a closed subset of X and so, $\overline{\bigcup_{i=1}^{n-1} f_i(X)} = \bigcup_{i=1}^{n-1} f_i(X)$.

To prove the other equivalences is routine. ∎

COROLLARY 9. ([Figallo et al., 2004, Proposition 3.1]) Let $(X, \{f_1, \ldots, f_{n-1}\})$ be an $l_n P$-space. Then X is the cardinal sum of a family of chains, each of which has at most $n-1$ elements.

Proof. By (lP11), X is the cardinal sum of the sets $[\{f_i(x) : 1 \leq i \leq n-1\}) \cup (\{f_i(x) : 1 \leq i \leq n-1\}]$, $x \in X$. From (lP3), (lP4), (lP5) and (l_nP8) we infer that for each $x \in X$, the set $[\{f_i(x) : 1 \leq i \leq n-1\}) \cup (\{f_i(x) : 1 \leq i \leq n-1\}] = \{f_i(x) : 1 \leq i \leq n-1\}$ is a maximal chain in X and so, the proof is complete. ∎

Now we are going to introduce the notion of modal subset of an $l_\theta P$-space. These subsets play a fundamental role in the characterization of the congruences on Lk_n-algebras, as we will show next.

DEFINITION 10. Let $(X, \{f_i\})_{i \in I}$ be an $l_\theta P$-space. A subset Y of X is modal if $Y = f_i^{-1}(Y)$ for all $i \in I$.

In order to reach our goal we will show the following lemmas.

LEMMA 11. *Let $(X, \{f_1, \ldots, f_{n-1}\})$ be an $l_n P$-space and let Y be a non empty set of X. Then the following conditions are equivalent:*

(i) *Y is semimodal,*

(ii) *Y is modal,*

(iii) *Y is a cardinal sum of maximal chains in X.*

Proof. (i) $\Rightarrow$ (ii): Suppose that $f_i(z) \in Y$. By $(l_n P8)$ we have that $f_{i_0}(z) = z$ for some i_0, $1 \leq i_0 \leq n-1$. Then, from the hypothesis and (lP5) we infer that $z = f_{i_0}(f_i(z)) \in Y$. Therefore, $f_i^{-1}(Y) \subseteq Y$ for all i, $1 \leq i \leq n-1$. The other inclusion follows immediately.

(ii) $\Rightarrow$ (iii): For each $y \in Y$, let $C_y = \{f_i(y) : 1 \leq i \leq n\}$. Then, taking into account the proof of Corollary 9 and $(l_n P8)$ it follows that C_y is a maximal chain in X to which y belongs. Therefore, $Y \subseteq \bigcup_{y \in Y} C_y$. On the other hand, $\bigcup_{y \in Y} C_y = \bigcup_{i=1}^{n-1} f_i(Y)$ and since Y is modal, we conclude that $\bigcup_{y \in Y} C_y \subseteq Y$, which completes the proof.

(iii) $\Rightarrow$ (i): It follows from (lP5) and the fact that a subset C of X is a chain if and only if $C = \{f_i(x) : 1 \leq i \leq n-1\}$, for some $x \in X$. ∎

LEMMA 12. *Let $(X, \{f_1, \ldots, f_{n-1}\})$ be an $l_n P$-space and let Y be a modal subset of X. Then $X \setminus Y$ is modal.*

Proof. It is a direct consequence of Definition 10. ∎

THEOREM 13. *Let A be an LM_n-algebra and let $L(A)$ be the $l_n P$-space associated with A. Then, it holds:*

(ii) *the lattice $\mathcal{C}_M(X(A))$ of all closed and modal subsets of $X(A)$ is isomorphic to the dual lattice $Con_{Lk_n}(A)$ of all Lk_n-congruences on A and the isomorphism is the function $\Theta_M : \mathcal{O}_M(X(A)) \longrightarrow Con_{Lk_n}(A)$, defined as in (A4).*

(ii) *The lattice $\mathcal{O}_M(X(A))$ of all open and modal subsets of $X(A)$ is isomorphic to the lattice $Con_{Lk_n}(A)$ of all Lk_n-congruences on A and the isomorphism is the function $\Theta_{OM} : \mathcal{O}_M(X(A)) \longrightarrow Con_{Lk_n}(A)$, defined as in Theorem 1.*

Proof. (i): It follows from (A4) and Lemma 11.

(ii): It is a consequence of Theorem 1 and Lemmas 11 and 12. ∎

In the sequel, we take into account the well-known fact that Priestley duality provides an isomorphism between the lattices $\mathcal{F}(L)$ of all filters of a bounded distributive lattice L and that of $\mathcal{C}_I(X(L))$ of all closed and increasing subsets of $X(L)$. Under this isomorphism, any $F \in \mathcal{F}(L)$ corresponds to the increasing closed subset $Y_F = \bigcap \{\sigma_L(a) : a \in F\}$, and any $Y \in \mathcal{C}_I(X(L))$ corresponds to the filter $F_Y = \{a \in L : Y \subseteq \sigma_L(a)\}$, and $\Theta(F) = \Theta(Y_F)$ and $\Theta(Y) = \Theta(F_Y)$, where $\Theta(Y)$ is defined as in (A4) for all $Y \in \mathcal{C}_I(X(L))$ and $\Theta(F)$ is the congruence associated with F.

PROPOSITION 14. *Let A be an LM_n–algebra and let $L_n(A)$ be the l_nP-space associated with A. Then the following conditions are equivalent for all $a,b \in A$, $a \leqslant b$:*

(i) $\Theta_{OM}(G) = \Theta(a,b)$ and $G \in \mathcal{O}_M(X(A))$,

(ii) G is the least element of $\mathcal{O}_M(X(A))$, ordered by inclusion, which contains $\sigma_A(b) \setminus \sigma_A(a)$,

(iii) $G = \bigcup_{i=1}^{n-1} f_i^{A^{-1}}(\sigma_A(b) \setminus \sigma_A(a))$,

(iv) $G = \sigma_A(\bigvee_{j=1}^{n-1} (\phi_i b \wedge \overline{\phi}_i a))$,

(v) $\Theta_{OM}(G) = \Theta([\bigwedge_{i=1}^{n-1} (\overline{\phi}_i b \vee \phi_i a)))$.

Proof. (i) $\Leftrightarrow$ (ii): It follows from Theorem 13 using the same argument as in Proposition 2.

(ii) $\Leftrightarrow$ (iii): By $(l_n\mathrm{IP8})$ we infer that $\sigma_A(b) \setminus \sigma_A(a) \subseteq \bigcup_{i=1}^{n-1} f_i^{A^{-1}}(\sigma_A(b) \setminus \sigma_A(a))$. Furthermore, by (IP2) and (IP5) we have that $\bigcup_{i=1}^{n-1} f_i^{A^{-1}}(\sigma_A(b) \setminus \sigma_A(a)) \in \mathcal{O}_M(X(A))$. On the other hand, if $H \in \mathcal{O}_M(X(A))$ and $\sigma_A(b) \setminus \sigma_A(a) \subseteq H$, since H is modal, we conclude that $\bigcup_{i=1}^{n-1} f_i^{A^{-1}}(\sigma_A(b) \setminus \sigma_A(a)) \subseteq H$. Therefore, $G = \bigcup_{i=1}^{n-1} f_i^{A^{-1}}(\sigma_A(b) \setminus \sigma_A(a))$ if and only if G verifies (ii).

(iii) $\Leftrightarrow$ (iv): Taking into account that σ_A is an LM_n–isomorphism we have that $\bigcup_{i=1}^{n-1} f_i^{A^{-1}}(\sigma_A(b) \setminus \sigma_A(a)) = \bigcup_{i=1}^{n-1} (f_i^{A^{-1}}(\sigma_A(b)) \cap (X(A) \setminus f_i^{A^{-1}}(\sigma_A(a)))) = \bigcup_{i=1}^{n-1} (\phi_i^{X(A)} \sigma_A(b) \cap \overline{\phi}_i^{X(A)} \sigma_A(a)) = \bigcup_{i=1}^{n-1} (\sigma_A(\phi_i b) \cap \sigma_A(\overline{\phi}_i a)) = \bigcup_{i=1}^{n-1} \sigma_A(\phi_i b \wedge \overline{\phi}_i a) = \sigma_A(\bigvee_{j=1}^{n-1} (\phi_i b \wedge \overline{\phi}_i a))$, from which the proof is complete.

(iv) ⇒ (v): Bearing in mind that $\Theta_{OM}(G) = \Theta_M(X(A) \setminus G)$ and that $X(A) \setminus G = \sigma_A(\bigwedge_{i=1}^{n-1} (\overline{\phi}_i b \vee \phi_i a))$, by Theorem 13 we infer that $\Theta_{OM}(G) = \Theta([\bigwedge_{i=1}^{n-1} (\overline{\phi}_i b \vee \phi_i a)))$.

(v) ⇒ (iv): Since $\sigma_A(\bigwedge_{i=1}^{n-1} (\overline{\phi}_i b \vee \phi_i a)) = \bigcap \{\sigma_A(x) : x \in [\bigwedge_{i=1}^{n-1} (\overline{\phi}_i b \vee \phi_i a))\}$, by Theorem 13 we have that $\Theta([\bigwedge_{i=1}^{n-1} (\overline{\phi}_i b \vee \phi_i a))) = \Theta_M(\sigma_A(\bigwedge_{i=1}^{n-1} (\overline{\phi}_i b \vee \phi_i a)))$. Besides, by Theorem 13 we infer that $\Theta_M(\sigma_A(\bigwedge_{i=1}^{n-1} (\overline{\phi}_i b \vee \phi_i a))) = \Theta_{OM}(\sigma_A(\bigvee_{j=1}^{n-1} (\phi_i b \wedge \overline{\phi}_i a)))$. Hence, $\Theta_{OM}(\sigma_A(\bigvee_{j=1}^{n-1} (\phi_i b \wedge \overline{\phi}_i a))) = \Theta([\bigwedge_{i=1}^{n-1} (\overline{\phi}_i b \vee \phi_i a)))$. From this last equality and the hypothesis we conclude that $\Theta_{OM}(G) = \Theta_{OM}(\sigma_A(\bigvee_{j=1}^{n-1} (\phi_i b \wedge \overline{\phi}_i a)))$ and so, by Theorem 13 we get $G = \sigma_A(\bigvee_{j=1}^{n-1} (\phi_i b \wedge \overline{\phi}_i a))$. ∎

PROPOSITION 15. *Let A be an LM_n-algebra and let $L(A)$ be the $l_n P$-space associated with A. Then the following conditions are equivalent:*

(i) *ϑ is a principal Lk_n-congruence,*

(ii) *$\vartheta = \Theta_{OM}(G)$, where $G = \bigcup_{i=1}^{n-1} f_i^{A^{-1}}(R)$ and R is a closed, open and convex subset of $X(A)$,*

(iii) *$\vartheta = \Theta_{OM}(G)$, where G is a closed, open and modal subset of $X(A)$.*

Proof.

(i) ⇔ (ii): It is an immediate consequence of Proposition 14 and (A7).

(ii) ⇒ (iii): From the hypothesis, (lP2) and (lP5) we have that G is closed, open and modal subset of $X(A)$.

(iii) ⇒ (ii): From Proposition 11, we have that G is the cardinal sum of maximal chains and so, G is a convex subset of $X(A)$. Hence, from the hypothesis by taking $R = G$ we conclude the proof. ∎

COROLLARY 16. *Let A be an Lk_n-algebra. Then, the intersection of two principal Lk_n-congruence on A is a principal one.*

Proof. It is a direct consequence of Proposition 15 and Theorem 13. ∎

4 Boolean congruences

Next, our attention is focus on determine the Boolean congruences and θ-congruences on LM_θ-algebras bearing in mind the topological duality for them established in Section 1. In order to do this, we will start studying certain subsets of $l_\theta P$-spaces which will be fundamental to reach our goal.

PROPOSITION 17. *Let A be an LM_θ-algebra and let $L_\theta(A)$ be the $l_\theta P$-space associated with A. Then for each $Y \subseteq X(A)$ holds:*

(i) $\Theta_{OS}(Y)$ *is a Boolean congruence on A if and only if Y is a closed and open subset of $X(A)$ such that Y and $X(A) \setminus Y$ are semimodal, where $\Theta_{OS}(Y)$ is defined as in Theorem 1.*

(ii) $\Theta_{O\theta}(Y)$ *is a Boolean θ-congruence on A if and only if Y is a closed and open subset of $X(A)$ such that Y and $X(A) \setminus Y$ are θ-subsets of $X(A)$, where $\Theta_{O\theta}(Y)$ is defined as in Theorem 1.*

Proof. It is a direct consequence of Theorem 1. ∎

Now, we will recall two concepts which will be used in Proposition 18. Let Y be a topological space and $y_0 \in Y$. A net in a space Y is a map $\varphi : D \to Y$ of some directed set $(D, \prec)$. Besides, we say that φ converges to y_0 (written $\varphi \to y_0$) if for all neighborhood $U(y_0)$ of y_0 there is $a \in D$ such that for all $a \prec b$, $\varphi(b) \in U(y_0)$.

PROPOSITION 18. *Let $(X, \{f_i\}_{i \in I})$ be an $l_\theta P$-space and let Y be a closed, open and semimodal subset of X. Then $X \setminus Y$ is semimodal.*

Proof. Suppose that there is (1) $x \in X \setminus Y$ such that $f_{i_0}(x) \in Y$ for some $i_0 \in I$. Since (2) Y is semimodal, we infer by (lP5) that (3) $f_i(x) \in Y$ for all $i \in I$. Taking into account that Y is a closed subset of X it follows by (1) that $X \setminus Y$ is a neighborhood of x. Then, by (2) we have that $(X \setminus Y) \cap \bigcup_{i \in I} f_i(Y) = \emptyset$ and so, $x \notin \overline{\bigcup_{i \in I} f_i(Y)}$. From this last assertion and (lP6) we conclude that $x \in \overline{\bigcup_{i \in I} f_i(X \setminus Y)}$, from which it follows that there exists a net $\{x_d\}_{d \in D} \subseteq X \setminus Y$ and (4) $f_{i_d}(x_d) \to x$. Therefore, there exists $d_0 \in D$ such that $\{f_{i_d}(x_d) : d_0 \prec d,\ d \in D\} \subseteq X \setminus Y$. From (2) and (lP5), $\{f_i(x_d) : d_0 \prec d,\ d \in D,\ i \in I\} \subseteq X \setminus Y$. On the other hand, by (4), (lP2) and (lP5) we have that $f_i(x_d) \to f_i(x)$ for all $i \in I$ and from the fact that $X \setminus Y$ is closed, we conclude that $f_i(x) \in X \setminus Y$ for all $i \in I$, which contradicts (3). ∎

PROPOSITION 19. *Let $(X, \{f_i\}_{i \in I})$ be an $l_\theta P$-space. Then for each $x \in X$, the set $[\{f_i(x)\}_{i \in I}] \cup (\{f_i(x)\}_{i \in I}]$ is convex. If I has least element 0 and greatest element 1, then for each $x \in X$ the set $[f_0(x), f_1(x)]$ is convex.*

Proof. Let $y, z, w \in X$ be such that $y \leq z \leq w$ and let $y, w \in [\{f_i(x)\}_{i \in I}) \cup (\{f_i(x)\}_{i \in I}]$. Then, by (lP3) and (lP5) we have that $f_i(y) = f_i(z) = f_i(w) = f_i(x)$ for all $i \in I$. This statement and (lP8) imply that $z \in [\{f_i(x)\}_{i \in I}) \cup (\{f_i(x)\}_{i \in I}]$ and so, $[\{f_i(x)\}_{i \in I}) \cup (\{f_i(x)\}_{i \in I}]$ is a convex set.

On the other hand, if I has least and greatest element, for each $x \in X$ we have that $[f_0(x), f_1(x)] = [\{f_i(x)\}_{i \in I}) \cup (\{f_i(x)\}_{i \in I}]$ and the proof is concluded. ∎

PROPOSITION 20. *Let $(X, \{f_i\}_{i \in I})$ be an $l_\varrho P$–space and let $Y \subseteq X$. Then the following conditions are equivalent:*

(i) *Y is modal,*

(ii) *$Y = \bigcup_{y \in Y} ([\{f_i(y)\}_{i \in I}) \cup (\{f_i(y)\}_{i \in I}])$. Besides, if I has least and greatest element, $Y = \bigcup_{y \in Y} [f_0(y), f_1(y)]$.*

Proof. (i) ⇒ (ii): By (lP8), $Y \subseteq \bigcup_{y \in Y} ([\{f_i(y)\}_{i \in I}) \cup (\{f_i(y)\}_{i \in I}])$. On the other hand, if $z \in [\{f_i(y)\}_{i \in I}) \cup (\{f_i(y)\}_{i \in I}]$, for some $y \in Y$, then by (lP3) and (lP5) we have that $f_i(z) = f_i(y)$ for all $i \in I$ and so, from the hypothesis we conclude that $z \in Y$.

(ii) ⇒ (i): From (lP3), (lP5) and (lP8), $f_i^{-1}(([\{f_i(y)\}_{i \in I}) \cup (\{f_i(y)\}_{i \in I}])) = [\{f_i(y)\}_{i \in I}) \cup (\{f_i(y)\}_{i \in I}]$ for all $i \in I$ and so, $Y = f_i^{-1}(Y)$ for all $i \in I$. In case that I has least and greatest element, it holds that $\bigcup_{y \in Y} ([\{f_i(y)\}_{i \in I}) \cup (\{f_i(y)\}_{i \in I}]) = \bigcup_{y \in Y} [f_0(y), f_1(y)]$ and hence, $Y = \bigcup_{y \in Y} [f_0(y), f_1(y)]$. ∎

COROLLARY 21. *Let $(X, \{f_i\}_{i \in I})$ be an $l_\varrho P$–space. Then the following conditions hold, for each modal subset Y of X:*

(i) *$X \setminus Y$ is modal,*

(ii) *Y is convex,*

(iii) *Y is increasing and decreasing.*

Proof. (i) and (iii): They are an immediate consequence of Proposition 20 and (lP11).

(ii): It follows from Propositions 20, 19 and (lP11). ∎

PROPOSITION 22. *Let $(X, \{f_i\}_{i \in I})$ be an $l_\varrho P$–space and let Y be a closed and open subset of X. Then the following conditions are equivalent:*

(i) *Y is a θ–subset,*

(ii) *Y is semimodal,*

(iii) *Y is modal.*

Proof. (i) $\Rightarrow$ (ii): It follows immediately.

(ii) $\Rightarrow$ (iii): From (lP8) we have that $Y \subseteq \bigcup_{y \in Y} (\overline{[\{f_i(y)\}_{i \in I}]} \cup (\{f_i(y)\}_{i \in I}])$.
On the other hand, if $z \in [\{f_i(y)\}_{i \in I}] \cup (\{f_i(y)\}_{i \in I}]$ for some $y \in Y$, by (lP3) and (lP5) we conclude that $f_i(z) = f_i(y)$ for all $i \in I$. Furthermore, from the hypothesis we infer that $f_i(z) \in Y$ for all $i \in I$ and then, Proposition 18 allows us to assert that $z \in Y$. Therefore, $Y = \bigcup_{y \in Y} ([\{f_i(y)\}_{i \in I}] \cup (\{f_i(y)\}_{i \in I}])$ and by Proposition 20 we have that Y is modal.

(iii) $\Rightarrow$ (i): It follows immediately that Y is semimodal. Furthermore, since Y is a closed and open subset of X, from Proposition 18 we get that $X \setminus Y$ is semimodal, which implies that (1) $\overline{\bigcup_{i \in I} f_i(Y)} \subseteq Y$ and (2) $\overline{\bigcup_{i \in I} f_i(X \setminus Y)} \subseteq X \setminus Y$. On the other hand, by (lP6) it holds (3) $X = \overline{\bigcup_{i \in I} f_i(Y)} \cup \overline{\bigcup_{i \in I} f_i(X \setminus Y)}$. From (2) and (3) we infer that $Y \subseteq \overline{\bigcup_{i \in I} f_i(Y)}$ and so, by (1) we conclude that $Y = \overline{\bigcup_{i \in I} f_i(Y)}$. ∎

COROLLARY 23. *Let $(X, \{f_i\}_{i \in I})$ be a $l_\theta P$-space. Then the following conditions hold for each closed, open and θ-subset Y of X:*

(i) $X \setminus Y$ *is a θ-subset,*

(ii) Y *and* $X \setminus Y$ *are convex subsets of* X.

Proof. (i): It follows from Proposition 22 and Corollary 21.

(ii): It is a direct consequence of Proposition 22 and Corollaries 21 and 23. ∎

PROPOSITION 24. *Let A be an LM_θ–algebra and let $L_\theta(A)$ be the $l_\theta P$-space associated with A. Then for each $Y \subseteq X(A)$ the following conditions are equivalent:*

(i) Y *is a closed, open and modal subset of* $X(A)$,

(ii) *there is* $b \in C(A)$ *such that* $Y = \sigma_A(b)$.

Proof. (i) $\Rightarrow$ (ii): From the hypothesis and item (iii) in Corollary 21, $Y \in D(X(A))$. Hence, there exists $a \in A$ such that (1) $Y = \sigma_A(a)$. Taking into account that $Y = f_i^{A^{-1}}(Y)$ for all $i \in I$, σ_A is an LM_θ–isomorphism and (A1) we conclude that $Y = \sigma_A(\phi_i a)$ for all $i \in I$ and so, by (1) we have that $a = \phi_i(a)$ for all $i \in I$. This statement and (L6) imply that $a \in C(A)$.

(ii) $\Rightarrow$ (i): From the hypothesis it follows that $Y \in D(X(A))$. Besides, for all $i \in I$, $\sigma_A(\phi_i b) = \phi_i^{D(X(A))}(\sigma_A(b))$ which by (A1) entails $\sigma_A(\phi_i b) = f_i^{A^{-1}}(Y)$. Since $b \in C(A)$ by (L6) we have that $\phi_i b = b$ for all $i \in I$. Therefore, $f_i^{A^{-1}}(Y) = Y$ for all $i \in I$ which completes the proof. ∎

Let $(X, \{f_i\}_{i \in I})$ be a $l_\theta P$-space. We will denote by $\mathcal{CO}_M(X)$ the Boolean lattice of all closed, open and modal subsets of X.

COROLLARY 25. *Let A be an LM_θ-algebra and let $L_\theta(A)$ be the $l_\theta P$-space associated with A. Then $\mathcal{CO}_M(X(A))$ is isomorphic to the Boolean lattice $C(A)$.*

Proof. Proposición 24 allows us to assert that the restriction of σ_A to $C(A)$ is a Boolean isomorphism. ∎

The above results allow us to obtain the description of Boolean congruences we were looking for.

THEOREM 26. *Let A be an LM_θ-algebra and let $L_\theta(A)$ be the $l_\theta P$-space associated with A. Then the lattice $\mathcal{CO}_M(X(A))$ is isomorphic to the lattice (dual lattice) $Con_{bLM_\theta}(A)$ of Boolean congruences on A, and the isomorphism Θ_{OM} (Θ_{CM}) is the restriction of Θ_{OS} (Θ_S) to $\mathcal{CO}_M(X(A))$, where these functions are defined as in Theorem 1 (in (A4)) respectively.*

Proof. If Y is a closed, open and modal subset of $X(A)$, then from Propositions 17, 18 and 22 we have that $\Theta_{OM}(Y)$ is a Boolean congruence on A. Conversely, let $\varphi \in Con_{bLk_\theta}(A)$. Then, by Theorem 1 and Proposition 17 we infer that there is a closed and open subset Y of $X(A)$ such that Y, $X(A) \setminus Y$ are semimodal and $\varphi = \Theta_{OS}(Y)$. These assertions and Proposition 22 imply that $Y \in \mathcal{CO}_M(X(A))$ and $\Theta_{OS}(Y) = \Theta_{OM}(Y)$ and so, by Theorem 1 we conclude the proof.

On the other hand, taking into account that $Y \in \mathcal{CO}_M(X(A))$ if and only if $X(A) \setminus Y \in \mathcal{CO}_M(X(A))$ and that $\Theta_{CM}(Y) = \Theta_{OM}(X(A) \setminus Y)$ we infer that Θ_{CM} establishes an isomorphism between $\mathcal{CO}_M(X(A))$ and the dual of $Con_{bLM_\theta}(A)$. ∎

COROLLARY 27. *Let A be an LM_θ-algebra and let $L_\theta(A)$ be the $l_\theta P$-space associated with A. If φ is a congruence on A then the following conditions are equivalent:*

(i) *φ is a Boolean congruence,*

(ii) *φ is a Boolean θ-congruence.*

Proof. (i) $\Rightarrow$ (ii): It is a direct consequence of Theorem 26 and Propositions 22 and 17.

(ii) $\Rightarrow$ (i): It follows immediately. ∎

COROLLARY 28. *Let A be an LM_θ-algebra and let $L_\theta(A)$ be the $l_\theta P$-space associated with A. Then each Boolean congruence on A is both a principal congruence and θ-congruence on A.*

Proof. Let φ be a Boolean LM_θ-congruence on A. Then, Theorem 26 implies that there is $G \in \mathcal{CO}_M(X(A))$ such that $\varphi = \Theta_{OM}(G)$. Furthermore, Proposition 22 and Corollary 21 imply that $\mathcal{CO}_M(X(A)) \subseteq \mathcal{O}_{CS}(X(A))$. Hence, we have that $G \in \mathcal{O}_{CS}(X(A))$ and so, $\Theta_{OM}(G) = \Theta_{OS}(G) = \varphi$. Since G is modal, by Corollary 21 we infer that G is convex. From this last assertion and Corollary 3 we conlude that φ is a principal LM_θ− congruence on A.

On the other hand, from Proposition 22 and Corollary 21 we have that $\mathcal{CO}_M(X(A)) \subseteq \mathcal{O}_{C\theta}(X(A))$ from which we get that $\Theta_{OM}(G) = \Theta_{O\theta}(G) = \varphi$. Hence, by Proposition 22 we conclude that G is a closed and open θ−subset of $X(A)$ and so, from Corollary 23 we infer that $X(A) \setminus G$ is a closed θ−subset of $X(A)$. This statement means that $X(A) \setminus G = \overline{\bigcup_{i \in I} f_i^A(X(A) \setminus G)}$. On the other hand since $X(A) \setminus G$ is modal, $X(A) \setminus G = \bigcap_{i \in I} f_i^{A^{-1}}(X(A) \setminus G)$. These last assertions imply that $G = X(A) \setminus \overline{\bigcup_{i \in I} f_i^A(\bigcap_{i \in I} f_i^{A^{-1}}(X(A) \setminus G))}$ and since G is convex, Corollary 5 allows us to conclude that φ is a principal θ−congruence on A. ∎

COROLLARY 29. *Let A be an LM_θ−algebra. Then the Boolean algebras $Con_{bLM_\theta}(A)$ and $C(A)$ are isomorphic and therefore, $|Con_{bLM_\theta}(A)| = |C(A)|$, where $|Z|$ denotes the cardinality of the set Z.*

Proof. It is a direct consequence of Corollary 25 and Theorem 26. ∎

COROLLARY 30. *Let A be an LM_θ−algebra let $L_\theta(A)$ be the $l_\theta P$-space associated with A. Then, Boolean congruences on A are permutable.*

Proof. Let $\varphi_1, \varphi_2 \in Con_{bLM_\theta}(A)$. Then, by Theorem 26 there are closed, open and modal subsets Y_1, Y_2 of $X(A)$ such that $\theta_S(Y_1) = \varphi_1$ and $\theta_S(Y_2) = \varphi_2$. Suppose now that $(x, y) \in \varphi_2 \circ \varphi_1$. Hence, there is $z \in A$ such that $(x, z) \in \varphi_1$ and $(z, y) \in \varphi_2$ and so, from Theorem 26 we have that $\sigma_A(x) \cap Y_1 = \sigma_A(z) \cap Y_1$ and $\sigma_A(y) \cap Y_2 = \sigma_A(z) \cap Y_2$. These statements imply that $\sigma_A(x) \cap (Y_1 \cap Y_2) = \sigma_A(y) \cap (Y_1 \cap Y_2)$. On the other hand, since $Y_1, Y_2 \in \mathcal{CO}_M(X(A))$, by Corollary 21 and Proposition 24 we infer that $(\sigma_A(x) \cap (Y_1 \cap Y_2)) \cup (\sigma_A(x) \cap (Y_2 \setminus Y_1)) \cup (\sigma_A(y) \cap (Y_1 \setminus Y_2)) \in D(X(A))$ and so, $w = \sigma_A^{-1}((\sigma_A(x) \cap (Y_1 \cap Y_2)) \cup (\sigma_A(x) \cap (Y_2 \setminus Y_1)) \cup (\sigma_A(y) \cap (Y_1 \setminus Y_2))) \in A$. Furthermore, we have that $\sigma_A(x) \cap Y_2 = \sigma_A(w) \cap Y_2$ and $\sigma_A(w) \cap Y_1 = \sigma_A(y) \cap Y_1$, hence $(x, w) \in \varphi_2$ and $(w, y) \in \varphi_1$. Therefore, $(x, y) \in \varphi_1 \circ \varphi_2$ from which we conclude that $\varphi_2 \circ \varphi_1 \subseteq \varphi_1 \circ \varphi_2$. The other inclusion follows similarly. ∎

Next, we will give another characterization of the Boolean congruences which will be useful in order to determine some properties of them.

LEMMA 31. *Let A be an LM_θ−algebra and let $L_\theta(A)$ be the $l_\theta P$-space associated with A. Then $\Theta([\phi_i a))$ is an congruence on A for all $a \in A$ and for all $i \in I$.*

Proof. Since $\sigma_A(\phi_i(a)) = f_i^{A^{-1}}(\sigma_A(a))$, by (lP5) we have that $\sigma_A(\phi_i(a))$ is a modal subset of $X(A)$ and so, it is semimodal. Then, by (A4) we infer that $\Theta_S(\sigma_A(\phi_i(a)))$ is a congruence on A. Bearing in mind the definition of $\Theta_S(\sigma_A(\phi_i(a)))$, we conclude that $\Theta_S(\sigma_A(\phi_i(a))) = \Theta([\phi_i(a))$ which completes the proof. ∎

PROPOSITION 32. *Let A be an LM_θ–algebra and $\varphi \in Con_{Lk_\theta}(A)$. Then the following conditions are equivalent:*

(i) *φ is a Boolean congruence on A,*

(ii) *there is $a \in A$ and $i \in I$ such that $\varphi = \Theta([\phi_i a))$.*

Proof. (i) $\Rightarrow$ (ii): From the hypothesis and Theorem 26 there is $Y \in \mathcal{CO}_M(X(A))$ such that $\varphi = \Theta_{CM}(Y)$. Besides, from Proposition 24 and (L6) there is $a \in A$ and $Y = \sigma_A(\phi_i(a))$ for some $i \in I$. Therefore, for all $b, c \in A$ we have that $(b, c) \in \Theta_{CM}(Y)$ if and only if $\sigma_A(b) \cap \sigma_A(\phi_i(a)) = \sigma_A(c) \cap \sigma_A(\phi_i(a))$. Hence, taking into account that σ_A is an isomorphism it follows that $(b, c) \in \Theta_{CM}(Y)$ if and only if $b \wedge \phi_i(a) = c \wedge \phi_i(a)$. From this last assertion we conclude that $\varphi = \Theta([\phi_i(a)))$.

(ii) $\Rightarrow$ (i): From Lemma 31 we infer that $\Theta([\phi_i a))$ is a congruence on A and $\Theta([\phi_i a)) = \Theta_S(\sigma_A(\phi_i(a)))$. Besides, from Proposition 24, $\sigma_A(\phi_i(a)) \in \mathcal{CO}_M(X(A))$ and so, by Theorem 26 we have that $\Theta_{CM}(\sigma_A(\phi_i(a))) = \Theta_S(\sigma_A(\phi_i(a)))$ is a Boolean congruence on A. ∎

COROLLARY 33. *Let A be an LM_θ–algebra. Then the Boolean congruences on A are normal and regular.*

Proof. Let φ be a Boolean congruence on A. Then, by Proposition 32 there is $a \in A$ and $i \in I$ such that $\varphi = \Theta([\phi_i a))$. Furthermore, for each $b \in A$ we have that $\overline{b}_\varphi = \{(b \wedge \phi_i a) \vee c : c \in (\overline{\phi}_i a]\}$ where $\overline{b}_\varphi$ stands for the equivalence class of b modulo φ. From this last assertion we infer that $\overline{0}_\varphi = (\overline{\phi}_i a]$ and therefore, $\overline{b}_\varphi = \{(b \wedge \phi_i a) \vee c : c \in \overline{0}_\varphi\}$ and $|\overline{b}_\varphi| = |\overline{0}_\varphi|$ for all $b \in A$, which allows us to conclude the proof. ∎

In what follows we will determine necessary and sufficient conditions for a principal congruence on an LM_θ–algebra be a Boolean one. These are also sufficient conditions for the fact that the intersection of two principal LM_θ–congruences be a principal one.

PROPOSITION 34. *Let A be an LM_θ–algebra and let $L_\theta(A)$ be the $l_\theta P$–space associated with A. Then the following conditions are equivalent for all $a, b \in A$ such that $a \leqslant b$:*

(i) $\Theta(a, b)$ *is a Boolean congruence on A,*

(ii) $\bigcup_{i \in I} f_i^{A^{-1}}(\sigma_A(b) \setminus \sigma_A(a))$ *is a closed subset of $X(A)$,*

(iii) $\bigcup_{i \in I} \sigma_A(\phi_i b \wedge \overline{\phi}_i a)$ a closed subset of $X(A)$.

Proof. (i) $\Rightarrow$ (ii): From the hypothesis and Proposition 2 we have that $\Theta(a,b) = \Theta_{OS}(G)$, where $G = (\sigma_A(b) \setminus \sigma_A(a)) \cup \bigcup_{i \in I} f_i^{A^{-1}}(\sigma_A(b) \setminus \sigma_A(a))$ and taking into account Theorem 26 we infer that G is a closed, open and modal subset of $X(A)$. This last assertion allows us to conclude that $G = \bigcup_{i \in I} f_i^{A^{-1}}(\sigma_A(b) \setminus \sigma_A(a))$ and so, the proof is completed.

(ii) $\Rightarrow$ (i): Let (1) $G = \bigcup_{i \in I} f_i^{A^{-1}}(\sigma_A(b) \setminus \sigma_A(a))$. Then from the hypothesis, (lP2) and (lP5) we have that (2) $G \in \mathcal{CO}_M(X(A))$, and as a consequence of Theorem 26 we infer that $\Theta_{OM}(G)$ is a Boolean congruence on A. On the other hand, from (2), Proposition 22 and Corollary 21 it follows that $G \in \mathcal{O}_{CS}(X(A))$ and so, (3) $\Theta_{OM}(G) = \Theta_{OS}(G)$. Besides, $\sigma_A(b) \setminus \sigma_A(a) \subseteq G$. Indeed, suppose that $(\sigma_A(b) \setminus \sigma_A(a)) \setminus G \neq \emptyset$. Since G is a closed subset of $X(A)$, by (lP6) we have that $((\sigma_A(b) \setminus \sigma_A(a)) \setminus G) \cap \bigcup_{i \in I} f_i^A(X(A)) \neq \emptyset$, from which we infer that there are (4) $x \in (\sigma_A(b) \setminus \sigma_A(a)) \setminus G$ and $y \in X(A)$ such that $x = f_{i_0}(y)$ for some $i_0 \in I$. Then by (lP5), $x = f_{i_0}(x)$ for some $i_0 \in I$. This statement, (1) and (4) imply that $x \in G$, which contradicts (4). On the other hand, it follows immediately that G is the least element of $\mathcal{O}_{CS}(X(A))$, ordered by inclusion such that $\sigma_A(b) \setminus \sigma_A(a) \subseteq G$. Hence, by Proposition 2 and (3) we conclude that $\Theta(a,b) = \Theta_{OM}(G)$. Therefore, $\Theta(a,b)$ is a Boolean congruence on A.

(ii) $\Leftrightarrow$ (iii): It is a direct consequence of the fact that σ_A is an isomorphism. ∎

LEMMA 35. Let $(X, \{f_i\}_{i \in I})$ be an $l_\theta P$-space. If $R \subseteq X$ is such that $R \subseteq \bigcup_{i \in I} f_i^{-1}(R)$, then $\bigcup_{i \in I} f_i^{-1}(R) = (R] \cup [R)$.

Proof. Suppose that $f_i(x) \in R$ for some $i \in I$. Hence, by (lP8) we have that $x \in (R]$ or $x \in [R)$. Conversely, let $x \in (R] \cup [R)$. Then, there is $y \in R$ such that $x \leqslant y$ or $y \leqslant x$ and so, by (lP3) we infer that $f_i(x) = f_i(y)$ for all $i \in I$. This assertion and the fact that from the hypothesis $f_{i_0}(y) \in R$ for some $i_0 \in I$ allow us to conclude that $x \in \bigcup_{i \in I} f_i^{-1}(R)$. ∎

LEMMA 36. Let $(X, \{f_i\}_{i \in I})$ be an $l_\theta P$-space and R be a closed and open subset of X. Then the following conditions are equivalent:

(i) $R \subseteq \bigcup_{i \in I} f_i^{-1}(R)$,

(ii) $\bigcup_{i \in I} f_i^{-1}(R)$ is closed.

Proof. (i) $\Rightarrow$ (ii): From the hypothesis and Lemma 35 we have that $\bigcup_{i \in I} f_i^{-1}(R) = (R] \cup [R)$. Since R is a closed subset of X by (A6) we infer that $(R] \cup [R)$ is closed, which completes the proof.

(ii) $\Rightarrow$ (i): From the hypothesis and the fact that R is an open subset of X we have that $R \setminus \bigcup_{i \in I} f_i^{-1}(R)$ is open. Suppose now that $R \setminus \bigcup_{i \in I} f_i^{-1}(R) \neq \emptyset$. Then, by (lP6) there are $x \in X$ and $i_0 \in I$ such that $f_{i_0}(x) \in R \setminus \bigcup_{i \in I} f_i^{-1}(R)$ and from (lP5) we conclude that $f_{i_0}(x) \in R$ and $f_i(x) \notin R$ for all $i \in I$, which is a contradiction. Therefore, $R \subseteq \bigcup_{i \in I} f_i^{-1}(R)$. ∎

PROPOSITION 37. *Let A be an LM_θ–algebra and let $L_\theta(A)$ be the $l_\theta P$-space associated with A. Then the following conditions are equivalent for all $a, b \in A$, $a \leq b$:*

(i) $\Theta(a,b) = \Theta_{OS}(G)$ *is a Boolean congruence on A,*

(ii) $G = \bigcup_{i \in I} \sigma_A(\phi_i b \wedge \overline{\phi}_i a)$ *is a closed subset of $X(A$,*

(iii) *there are $i_j \in I$, $1 \leq j \leq n$ such that $G = \sigma_A(\bigvee_{j=1}^n (\phi_{i_j} b \wedge \overline{\phi}_{i_j} a))$ and $\sigma_A(b) \setminus \sigma_A(a) \subseteq G$,*

(iv) *there are $i_j \in I$, $1 \leq j \leq n$ such that $\sigma_A(b) \setminus \sigma_A(b) \subseteq \sigma_A(\bigvee_{j=1}^n (\phi_{i_j} b \wedge \overline{\phi}_{i_j} a))$ and $\Theta_{OS}(G) = \Theta([\bigwedge_{j=1}^n (\overline{\phi}_{i_j} b \vee \phi_{i_j} a)))$.*

Proof. (i) $\Rightarrow$ (ii): From the hypothesis and Proposition 2 we have that $G = (\sigma_A(b) \setminus \sigma_A(a)) \cup \bigcup_{i \in I} f_i^{A^{-1}}(\sigma_A(b) \setminus \sigma_A(a))$. On the other hand, since $\Theta_{OS}(G)$ is a Boolean congruence, by Proposition 34 we infer that $\bigcup_{i \in I} f_i^{A^{-1}}(\sigma_A(b) \setminus \sigma_A(a))$ is a closed subset of $X(A)$. Hence, by Lemma 36 we have that $\sigma_A(b) \setminus \sigma_A(a) \subseteq \bigcup_{i \in I} f_i^{A^{-1}}(\sigma_A(b) \setminus \sigma_A(a))$. Therefore, $G = \bigcup_{i \in I} f_i^{A^{-1}}(\sigma_A(b) \setminus \sigma_A(a))$ and taking into account that σ_A is an isomorphism we conclude the proof.

(ii) $\Rightarrow$ (iii): The hypothesis implies that $G = \bigcup_{i \in I} f_i^{A^{-1}}(\sigma_A(b) \setminus \sigma_A(a))$ is a closed subset of $X(A)$ and so, from Lemma 36 we have that $\sigma_A(b) \setminus \sigma_A(a) \subseteq G$. To complete the proof note that $\{f_i^{A^{-1}}(\sigma_A(b) \setminus \sigma_A(a)) : i \in I\}$ is an open covering of G. Then, a simple compactness argument shows that $G = \bigcup_{j=1}^n f_{i_j}^{A^{-1}}(\sigma_A(b) \setminus \sigma_A(a))$, and since σ_A is an isomorphism, we infer that $G = \sigma_A(\bigvee_{j=1}^n (\phi_{i_j} b \wedge \overline{\phi}_{i_j} a))$.

(iii) $\Rightarrow$ (ii): From the hypothesis, $G = \bigcup_{j=1}^{n} f_{ij}^{A^{-1}}(\sigma_A(b) \setminus \sigma_A(a))$ and $\sigma_A(b) \setminus \sigma_A(a) \subseteq \bigcup_{j=1}^{n} f_{ij}^{A^{-1}}(\sigma_A(b) \setminus \sigma_A(a))$. Then, by (lP5) we infer that $\bigcup_{i \in I} f_i^{A^{-1}}(\sigma_A(b) \setminus \sigma_A(a)) = \bigcup_{j=1}^{n} f_{ij}^{A^{-1}}(\sigma_A(b) \setminus \sigma_A(a))$. Therefore, $G = \bigcup_{i \in I} f_i^{A^{-1}}(\sigma_A(b) \setminus \sigma_A(a))$ is a closed subset of $X(A)$ and bearing in mind that σ_A is an isomorphismo we conclude that $G = \bigcup_{i \in I} \sigma_A(\phi_i b \wedge \overline{\phi}_i a)$.

(ii) $\Rightarrow$ (i): It is easy to check that G is an open and modal subset of $X(A)$. Therefore, $G \in \mathcal{CO}_M(X(A))$. Then, taking into account that $\mathcal{CO}_M(X(A)) \subseteq \mathcal{O}_{CS}(X(A))$ we have that $\Theta_{OM}(G) = \Theta_{OS}(G)$ and so, by Theorem 26 we conclude that $\Theta_{OS}(G)$ is a Boolean congruence on A. Furthermore, since G is a closed subset of $X(A)$ by Lemma 36, $\sigma_A(b) \setminus \sigma_A(a) \subseteq \bigcup_{i \in I} f_i^{A^{-1}}(\sigma_A(b) \setminus \sigma_A(a))$ which implies that $G = (\sigma_A(b) \setminus \sigma_A(a)) \cup \bigcup_{i \in I} f_i^{A^{-1}}(\sigma_A(b) \setminus \sigma_A(a))$. This equality and Proposition 2 imply that $\Theta(a,b) = \Theta_{OS}(G)$ and so, the proof is complete.

(iii) $\Leftrightarrow$ (iv): It follows as a consequence of the fact that $\Theta_{OM}(\sigma_A(\bigvee_{j=1}^{n}(\phi_{i_j} b \wedge \overline{\phi}_{i_j} a))) = \Theta_S(X(A) \setminus \sigma_A(\bigvee_{j=1}^{n}(\phi_{i_j} b \wedge \overline{\phi}_{i_j} a))) = \Theta_S(\sigma_A(\bigwedge_{j=1}^{n}(\overline{\phi}_{i_j} b \vee \phi_{i_j} a)))$ and that $\Theta_S(\sigma_A(\bigwedge_{j=1}^{n}(\overline{\phi}_{i_j} b \vee \phi_{i_j} a))) = \Theta([\bigwedge_{j=1}^{n}(\overline{\phi}_{i_j} b \vee \phi_{i_j} a)))$. ■

Finally, we will complete this section establishing a characterization of the Boolean congruences on n-valued Łukasiewicz algebras.

THEOREM 38. *Let A be an LM_n-algebra and let $L(A)$ be the $l_n P$-space associated with A. Then the lattice $\mathcal{CO}_M(X(A))$ is isomorphic to the lattice $Con_{bLk_n}(A)$ of the Boolean congruences on A and the isomorphism is the function Θ_{OM} defined as in Theorem 26.*

Proof. It is a direct consequence of Theorem 26 and Corollary 25. ■

COROLLARY 39. *Let A be an LM_n-algebra. Then Boolean and principal congruences coincide.*

Proof. It is a direct consequence of Proposition 15 and Theorem 38. ■

BIBLIOGRAPHY

[Adams, 1987] M. E. Adams. Principal congruences in De Morgan algebras. *Edimburgh Mathematical Society*, 30:415–421, 1987.

[Balbes and Dwinger, 1974] R. Balbes and P. Dwinger. *Distributive Lattices*. Univ. of Missouri Press, 1974.

[Boicescu et al., 1991] V. Boicescu, A. Filipoiu, G. Georgescu, and S. Rudeanu. *Łukasiewicz–Moisil Algebras*. North-Holland, 1991.

[Boicescu, 1984] V. Boicescu. *Contributions to the study of Łukasiewicz algebras* (Romanian). PhD thesis, Univ. of Bucharest, 1984.

[Burris and Sankappanavar, 1981] S. Burris and H. P. Sankappanavar. *A Course in Universal Algebra*, volume 78 of *Graduate Texts in Mathematics*. Springer-Verlag, Berlin, 1981.

[Cignoli et al., 2000] R. Cignoli, I. D'Ottaviano, and D. Mundici. *Algebraic Foundations of Many-valued Reasoning*. Kluwer, 2000.

[Cignoli, 1970] R. Cignoli. *Moisil Algebras*. Notas de Lógica Matemática 27, 1970. Instituto de Matemática, Universidad Nacional del Sur.

[Cignoli, 1982] R. Cignoli. Proper n−valued Łukasiewicz algebras as s−algebras of Łukasiewicz valued propositional calculi. *Studia Logica*, 41:3–16, 1982.

[Figallo et al., 2004] A.V. Figallo, I. Pascual, and A. Ziliani. Notes on monadic n−valued Łukasiewicz algebras. *Mathematica Bohemica*, 129(3):255–271, 2004.

[Figallo et al., 2010] A.V. Figallo, I. Pascual, and A. Ziliani. A duality for θ−valued Łukasiewicz–Moisil algebras and applications. *J. Mult-Valued Logic & Soft Computing*, 16:303–322, 2010.

[Filipoiu, 1981] A. Filipoiu. θ valued Łukasiewicz-Moisil algebras and logics (Romanian). PhD thesis, Univ. of Bucharest, 1981.

[Iorgulescu, 1984] A. Iorgulescu. $(1+\theta)$−valued Łukasiewicz–Moisil algebras with negation (Romanian). PhD thesis, Univ. of Bucharest, 1984.

[Lane, 1971] S. Mac Lane. *Categories for the Working Mathematician*. Springer–Verlag, Berlin, 1971.

[Marek and Traczyck, 1969] W. Marek and T. Traczyck. Generalized Łukasiewicz algebras. *Bull. Acad. Polonaise Sci. sér. Math. Asttronom. Phys.*, 17:789–792, 1969.

[Moisil, 1963] Gr. Moisil. Le algèbre di Łukasiewicz. *Acta Logica (Bucharest)*, 6:97–135, 1963.

[Moisil, 1972] Gr. Moisil. *Łukasiewiczian algebras*. Preprint, 1972. Computing Center, Univ. Bucharest, 311–324.

[Priestley, 1970] H. Priestley. Representation of distributive lattices by means of ordered Stone spaces. *Bull. London Math. Soc.*, 2:186–190, 1970.

[Priestley, 1972] H. Priestley. Ordered topological spaces and the representation of distributive lattices. *Proc. London Math. Soc.*, 3:507–530, 1972.

[Priestley, 1984] H. Priestley. Ordered sets duality for distributive lattices. *Ann. Discrete Math.*, 23:39–60, 1984.

Aldo V. Figallo
Departamento de Matemática
Universidad Nacional del Sur
Bahía Blanca, Argentina
Instituto de Ciencias Básicas.
Universidad Nacional de San Juan
San Juan, Argentina
E-mail: avfigallo@gmail.com

Inés Pascual
Instituto de Ciencias Básicas
Universidad Nacional de San Juan
San Juan, Argentina
E-mail: ipascualdiz@gmail.com

Alicia Ziliani
Departamento de Matemática
Universidad Nacional del Sur
Bahía Blanca, Argentina
E-mail: aziliani@gmail.com

Analytical Tableaux for da Costa's Paraconsistent Predicate Calculi $\mathbf{C}_n^*$

ITALA M. LOFFREDO D'OTTAVIANO AND MILTON AUGUSTINIS DE CASTRO

ABSTRACT. In this paper, based on [Castro, 2004] and [D'Ottaviano and Castro, 2005], we introduce a hierarchy of syntactical tableaux systems $\mathbf{TNDC}_n^*, 1 \leq n < \omega$, for da Costa's hierarchy of predicate paraconsistent logics $\mathbf{C}_n^*, 1 \leq n < \omega$. As in our paper of 2005, in our tableaux formulation for the $\mathbf{TNDC}_n^*$, we introduce da Costa's 'ball' operator 'o', the generalized operators 'k', '(k)' and the negations '$\sim_k$', for $k \geq 1$, as primitive operators. It is necessary to deal with specific problems, concerning relationships between these generalized distinct operators; and relationships between the different systems of the hierarchy $\mathbf{TNDC}_n^*$. Another peculiarity is that we define two conditions for the closure of the branches. We prove a generalized version of the Cut Rule for the $\mathbf{TNDC}_n^*$ and also prove that these systems are logically equivalent to the corresponding da Costa's calculi. The $\mathbf{TNDC}_n^*$ are introduced from a denumerable (infinite) set of primitive operators and this allows us to capture $\mathbf{C}_n^*$ as paraconsistent extensions of classical logic. As far as we know, besides introducing every one of the operators of the three da Costa's families of operators 'k', '(k)' and the negations '$\sim_k$', for $k \geq 1$, as primitive, this is the first paper in the literature in which the hierarchy of logics $\mathbf{C}_n^*, 1 \leq n < \omega$, receive a tableaux approach. Given that our tableaux are deterministic automated procedures for the analysis of the validity of formulas of the systems of da Costa's hierarchy $\mathbf{C}_n^*, 1 \leq n < \omega$, our method may still be useful for the study of other paraconsistent logics and other non-classical logics in general.

Introduction

Problems concerning the decidability of da Costa's hierarquy of paraconsistent propositional logics $\mathbf{C}_n, 1 \leq n \leq \omega$, have motivated several authors, among them Raggio, da Costa, Alves, Loparić, Fidel, Sette, Béziau, Moura, Lima-Marques, Carnielli, Marcos, Coniglio, Bueno-Soler, Castro and D'Ottaviano. These logicians introduced sequent calculi, natural deduction systems and tableaux systems equivalent to da Costa's propositional logics; the known quasi-matrices semantics, possible-translations semantics, and algebraic possible-translations semantics[1].

---

[1] See, for instance, [D'Ottaviano, 1990], [D'Ottaviano and Castro, 2005], and [Carnielli et al., 2009].

In this paper, we introduce a new hierarchy of predicate analytical tableaux systems, the $\mathbf{TNDC}_n^*, 1 \leq n < \omega$, for da Costa's historical hierarchy of predicate calculi for the study of inconsistent but non-trivial theories: the hierarchy of first-order paraconsistent predicate calculi $\mathbf{C}_n^*, 1 \leq n < \omega$ (see [da Costa, 1964] and [da Costa, 1974]).

In our hierarchy $\mathbf{TNDC}_n^*, 1 \leq n < \omega$, as in the hierarchy of propositional analytical tableaux systems $\mathbf{TNDC}_n, 1 \leq n < \omega$, introduced by [Castro, 2004] and [D'Ottaviano and Castro, 2005], the 'ball' operator 'o' is introduced as primitive, as well as the generalized operators 'k' and '(k)', and the negations '$\sim_k$', for $k \geq 1$, are primitive operators. As in the case of our previous paper, it is necessary to deal with specific problems concerning relationships between the generalized distinct primitive operators; and with relationships between the different systems of the hierarchy $\mathbf{TNDC}_n^*, 1 \leq n < \omega$. We also define two conditions for the closure of the branches of the tableaux, looking for reflecting the meaning of the paraconsistent negation: either they are closed by the strong negation '$\sim_n$', as usual, or they are closed by the paraconsistent negation '$\neg$' and additional conditions.

We prove a general version of the Cut Rule (Theorem) for $\mathbf{TNDC}_n^*, 1 \leq n < \omega$, and also prove that these systems are logically equivalent to the corresponding systems $\mathbf{C}_n^*, 1 \leq n < \omega$, respectively.

As far as we know in the literature, the first paraconsistent system in which da Costa's 'ball' operator 'o' was treated as a primitive "consistency" operator was introduced by [D'Ottaviano and Epstein, 1988], in a modified version of the system $\mathbf{J}_3$ of [D'Ottaviano and da Costa, 1970], also presented in [Epstein, 1990] and [Epstein, 1995] (Chapter IX, written in collaboration with D'Ottaviano).

Castro presented, for the first time, our conception concerning the primitiveness of the operators of the families 'k', '(k)' and the negations '$\sim_k$' for the construction of the hierarchy of tableaux systems $\mathbf{TNDC}_n, 1 \leq n < \omega$, in his communication during the "II World Congress on Paraconsistency (II WCP)", held in Juquehy, Brazil, in 2000 (see [Castro, 2000]).

Carnielli, Coniglio and Marcos have studied an ample class of paraconsistent axiomatic propositional systems, the so-called *Logics of Formal Inconsistency* (LFIs), in which a unary "consistency operator" is introduced as a primitive operator of the object language (see [Carnielli and Marcos, 2002] and [Carnielli et al., 2007]). The system $\mathbf{J}_3$ is a particular logic of LFIs; da Costa's propositional systems $\mathbf{C}_n^*, 1 \leq n < \omega$, are seen as a primary sub-class of LFIs, such that, in C_1, the primitive "consistency operator" corresponds to da Costa's defined "ball" operator "o". It was also at the "II World Congress on Paraconsistency" (celebrated in Juquehy, 2000) that Carnielli presented for the first time the Logics of Formal Inconsistency, in a joint work with João Marcos (see [Carnielli and Marcos, 2000]). Since then, the LFIs were extensively studied for several logicians worldwide.

We remind that [Marconi, 1980] is the first tableaux system for da Costa's calculus C_1, the first logic of the hierarchy of propositional calculi $\mathbf{C}_n, 1 \leq n \leq \omega$. He introduces a variant of semantic tableaux systems, *à la* Beth (see [Beth, 1959]), in order to prove the completeness and decidability of da Costa's propositional system C_1. He also claims that his method could be extended for the systems $\mathbf{C}_n, 2 \leq n < \omega$.

[Carnielli and Lima-Marques, 1992] introduce a semantic tableaux system, *à la* Smullyan (see [Smullyan, 1968]), for Alves's paraconsistent propositional logic[2] C_1^1, and for the paraconsistent predicate logic with equality $C_1^{1=}$ (see [Alves, 1976]), namely the systems $\mathbf{TC}_1$ and $\mathbf{TC}_1^=$ respectively, and show that these systems are complete and decidable.

[Buchsbamum and Pequeno, 1993] introduce a syntactical tableaux system, also *à la* Smullyan, for da Costa's $\mathbf{C}_1^*$, the system $\mathbf{SC}_1^*$, showing that $\mathbf{SC}_1^*$ is complete.

In the system $\mathbf{SC}_1^*$ of Buchsbaum and Pequeno we do not have an explicit rule that determines *a priori* when the definition of the operator 'o' must be used or must not be used during the derivations; on account of this it is possible to occur open branches that must be rebuilt, in a distinct way, from the mentioned occurrence of the operator 'o'.

Also in Carnielli and Lima-Marques's systems $\mathbf{TC}_1$ and $\mathbf{TC}_1^=$ there are not specific rules that determine *a priori* when to use the definition of the operator 'o', what may make necessary to rebuilding branches. Particularly, in these systems infinite loops may occur, 'postponing indefinitely', according to the own authors, the analysis of formulae that involve the operator of primitive negation and, as a natural consequence, the operator 'o'; Carnielli and Lima-Marques prove the decidability of $\mathbf{TC}_1$ and $\mathbf{TC}_1^=$, showing how to deal with the infinite loops. [Carnielli et al., 2007] improve the system $\mathbf{TC}_1$, introducing a new semantic tableaux system for $\mathbf{C}_1$, trying to avoid the presence of infinite loops in the derivations of the branches: in this new tableaux system da Costa's 'ball' operator 'o' is maintained as a defined operator and, as the nodes of the branches in the derivations are not univocally determined, it is possible (as in Buchsbaum and Pequeno's $\mathbf{SC}_1^*$ system) to occur open branches that may be rebuilt in a distinct way.

Due to the "primitiveness" of the 'ball' operator 'o' and of the other above mentioned denumerable da Costa's operators, in every one of our tableaux systems $\mathbf{TNDC}_n^*, 1 \leq n < \omega$, we emphasize that there are specific rules to objectively deal with the operator 'o', as well as with the operators 'k' and '(k)', and the negations '$\sim_k$', for $k \geq 1$. The branches of the tableaux are univocally and automatically generated and infinite loops do not occur.

Our tableaux are deterministic automated procedures for the analysis of the validity of formulas of da Costa's logical systems $\mathbf{C}_n^*, 1 \leq n < \omega$.

As far as we know this is the first paper in the literature in which every one of the operators of the three famílies of operators 'k' and '(k)', and the negations '$\sim_k$', for $k \geq 1$, introduced as defined operators by da Costa, are considered as primitive operators.

By defining two conditions for the closure of the branches of the tableaux $\mathbf{TNDC}_n^*, 1 \leq n < \omega$, we look for reflecting the meaning of da Costa's negations in his original hierarchies of paraconsistent systems. This also allows us to capture da Costa's systems $\mathbf{C}_n^*, 1 \leq n < \omega$, as paraconsistent extensions of classical first-order predicate logic.

---

[2] C_1^1, a system slightly stronger than C_1, is obtained by replacing the schema of axioms $\neg\neg A \supset A$ of da Costa's C_1 by the schema $\neg\neg A \equiv A$.

Furthermore this is the first paper in which all the systems of da Costa's hierarchy of paraconsistent logics $\mathbf{C}_n^*$, $1 \leq n < \omega$, receive a tableaux approach, and the method may be useful for the study of other paraconsistent logis and other non-classical logics, in general.

1 Da Costa's paraconsistent first-order logics C_n^*

In this section, we briefly recall some known results on da Costa's paraconsistent systems $\mathbf{C}_n^*$, $1 \leq n \leq \omega$.

The language $\mathbf{L}^*$ of da Costa's systems has as primitive connectives $\neg, \vee, \&$ and $\supset$, and as primitive quantifiers $\forall$ and $\exists$, as well as a denumerable family of individual variables, predicate symbols and auxiliary symbols (see [da Costa, 1964] and [da Costa, 1974]). The notions of formula, free and bound variables in a formula, sentence, theorem, as well as the general conventions and notations, are the standard ones, as in [Kleene, 1952].

Let A and B be formulae. The following operators, as in da Costa's original papers, are added, by definition, to the language $\mathbf{L}^*$.

DEFINITION 1. $A^0 =_{df} \neg(A \& \neg A)$.

DEFINITION 2. $A^k =_{def} A^{00...0}$ ("o" k times, for $k \geq 1$)[3].

DEFINITION 3. $A^{(k)} =_{df} A^1 \& A^2 \& \ldots \& A^k$, for $k \geq 1$.

DEFINITION 4. $\sim_k A =_{df} \neg A \& A^{(k)}$, for $k \geq 1$.

DEFINITION 5. $(A \equiv B) =_{df} (A \supset B) \& (B \supset A)$.

In $\mathbf{C}_n^*$, $1 \leq n < \omega$, the operator "o" is usually named *"ball operator"* and $A^{(n)}$ may be read as "A is a *well-behaved formula*" or "A is *regular*"; for every $\mathbf{C}_n^*$, $1 \leq n \leq \omega$, the primitive negation "$\neg$" is the basic paraconsistent negation, or *weak negation* of the system; and for every $\mathbf{C}_n^*$, $n \geq 1$, the connective "$\sim_n$" is called *strong negation*.

For every first-order predicate calculus $\mathbf{C}_n^*$, $1 \leq n < \omega$, the schemata of *axioms* and the *deduction rules* are the following.

Axiom 1: $A \supset (B \supset A)$
Axiom 2: $(A \supset B) \supset ((A \supset (B \supset C)) \supset (A \supset C))$
Axiom 3: $A \& B \supset A$
Axiom 4: $A \& B \supset B$
Axiom 5: $A \supset (B \supset A \& B)$
Axiom 6: $A \supset A \vee B$
Axiom 7: $A \supset B \vee A$
Axiom 8: $(A \supset C) \supset ((B \supset C) \supset (A \vee B \supset C))$
Axiom 9: $\neg\neg A \supset A$
Axiom 10: $A \vee \neg A$
Axiom 11: $B^{(n)} \supset ((A \supset B) \supset ((A \supset \neg B) \supset \neg A))$
Axiom 12: $A^{(n)} \& B^{(n)} \supset (A \& B)^{(n)}$

---

[3] As an easy recursive alternative formulation, we could define: $A^1 =_{df} \neg(A \& \neg A)$; $A^{k+1} =_{df} (A^k)^1 = \neg(A^k \wedge \neg A^k)$.

Axiom 13: $A^{(n)} \& B^{(n)} \supset (A \vee B)^{(n)}$
Axiom 14: $A^{(n)} \& B^{(n)} \supset (A \supset B)^{(n)}$
Axiom 15: $\forall x A(x) \supset A(t)$
Axiom 16: $A(t) \supset \exists x A(x)$
Axiom 17: $\forall x (A(x))^{(n)} \supset (\forall x A(x))^{(n)}$
Axiom 18: $\forall x (A(x))^{(n)} \supset (\exists x A(x))^{(n)}$
Axiom 19: $A \equiv B$, where A and B are congruent[4] formulae

Rule of *Modus Ponens* (MP) $\dfrac{A, A \supset B}{B}$

Rule II $\dfrac{C \supset A(x)}{C \supset \forall x A(x)}$

Rule III $\dfrac{A(x) \supset C}{\exists x A(x) \supset C}$,

where the variable x, the term t and the formulae $A(x)$ and C satisfy the usual restrictions[5].

Finally, the *system* $\mathbf{C}^*_\omega$ is defined by:

Axiom 1 to Axiom 10, Axiom 15-16, Axiom 19
and the Rules MP, II and III.

The classical first-order predicate calculus may be considered as the system $\mathbf{C}^*_0$ of this hierarchy.

Some very well known results[6] are necessary for the development of this paper: the *Deduction Theorem* is valid for the $\mathbf{C}^*_n, 1 \leq n \leq \omega$; in every $\mathbf{C}^*_n, n \geq 1$, the strong negation $\sim_n$ has all the properties of classical negation; every system in the hierarchy $\mathbf{C}^*_n, 1 \leq n \leq \omega$, is strictly stronger than those which follow it; every $\mathbf{C}^*_n, n \geq 1$, is finitely trivializable; the *Replacement Theorem*[7] is not valid in $\mathbf{C}^*_n, n \geq 1$. Furthermore, every system of the hierarchy $\mathbf{C}^*_n, 1 \leq n \leq \omega$, is consistent, *paraconsistent lato sensu*[8] and *undecidable*. The proof of the following theorem is trivial.

THEOREM 6. *In every* $\mathbf{C}^*_n, 1 \leq n < \omega$, *we have that:*

1. $(A \supset B) \supset (\sim_n A \vee B)$

2. $A \& \sim_n A \supset B$

3. $\sim_n \sim_n A \supset A$

4. $(A \supset B) \supset (\sim_n B \supset \sim_n A)$

5. $(A \vee A) \supset A$

---

[4]Here, two formulas are *congruent* if they differ only in their bound variables, which are bound to the corresponding quantifiers (see [Kleene, 1952], p. 153).
[5]See Kleene (1952, p. 81).
[6]See, for instance, [D'Ottaviano, 1990], [Castro, 2004] and [da Costa et al., 2006]).
[7]See [Kleene, 1952], pp. 115-116.
[8]In every $\mathbf{C}^*_n, 1 \leq n < \omega$, the formulae $A \supset (\neg A \supset B)$ and $\neg A \supset (A \supset B)$ are not valid.

6. $(\forall x(A(x))^{(n)}) \supset (\neg\forall x A(x) \supset \exists x \neg(A(x)))$

7. $(\forall x(A(x))^{(n)}) \supset (\neg\exists x(A(x)) \supset \forall x \neg(A(x)))$

8. $\neg\forall x A(x) \supset (\sim_n (\forall x(A(x))^{(n)}) \vee (\exists x \neg(A(x))))$

9. $\neg\exists x(A(x)) \supset (\sim_n (\forall x(A(x))^{(n)}) \vee (\forall x \neg(A(x))))$.

2 Tableaux systems for $\mathbf{C}_n^*, 1 \leq n < \omega$

In this section, we introduce analytical tableaux versions, à la [Smullyan, 1968], for the systems $\mathbf{C}_n^*, 1 \leq n < \omega$, named $\mathbf{TNDC}_n^*$. We adapt the notion of tableau sequence presented by [van Fraassen, 1971].

The language of the systems $\mathbf{TNDC}_n^*, 1 \leq n < \omega$, is the language $\mathbf{L}^*$ of the $\mathbf{C}_n^*, 1 \leq n < \omega$, excepting that we consider the symbol "o" (the ball operator), the symbols "k", "(k)" and the negations "$\sim_k$", for $k \geq 1$, as primitive symbols. Therefore, the language $\mathbf{L}^*$ contains a (infinite) denumerable set of primitive connectives.

The tableaux method is based on expansion rules, which allow us to analyze the formulae of $\mathbf{L}^*$. Essentially, the expansion rules allow us to expand a sequence of formulae into another sequence of formulae.

We need here to present the following definitions, necessary for the introduction of our hierarchy of tableaux systems, though having them been already introduced by [D'Ottaviano and Castro, 2005].

DEFINITION 7. For every tableaux system $\mathbf{TNDC}_n^*, 1 \leq n < \omega$, a *tableau sequence* for a given formula S, or simply a *tableau*, is a sequence of expressions $A_1, A_2, \ldots, A_k$, such that the formula S is put at the origin of the tableau, as the initial expression A_1; and every expression $A_i, 1 < i \leq k$, corresponds to a finite disjunction A_i^1 or ... or $A_i^{m_i}, m_i \geq 1$, where every $A_i^j, 1 \leq j \leq m_i$, is generated from the preceding expression(s) A_p^j, by applying one of the Expansion Rules 2.4 of the system. We call each A_i^j a *disjunct* of the expression A_i.

DEFINITION 8. For every system $\mathbf{TNDC}_n^*, 1 \leq n < \omega$, a branch j of a tableau sequence, $1 \leq j \leq m$ (with $m = max\ m_i, 1 \leq i \leq k$), corresponds to a sequence of expressions $A_i^s, 1 \leq i \leq k$ and $1 \leq s \leq m$, with A_1^1 the first expression and A_k^j the last one (the superior index s is equal to $1(s = 1)$, for $1 \leq i \leq i'$, for some $i' \leq k; s = j$, for $1 \leq i'' \leq k$, for some $i'' > i'$; and for $i' < i < i'', s$ assumes values between 1 and j).

REMARK 9. We observe that the tableau sequence has the structure of a *tree*, if we leave out the disjunction, and write the results of applying some rule under the disjunct to which the rule was applied. Thus, by thinking the disjunction as indicating a branching, the tableau sequence has the structure of an ordered dyadic tree à la [Smullyan, 1968].

For simplicity, the expressions of a given tableau branch j will be identified as of type A_i^j, with $1 \leq i \leq k$ and fixed $j, 1 \leq j \leq m_i$.

DEFINITION 10. A *node* corresponds to every expression A_i^j of every branch of a tableau, with $1 \leq i \leq k$ and $1 \leq j \leq m_i$.

Let the letters $\alpha, \beta, \gamma, \ldots, \psi$, if necessary also with indexes, stand for formulae of $\mathbf{L}^*$.

Now, suppose that $\mathbf{T}$ is a tableau being constructed from an initial formula A. Given a certain branch j, let A_{i-1}^j be the last expression of the branch. Then we may extend $\mathbf{T}$ by one of the following six operations:

(i) If some formula α occurs on the branch of the last expression A_{i-1}^j then, if δ_i^j and δ_{i+1}^j are generated from α by one of the Rules of Conjunctive Type $\mathbf{C}$ of the system, we may simultaneously adjoin the formulae δ_i^j and δ_{i+1}^j as the next expressions, on the branch j, after A_{i-1}^j;

(ii) If some formula β occurs on the branch of the last expression A_{i-1}^j then, if β_1 or β_2 is generated from β either by one of the Rules of Disjunctive Type $\mathbf{D}$ or by one of the Rules of Type $\mathbf{G}$ of the system, we may simultaneously adjoin the formula β_1 to the left of A_{i-1}^j, as the node δ_i^j, and the formula β_2 as the next expression to the right of A_{i-1}^j, as the node δ_i^{j+1};

(iii) If some formula γ occurs on the branch of the last expression A_{i-1}^j then, if the formula δ_i^j is generated from γ by one of the Rules of Special Type $\mathbf{S}_1$ Rules of Special Type $\mathbf{S}_3$, or Rules of Special Type $\mathbf{H}$ of the system, we may adjoin the formula δ_i^j, on the branch j, as the next expression after A_{i-1}^j;

(iv) If the formulae $\alpha_1, \ldots, \alpha_m$ occur on the branch of the last expression A_{i-1}^j then, if the formula δ_i^j is generated from $\{\alpha_1, \ldots, \alpha_m\}$ by one of the Rules of Special Type $\mathbf{S}_2$ of the system, we may adjoin, on the branch j, the formula δ_i^j as the next expression after A_{i-1}^j;

(v) If some formula α occurs on the branch of the last expression A_{i-1}^j then, if $\delta_i^j(t)$ is generated from α by one of the Rules of Type $\mathbf{E}$ of the system, we may adjoin the formulae $\delta_i^j(t)$ as the next expression, on the branch j, after A_{i-1}^j;

(vi) If some formula α occurs on the branch of the last expression A_{i-1}^j then, if $\delta_i^j(c)$ is generated from α by one of the Rules of Type $\mathbf{F}$ of the system, we may adjoin the formulae $\delta_i^j(c)$ as the next expression, on the branch j, after A_{i-1}^j.

[D'Ottaviano and Castro, 2005] introduce a hierarchy of syntactical tableaux systems $\mathbf{TNDC}_n, 1 \leq n < \omega$, in which every system $\mathbf{TNDC}_n$ is equivalent to da Costa's corresponding system $\mathbf{C}_n, 1 \leq n < \omega$. Here, besides the Expansion Rules of $\mathbf{TNDC}_n$, we have specific rules to deal with quantifiers.

The Expansion Rules of $\mathbf{TNDC}_n^*$ are introduced below.

Expansion Rules 2.4 The *expansion rules* of the tableaux systems **TNDC**$_n^*$, $1 \leq n < \omega$, are the following.

2.4.1 Rules of Conjunctive Type C:

$$\frac{\alpha}{\delta_i^j}$$
$$\delta_{i+1}^j$$

α	δ_i^j	δ_{i+1}^j	Name of the Rule
$A \& B$	A	B	$E\&$
$A^{(k)}$	A^k	$A^{(k-1)}$	$E(k), k > 1$
$\neg(A^k)$	A^{k-1}	$\neg(A^{k-1})$	$Ek\neg, k \geq 1$, where A^0 is A
$\neg(A^{(k)})$	A	$\neg A$	$E(k)\neg, k \geq 1$
$\sim_n \neg A$	$\neg\neg A$	$A^{(n)}$	$E \sim_n \neg$
$\sim_n (A^k)$	$\neg(A^k)$	$(A^k)^{(n)}$	$Ek \sim_n, k \geq 1$
$\sim_n (A \vee B)$	$\sim_n A$	$\sim_n B$	$DND \sim_n$
$\sim_n (A \supset B)$	A	$\sim_n B$	$DNI \sim_n$
$\sim_n (A^{(k)})$	A	$\neg A$	$E(k) \sim_n, k \geq 1$
$\sim_k A$	$\neg A$	$A^{(k)}$	$E \sim_k, k < n$ (i)

2.4.2 Rules of Disjunctive Type D:

$$\frac{\beta}{\delta_i^j \mid \delta_i^{j+1}}$$

β	δ_i^j	δ_i^{j+1}	Name of the Rule
$A \vee B$	A	B	$E\vee$
$A \supset B$	$\sim_n A$	B	$E \supset$
$\neg(A\&B)$	$\neg A$	$\neg B$	$DNC\neg$, where B is distinct of $\neg A$ (ii)
$\neg(A \vee B)$	$\neg(A^{(n)} \& B^{(n)})$	$\neg A \& \neg B$	$DND\neg$
$\neg(A \supset B)$	$\neg(A^{(n)} \& B^{(n)})$	$A \& \neg B$	$DNI\neg$
$\sim_n (A\&B)$	$\sim_n A$	$\sim_n B$	$DNC \sim_n$

2.4.3 Rules of Special Type S_1:

$$\frac{\gamma}{\delta_i^j}$$

γ	δ_i^j	Name of the Rule
$\neg\neg A$	A	$E\neg\neg$
$\neg \sim_k A$	A	$E\neg \sim_k, k \geq 1$
$\sim_n \sim_k A$	A	$E \sim_n \sim_k, k \geq 1$
$\sim_k A$	$\sim_{k-1} A$	$R \sim_k, k > n$
A^k	$\neg(A^{k-1} \& \neg A^{k-1})$	$Rk, k \geq 1$, where A^0 is A (iii)
$A^{(1)}$	A^1	$E(1)$

2.4.4 Rules of Special Type S_2:

$$\frac{\alpha_1 \\ \vdots \\ \alpha_m}{\delta_i^j}$$

$\alpha_1, \ldots, \alpha_m$	δ_i^j	Name of the Rule
$\{\neg A, A^1, \ldots, A^k\}$	$\sim_k A$	$I \sim_k, 0 < k < n$
$\{A^1, A^2, \ldots, A^k\}$	$A^{(k)}$	$I(k), 0 < k < n$ (i)

2.4.5 Rules of Special Type S_3 (iv):

$$\frac{\varepsilon}{\delta_i^j}$$

ε	δ_i^j	Name of the Rule
$A^{00\ldots 0}$	A^k	$E\circ$ (with "$\circ$" k-times)
$\neg(A^{k-1} \& \neg A^{k-1})$	A^k	$Ik, k \geq 1$, where A^0 is A (i)
$(A^s)^k$	A^{s+k}	$Is + k$, for $s, k \geq 1$
$A^1 \& A^2 \& \ldots \& A^k$	$A^{(k)}$	$I'(k), k \geq 1$ (v)
$\neg A \& A^{(k)}$	$\sim_k A$	$I' \sim_k, k \geq 1$ (v)

(i) This Rule must be applied only once, on each branch and for each formula.

(ii) If A is of type $(C^{k-1} \& \neg (C^{k-1}))$, then B must be distinct of C^k.

(iii) This Rule must be applied only when there is no possibility of applying any other Rule; it can be applied in subformulae of formulas that occur in the nodes and, in these cases, it must be applied "from outside to inside", that is, from the connective of largest scope to the connective of smallest scope.

(iv) The Rules of Special Type S_3 must be immediately applied, in every case, after applying the first Rule in the initial node of the tableau; they can be applied to subformulae of formulas that occur in the nodes and, in these cases, they must be applied "from outside to inside".

(v) These Rules, under conditions (iv), can only be applied to proper subformulae of formulas that occur in the nodes and, in these cases, they must be applied "from outside to inside".

REMARK 11. We observe that A^0, which corresponds to the formula A with superior index "0" (numeral 0), coincides with the formula A. This formula is distinct of the formula A^0 ("A-ball").

In the Rules of Special Type S_2 we use the notation of set in order to indicate that it is not important the order in which the formulas occur in the nodes of a branch.

Also, the only Rules that can be applied to subformulae, are the Rules of Special Type S_2 and the Rule Rk.

Following we introduce the specific rules for quantifiers.

2.4.6 *Rules of Type* **E**: $\dfrac{\alpha}{\delta_i^j(t)}$, where t is a given term free for a given variable occurring in the formula α.

α	δ_i^j	Name of the Rule
$\forall xA$	$A(t)$	E-$\forall$, with proviso
$\sim_n \exists xA$	$\sim_n(A(t))$	$\sim_n \exists$, with proviso

2.4.7 *Rules of Type* **F** (i): $\dfrac{\alpha}{\delta_i^j(c)}$, where c is a constant that does not occur in the branch; or c has not been previously introduced by Rule of Type **F**, and does not occur in α, and no constant of α has been previously introduced by Rule of Type **F**.

α	δ_i^j	Name of the Rule
$\exists xA$	$A(c)$	E-$\exists$, with proviso
$\sim_n \forall xA$	$\sim_n(A(c))$	$\sim_n \forall$, with proviso

2.4.8 *Rules of Special Type* **G**: $\dfrac{\beta}{\delta_i^j \mid \delta_i^{j+1}}$

β	δ_i^j	δ_i^{j+1}	Name of the Rule
$\neg \forall xA$	$\sim_n(\forall x(A)^{(n)})$	$\exists x \neg A$	$\neg \forall$
$\neg \exists xA$	$\sim_n(\forall x(A)^{(n)})$	$\forall x \neg A$	$\neg \exists$

2.4.9 *Rules of Special Type* **H**: $\dfrac{\gamma}{\delta_i^j}$

γ	δ_i^j	Name of the Rule
A_1	A_2	tA, where A_2 is congruent to A_1, and the new variables must previously occur in the branch
$\sim_k A_1$	$\sim_k A_2$	sA, $k \geq 1$ (ii)
$\neg A_1$	$\neg A_2$	wA, (ii)

(i) The Rules of Special Type **F** must be immediately applied, in every case, after applying the first Rule in the initial node of the tableau.

(ii) Where **A**$_2$ is congruent to **A**$_1$ and **A**$_2$ occurs previously in the branch.

REMARK 12. In the application of the Expansion Rules, priority must be given to the Rules of Type $C, E-\exists, \sim_n \exists, \neg\forall$ and to the Rules of Special Type.

Definition 2.5 For every system $\mathbf{TNDC}_n^*, 1 \leq n < \omega$, a branch $A_1^j, \ldots, A_s^j$ of a tableau is called a *closed branch* if there exist nodes $A_i^j, 1 \leq r \leq s$, that correspond either to formulae B and $\sim_n B$, or to formulae $B, \neg B$ and $B^1, B^2, \ldots, B^n$.

The next definition gives us the closure criterion for the tableaux.

DEFINITION 13. Given a formula S, a *tableau* for S is *closed* if all its branches are closed; otherwise, it is *open*.

DEFINITION 14. A *set of formulae* Γ is said to be *closed* if, and only if, there exists a finite subset Γ_0 of Γ, such that there exists a closed tableau for the conjunction of the formulae of Γ_0; otherwise, it is *open*.

In what follows, we use Γ, A as an abbreviation for $\Gamma \cup \{A\}$[9].

DEFINITION 15. For every tableaux system $\mathbf{TNDC}_n^*, 1 \leq n < \omega$, a formula S is said to be an *analytical consequence* of a set Γ of formulae if, and only if, $\Gamma, \sim_n S$ is closed. We also say that Γ, by the Expansion Rules, *generates* S.

This is denoted by: $\Gamma \vdash_{\mathbf{TNDC}_n^*} S$.

We observe that, for a formula S to be provable in $\mathbf{TNDC}_n^*, 1 \leq n < \omega$, a closed tableau must be generated from the strong negation $\sim_n S$ f the formula.

DEFINITION 16. For every tableaux system $\mathbf{TNDC}_n^*, 1 \leq n < \omega$, a formula S is said to be *provable* if, and only if, there is a closed tableau for $\sim_n S$, that is, if $\{\sim_n S\}$ is closed.

This is denoted by: $\vdash_{\mathbf{TNDC}_n^*} S$.

DEFINITION 17. For a given tableau $\mathbf{T}$ in $\mathbf{TNDC}_n^*, 1 \leq n < \omega$, a branch j is *complete* if, and only if, there is not any Expansion Rule that can still be applied on some node of j.

DEFINITION 18. A tableau $\mathbf{T}$, in $\mathbf{TNDC}_n^*, 1 \leq n < \omega$, is *complete* if, and only if, every branch j of $\mathbf{T}$ is either closed or complete.

---

[9] Γ, A, B is the same as Γ, B, A.

EXAMPLES 19. Following, we present some examples of proofs in the systems $\mathbf{TNDC}_n^*$, $1 \leq n < \omega$. The rules used are indicated to the right of each step of the proof; the numbers on the left side are added only to facilitate mentioning the tableau.

a) $\not\vdash_{TNDC_n}{}^* \neg\forall x \neg A(x) \equiv \exists x A(x)$

a.1) $\vdash_{TNDC_n}{}^* \neg\forall x \neg A(x) \supset \exists x A(x)$

1	$\sim_n(\neg\forall x \neg A(x) \supset \exists x A(x))$	
2	$\neg\forall x \neg A(x)$	1, DNI$\sim_n$
3	$\sim_n(\exists x A(x))$	1, DNI$\sim_n$

4	$\sim_n(\forall x(\neg A)^{(n)})$	2, $\neg\forall$		5	$\exists x \neg\neg A$	2, $\neg\forall$
6	$\sim_n((\neg A(c))^{(n)})$	4, $\sim_n \forall$		i+1	$\neg\neg A(c)$	5, E-$\exists$
7	$(\neg A(c))^{(n)}$	6, E(k)$\sim_n$		i+2	$A(c)$	i+1, E$\neg\neg$
8	$\neg((\neg A(c))^{(n)})$	6, E(k)$\sim_n$		i+3	$\sim_n A(c)$	3, $\sim_n \exists$
9	$\neg A(c)$	8, E(k) $\neg$			*	
10	$\neg(\neg A(c))$	8, E(k) $\neg$				
11	$A(c)$	10, E-$\neg\neg$				
12	$(\neg A(c))^n$	7, E(k)				
13	$(\neg A(c))^{(n-1)}$	7, E(k)				
14	$(\neg A(c))^{n-1}$	13, E(k)				
15	$(\neg A(c))^{(n-2)}$	13, E(k)				
16	$(\neg A(c))^{n-2}$	15, E(k)				
17	$(\neg A(c))^{(n-3)}$	15, E(k)				
⋮	⋮					
i-1	$(\neg A(c))^{(1)}$	i-2, E(k)				
i	$(\neg A(c))^1$	i-1, E(1)				
	*					

The tableau closes in the first branch by the formulae $\neg A(c), \neg(\neg A(c))$, $(\neg A(c))^n, (\neg A(c))^{n-1}, (\neg A(c))^{n-2}, \ldots, (\neg A(c))^1$ which occur on the nodes 9, 10, 12, 14, 16, ..., i; and in the second branch by the formulae $A(c)$ and $\sim_n A(c)$ which occur on the nodes $i+2$ and $i+3$.

Analytical Tableaux for da Costa's Paraconsistent Predicate Calculi C_n^*

a.2) $\nvdash_{TNDC_n^*} \exists x A(x) \supset \neg \forall x \neg A(x)$

1	$\sim_n(\exists x A(x) \supset \neg \forall x \neg A(x))$	
2	$\exists x A(x)$	1, DNI$\sim_n$
3	$\sim_n(\neg \forall x \neg A(x))$	1, DNI$\sim_n$
4	$A(c)$	2, E-$\exists$
5	$\neg\neg\forall x \neg A(x)$	3, E$\sim_n\neg$
6	$(\forall x \neg A(x))^{(n)}$	3, E$\sim_n\neg$
7	$\forall x \neg A(x)$	5, E$\neg\neg$
9	$\neg A(c)$	7, E$\forall$
10	$(\forall x \neg A(x))^n$	6, E(k)
11	$(\forall x \neg A(x))^{(n-1)}$	6, E(k)
12	$(\forall x \neg A(x))^{n-1}$	11, E(k)
13	$(\forall x \neg A(x))^{(n-2)}$	11, E(k)
14	$(\forall x \neg A(x))^{n-2}$	13, E(k)
15	$(\forall x \neg A(x))^{(n-3)}$	13, E(k)
$\vdots$	$\vdots$	
i-1	$(\forall x \neg A(x))^{(1)}$	i-2, E(k)
i	$(\forall x \neg A(x))^1$	i-1, E(k)
i+1	$\neg((\forall x \neg A(x))^{n-1} \& \neg(\forall x \neg A(x))^{n-1})$	10, Rk
i+2	$\neg((\forall x \neg A(x))^{n-2} \& \neg(\forall x \neg A(x))^{n-2})$	12, Rk
i+3	$\neg((\forall)\neg A(x))^{n-3} \& \neg(\forall x \neg A(x))^{n-3})$	14, Rk
$\vdots$	$\vdots$	
i+j	$\neg((\forall x \neg A(x)) \& \neg(\forall x \neg A(x)))$	i-1, Rk
i+j+1	$(\forall x \neg A(x))^n$	i+1, Ik
i+j+2	$(\forall x \neg A(x))^{n-1}$	i+2, Ik
i+j+3	$(\forall x \neg A(x))^{n-2}$	i+3, Ik
i+j+4	$(\forall x \neg A(x))^{n-3}$	i+4, Ik
$\vdots$	$\vdots$	
i+j+s	$(\forall x \neg A(x))^1$	i+j, Ik
i+j+s+1	$\neg((\forall x \neg A(x))^{n-1} \& \neg(\forall x \neg A(x))^{n-1})$	i+j+1, Rk
i+j+s+2	$\neg((\forall x \neg A(x))^{n-2} \& \neg(\forall x \neg A(x))^{n-2})$	i+j+2, Rk
i+j+s+3	$\neg((\forall)\neg A(x))^{n-3} \& \neg(\forall x \neg A(x))^{n-3})$	i+j+3, Rk
$\vdots$	$\vdots$	
i+j+s+p	$\neg((\forall x \neg A(x)) \& \neg(\forall x \neg A(x)))$	i+j+s, Rk
i+j+s+p+1	$(\forall x \neg A(x))^{(1)}$	i, I(k)
i+j+s+p+2	$(\forall x \neg A(x))^{(2)}$	i-2, i, I(k)
i+j+s+p+2	$(\forall x \neg A(x))^{(3)}$	i-4, i, I(k)
i+j+s+p+u	$(\forall x \neg A(x))^{(n-1)}$	12, 14, 16, ..., i, I(k)
$\vdots$	$\vdots$	
i+j+s+p+u+1	$(\forall x \neg A(x))^{(n-2)}$	i+j+s+p+u, E(k)
i+j+s+p+u+2	$(\forall x \neg A(x))^{n-2}$	i+j+s+p+u+1, E(k)
i+j+s+p+u+3	$(\forall x \neg A(x))^{(n-3)}$	i+j+s+p+u+2, E(k)
i+j+s+p+u+4	$(\forall x \neg A(x))^{n-3}$	i+j+s+p+u+3, E(k)
$\vdots$	$\vdots$	
i+j+s+p+u+z	$(\forall x \neg A(x))^{(1)}$	i+j+s+p+u+z-1, E(k)
i+j+s+p+u+z+1	$(\forall x \neg A(x))^1$	i+j+s+p+u+z, E(k)

The tableau is complete but is not closed.

b) $\vdash_{TNDC_{155}^{*}} ((A \supset B) \supset A) \supset A$

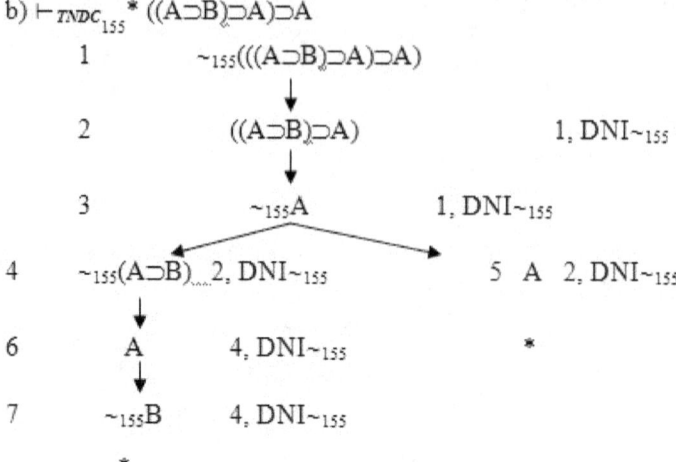

The tableau is closed by the formulas $\sim_{155} A$ and A, that occur in the nodes 3 and 6 of the first branch, and in the nodes 3 and 5 of the second branch.

3 The Cut Rule for the systems $\mathbf{TNDC}_n^*$

In this section we prove a special version of Cut Rule for the systems $\mathbf{TNDC}_n^*$.

LEMMA 20. *If Γ, A is closed, then Γ, A, B is closed.*

Proof. Immediate, from Definition 14. ∎

THEOREM 21. *(Cut Rule) For every system $\mathbf{TNDC}_n^*$, there exists a closed tableau for a set Γ of formulae if, and only if, for a given formula S there exist closed tableaux either for $\Gamma \cup \{S\}$ and $\Gamma \cup \{\sim_n S\}$, or for $\Gamma \cup \{S\}$ and $\Gamma \cup \{\neg S, S^1, S^2, \ldots, S^n\}$.*

Proof. If there exists a closed tableau for Γ, by Lemma 3.1, it is immediate that there are closed tableaux for Γ, S and $\Gamma, \sim_n S$, and for Γ, S and $\Gamma, \neg S, S^1, S^2, \ldots, S^n$.

Now, suppose that either there exist closed tableaux for Γ, S and for $\Gamma, \sim_n S$, or there exist closed tableaux for Γ, S and $\Gamma, \neg S, S^1, S^2, \ldots, S^n$. The proof that there exists a closed tableau for Γ is done by induction on the complexity of the formula S, as we have similarly presented in [D'Ottaviano and Castro, 2005], Section 4, pp 85-91.

1 Let S be an atomic formula A.

Suppose that there are closed tableaux for Γ, A and for $\Gamma, \sim_n A$. In the cases when either $A \in \Gamma$ or $\sim_n A \in \Gamma$, it is immediate that Γ is closed. Hence, we have only to analyze the case when $A \notin \Gamma$ and $\sim_n A \notin \Gamma$. If either Γ, A and $\Gamma, \sim_n A$ is closed only on account of formulae of Γ, then Γ is closed and we have nothing to prove; the same reasoning is applicable to the case when we have that Γ, A and $\Gamma, \neg A, A^1, A^2, \ldots, A^n$ are closed.

1.1 Suppose that there are closed tableaux for Γ, A and for $\Gamma, \sim_n A$. Observe that from A atomic we can not generate any formula, and from $\sim_n A$ we also can not generate any formula.

If Γ, A is closed then, by Definition 14, there is a tableau **T** such that its branches are closed either by $\sim_n A$, or by $\neg A, A^1, A^2, \ldots, A^n$. As $\Gamma, \sim_n A$ is also closed, then there is a closed tableau **T'** such that its branches are closed by A, or by $\sim_n \sim_n A$, or by $\neg \sim_n A$ and $(\sim_n A)^1, (\sim_n A)^2, \ldots, (\sim_n A)^n$; that is, by Rules $E \sim_n \sim_k$ and $E \neg \sim_k$, the formula A appears in all the branches of **T'**.

Therefore, in the tableaux **T** and **T'** the formulae $\sim_n A, \neg A, A^1, A^2, \ldots, A^n$ (in **T**), and A (in **T'**), respectively, are directly generated, by the Expansion Rules, from Γ, because, neither $\sim_n A, \neg A, A^1, A^2, \ldots, A^n$ could be generated from A, nor A could be generated from $\sim_n A$.

Hence, there is a closed tableau for Γ and so, by Definition 2.7, Γ is closed.

1.2 Suppose that there are closed tableaux for Γ, A and for $\Gamma, \neg A, A^1, A^2, \ldots, A^n$. Observe that it is not possible to generate any formula from A and from $\neg A, A^1, A^2, \ldots, A^n$ it is only possible to generate $\sim_k A$ and $A^{(k)}, k < n$ (by Rules $I \sim_k$ and $I(k)$).

If Γ, A is closed then, by Definition 14, there is a tableau **T** such that its branches are closed either by $\sim_n A$, or by $\neg A$ and $A^1, A^2, \ldots, A^n$.

As $\Gamma, \neg A, A^1, A^2, \ldots, A^n$ is also closed, then there is a closed tableau **T'** such that its branches are closed by A; or by $\sim_n \neg A$; or by $\neg\neg A, (\neg A)^1, (\neg A)^2, \ldots, (\neg A)^n$; or by $\sim_n (A^i)$, for every $i, 1 \leq i \leq n$; or by $\neg(A^i), (A^i)^1, (A^i)^2, \ldots, (A^i)^n$, for every $i, 1 \leq i \leq n$. So, by Rules $E \sim_n \neg, E\neg\neg, Ek \sim_n, E\&$ and $Ek\neg$, the formula A appears in all the branches of **T'**.

Therefore, in the tableaux **T** and **T'** the formulae $\sim_n A$, or $\neg A, A^1, A^2, \ldots, A^n$ (in **T**) and A (in **T'**), respectively, are directly generated, by the Expansion Rules, from Γ.

Hence, there exists a closed tableau for Γ and so, by Definition 14, Γ is closed.

2 Suppose that the result holds for formulae S of complexity $p, p0$.

3 Let S be a formula of complexity $p + 1$.

The cases where S is of type $\neg B$, with B of complexity p; S is of type $B^k, k \geq 1$; S is of type $B^{(k)}$, with $k \geq 1$; S is of type $\sim_k B$, with $k \geq 1$: S is of type $(B\&C)$; S is of type $(B \vee C)$ and S is of type $(B \supset C)$ can be proved exactly as in 14.

3.1 Let S be of type $\forall x B$, with B of complexity p.

3.1.1 Let $\Gamma, \forall x B$ and $\Gamma, \sim_n \forall x B$ be closed, considering that $\forall x B$ and $\sim_n \forall x B$ are not formulae of Γ.

If $\Gamma, \sim_n \forall x B$ is closed, then, by Rule $\sim_n \forall, \Gamma, \sim_n B(c)$ is closed (with c not occurring in the considered branch of the tableaux; or c has not been previously introduced by Rule of Type **F**, and does not occur in $\exists x B$, and no constant of $\exists x B$ has been previously introduced by Rule of Type **F**).

As $\Gamma, \sim_n \forall x B$ is also closed then, by Rule $E - \forall, \Gamma, B(c)$ is closed.

So, as $\Gamma, B(c)$ and $\Gamma, \sim_n B(c)$ are closed, hence by induction hypothesis, Γ is closed.

3.1.2 Let $\Gamma, \forall x B$ and $\Gamma, \neg \forall x B, (\forall x B)^1, (\forall x B)^2, \ldots, (\forall x B)^n$ be closed, also considering that $\forall x B, \neg \forall x B$ and $(\forall x B)^i$, for every $i, 1 \leq i \leq n$, are not formulae of Γ. We observe that, from $((\forall x B)^i$, for $1 \leq i \leq k$, by Rule $Rk, k \geq 1$, it is only possible to generate the formula $\neg((\forall x B)^{i-1} \& \neg((\forall x B)^{i-1}))$.

If $\Gamma, \neg \forall x B, (\forall x B)^1, (\forall x B)^2, \ldots, (\forall x B)^n$ is closed then, by Rule $\neg \forall$, there exist closed tableaux for $\Gamma, \sim_v (\forall x (B)^{(n)}), (\forall x B)^1, (\forall x B)^2, \ldots, (\forall x B)^n$ and for $\Gamma, \exists x \neg B, (\forall x B)^1, (\forall x B)^2, \ldots, (\forall x B)^n$. Therefore,

(i) $\Gamma, \sim_n \ (\forall x (B)^{(n)}), (\forall x B)^1, (\forall x B)^2, \ldots, (\forall x B)^n$ is closed, that is, $\Gamma, (\forall x B)^1, (\forall x B)^2, \ldots, (\forall x B)^n, \sim_n (\forall x (B)^{(n)})$ is closed.

As $\Gamma, \forall x B$ is also closed, by Lemma 3.1, we have that $\Gamma, \forall x B, (\forall x B)^1, (\forall x B)^2, \ldots, (\forall x B)^n, \forall x (B)^{(n)}$ is closed.

Hence, by induction hypothesis, we have that $\Gamma, \forall x B, (\forall x B)^1, (\forall x B)^2, \ldots, (\forall x B)^n$ is closed.

(ii) If $\Gamma, \exists x \neg B, (\forall x B)^1, (\forall x B)^2, \ldots, (\forall x B)^n$ is closed, then, by Rule $E - \exists, \Gamma, \neg B(c), (\forall x B)^1, (\forall x B)^2, \ldots, (\forall x B)^n$ is closed, with c not occurring in the considered branch of the tableaux; or c has not been previously introduced by Rule of Type **F**, and does not occur in $\exists x \neg B$, and no constant of $\exists x \neg B$ has been previously introduced by Rule of Type **F**. That is, $\Gamma, (\forall x B)^1, (\forall x B)^2, \ldots, (\forall x B)^n, \neg B(c)$ is closed.

(iii) As, by (i), $\Gamma, \forall x (B), (\forall x B)^1, (\forall x B)^2, \ldots, (\forall x B)^n$ is closed then, by Rule $E - \forall, \Gamma, B(c), (\forall x B)^1, (\forall x B)^2, \ldots, (\forall x B)^n$ is also closed, that is, $\Gamma, (\forall x B)^1, (\forall x B)^2, \ldots, (\forall x B)^n, B(c)$ is closed. So, by (ii), by induction hypothesis, $\Gamma, (\forall x B)^1, (\forall x B)^2, \ldots, (\forall x B)^n$ is closed.

If there is a branch that is closed on account of Γ and one of the formulae $(\forall x B)^1, (\forall x B)^2, \ldots, (\forall x B)^n$, then it is closed either by $\sim_n ((\forall x B)^i)$; or by $\neg((\forall x B)^i), ((\forall x B)^i)^1, ((\forall x B)^i)^2, \ldots, ((\forall x B)^i)^n$, for $1 \leq i \leq n$. Then, by applying Rules $Ek \sim_n, E\&$ and $Ek\neg$, in these previous mentioned formulae, the formula $\forall x B$ appears in this branch. As $\forall x B$ can not be generated from

$\neg \forall x B, \sim_n (\forall x(B)^{(n)})$ or $\exists x \neg B$, by the Expansion Rules, Γ generates $\forall x B$. Therefore, as $\Gamma, \forall x B$ is closed and Γ generates $\forall x B$, we have that Γ is closed.

3.2 Let S be of type $\exists x B$, with B of complexity p.

3.2.1. Let $\Gamma, \exists x B$ and $\Gamma, \sim_n \exists x B$ be closed, considering that $\exists x B$ and $\sim_n \exists x B$ are not formulae of Γ.

If $\Gamma, \exists x B$ is closed, then, by Rule $E - \exists, \Gamma, B(c)$ is closed (with c not occurring in the considered branch of the tableaux; or c has not been previously introduced by Rule of Type **F**, and does not occur in $\exists x B$, and no constant of $\exists x B$ has been previously introduced by Rule of Type **F**).

If $\Gamma, \sim_n \exists x B$ is closed, then, by Rule $\sim_n \exists, \Gamma, \sim_n B(c)$ is closed.

Hence, as $\Gamma, B(c)$ and $\Gamma, \sim_n B(c)$ are closed, then by induction hypothesis, Γ is closed.

3.2.2 Let $\Gamma, \exists x B$ and $\Gamma, \neg \exists x B, (\exists x B)^1, (\exists x B)^2, \ldots, (\exists x B)^n$ be closed, considering also that $\exists x B, \neg \exists x B$ and $(\exists x B)^i$, for any $i, 1 \leq i \leq n$, are not formulae of Γ. We observe that, from $(\exists x B)^i$, for $1 \leq i \leq k$, by Rule $Rk, k \geq 1$, it is only possible to generate the formula $\neg((\exists x B)^{i-1} \& \neg((\exists x B)^{i-1}))$.

If $\Gamma, \exists x B$ is closed, then, by Rule $E - \exists, \Gamma, B(c)$ is closed, with proviso.

If $\Gamma, \neg \exists x B, (\exists x B)^1, (\exists x B)^2, \ldots, (\exists x B)^n$ is closed then, by Rule $\neg \exists$, there exist closed tableaux for $\Gamma, \sim_n (\forall x(B)^{(n)}), (\exists x B)^1, (\exists x B)^2, \ldots, (\exists x B)^n$ and for $\Gamma, \forall x \neg B, (\exists x B)^1, (\exists x B)^2, \ldots, (\exists x B)^n$. Therefore,

(i) $\Gamma, \sim_n (\forall x(B)^{(n)}), (\exists x B)^1, (\exists x B)^2, \ldots, (\exists x B)^n$ is closed then, by Lemma 3.1, we have that $\Gamma, \exists x B, (\exists x B)^1, (\exists x B)^2, \ldots, (\exists x B)^n, \sim_n (\forall x(B)^{(n)})$ is closed.

As $\Gamma, \exists x B$ is also closed, again by Lemma 3.1, we have that $\Gamma, \exists x B, (\exists x B)^1, (\exists x B)^2, \ldots, (\exists x B)^n, \forall x(B)^n$ is closed.

Hence, by induction hypothesis, we have that $\Gamma, \exists x B, (\exists x B)^1, (\exists x B)^2, \ldots, (\exists x B)^n$ is closed. So, by applying Rule $E - \exists, \Gamma, B(c), (\exists x B)^1, (\exists x B)^2, \ldots, (\exists x B)^n$ is closed (with proviso). That is, $\Gamma, (\exists x B)^1, (\exists x B)^2, \ldots, (\exists x B)^n, B(c)$ is closed.

(ii) If $\Gamma, \forall x \neg B, (\exists x B)^1, (\exists x B)^2, \ldots, (\exists x B)^n$ is closed, then, by Rule $E - \forall, \Gamma, \neg B(c), (\exists x B)^1, (\exists x B)^2, \ldots, (\exists x B)^n$ is closed, with proviso. That is, $\Gamma, (\exists x B)^1, (\exists x B)^2, \ldots, (\exists x B)^n, \neg B(c)$ is closed and so, by Lemma 20, $\Gamma, (\exists x B)^1, (\exists x B)^2, \ldots, (\exists x B)^n, \neg B(c), (B(c))^1, (B(c))^2, \ldots, (B(c))^n$ is closed.

(iii) As, by (i) and (ii), $\Gamma, (\exists x B)^1, (\exists x B)^2, \ldots, (\exists x B)^n, B(c)$ and $\Gamma, (\exists x B)^1, (\exists x B)^2, \ldots, (\exists x B)^n, \neg B(c), (B(c))^1, (B(c))^2, \ldots, (B(c))^n$ are closed then, by induction hypothesis, $\Gamma, (\exists x B)^1, (\exists x B)^2, \ldots, (\exists x B)^n$ is closed.

If there is a branch that is closed on account of Γ and one of the formulae $(\exists x B)^1, (\exists x B)^2, \ldots, (\exists x B)^n$, then it is closed by $\sim_n ((\exists x B)^i)$; or by $\neg((\exists x B)^i), ((\exists x B)^i)^1, ((\exists x B)^i)^2, \ldots, ((\exists x B)^i)^n$, for $1 \leq i \leq n$. Then, by Rules $Ek \sim_n, E\&$ and $Ek\neg$, the formula $\exists x B$ appears in this branch. As $\exists x B$ can not be generated from $\neg \exists x B$ or $\sim_n (\exists x(B)^{(n)})$ or $\forall x \neg B$, by the Expansion Rules, Γ generates $\exists x B$. Therefore, as $\Gamma, \exists x B$ is closed and Γ generates $\exists x B$, we have that Γ is closed.

Hence, by Cases 1-3, we have proved that if either Γ, S and $\Gamma, \sim_n S$ are closed, or Γ, S and $\Gamma, S, \neg S, S^1, S^2, \ldots, S^n$ are closed, then Γ is closed. ∎

4 The logical equivalence between the systems of the hierarchy $\mathbf{TNDC}_n^*$ and the corresponding da Costa's systems $\mathbf{C}_n^*, 1 \leq n < \omega$

Now, based on the Cut Rule for $\mathbf{TNDC}_n^*, 1 \leq n < \omega$, we can prove the "equivalence" between the systems $\mathbf{TNDC}_n^*$ and the corresponding da Costa's paraconsistent systems $\mathbf{C}_n^*, 1 \leq n < \omega$, in the sense that, for every set of formulas $\Gamma \cup \{S\}$ (in the language of $\mathbf{C}_n^*$), $\Gamma \vdash_{\mathbf{C}_n^*} S$ if and only if, $\Gamma \vdash_{\mathbf{TNDC}_n^*} S$. In particular, a formula S is a theorem of $\mathbf{C}_n^*$ if, and only if, S is provable in $\mathbf{TNDC}_n^*$; that is, $\vdash_{\mathbf{C}_n^*} S$ if, and only if, $\vdash_{\mathbf{TNDC}_n^*} S$.

THEOREM 22. *If* $\Gamma \vdash_{\mathbf{C}_n^*} S$, *then* $\Gamma \vdash_{\mathbf{TNDC}_n^*} S$, *for every* $n, 1 \leq n < \omega$.

Proof. Suppose that $\Gamma \vdash_{\mathbf{C}_n^*} S$.

If $S \in \Gamma$ then, for every $n, 1 \leq n < \omega$, it is immediate that $\Gamma \vdash_{\mathbf{TNDC}_n^*} S$. So, let us suppose that S is not in Γ.

1 Let S be an axiom schema of $\mathbf{C}_n^*, 1 \leq n < \omega$.

Let us prove that $\Gamma \vdash_{\mathbf{TNDC}_n^*} S$, that is, we have to prove that $\Gamma, \sim_n S$ is closed in $\mathbf{TNDC}_n^*, 1 \leq n < \omega$.

Here, we present the proofs for the Axiom schemata 15 to 19 and for the Deduction Rules. The cases of Axioms 1 to 14 were proved by [D'Ottaviano and Castro, 2005].

1.1 Let S be Axiom 15, that is, S is $\forall x A \supset A(t)$, where t and the formulae $A(x)$ satisfy the usual restrictions[10]. We shall generate a closed tableau, whose initial node, $\Gamma, \sim_n S$, constitutes the step 1 below.

1	$\Gamma, \sim_n(\forall x A \supset A(t))$	
2	$\Gamma, \forall x A$	1, DNI$\sim_n$
3	$\Gamma, \sim_n(A(t))$	1, DNI$\sim_n$
4	$\Gamma, A(t)$	2, E$\forall$
	*	

---

[10]See [Kleene, 1952], p. 81.

In this case, the tableau closes by the formulae $\sim_n (A(t))$ and $A(t)$, occurring in the nodes 3 and 4, respectively.

1.2 Let S be Axiom 16, that is, S is $A(t) \supset (\exists x A)$, where t is a term which is free for x in $A(x)$. We shall generate a closed tableau, whose initial node, $\Gamma, \sim_n S$, constitutes the step 1 below.

1	$\Gamma, \sim_n(A(t) \supset (\exists xA))$	
2	$\Gamma, A(t)$	1, DNI$\sim_n$
3	$\Gamma, \sim_n(\exists xA)$	1, DNI$\sim_n$
4	$\Gamma, \sim_n(A(t))$	3, $\sim_n\exists$
	*	

1.3 Let S be Axiom 17, that is, S is $\forall \underline{x}(A)^{(n)} \supset (\forall xA)^{(n)}$.

1	$\Gamma, \sim_n(\forall x(A)^{(n)} \supset (\forall xA)^{(n)})$	
2	$\Gamma, \forall x(A)^{(n)}$	1, DNI$\sim_n$
3	$\Gamma, \sim_n((\forall xA)^{(n)})$	1, DNI$\sim_n$
4	$\Gamma, (\forall xA)$	3, E(k)$\sim_n$
5	$\Gamma, \neg(\forall \underline{x}A)$	3, E(k)$\sim_n$

6	$\Gamma, \sim_n(\forall x(A)^{(n)})$ 5, $\neg\forall$		7	$\Gamma, \exists x\neg A$	5, $\neg\forall$
	*		8	$\Gamma, \neg A(c)$	7, E-$\exists$
			9	$\Gamma, A(c)$	4, E-$\forall$
			10	$\Gamma, (A(c))^{(n)}$	2, E-$\forall$
			11	$\Gamma, (A(c))^n$	10, E(k)
			12	$\Gamma, (A(c))^{(n-1)}$	10, E(k)
			13	$\Gamma, (A(c))^{n-1}$	12, E(k)
			14	$\Gamma, (A(c))^{(n-2)}$	12, E(k)
			15	$\Gamma, (A(c))^{n-2}$	14, E(k)
			⋮	⋮	
			i-1	$\Gamma, (A(c))^{(1)}$	i-2, E(k)
			i	$\Gamma, (A(c))^1$	i-1, E(1)
				*	

1.4 Let S be Axiom 18, that is, S is $\forall \underline{x}(A)^{(n)} \supset (\exists xA)^{(n)}$.

1	$\Gamma, \sim_n(\forall x(A)^{(n)} \supset (\exists xA)^{(n)})$	
2	$\Gamma, \forall x(A)^{(n)}$	1, DNI$\sim_n$
3	$\Gamma, \sim_n((\exists xA)^{(n)})$	1, DNI$\sim_n$
4	$\Gamma, (\exists xA)$	3, E(k)$\sim_n$
5	$\Gamma, \neg(\exists xA)$	3, E(k)$\sim_n$

6	$\Gamma, \sim_n(\forall x(A)^{(n)})$ 5, $\neg\exists$		7	$\Gamma, \forall x\neg A$	5, $\neg\exists$
	*		8	$\Gamma, A(c)$	4, E-$\exists$
			9	$\Gamma, \neg A(c)$	7, E-$\forall$
			10	$\Gamma, (A(c))^{(n)}$	2, E-$\forall$
			11	$\Gamma, (A(c))^n$	10, E(k)
			12	$\Gamma, (A(c))^{(n-1)}$	10, E(k)
			13	$\Gamma, (A(c))^{n-1}$	12, E(k)
			14	$\Gamma, (A(c))^{(n-2)}$	12, E(k)
			15	$\Gamma, (A(c))^{n-2}$	14, E(k)
			⋮	⋮	
			i-1	$\Gamma, (A(c))^{(1)}$	i-2, E(k)
			i	$\Gamma, (A(c))^1$	i-1, E(1)
				*	

1.5 Let S be Axiom 19, that is, S is $A \equiv B$, where A and B are congruent formulae.

Immediate, from Rules $E - \forall, \sim_n \forall, E - \exists, \sim_n \exists$ and Special Rules **H**.

2 Now, let us consider that the formula S is a consequence of preceding formulae in a proof in $\mathbf{C}_n^*, 1 \leq n < \omega$, by *Modus Ponens*; that is, we have that $\Gamma \vdash_{\mathbf{C}_n^*} S$ is a consequence of $\Gamma \vdash_{\mathbf{C}_n^*} A$ and $\Gamma \vdash_{\mathbf{C}_n^*} A \supset S$. Then, as we have that $\Gamma \vdash_{\mathbf{TNDC}_n^*} A$ and $\Gamma \vdash_{\mathbf{TNDC}_n^*} A \supset S$, the sets $\Gamma \cup \{\sim_n A\}$ and $\Gamma \cup \{\sim_n (A \supset S)\}$ are closed in $\mathbf{TNDC}_n^*$ and so, by Rule $DNI \sim_n, \Gamma \cup \{\sim_n A\}$ and $\Gamma \cup \{A, \sim_n S\}$ are closed. Hence, $\Gamma, \sim_n S, A$ and $\Gamma, \sim_n S, \sim_n A$ are closed and, so, by the Cut Rule, $\Gamma \cup \{\sim_n S\}$ is closed. Therefore, Γ generates S in $\mathbf{TNDC}_n^*, 1 \leq n < \omega$, that is, $\Gamma \vdash_{\mathbf{TNDC}_n^*} S$.

3 Let us consider that the formula S is a consequence of a preceding formula in a proof in $\mathbf{C}_n^*, 1 \leq n < \omega$, by an application of Rule II; that is, we have that $\Gamma \vdash_{\mathbf{C}_n^*} C \supset \forall x A(x)$ is a consequence of $\Gamma \vdash_{\mathbf{C}_n^*} C \supset A(x)$, where C^{11} is a formula which does not contain x free. Let us suppose that $\Gamma \nvdash_{\mathbf{TNDC}_n^*} C \supset \forall x A(x)$, then, as we have that $\Gamma \vdash_{\mathbf{TNDC}_n^*} C \supset A(x)$, the set $\Gamma \cup \{\sim_n (C \supset A(x))\}$ is closed and $\Gamma \cup \{\sim_n (C \supset \forall x A(x))\}$ is not closed in $\mathbf{TNDC}_n^*$; and so, by Rule $DNI \sim_n, \Gamma \cup \{C, \sim_n A(x)\}$ is closed and $\Gamma \cup \{C, \sim_n (\forall x A(x))\}$ is not closed and so, by Rule $\sim_n \forall, \Gamma \cup \{C, \sim_n A(c)\}$ is not closed, where c is a constant that does not occur in the branch; or c has not been previously introduced by Rule of Type **F**, and does not occur in $\forall x A(x)$, and no constant of $\forall x A$ has been previously introduced by Rule of Type **F**. From $\Gamma \cup \{C, \sim_n A(x)\}$ being closed, we obtain that $\Gamma \cup \{C, \sim_n A(c)\}^{12}$ is closed. Hence, we obtain that $\Gamma \cup \{C, \sim_n A(c)\}$ is closed and $\Gamma \cup \{C, \sim_n A(c)\}$ is not closed.

4 Let us consider that the formula S is a consequence of a preceding formula in a proof in $\mathbf{C}_n^*, 1 \leq n < \omega$, by an application of Rule III; that is, we have that $\Gamma \vdash_{\mathbf{C}_n^*} \exists x A(x) \supset C$ is a consequence of $\Gamma \vdash_{\mathbf{C}_n^*} A(x) \supset C$, where C is a formula which does not contain x free. Let us suppose that $\Gamma \nvdash_{\mathbf{TNDC}_n^*} \exists x A(x) \supset C$; then, as we have that $\Gamma \vdash_{\mathbf{TNDC}_n^*} A(x) \supset C$, the set $\Gamma \cup \{\sim_n (A(x) \supset C)\}$ is closed and $\Gamma \cup \{\sim_n (\exists x A(x) \supset C)\}$ is not closed in $\mathbf{TNDC}_n^*$; and so, by Rules $DNI \sim_n$ and $E - \exists, \Gamma \cup \{A(x), \sim_n C\}$ is closed and $\Gamma \cup \{A(c), \sim_n C\}$ is not closed. Hence, there exists at least one constant c (in the domain), such that $\Gamma \cup \{A(x), \sim_n C\}$ is not closed, but it is an absurd. ∎

REMARK 23. The Deduction Theorem is provable in the $\mathbf{TNDC}_n^*, 1 \leq n < \omega$, with the necessary restrictions, as in $\mathbf{TNDC}_n^*, 1 \leq n < \omega^{13}$.

THEOREM 24. *If* $\Gamma \vdash_{\mathbf{TNDC}_n^*} S$, *then* $\Gamma \vdash_{\mathbf{C}_n^*} S$.

---

[11]We observe that x, in $C \supset A(x)$, denotes necessarily some constant in the domain.
[12]Where c is a constant that does not occur in the branch.
[13]D'Ottaviano and Castro (2005) prove the Deduction Theorem for the $\mathbf{TNDC}_n^*, 1 \leq n < \omega$.

Proof. Suppose that $\Gamma \vdash_{\mathbf{TNDC}_n^*} S$. If $S \in \Gamma$, then $\Gamma \vdash_{\mathbf{C}_n^*} S$ is immediate. So, let us suppose that S is not in Γ.

In order to prove the theorem, let us consider S as a formula generated from Γ by the expansion rules of $\mathbf{TNDC}_n^*, 1 \leq n < \omega$.

We shall transform every Expansion Rule of $\mathbf{TNDC}_n^*, 1 \leq n < \omega$, into a correspondent valid proof in $\mathbf{C}_n^*, 1 \leq n < \omega$. That is, the rules of type $C, D, S_1, S_2, S_3, E, F, G$ and H will be transformed into the proofs of $\alpha \vdash_{\mathbf{C}_n^*} (\delta_i^j) \& (\delta_{i+1}^j); \beta \vdash_{\mathbf{C}_n^*}; (\delta_i^j) \vee (\delta_{i+1}^j); \gamma \vdash_{\mathbf{C}_n^*} \delta_i^j; \alpha_1, \ldots, \alpha_n \vdash_{\mathbf{C}_n^*} \delta_i^j; \varepsilon \vdash_{\mathbf{C}_n^*} \delta_i^j; \alpha \vdash_{\mathbf{C}_n^*} \delta_i^j(t); \alpha \vdash_{\mathbf{C}_n^*} \delta_i^j(c); \beta \vdash_{\mathbf{C}_n^*} (\delta_i^j) \vee (\delta_{i+1}^j)$ and $\gamma \vdash_{\mathbf{C}_n^*} \delta_i^j$, respectively.

We shall only present the complete proofs relative to the Expansion Rules involving the quantifiers. The other cases of Expansion Rules were proved by [D'Ottaviano and Castro, 2005].

1. Let S be of type $A(t)$, generated, in $\mathbf{TNDC}_n^*, 1 \leq n < \omega$, from $\forall x A(x)$, by Rule $E - \forall$, where t is one term free for some variable occurring in the formula $\forall x A(x)$. We have to prove that $\forall x A \vdash_{\mathbf{C}_n^*} A(t)$.

From Axiom 15 and the Deduction Theorem, the proof is immediate.

2. Let S be of type $A(c)$, generated, in $\mathbf{TNDC}_n^*, 1 \leq n < \omega$, from $\exists x A(x)$, by Rule $E - \exists$, with c not occurring in the considered branch of the tableaux; or c has not been previously introduced by Rule of Type **F**, and does not occur in $\exists x A(x)$, and no constant of $\exists x A$ has been previously introduced by Rule of Type **F**. We have to prove that $\exists x A(x) \vdash_{\mathbf{C}_n^*} A(c)$.

1. $A(c) \vdash_{C_n^*} A(c)$ property of $\vdash_{C_n^*}$
2. $\vdash_{C_n^*} A(c) \supset A(c)$ 1, Deduction Theorem
3. $\vdash_{C_n^*} \exists x A(x) \supset A(c)$ 2, Rule III
4. $\exists x A(x) \vdash_{C_n^*} \exists x A(x)$ property of $\vdash_{C_n^*}$
5. $\exists x A(x) \vdash_{C_n^*} A(c)$ 3, 4, MP

3. Let S be of type $\sim_n A(t)$, generated, in $\mathbf{TNDC}_n^*, 1 \leq n < \omega$, from $\sim_n \exists x A(x)$, by Rule $\sim_n \exists$, where t is one term free for some variable occurring in the formula $\exists x A(x)$. We have to prove that $\sim_n \exists x A(x) \vdash_{\mathbf{C}_n^*} \sim_n A(t)$.

1	$\sim_n \exists xA(x), A(x) \vdash_{C_n^*} \sim_n \exists xA(x)$	property of $\vdash_{C_n^*}$
2	$\sim_n \exists xA(x), A(x) \vdash_{C_n^*} A(x)$	property of $\vdash_{C_n^*}$
3	$\vdash_{C_n^*} A(x) \supset \exists xA(x)$	Axiom 16
4	$\sim_n \exists xA(x), A(x) \vdash_{C_n^*} \exists xA(x)$	2, 3, MP
5	$\sim_n \exists xA(x), A(x) \vdash_{C_n^*} \exists xA(x) \supset (\sim_n \exists xA(x) \supset (\exists xA(x) \& \sim_n \exists xA(x)))$	Axiom 5
6	$\sim_n \exists xA(x), A(x) \vdash_{C_n^*} \sim_n \exists xA(x) \supset (\exists xA(x) \& \sim_n \exists xA(x))$	4, 5, MP
7	$\sim_n \exists xA(x), A(x) \vdash_{C_n^*} \exists xA(x) \& \sim_n \exists xA(x)$	1, 6, MP
8	$\vdash_{C_n^*} \exists xA(x) \& \sim_n \exists xA(x) \supset \sim_n A(x)$	Theorem 1.12 (ii)
9	$\sim_n \exists xA(x), A(x) \vdash_{C_n^*} \sim_n A(x)$	7, 8, MP
10	$\sim_n \exists xA(x) \vdash_{C_n^*} A(x) \supset \sim_n A(x)$	9, Deduction Theorem
11	$\vdash_{C_n^*} (A(x) \supset \sim_n A(x)) \supset (\sim_n A(x) \vee \sim_n A(x))$	Theorem 1.12 (i)
12	$\sim_n \exists xA(x) \vdash_{C_n^*} \sim_n A(x) \vee \sim_n A(x)$	10, 11, MP
13	$\vdash_{C_n^*} \sim_n A(x) \vee \sim_n A(x) \supset \sim_n A(x)$	Theorem 1.12 (v)
14	$\sim_n \exists xA(x) \vdash_{C_n^*} \sim_n A(x)$	12, 13, MP
15	$\vdash_{C_n^*} \sim_n \exists xA(x) \supset \sim_n A(x)$	14, Deduction Theorem
16	$\vdash_{C_n^*} \sim_n \exists xA(x) \supset \forall x \sim_n A(x)$	15, Rule II
17	$\sim_n \exists xA(x) \vdash_{C_n^*} \sim_n \exists xA(x)$	property of $\vdash_{C_n^*}$
18	$\sim_n \exists xA(x) \vdash_{C_n^*} \forall x \sim_n A(x)$	16, 17, MP
19	$\vdash_{C_n^*} \forall x \sim_n A(x) \supset \sim_n A(t)$	Axiom 15
20	$\sim_n \exists xA(x) \vdash_{C_n^*} \sim_n A(t)$	18, 19, MP

4. Let S be of type $\sim_n A(c)$, generated, in $\mathbf{TNDC}_n^*, 1 \leq n < \omega$, from $\sim_n \forall x A(x)$, by Rule $\sim_n \forall$, with c not occurring in the considered branch of the tableaux; or c has not been previously introduced by Rule of Type $\mathbf{F}$, and does not occur in $\forall x A(x)$, and no constant of $\forall x A(x)$ has been previously introduced by Rule of Type $\mathbf{F}$. We have to prove that $\sim_n \forall x A(x) \vdash_{\mathbf{C}_n^*} \sim_n A(c)$.

1. $\sim_n\exists x \sim_n A(x), \sim_n A(x) \vdash_{C_n^*} \sim_n\exists x \sim_n A(x)$ property of $\vdash_{C_n^*}$
2. $\sim_n\exists x \sim_n A(x), \sim_n A(x) \vdash_{C_n^*} \sim_n A(x)$ property of $\vdash_{C_n^*}$
3. $\vdash_{C_n^*} \sim_n A(x) \supset \exists x \sim_n A(x)$ Axiom 16
4. $\sim_n\exists x \sim_n A(x), \sim_n A(x) \vdash_{C_n^*} \exists x \sim_n A(x)$ 2, 3, MP
5. $\sim_n\exists x \sim_n A(x), \sim_n A(x) \vdash_{C_n^*} \exists x \sim_n A(x) \supset (\sim_n\exists x \sim_n A(x) \supset (\exists x \sim_n A(x) \& \sim_n\exists x \sim_n A(x)))$
     Axiom 5
6. $\sim_n\exists x \sim_n A(x), \sim_n A(x) \vdash_{C_n^*} \sim_n\exists x \sim_n A(x) \supset (\exists x \sim_n A(x) \& \sim_n\exists x \sim_n A(x))$ 4, 5, MP
7. $\sim_n\exists x \sim_n A(x), \sim_n A(x) \vdash_{C_n^*} \exists x \sim_n A(x) \& \sim_n\exists x \sim_n A(x)$ 1, 6, MP
8. $\vdash_{C_n^*} \exists x \sim_n A(x) \& \sim_n\exists x \sim_n A(x) \supset \sim_n\sim_n A(x)$ Theorem 1.12 (ii)
9. $\sim_n\exists x \sim_n A(x), \sim_n A(x) \vdash_{C_n^*} \sim_n\sim_n A(x)$ 7, 8, MP
10. $\sim_n\exists x \sim_n A(x) \vdash_{C_n^*} \sim_n A(x) \supset \sim_n\sim_n A(x)$ 9, Deduction Theorem
11. $\vdash_{C_n^*} (\sim_n A(x) \supset \sim_n\sim_n A(x)) \supset (\sim_n\sim_n A(x) \vee \sim_n\sim_n A(x))$ Theorem 1.12 (i)
12. $\sim_n\exists x \sim_n A(x) \vdash_{C_n^*} \sim_n\sim_n A(x) \vee \sim_n\sim_n A(x)$ 10, 11, MP
13. $\vdash_{C_n^*} \sim_n\sim_n A(x) \vee \sim_n\sim_n A(x) \supset \sim_n\sim_n A(x)$ Theorem 1.12 (ix)
14. $\sim_n\exists x \sim_n A(x) \vdash_{C_n^*} \sim_n\sim_n A(x)$ 12, 13, MP
15. $\vdash_{C_n^*} \sim_n\sim_n A(x) \supset A(x)$ Theorem 1.12 (iii)
16. $\sim_n\exists x \sim_n A(x) \vdash_{C_n^*} A(x)$ 14, 15, MP
17. $\vdash_{C_n^*} \sim_n\exists x \sim_n A(x) \supset A(x)$ 16, Deduction Theorem
18. $\vdash_{C_n^*} \sim_n\exists x \sim_n A(x) \supset \forall x A(x)$ 17, Rule II
19. $\vdash_{C_n^*} (\sim_n\exists x \sim_n A(x) \supset \forall x A(x)) \supset (\sim_n\forall x A(x) \supset \sim_n\sim_n\exists x \sim_n A(x))$ Theorem 1.12 (iv)
20. $\vdash_{C_n^*} \sim_n\forall x A(x) \supset \sim_n\sim_n\exists x \sim_n A(x)$ 18, 19, MP

21. $\vdash_{C_n^*} \sim_n\sim_n\exists x \sim_n A(x) \supset \exists x \sim_n A(x)$ Theorem 1.12 (iii)
22. $\vdash_{C_n^*} \sim_n\forall x A(x) \supset \exists x \sim_n A(x)$ 20, 21, transitivity
23. $\sim_n\forall x A(x) \vdash_{C_n^*} \sim_n\forall x A(x)$ property of $\vdash_{C_n^*}$
24. $\sim_n\forall x A(x) \vdash_{C_n^*} \exists x \sim_n A(x)$ 22, 23, MP
25. $\sim_n A(c) \vdash_{C_n^*} \sim_n A(c)$ property of $\vdash_{C_n^*}$
26. $\vdash_{C_n^*} \sim_n A(c) \supset \sim_n A(c)$ 25, Deduction Theorem
27. $\vdash_{C_n^*} \exists x \sim_n A(x) \supset \sim_n A(c)$ 26, Rule III[1]
28. $\sim_n\forall x A(x) \vdash_{C_n^*} \sim_n A(c)$ 24, 27, MP

---

[1] Note that c is a closed term and so, it is not free in **C**.

5. Let S be of type $\sim_n(\forall \underline{x}(A(x))^{(n)}) \vee (\exists x \neg (A(x)))$, generated, in $TNDC_n{}^*$, $1 \leq n < \omega$, from $\neg \forall x A(x)$, by Rule $\neg \forall$. We have to prove that $\neg \forall x A(x) \vdash_{C_n^*} \sim_n(\forall \underline{x}(A(x))^{(n)}) \vee (\exists x \neg A(x))$.

1    $\vdash_{C_n^*} \neg \forall \underline{x} A(x) \supset (\sim_n (\forall x(A(x))^{(n)}) \vee (\exists x \neg (A(x))))$ Theorem 1.12 (viii)

2    $\neg \forall \underline{x} A(x) \vdash_{C_n^*} \neg \forall \underline{x} A(x)$ property of $\vdash_{C_n^*}$

3    $\neg \forall \underline{x} A(x) \vdash_{C_n^*} \sim_n(\forall x(A(x))^{(n)}) \vee (\exists x \neg A(x))$ 1, 2, MP

6. Let S be of type $\sim_n(\forall \underline{x}(A(x))^{(n)}) \vee (\forall x \neg A(x))$, generated, in $TNDC_n{}^*$, $1 \leq n < \omega$, from $\neg \exists x A(x)$, by Rule $\neg \exists$. We have to prove that $\neg \exists \underline{x} A(x) \vdash_{C_n^*} \sim_n(\forall x(A(x))^{(n)}) \vee (\forall x \neg A(x))$.

1    $\vdash_{C_n^*} \neg \exists x(A(x)) \supset (\sim_n(\forall x(A(x))^{(n)}) \vee (\forall x \neg (A(x))))$ Theorem 1.12 (ix)

2    $\neg \exists \underline{x} A(x) \vdash_{C_n^*} \neg \exists \underline{x} A(x)$ property of $\vdash_{C_n^*}$

3    $\neg \exists \underline{x} A(x) \vdash_{C_n^*} \sim_n(\forall x(A(x))^{(n)}) \vee (\forall x \neg A(x))$ 1, 2, MP   □

∎

Hence, by Theorem 22 and Theorem 24, we have the "equivalence" between the correspondent systems of both hierarchies $\mathbf{C}_n^*$ and $\mathbf{TNDC}_n^*$, $1 \leq n < \omega$.

THEOREM 25. $\Gamma \vdash_{C_*^n} S$ if, and only if, $\Gamma \vdash_{\mathbf{TNDC}_n^*} S$, for every n, $1 \leq n < \omega$

As every system $\mathbf{TNDC}_n^*$, $1 \leq n < \omega$, is equivalent to the corresponding $\mathbf{C}_n^*$, $1 \leq n < \omega$, the syntactical and semantic results concerning the $\mathbf{TNDC}_n^*$ are immediate.

So, the soundness and completeness of our tableaux systems can be proved.

Besides, the decidability of the monadic first-order predicate systems $\mathbf{TNDC}_n^*$, $1 \leq n < \omega$, could also be proved, from the characteristics of the Expansion Rules of the systems: for every formula S we have to check, in a finite number of steps, either if $\sim_n S$ is closed, or if $\sim_n S$ is not closed; for every tableau for $\sim_n S$, in the case when $\sim_n S$ is not closed, we have to generate at least a finite, open and complete branch. We intend to develop this proof in a future paper.

Finally, we observe that, in spite of having carefully studied da Costa's calculi $\mathbf{C}_\omega$ and $\mathbf{C}_\omega^*$, we could not obtain tableaux systems for such systems.

BIBLIOGRAPHY

[Alves, 1976] Alves, E.H. (1976) *Lógica e inconsistência: um estudo dos cálculos* $\mathbf{C}_n^*, 1 \leq n \leq \omega$ *(Logic and inconsistency: a study of the calculi* $\mathbf{C}_n^*, 1 \leq n \leq \omega$, in Portuguese), Master's Dissertation, FFLCH-USP, São Paulo, 1976.

[Beth, 1959] Beth, H.W. (1959) *The foundations of mathematics*, North Holland, Amsterdan.

[Buchsbamum and Pequeno, 1993] Buchsbamum, A., Pequeno, T. (1993) A reasoning method for a paraconsistent logic, *Studia Logica*, vol. 52, pp. 281-289.

[Carnielli and Marcos, 2000] Carnielli, W.A., Marcos, J. (2000) The C-Systems: Paleontology and Futurology. In: II World Congress on Paraconsistency. Juquehy: *Proceedings*.

[Carnielli et al., 2007] Carnielli, W.A., Coniglio, M.E., Marcos, J. (2007) Logics of formal inconsistency. In: Gabbay D.V., Guenthner F. (org.), *Handbook of Philosophical Logic*, 2 ed., Dordrecht, Springer, v. 14, pp. 1-93.

[Carnielli et al., 2009] Carnielli, W.A., Coniglio, M.E., D'Ottaviano, I.M.L. (2009) New dimensions on translations between logics. *Logica Universalis* (Print), v. 3, pp. 1-19.

[Carnielli and Lima-Marques, 1992] Carnielli, W.A., Lima-Marques, M. (1992) Reasoning under inconsistent knowledge, *Journal of Applied Non-Classical Logics*, vol. 2, no. 1, pp. 49-79.

[Carnielli and Marcos, 2002] Carnielli, W.A., Marcos, J. (2002) A taxonomy of C-systems. In: Carnielli, W.A., Coniglio, M.E., D'Ottaviano, I.M.L. (eds). *Paraconsistency: the logical way to the inconsistent*. New York: Marcel Dekker, pp. 01-94. (Lectures Notes in Pure and Applied Mathematics, v.228)

[Castro, 2000] Castro, M.A. (2000) Analytical tableaux for da Costa's hierarchy of paraconsistent logics $\mathbf{C}_n^*, 1 \leq n < \omega$ In: II World Congress on Paraconsistency. Juquehy: *Proceedings*.

[Castro, 2004] Castro M.A. (2004) *Hierarquias de sistemas de dedução natural e de sistemas de tableaux analíticos para os sistemas $\mathbf{C}_n$ de da Costa (Hierarchies of natural deduction systems and of analitycal tableaux systems for da Costa's systems $\mathbf{C}_n$, in Portuguese)*, PhD Thesis, IFCH-UNICAMP, Campinas, SP.

[da Costa, 1963] da Costa N.C.A. (1963) *Sistemas formais inconsistentes (Inconsistent formal systems)*, in Portuguese, Thesis, UFPr, Curitiba, PR.

[da Costa, 1964] da Costa N.C.A. (1964), Calculs des prédicats pour les systèmes formels inconsistants, *Comptes Rendus de l'Académie de Sciences de Paris*, t. 258, pp. 27-29.

[da Costa, 1974] da Costa N.C.A. (1974) On the theory of inconsistent formal systems, *Notre Dame Journal of Formal Logic*, vol. 15, pp. 497-510.

[da Costa et al., 2006] da Costa N.C.A., Krause D., Bueno O. (2006) Paraconsistent logics and paraconsistency, In: Jacquette D., Gabbay D.M., Thagard P., Woods J., Org., *Philosophy of Logic*, 5. 1 ed. : Elsevier, v. 5, pp. 791-911.

[D'Ottaviano, 1990] D'Ottaviano I. M. L. (1990) On the development of paraconsistent logic and da Costa's work, *The Journal of Non-Classical Logic*, vol. 7, no. 1/2, pp. 89-152.

[D'Ottaviano and Castro, 2005] D'Ottaviano, I.M.L., Castro, M.A. de (2005) Analytical tableaux for da Costa's hierarchy of paraconsistent logics $\mathbf{C}_n$, *Journal of Applied Non-Classical Logics*, Paris, vol. 15, n. 1, pp. 69-103.

[D'Ottaviano and da Costa, 1970] D'Ottaviano, I.M.L., Da Costa, N.C.A. (1970) Sur un problème de Jaśkowski. *Comptes Rendus de l'Académie de Sciences de Paris*, n. 21, v. 270 A, pp.1349-1353.

[D'Ottaviano and Epstein, 1988] D'Ottaviano, I.M.L., Epstein, R.L. (1988) A many-valued paraconsistent logic, *Reports on Mathematical Logic*, vol. 22, pp. 89-103.

[Epstein, 1990] Epstein, R.L. (1990) *The Semantic Foundations of Logic*. Vol. 1. *Propositional Logics*. Nijhoff International Philosophy Series, 35. Kruwer Academic Publishers Group, Dordrecht.

[Epstein, 1995] Epstein, R.L. (1995) *The Semantic Foundations of Logic*. Vol. 1. *Propositional Logics*. Second Edition. The Claredon Press, Oxford University Press, New York.

[Kleene, 1952] Kleene, S.C. (1952) *Introduction to metamathematics*, Van Nostrand, New York.

[Marconi, 1980] Marconi, D. (1980) A decision method for the calculus $\mathbf{C}_1$, Arruda, A.I., da Costa, N.C.A., Sette, A.M., Eds. *Proceedings of the 3rd Brazilian Conference on Mathematical Logic*, São Paulo, Sociedade Brasileira de Lógica, pp. 211-223.

[Smullyan, 1968] Smullyan, R.M. (1968) *First-order logic*, Springer Verlag, New York.

[van Fraassen, 1971] van Fraasen, B.C. (1971) *Formal semantics and logic*, The Macmillan Company, New York.

Itala M. Loffredo D'Ottaviano and Milton Augustinis de Castro
Centre for Logic, Epistemology and the History of Science (CLE)
and Department of Philosophy
State University of Campinas (UNICAMP)
Campinas, Brazil
E-mail: {itala, milton}@cle.unicamp.br

Analytic Calculi for Basic Logics of Formal Inconsistency

ARNON AVRON, BEATA KONIKOWSKA AND ANNA ZAMANSKY

ABSTRACT.
This paper makes a substantial step towards automatization of paraconsistent reasoning by exemplifying a method for a systematic generation of analytic calculi for thousands of Logics of Formal (In)consistency. The method relies on the availability of non-deterministic three-valued semantics for these logics.

1 Introduction

A *paraconsistent logic* is a logic which allows non-trivial inconsistent theories. One of the oldest and best known approaches to paraconsistency is da Costa's one ([da Costa, 1974; da Costa *et al.*, 1995; D'Ottaviano, 1990]), which seeks to allow the use of classical logic whenever it is safe to do so, but behaves completely differently when contradictions are involved. This approach has led to the introduction of the family of *Logics of Formal (In)consistency (LFIs)* by W.A. Carnielli ([Carnielli *et al.*, 2007; Carnielli and Marcos, 2002]). This family is based on two key ideas. The first is that propositions should be divided into two sorts: the "normal" (or consistent), and the "abnormal" (or inconsistent) ones. While classical logic can be applied freely to normal propositions, its use is restricted for the abnormal ones. The second idea is to reflect this classification within the language used. In the important class of LFIs called *C-systems* ([Carnielli and Marcos, 2002]), this is done by employing a specific formula constructor $C(\varphi)$ which is available in the language, where the intuitive meaning of $C(\varphi)$ is "φ is consistent". Usually, this is done by simply introducing a special unary connective $\circ$, and taking $C(\varphi) = \circ\varphi$.

For a long time, the class of C-systems (which is the most important and useful class of LFIs) had a major shortcoming: with few exceptions (see [Béziau, 2001; Carnielli and Lima-Marques, 1992; Carnielli *et al.*, 2007; Neto and Finger, 2007] and recently [Gentilini, 2011]), no analytic calculi were available for most of them. Moreover, each of the calculi that have been suggested was tailored to some specific LFI, and the rules of those calculi were introduced in a sort of an ad-hoc manner, and so they do not have any uniform structure. Therefore even a slight modification of any of these LFIs practically means starting the search for a corresponding analytic calculus all over again.

In this paper we present, by way of example, a systematic way to generate cut-free sequent calculi for a very large family of LFIs, for which three-valued non-deterministic semantics were provided in [Avron, 2007b] and [Avron, 2005].

The method exploits an algorithm given in [Avron et al., 2006] for constructing an analytic Gentzen-type system for a logic which has a non-deterministic finite-valued characterization, assuming that its language is sufficiently expressive (which is the case here). The resulting sequent calculus automatically enjoys cut-admissibility, and its rules have a uniform form, closely related to the form used in calculi for classical logic and other well known calculi. We demonstrate the method by constructing a cut-free Gentzen-type calculus for one particular LFI: $\mathbf{Bk_{o_\vee}}$. However, the same method can be applied to any of the other LFIs with three-valued non-deterministic semantics which were studied in the papers cited above.

2 Preliminaries

In what follows, $\mathcal{L}$ is a propositional language, and $Frm_\mathcal{L}$ is its set of wffs. The metavariables T, S range over theories of $\mathcal{L}$-formulas, and Γ, Δ range over finite theories of $\mathcal{L}$-formulas.

2.1 The systems B, Bk and Bk$_{o_\vee}$

Let $\mathcal{L}_{cl}^+ = \{\wedge, \vee, \supset\}$, and $\mathcal{L}_C = \{\wedge, \vee, \supset, \neg, \circ\}$.

DEFINITION 1. Let HCL^+ be some standard Hilbert-style system which has MP as the only inference rule, and is sound and strongly complete for the $\mathcal{L}_{cl}^+$-fragment of classical propositional logic. The system **B** over $\mathcal{L}_C$ is obtained by adding to HCL^+ the following axioms:

(t) $\neg \varphi \vee \varphi$ \qquad (b) $\circ \varphi \supset (\varphi \wedge \neg \varphi \supset \psi)$

The system **Bk** is obtained from **B** by adding the following axiom:

(k) $\circ \varphi \vee (\varphi \wedge \neg \varphi)$

REMARK 2. The system **B** is frequently taken as the most basic LFI (see e.g. [Carnielli et al., 2007; Carnielli and Marcos, 2002], where it is called **mbC**). However, we find it much more appropriate to take **Bk** as the most basic C-system. The reason is that the intended meaning of $\circ \varphi$ is "φ is consistent". Now axiom (b) reflects this by saying that no formula is both consistent and contradictory. This is complemented by axiom (k), which intuitively says that every formula is either consistent or contradictory. This last principle seems to be as essential for the intended meaning of $\circ \varphi$ as the one expressed by axiom (b). We note that the intuitive content of both (b) and (k) we have just explained is well reflected in the three-valued non-deterministic semantics used in this paper.

There is an extensive family of LFIs that extend **Bk** by various additional schemata ([Carnielli et al., 2007; Carnielli and Marcos, 2002]). For thousands of these LFIs, three-valued non-deterministic semantics has been provided in [Avron, 2007b; Avron, 2005]. Using this semantics, cut-free sequent systems can be constructed for all of them with the method of [Avron et al., 2006]. This method will be exemplified here for the following extension of **Bk**:

DEFINITION 3. The system **Bko**$_\vee$ is obtained from **Bk** by adding the axiom:

$$(\circ\varphi \vee \circ\psi) \supset \circ(\varphi \vee \psi)$$

REMARK 4. It is useful to split (**o**$_\vee$) into the following two axioms:

$$(\mathbf{o}_\vee^1) \quad \circ\varphi \supset \circ(\varphi \vee \psi) \qquad\qquad (\mathbf{o}_\vee^2) \quad \circ\psi \supset \circ(\varphi \vee \psi)$$

We shall refer by **Bko**$_\vee^1$ and **Bko**$_\vee^2$ to the systems obtained from **Bk** by adding ($\mathbf{o}_\vee^1$) and ($\mathbf{o}_\vee^2$), respectively.

2.2 Non-deterministic Matrices

Our main semantic tool in what follows will be the following generalization of the concept of a many-valued matrix introduced in [Avron and Lev, 2005] (for a comprehensive survey on non-deterministic matrices, see also [Avron and Zamansky, 2011]):

DEFINITION 5.

1. A non-deterministic matrix (Nmatrix) for a language $\mathcal{L}$ is a tuple $\mathcal{M} = \langle \mathcal{V}, \mathcal{D}, \mathcal{O} \rangle$, where: $\mathcal{V}$ is a non-empty set of truth values, $\mathcal{D}$ (designated truth values) is a non-empty proper subset of $\mathcal{V}$ and $\mathcal{O}$ includes an interpretation function $\tilde{\diamond}_\mathcal{M} : \mathcal{V}^n \to P^+(\mathcal{V})$ for every n-ary connective $\diamond$. We say that $\mathcal{M}$ is *finite* if so is $\mathcal{V}$.

2. Let $\mathcal{M} = \langle \mathcal{V}, \mathcal{D}, \mathcal{O} \rangle$ be an Nmatrix. Let F be some set of $\mathcal{L}$-formulas closed under subformulas. An $\mathcal{M}$-valuation on F is a function $v : F \to \mathcal{V}$ which satisfies the following condition for every n-ary connective $\diamond$ of $\mathcal{L}$ and every $\psi_1, \ldots, \psi_n \in F$:

$$v(\diamond(\psi_1, \ldots, \psi_n)) \in \tilde{\diamond}_\mathcal{M}(v(\psi_1), \ldots, v(\psi_n))$$

A *full $\mathcal{M}$-valuation* is an $\mathcal{M}$-valuation on $Frm_\mathcal{L}$.

3. Let F be as above, and let $\psi \in F$. An $\mathcal{M}$-valuation v on F *satisfies* ψ, denoted by $v \models_\mathcal{M} \psi$, if $v(\psi) \in \mathcal{D}$. v satisfies a set $\Gamma \subseteq F$ of formulas, denoted by $v \models_\mathcal{M} \Gamma$, if it satisfies every formula of Γ.

4. Let F be as above, and let v be an $\mathcal{M}$-valuation on F. A sequent $\Gamma \Rightarrow \Delta$ such that $\Gamma \cup \Delta \subseteq F$ is *true in* v if either there is some $\psi \in \Delta$, such that $v \models_\mathcal{M} \psi$, or for every $\psi \in \Gamma$, $v \not\models_\mathcal{M} \psi$. A sequent is *valid in* $\mathcal{M}$ if it is true in every full $\mathcal{M}$-valuation.

5. $\vdash_\mathcal{M}$, the consequence relation induced by $\mathcal{M}$, is defined by: $T \vdash_\mathcal{M} \psi$ if $v \models_\mathcal{M} \psi$ for every full $\mathcal{M}$-valuation v such that $v \models_\mathcal{M} T$.

Notation. Below we shall frequently write just $\diamond$ instead of $\tilde{\diamond}_\mathcal{M}$, relying on the context to determine whether we mean the connective or its interpretation in some Nmatrix $\mathcal{M}$.

It should be noted that Nmatrices enjoy many of the attractive properties of usual (deterministic) finite-valued matrices, like compactness and semantic analyticity ([Avron and Zamansky, 2011]). The latter, in turn, implies the decidability of the logics induced by them.

The following notion of *simple refinements* is going to be useful in the sequel:

DEFINITION 6. Let $\mathcal{M}_1 = \langle \mathcal{V}_1, \mathcal{D}_1, \mathcal{O}_1 \rangle$ and $\mathcal{M}_2 = \langle \mathcal{V}_2, \mathcal{D}_2, \mathcal{O}_2 \rangle$. $\mathcal{M}_2$ is a *simple refinement* of $\mathcal{M}_1$ if $\mathcal{V}_1 = \mathcal{V}_2$, $\mathcal{D}_1 = \mathcal{D}_2$ and for every n-ary connective $\diamond$ and every $a_1, \ldots, a_n \in \mathcal{V}_1$, $\tilde{\diamond}_{\mathcal{M}_2}(a_1, \ldots, a_n) \subseteq \tilde{\diamond}_{\mathcal{M}_1}(a_1, \ldots, a_n)$.

PROPOSITION 7. ([Avron, 2007b]) *If $\mathcal{M}_2$ is a simple refinement of $\mathcal{M}_1$, then $\vdash_{\mathcal{M}_1} \subseteq \vdash_{\mathcal{M}_2}$.*

3 Construction of Analytic Calculi

Our method for a systematic construction of Gentzen-type calculi is suitable for all the LFIs which have been given a semantic characterization in terms of three-valued Nmatrices in [Avron, 2005; Avron, 2007a; Avron, 2007b]. It relies on the algorithm provided in [Avron et al., 2006] for constructing cut-free Gentzen-type systems for logics which have a characteristic finite-valued Nmatrix $\mathcal{M}$, and the language of which is *sufficiently expressive* with respect to $\mathcal{M}$. The latter means that from any formula φ we can construct in a uniform way a finite set of formulas $S(\varphi)$ such that, for any valuation v in $\mathcal{M}$, the truth-value of $v(\varphi)$ is uniquely determined by the subset of formulas in $S(\varphi)$ that v satisfies. (Note there is no need to know the exact truth-values assigned by v to formulas in $S(\varphi)$, but only whether they are designated or not). We will shortly see that the language of the LFIs studied in [Avron, 2005; Avron, 2007a; Avron, 2007b] is sufficiently expressive with respect to all the three-valued Nmatrices used in those papers. Hence we can indeed exploit the algorithm of [Avron et al., 2006] in order to construct cut-free Gentzen-type systems for those logics.

Below we demonstrate the general method by way of example. We start by defining non-deterministic three-valued semantics for $\mathbf{Bko}_\vee$, and then introduce its corresponding Gentzen-type system.

3.1 The Non-deterministic Semantics

We begin with defining the Nmatrix $\mathcal{M}_0^3$ for the most basic system $\mathbf{B}$:

DEFINITION 8. The Nmatrix $\mathcal{M}_0^3 = (\{t, f, \top\}, \{t, \top\}, \mathcal{O})$ for $\mathcal{L}_C$ is defined as follows:

a	$\neg a$	$\circ a$
t	$\{f\}$	$\{t, \top, f\}$
$\top$	$\{t, \top\}$	$\{f\}$
f	$\{t, \top\}$	$\{t, \top, f\}$

$\wedge$	t	$\top$	f
t	$\{t, \top\}$	$\{t, \top\}$	$\{f\}$
$\top$	$\{t, \top\}$	$\{t, \top\}$	$\{f\}$
f	$\{f\}$	$\{f\}$	$\{f\}$

$\vee$	t	$\top$	f
t	$\{t, \top\}$	$\{t, \top\}$	$\{t, \top\}$
$\top$	$\{t, \top\}$	$\{t, \top\}$	$\{t, \top\}$
f	$\{t, \top\}$	$\{t, \top\}$	$\{f\}$

$\supset$	t	$\top$	f
t	$\{t, \top\}$	$\{t, \top\}$	$\{f\}$
$\top$	$\{t, \top\}$	$\{t, \top\}$	$\{f\}$
f	$\{t, \top\}$	$\{t, \top\}$	$\{t, \top\}$

PROPOSITION 9. ([Avron, 2007b]) $T \vdash_{\mathcal{M}_0^3} \psi$ *iff* $T \vdash_{\mathbf{B}} \psi$.

To obtain an Nmatrix which is characteristic for the system **Bk**, it is useful to split **(k)** into the following two axioms:

$$(\mathbf{k_1}) \quad \circ \varphi \vee \varphi \qquad\qquad (\mathbf{k_2}) \quad \circ \varphi \vee \neg \varphi$$

The basic Nmatrix $\mathcal{M}_0^3$ is then refined according to the semantic conditions induced by the **(k)**-schemata (see e.g. [Avron, 2005], where $(\mathbf{k_1})$ and $(\mathbf{k_2})$ are called $(\mathbf{d_1})$ and $(\mathbf{d_2})$, respectively):

DEFINITION 10. The Nmatrix $\mathcal{M}^3$ is the weakest refinement of $\mathcal{M}_0^3$, in which the following conditions hold:

$$C(\mathbf{k_1}): \quad \circ t = \{t, \top\}$$

$$C(\mathbf{k_2}): \quad \circ f = \{t, \top\}$$

In other words, $\mathcal{M}^3$ is similar to $\mathcal{M}_0^3$, except for its interpretation of $\circ$, which is as follows: $\circ t = \circ f = \{t, \top\}$, $\circ \top = \{f\}$.

PROPOSITION 11. ([Avron, 2005]) $T \vdash_{\mathcal{M}^3} \psi$ iff $T \vdash_{\mathbf{Bk}} \psi$.

To obtain an Nmatrix which is characteristic for the system $\mathbf{Bko_\vee}$, the Nmatrix $\mathcal{M}^3$ is further refined according to the semantic conditions induced by the **(o)**-schemata described in Remark 4 ([Avron, 2007b]):

DEFINITION 12. Let:

$$C(\mathbf{o_\vee^1}): \quad t \vee t = t \vee f = t \vee \top = f \vee t = f \vee \top = \{t\}$$

$$C(\mathbf{o_\vee^2}): \quad t \vee t = \top \vee t = f \vee t = t \vee f = \top \vee f = \{t\}$$

The Nmatrices $\mathcal{M}_{\mathbf{o_\vee^1}}^3, \mathcal{M}_{\mathbf{o_\vee^2}}^3$ and $\mathcal{M}_{\mathbf{o_\vee}}^3$ are the weakest refinements of $\mathcal{M}^3$, in which $C(\mathbf{o_\vee^1})$, $C(\mathbf{o_\vee^2})$ and both of them hold, respectively. In other words, $\mathcal{M}_{\mathbf{o_\vee^1}}^3, \mathcal{M}_{\mathbf{o_\vee^2}}^3$ and $\mathcal{M}_{\mathbf{o_\vee}}^3$ are similar to $\mathcal{M}^3$, except for their interpretations of $\vee$, which are as follows:

$\mathcal{M}_{\mathbf{o_\vee^1}}^3$:

$\vee$	t	$\top$	f
t	$\{t\}$	$\{t\}$	$\{t\}$
$\top$	$\{t, \top\}$	$\{t, \top\}$	$\{t, \top\}$
f	$\{t\}$	$\{t\}$	$\{f\}$

$\mathcal{M}_{\mathbf{o_\vee^2}}^3$:

$\vee$	t	$\top$	f
t	$\{t\}$	$\{t, \top\}$	$\{t\}$
$\top$	$\{t\}$	$\{t, \top\}$	$\{t\}$
f	$\{t\}$	$\{t, \top\}$	$\{f\}$

$\mathcal{M}_{\mathbf{o_\vee}}^3$:

$\vee$	t	$\top$	f
t	$\{t\}$	$\{t\}$	$\{t\}$
$\top$	$\{t\}$	$\{t, \top\}$	$\{t\}$
f	$\{t\}$	$\{t\}$	$\{f\}$

As an example, let us explain how $C(\mathbf{o_\vee^1})$ is derived. For this, assume that v is a valuation in $\mathcal{M}^3$. If $v(\varphi) = \top$ then v certainly satisfies $\circ \varphi \supset \circ(\varphi \vee \psi)$. Otherwise it satisfies this sentence iff it satisfies $\circ(\varphi \vee \psi)$, which is the case iff $v(\varphi \vee \psi) \neq \top$. This again necessarily holds if $v(\varphi) = v(\psi) = f$. In the remaining five cases all we know is that $v(\varphi \vee \psi) \in \{t, \top\}$. Hence to ensure it is indeed not $\top$, we have to force it to take the value t in these cases. This requires five basic semantic conditions, which can conveniently be grouped as follows: (i) $t \vee t = t \vee f = t \vee \top = \{t\}$ (i.e. $t \vee x = \{t\}$ for $x \in \{t, \top, f\}$), and (ii) $f \vee t = f \vee \top = \{t\}$.

PROPOSITION 13.

1. $T \vdash_{\mathcal{M}^3_{\circ^1_\vee}} \psi$ iff $T \vdash_{\mathbf{Bko}^1_\vee} \psi$.

2. $T \vdash_{\mathcal{M}^3_{\circ^2_\vee}} \psi$ iff $T \vdash_{\mathbf{Bko}^2_\vee} \psi$.

3. $T \vdash_{\mathcal{M}^3_{\circ_\vee}} \psi$ iff $T \vdash_{\mathbf{Bko}_\vee} \psi$.

Proof. The proof is practically identical to the proof of Theorem 3 in [Avron, 2007b], with some slight modifications (due to the fact that the axiom (o) has been split here into simpler axioms). ∎

3.2 The Corresponding Gentzen-type Systems

Before applying the algorithm of [Avron et al., 2006] to obtain analytic systems for (**Bk** and) **Bko**$_\vee$, we first need the following proposition and its proof:

PROPOSITION 14. $\mathcal{L}_C$ *is sufficiently expressive for every simple refinement of $\mathcal{M}^3_0$ (and so for every simple refinement of $\mathcal{M}^3$).*

Proof. Let $S(\psi) = \{\psi, \neg\psi\}$ for every formula ψ. Then the following holds in $\mathcal{M}^3_0$ and any simple refinement of it:

- $v(\psi) = t$ iff v does not satisfy $\neg\psi$ (i.e., $v(\neg\psi) \notin \mathcal{D}$).
- $v(\psi) = f$ iff v does not satisfy ψ (i.e., $v(\psi) \notin \mathcal{D}$).
- $v(\psi) = \top$ iff v satisfies both ψ and $\neg\psi$ (i.e., $v(\psi) \in \mathcal{D}$ and $v(\neg\psi) \in \mathcal{D}$).

It follows that $S(\psi)$ can be used to characterize in $\mathcal{M}^3_0$ all its three truth-values, and so $\mathcal{L}_C$ is sufficiently expressive for every simple refinement of $\mathcal{M}^3_0$. ∎

Now the method of [Avron et al., 2006] for constructing a cut-free, sound and complete Gentzen-type system for a given finite Nmatrix $\mathcal{M}$ involves two stages. In the first (and more important) stage, every entry of every truth-table of $\mathcal{M}$ is translated into a rule. In the second stage, certain streamlining principles are used to combine and simplify rules in order to get an optimal set of rules. The process can be significantly simplified in the present case, because just three truth values are used (and also because of the simplicity of the set $S(\psi)$ used here). All we need are the following six facts about valuations in $\mathcal{M}^3_0$ (corresponding to the nontrivial subsets of the set of truth values we employ) :

- $v(\psi) = t$ iff $\neg\psi \Rightarrow$ is true in v.
- $v(\psi) = f$ iff $\psi \Rightarrow$ is true in v.
- $v(\psi) = \top$ iff $\Rightarrow \psi$ and $\Rightarrow \neg\psi$ are both true in v.
- $v(\psi) \in \{f, \top\}$ iff $\Rightarrow \neg\psi$ is true in v.
- $v(\psi) \in \{t, \top\}$ iff $\Rightarrow \psi$ is true in v.
- $v(\psi) \in \{t, f\}$ iff $\psi, \neg\psi \Rightarrow$ is true in v.

To see how these facts are used to derive rules, take for example the truth table for $\vee$ in $\mathcal{M}_0^3$. The entry $f \vee f = \{f\}$ is first translated into: if $\varphi \Rightarrow$ is true and $\psi \Rightarrow$ is true then $\varphi \vee \psi \Rightarrow$ is true. By adding context, we get the rule:

$$\frac{\Gamma, \varphi \Rightarrow \Delta \quad \Gamma, \psi \Rightarrow \Delta}{\Gamma, \varphi \vee \psi \Rightarrow \Delta}$$

In a similar way we can derive the rules which correspond to the other eight entries of the truth table for $\vee$. The streamlining principles can then be used to combine all the eight rules into the single rule:

$$\frac{\Gamma \Rightarrow \Delta, \varphi, \psi}{\Gamma \Rightarrow \Delta, \varphi \vee \psi}$$

Alternatively, instead of first deriving eight different rules and then combining them, we can directly derive the last rule by observing that the relevant eight entries taken together mean that if either $v(\varphi) \in \{t, \top\}$ or $v(\psi) \in \{t, \top\}$ then $v(\varphi \vee \psi) \in \{t, \top\}$. This directly translates into: if $\Rightarrow \varphi, \psi$ is true then $\Rightarrow \varphi \vee \psi$ is true, and by adding context we get the last rule.

Using this method, we get the following systems for $\vdash_{\mathcal{M}_0^3}$ and $\vdash_{\mathcal{M}^3}$:

DEFINITION 15. The sequent system $\mathbf{G}_k$ consists of the following:

Axioms of $\mathbf{G}_k$: $\psi \Rightarrow \psi$

Rules of $\mathbf{G}_k$: Cut, Weakening, and the following logical rules:

$(\wedge \Rightarrow) \quad \dfrac{\Gamma, \psi, \phi \Rightarrow \Delta}{\Gamma, \psi \wedge \phi \Rightarrow \Delta}$ $\qquad$ $(\Rightarrow \wedge) \quad \dfrac{\Gamma \Rightarrow \Delta, \psi \quad \Gamma \Rightarrow \Delta, \phi}{\Gamma \Rightarrow \Delta, \psi \wedge \phi}$

$(\vee \Rightarrow) \quad \dfrac{\Gamma, \psi \Rightarrow \Delta \quad \Gamma, \phi \Rightarrow \Delta}{\Gamma, \psi \vee \phi \Rightarrow \Delta}$ $\qquad$ $(\Rightarrow \vee) \quad \dfrac{\Gamma \Rightarrow \Delta, \psi, \phi}{\Gamma \Rightarrow \Delta, \psi \vee \phi}$

$(\supset \Rightarrow) \quad \dfrac{\Gamma \Rightarrow \psi, \Delta \quad \Gamma, \phi \Rightarrow \Delta}{\Gamma, \psi \supset \phi \Rightarrow \Delta}$ $\qquad$ $(\Rightarrow \supset) \quad \dfrac{\Gamma, \psi \Rightarrow \phi, \Delta}{\Gamma \Rightarrow \psi \supset \phi, \Delta}$

$\qquad\qquad\qquad\qquad\qquad\qquad\qquad\qquad$ $(\Rightarrow \neg) \quad \dfrac{\Gamma, \psi \Rightarrow \Delta}{\Gamma \Rightarrow \Delta, \neg \psi}$

$(\circ \Rightarrow) \quad \dfrac{\Gamma \Rightarrow \psi, \Delta \quad \Gamma \Rightarrow \neg \psi, \Delta}{\Gamma, \circ \psi \Rightarrow \Delta}$ $\qquad$ $(\Rightarrow \circ) \quad \dfrac{\Gamma, \psi, \neg \psi \Rightarrow \Delta}{\Gamma \Rightarrow \circ \psi, \Delta}$

DEFINITION 16. $\mathbf{G}_0$ is the system obtained from $\mathbf{G}_k$ by deleting the rule $(\Rightarrow \circ)$.

Now from the results of [Avron et al., 2006] we directly get the following facts (first proved in [Avron, 2005]):

PROPOSITION 17.

1. $\mathcal{M}_0^3$ is a characteristic Nmatrix for $\mathbf{G}_0$.

2. $\mathcal{M}^3$ is a characteristic Nmatrix for $\mathbf{G_k}$.

3. Both $\mathbf{G_0}$ and $\mathbf{G_k}$ enjoy cut-admissibility.

The method we have just used for $\mathcal{M}_0^3$ can be separately applied to each of its simple refinements that were considered in [Avron, 2005; Avron, 2007b; Avron, 2007a]. In this way we can obtain a cut-free Gentzen-type formulation for each of the thousands of LFIs considered there. However, it would be much easier to do this in a modular way, by translating the semantic effect of each extra axiom into rules (and using the streamlining principles to simplify the results). Below we exemplify this process for the axioms $(\mathbf{o}_\vee^1)$, $(\mathbf{o}_\vee^2)$, and $(\mathbf{o}_\vee)$.

Consider first the schema $(\mathbf{o}_\vee^1)$. As explained above, its validity is equivalent to the combination of the following two conditions: (i) $t \vee x = \{t\}$ and (ii) $f \vee t = f \vee \top = \{t\}$. Now (i) can be reformulated as follows: if $\neg \psi \Rightarrow$ is true, then $\neg(\varphi \vee \psi) \Rightarrow$ is true. By adding context, we obtain:

$$\frac{\Gamma, \neg \varphi \Rightarrow \Delta}{\Gamma, \neg(\varphi \vee \psi) \Rightarrow \Delta}$$

In turn, (ii) can be reformulated as follows: if $\varphi \Rightarrow$ and $\Rightarrow \psi$ are true, then so is $\neg(\varphi \vee \psi) \Rightarrow$. Again, by adding context we get the following rule:

$$\frac{\Gamma, \varphi \Rightarrow \Delta \quad \Gamma \Rightarrow \Delta, \psi}{\Gamma, \neg(\varphi \vee \psi) \Rightarrow \Delta}$$

Taken together, these two Gentzen-type rules correspond to the schema $(\mathbf{o}_\vee^1)$.

Next, two analogous rules for $(\mathbf{o}_\vee^2)$ can be derived in a similar way. Finally, a set of four rules which corresponds (as a whole) to $\mathbf{o}_\vee$ is obtained by collecting together the rules for $(\mathbf{o}_\vee^1)$ and $(\mathbf{o}_\vee^2)$.

DEFINITION 18. The system $\mathbf{Gko}_\vee$ is obtained from $\mathbf{Gk}$ by adding to it the following four rules:

$$\frac{\Gamma, \neg \varphi \Rightarrow \Delta}{\Gamma, \neg(\varphi \vee \psi) \Rightarrow \Delta} \; (\mathbf{o}_\vee^{11}) \quad \frac{\Gamma, \varphi \Rightarrow \Delta \quad \Gamma \Rightarrow \Delta, \psi}{\Gamma, \neg(\varphi \vee \psi) \Rightarrow \Delta} \; (\mathbf{o}_\vee^{12})$$

$$\frac{\Gamma, \neg \psi \Rightarrow \Delta}{\Gamma, \neg(\varphi \vee \psi) \Rightarrow \Delta} \; (\mathbf{o}_\vee^{21}) \quad \frac{\Gamma, \psi \Rightarrow \Delta \quad \Gamma \Rightarrow \Delta, \varphi}{\Gamma, \neg(\varphi \vee \psi) \Rightarrow \Delta} \; (\mathbf{o}_\vee^{22})$$

The systems $\mathbf{Gko}_\vee^i$ ($i = 1, 2$) are obtained similarly, using $(\mathbf{o}_\vee^{i1})$ and $(\mathbf{o}_\vee^{i2})$.

EXAMPLE 19. Below we show that $(\mathbf{o}_\vee^1)$ indeed has a cut-free proof in $\mathbf{Bko}_\vee^1$, even though none of the Gentzen-type rules $(\mathbf{o}_\vee^{11})$ and $(\mathbf{o}_\vee^{12})$ corresponding to this axiom even mentions the connective $\circ$:

$$\cfrac{\cfrac{\neg(\varphi \vee \psi), \varphi \Rightarrow \varphi \quad \cfrac{\varphi, \psi \Rightarrow \varphi \quad \psi \Rightarrow \varphi, \psi}{\neg(\varphi \vee \psi), \psi \Rightarrow \varphi} \, (\mathbf{o}_\vee^{12})}{\cfrac{\neg(\varphi \vee \psi), \varphi \vee \psi \Rightarrow \varphi}{\cfrac{\circ \varphi, \neg(\varphi \vee \psi), \varphi \vee \psi \Rightarrow}{\cfrac{\circ \varphi \Rightarrow \circ(\varphi \vee \psi)}{\Rightarrow \circ \varphi \supset \circ(\varphi \vee \psi)} \, (\Rightarrow \supset)} \, (\Rightarrow \circ)} \, (\vee \Rightarrow)} \quad \cfrac{\neg \varphi, \varphi \vee \psi \Rightarrow \neg \varphi}{\neg(\varphi \vee \psi), \varphi \vee \psi \Rightarrow \neg \varphi} \, (\mathbf{o}_\vee^{11})}{} \, (\circ \Rightarrow)$$

THEOREM 20.

1. $\mathcal{M}^3_{o^1_\vee}$, $\mathcal{M}^3_{o^2_\vee}$ and $\mathcal{M}^3_{o_\vee}$ are characteristic Nmatrices for $\mathbf{G}_{\mathbf{ko}^1_\vee}$, $\mathbf{G}_{\mathbf{ko}^2_\vee}$ and $\mathbf{G}_{\mathbf{ko}_\vee}$ respectively.

2. $\mathbf{G}_{\mathbf{ko}^1_\vee}$, $\mathbf{G}_{\mathbf{ko}^2_\vee}$ and $\mathbf{G}_{\mathbf{ko}_\vee}$ enjoy cut-admissibility.

Proof. It is easy to see that these calculi are obtained for the corresponding Nmatrices by the algorithm of [Avron et al., 2006]. Thus the theorem follows from Proposition 13 and the results of [Avron et al., 2006]. ∎

4 Conclusions and Further Research

In this paper we have exemplified a *uniform* way to *systematically* construct analytic calculi for a large family of LFIs, each having a semantic characterization in terms of a three-valued Nmatrix. We believe that these results will help in producing efficient tools for automated reasoning with inconsistency, eventually making Logics of Formal (In)consistency a more appealing formalism for reasoning under uncertainty.

The most immediate directions for further research are the following.

- There are two axioms that are included in some of the most important C-systems, but cannot be handled in the framework developed in this paper. These are the following famous axioms from [Carnielli and Marcos, 2002; Carnielli et al., 2007]:

  (l) $\neg(\varphi \wedge \neg\varphi) \supset \circ\varphi$ (d) $\neg(\neg\varphi \wedge \varphi) \supset \circ\varphi$

  Systems with these axioms include, for instance, da Costa's historic system C_1, which can be shown to be equivalent to $\mathbf{Bk}[\{(\mathbf{a}), (\mathbf{c}), (\mathbf{i}), (\mathbf{l})\}]$.

  It is quite obvious how to translate (l) (say) into a Gentzen-type rule: simply substitute in ($\circ \Rightarrow$) the formula $\neg(\psi \wedge \neg\psi)$ for $\circ\psi$ (doing the same for ($\Rightarrow \circ$) results in a derivable rule of **B**). In the cases in which cut-free systems for logics with (l) have already been given in the literature (like in [Béziau, 2001; Gentilini, 2011]), this simple procedure leads to systems which are equivalent to those given before. However, it is not known yet whether this will be true in general, and whether the crucial modularity of our framework is preserved by this procedure.

  Now the main obstacle in extending the method of this paper to systems with (l) or (d) is that such systems have no semantic characterization in terms of finite-valued Nmatrices (this was shown in [Avron, 2007b]). However, they do have infinitely-valued characterizations of this type (which still suffice for guaranteeing their decidability). It is not clear yet whether these characterizations can be used for the same purposes that the three-valued framework has been used here.

- It is obvious that for building LFI-based theorem provers for real-life applications, the results of this paper need to be extended to the first-order case. To the best of our knowledge, currently there are no known analytic systems available on the first-order level. However, in [Avron and Zamansky, 2007] finite non-deterministic semantics was provided for many first-order LFIs. It is quite likely that it can be exploited along the lines of the approach presented in this paper.

BIBLIOGRAPHY

[Avron and Lev, 2005] A. Avron and I. Lev. Non-deterministic Multi-valued Structures. *Journal of Logic and Computation*, 15:241–261, 2005.

[Avron and Zamansky, 2007] A. Avron and A. Zamansky. Many-valued non-deterministic semantics for first-order logics of formal (in)consistency. In S. Aguzzoli, A. Ciabattoni, B. Gerla, C. Manara, and V. Marra, editors, *Algebraic and Proof-theoretic Aspects of Non-classical Logics*, number 4460 in LNAI, pages 1–24. Springer, 2007.

[Avron and Zamansky, 2011] A. Avron and A. Zamansky. Non-deterministic semantics for logical systems - A survey. In D. Gabbay and F. Guenther, editors, *Handbook of Philosophical Logic*, volume 16, pages 227–304. Springer, 2011.

[Avron et al., 2006] A. Avron, J. Ben-Naim, and B. Konikowska. Cut-free ordinary sequent calculi for logics having generalized finite-valued semantics. *Logica Universalis*, 1:41–69, 2006.

[Avron, 2005] A. Avron. Non-deterministic Matrices and Modular Semantics of Rules. In J. Y. Beziau, editor, *Logica Universalis*, pages 149–167. Birkhűser Verlag, 2005.

[Avron, 2007a] A. Avron. Non-deterministic semantics for families of paraconsistent logics. In J. Y. Beziau, W. A. Carnielli, and D. M. Gabbay, editors, *Handbook of Paraconsistency*, volume 9 of *Studies in Logic*, pages 285–320. College Publications, 2007.

[Avron, 2007b] A. Avron. Non-deterministic semantics for logics with a consistency operator. *Journal of Approximate Reasoning*, 45:271–287, 2007.

[Béziau, 2001] J.Y. Béziau. From Paraconsistent Logic to Universal Logic. *Sorites*, 12:5–32, 2001.

[Carnielli and Lima-Marques, 1992] W.A. Carnielli and M. Lima-Marques. Reasoning under inconsistent knowledge. *Journal of Applied Non-classical Logics*, 2(1):49–79, 1992.

[Carnielli and Marcos, 2001] W.A. Carnielli and J. Marcos. Tableau systems for logics of formal inconsistency. In *Proceedings of the 2001 International Conference on Artificial Intelligence*, volume 2, pages 848–852. CSREA Press, 2001.

[Carnielli and Marcos, 2002] W. A. Carnielli and J. Marcos. A taxonomy of C-systems. In W. A. Carnielli, M. E. Coniglio, and I. D'Ottaviano, editors, *Paraconsistency: The Logical Way to the Inconsistent*, number 228 in Lecture Notes in Pure and Applied Mathematics, pages 1–94. Marcel Dekker, 2002.

[Carnielli et al., 2007] W. A. Carnielli, M. E. Coniglio, and J. Marcos. Logics of formal inconsistency. In D. M. Gabbay and F. Guenthner, editors, *Handbook of Philosophical Logic*, volume 14, pages 15–107. Springer, 2007. Second edition.

[da Costa et al., 1995] N. C. A. da Costa, J.-Y. Béziau, and O.A.S. Bueno. Aspects of paraconsistent logic. *Bulletin of the IGPL*, 3:597–614, 1995.

[da Costa, 1974] N. C. A. da Costa. On the theory of inconsistent formal systems. *Notre Dame Journal of Formal Logic*, 15:497–510, 1974.

[D'Ottaviano, 1990] I. D'Ottaviano. On the development of paraconsistent logic and da Costa's work. *Journal of Non-classical Logic*, 7(1–2):89–152, 1990.

[Gentilini, 2011] P. Gentilini. Proof theory and mathematical meaning of paraconsistent C-systems. *Journal of Applied Logic*, 9:171–202, 2011.

[Neto and Finger, 2007] A. Neto and M. Finger. A KE tableau for a logic for formal inconsistency. In *Proceedings of TABLEAUX'07 position papers and Workshop on Agents, Logic and Theorem Proving*, volume LSIS.RR.2007.002, 2007.

Arnon Avron
School of Computer Science
Tel-Aviv University
Tel-Aviv, Israel
E-mail: aa@math.tau.ac.il

Beata Konikowska
Institute of Computer Science
Polish Academy of Sciences
Warsaw, Poland
E-mail: Beata.Konikowska@ipipan.waw.pl

Anna Zamansky
Institute for Discrete Mathematics and Geometry
Vienna Technical University
Vienna, Austria
E-mail: annaz@logic.at

The Value of the Two Values
João Marcos

ABSTRACT. Bilattices have proven again and again to be extremely rich structures from a logical point of view. As a matter of fact, even if one fixes the canonical notion of many-valued entailment and consider the smallest non-trivial bilattice, distinct logics may be defined according to the chosen ontological, epistemological or informational reading of the underlying truth-values. This note will explore the consequence relations of two very natural variants of Belnap's well-known 4-valued logic, and delve into their interrelationship. The strategy will be that of reformulating those logics using only two 'logical values', by way of uniform classic-like semantical and proof-theoretical frameworks, with the help of which such logics may be more easily compared to each other.

1 Introduction

Consider the order-bilattice $(\mathcal{V}, \leq_1, \leq_2)$ where $\mathcal{V} = \{t, \top, \bot, f\}$, the 'truth order' $\leq_1$ has t as its greatest element and f as its least element, as well as intermediate mutually incomparable elements $\top$ and $\bot$, and the 'information order' $\leq_2$ has $\top$ as its greatest element and $\bot$ as its least element, as well as intermediate mutually incomparable elements t and f (see Figure 1). Construing $\mathcal{V}$ as a set of 'truth-values', one may consider the algebraic structures $\mathcal{L}_i = (\mathcal{V}, \wedge_i, \vee_i, \neg_i)$, for $i = 1, 2$, where $\wedge_i$ (resp. $\vee_i$) denotes the meet $\sqcap_i$ (resp. the join $\sqcup_i$) under $\leq_i$, and $\neg_i$ is an order-reversing involution for $\leq_i$ having the intermediate elements, in each case, as fixed-points. It is easy to see that these algebraic structures are 'interlaced', i.e., the operators of $\mathcal{L}_1$ (resp. $\mathcal{L}_2$) are all monotone with respect to $\leq_2$ (resp. $\leq_1$). Even stronger than that, all distributive laws hold between the two meets and the two joins. Morever, $\neg_1$ (called 'negation') and $\neg_2$ (called 'conflation') obviously commute, that is, $\neg_1 \neg_2 x = \neg_2 \neg_1 x$. Coalescing $\mathcal{L}_1$ and $\mathcal{L}_2$ into a single diagram suggests a strongly symmetric bidimensional structure $\mathcal{B} = (\mathcal{V}, \wedge_1, \wedge_2, \vee_1, \vee_2, \neg_1, \neg_2)$.

Let $\Gamma \cup \Delta$ be a collection of formulas from the term algebra $\mathcal{T}_\mathcal{B}$ freely generated by a denumerable collection of atoms over the connectives (operator symbols) from the structure $\mathcal{B}$, and let Hom denote the set of all homomorphisms, called 'valuations', from $\mathcal{T}_\mathcal{B}$ into $\mathcal{V}$. This is known as a 'truth-functional interpretation', and the meaning of each operator is said to be fixed by a 'truth-table'. There are various notions of entailment that might be associated to the above structure, so that one could talk about an 'inference' that holds, or does not hold, between two given collections of formulas, Γ and Δ. For instance, one could explore the underlying orders once again (they have already been used in defining truth-tables for the connectives), and define, for each dimension i of the above structure its own notion of 'o-entailment' $\models_i^o$, according to which

$\Gamma \models_i^o \Delta$ iff $\bigsqcap_i \mathsf{w}(\Gamma) \leq_i \bigsqcup_i \mathsf{w}(\Delta)$, for every $\mathsf{w} \in \mathsf{Hom}$. A different notion of entailment that is canonically found in the literature on many-valued logics, that will here be called 'p-entailment', assumes some partition of the truth-values $\mathcal{V}_j$ into sets $\mathcal{D}_j$ (called 'designated values') and $\mathcal{U}_j$ (called 'undesignated values'). In that case, the inference $\Gamma \models_j^p \Delta$ is said to hold iff, for every $\mathsf{w} \in \mathsf{Hom}$, either $\mathsf{w}(\Gamma) \cap \mathcal{U}_j \neq \varnothing$ or $\mathsf{w}(\Delta) \cap \mathcal{D}_j \neq \varnothing$. We will almost exclusively be talking about p-entailment, from this point on, and omit accordingly the superscript from $\models^p$ whenever we see no risk of misunderstanding.

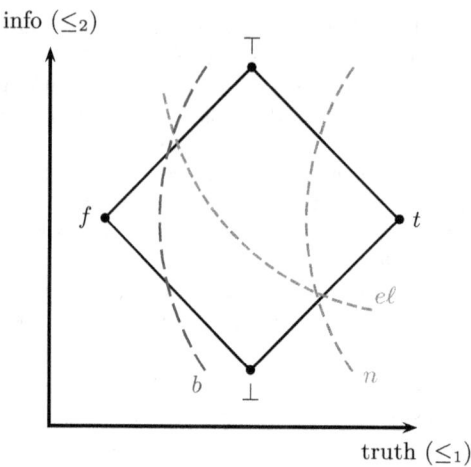

Figure 1. A representative bilattice, sliced in 3 different ways.

It is clear that the canonical many-valued entailment relation is remarkably sensitive to the choice of designated / undesignated values. There are at least three non-trivial such choices that could be made from the viewpoint of the truth order:

[$\mathcal{V}_b$] $\mathcal{D}_b = \{t, \top, \bot\}$ and $\mathcal{U}_b = \{f\}$
[$\mathcal{V}_{el}$] $\mathcal{D}_{el} = \{t, \top\}$ and $\mathcal{U}_{el} = \{\bot, f\}$
[$\mathcal{V}_n$] $\mathcal{D}_n = \{t\}$ and $\mathcal{U}_n = \{\top, \bot, f\}$

Choice [$\mathcal{V}_{el}$] has in fact been intensely investigated in the literature, and the corresponding p-entailment relation, $\models_{el}$, is essentially the same as each of the associated o-entailments. It is known to be adequate for the so-called 'Logic of De Morgan lattices', and some presentations of it are intimately related to a formalization of the so-called 'first-degree entailment'. It is also both paraconsistent and paracomplete. On the other hand, the p-entailment relation $\models_b$, that corresponds to [$\mathcal{V}_b$], is paraconsistent but not paracomplete, and the exact opposite is the case for the p-entailment relation $\models_n$, that corresponds to [$\mathcal{V}_n$]. A reasonable rationale for the choice [$\mathcal{V}_b$], according to the ordinary 'truth-degree interpretation', is that one might be dealing with vague states-of-affairs in which some values should not be ascertained to be 'false', yet they are 'not quite true'. Analogously, for [$\mathcal{V}_n$], there may be other kinds of inexact states-of-affairs in which some values should not be ascertained to be 'true', yet they are 'not quite false' (see Figure 1, again).

The present study will show in more detail what do such entailment relations have in common, and how do they differ from each other. The comparison will be made simpler when the logics involved are recast in terms of semantics and proof-systems that mention only *two* truth-values or *two* syntactic labels, as it happens in classical logic. To enhance the readability of the next sections, comments on the history, the scope and the challenges for our present approach will be left for Section 5. It should be clear at this point, however, that the task that interests us here is the one that is of concern for the logic-designer, much more than for the logic-user, to wit, the task of finding a common *coherent* framework in which it all the above logics can be simultaneously formulated and have their properties contrasted.

2 A Closer Look at the Logical Operators

From the semantical point of view, a logical operator & called *conjunction* is often used to internalize at the object-language level a collection of properties commonly attributed to the metalinguistic 'and', such as:

[and$_1$] $w(\alpha \& \beta) \in \mathcal{D}$ if $w(\alpha) \in \mathcal{D}$ and $w(\beta) \in \mathcal{D}$
[and$_2$] $w(\alpha \& \beta) \in \mathcal{D}$ only if $w(\alpha) \in \mathcal{D}$ and $w(\beta) \in \mathcal{D}$

Similarly, a logical operator $\|$ called *disjunction* is often used to internalize properties commonly attributed to the metalinguistic 'or', such as:

[or$_1$] $w(\alpha \| \beta) \in \mathcal{D}$ if $w(\alpha) \in \mathcal{D}$ or $w(\beta) \in \mathcal{D}$
[or$_2$] $w(\alpha \| \beta) \in \mathcal{D}$ only if $w(\alpha) \in \mathcal{D}$ or $w(\beta) \in \mathcal{D}$

Obviously, the use of a classical metalanguage, together with the above assumed partition of the truth-values into exactly two classes, allows us to immediately rewrite [and$_1$] and [or$_1$] as:

[and$_1$] $w(\alpha \& \beta) \in \mathcal{U}$ only if $w(\alpha) \in \mathcal{U}$ or $w(\beta) \in \mathcal{U}$
[or$_1$] $w(\alpha \| \beta) \in \mathcal{U}$ only if $w(\alpha) \in \mathcal{U}$ and $w(\beta) \in \mathcal{U}$

As it turns out, according to $\models_{e\ell}$, each operator $\wedge_i$ enjoys properties [and$_1$] and [and$_2$], and each operator $\vee_i$ enjoys properties [or$_1$] and [or$_2$], for $i = 1, 2$. However, according to either $\models_b$ or $\models_n$, this only holds good for $i = 1$, that is, for the logical operators defined according to the truth-order $\leq_1$. Indeed, for both the latter entailment relations, on what concerns the operators defined according to the information order $\leq_2$, it can easily be checked that $\wedge_2$ enjoys property [and$_1$] but fails property [and$_2$], while $\vee_2$ enjoys property [or$_2$] but fails property [or$_1$]. One can also say that $\neg_2$ only behaves like a real *negation* according to $\models_{e\ell}$, but not according to $\models_b$ nor according to $\models_n$. Indeed, for the latter entailment relations conflation behaves more like a kind of *identity* relation, in that the formulas φ and $\neg_2 \varphi$ are always interderivable (being equivalent, yet not congruent). This much for the similarities between $\models_b$ and $\models_n$. The mixed language of the structure $\mathcal{B}$, where a lot of surprising interactions occur between the original separate algebraic structures, produces other differences and dualities between $\models_b$ and $\models_n$ that go, as we shall see, much beyond the mere contrast between paraconsistent × paracomplete behavior.

It's not overemphasizing to insist here on the difficulty of the general task of *comparing* two given finite-valued logics just by having a quick look at their truth-tables. It is far from obvious, in fact, how examples of inferences that help in either distinguishing or likening two given logics are even to be found, without exercising some ingenuity. In all such cases, however, the task consists in, given logics $\mathcal{L}_x$ and $\mathcal{L}_y$, finding appropriate collections of formulas, Γ and Δ, such that: (1) $\Gamma \models_x \Delta$ (verified by performing a check ranging over all valuations w_x of $\mathcal{L}_x$) and (2) $\Gamma \not\models_y \Delta$ (verified, in the canonical reading of entailment, by finding a valuation w_y of $\mathcal{L}_y$ such that w_y satisfies all formulas in Γ and simultaneously falsifies all formulas in Δ). Nonetheless, serious difficulties hinder the automation, or even the very accomplishment, of such a task.

Let's assume, for the sake of the argument, that the language of $\mathcal{L}_x$ is at least as rich as the language of $\mathcal{L}_y$ — lest the comparison task is made more difficult by the requirement of some previous massage of the formulas being done by way of previous translations and reinterpretations of the underlying languages. Even based on the same initial language, however, it is not obvious how two distinct given many-valued logics compare to each other. One might think, for instance, of transferring the problem from semantics into proof-theory, with the following reasoning: If both $\mathcal{L}_x$ and $\mathcal{L}_y$ are formulated in terms of rules governing the behavior of their operators, one could in principle use the rules of $\mathcal{L}_x$ to check the derivability of the rules of $\mathcal{L}_y$, or perform an induction on the derivations of $\mathcal{L}_x$ to check whether each rule of $\mathcal{L}_y$ is admissible. But is there a common proof-theoretical language in which all such rules may be expressed? The most well-known automated approaches to finite-valued logics would suggest that this is *not* the case. It is common and straightforward, for instance, to extract signed tableau systems, in each case, from the received truth-tables, by way of a simple trick that transforms the truth-values, or the collections of truth-values, into syntactic signs that are used, say, in front of the formulas. An obvious difficulty for the logic comparison, in that case, appears when $\mathcal{L}_x$ is $\sharp(\mathcal{V}_x)$-valued and $\mathcal{L}_y$ is $\sharp(\mathcal{V}_y)$-valued, for $\sharp(\mathcal{V}_x) \neq \sharp(\mathcal{V}_y)$. In that case, the rules extracted for each logic will be written in different languages, as they will allow in principle for different collections of signs, and will hardly be comparable. At least this specific difficulty would seem to be circumvented, however, in our current case study, where we are concerned about the comparison of different 4-valued logics. One might argue, however, that in this case, where the availability of a common language would not seem to a problem, an even more insidious difficulty slips in. The problem could be described as one that touches upon *coherence* of the whole approach. Indeed, even if we transform the truth-values $\top$ and $\bot$ into signs to be used in expressing the rules of both the logic behind choice $[\mathcal{V}_b]$ and the logic behind choice $[\mathcal{V}_n]$, they could hardly be claimed to have the same significance: While both signs refer to gradations of truth from the viewpoint of $\models_b$, they refer to gradations of falsity from the viewpoint of $\models_n$. Using the same signs, for different logics, as symbols that refer to different entities in their corresponding interpretations, would just confuse the metatheory for the logic-designer and at the same time risk making equivocal the conversation between the users of each logic. It would be as if, for instance, the logic-users employed the same connective symbol to denote, in each case, a different operator — they would certainly experience a lot of difficulty thus in talking to each other about it. That the logic-designer should insist in lifting such misunderstandings to his own high-level framework is likely to make him prone to schizophrenia.

The remainder of the paper will show that there is indeed a strategy that lends itself for a straightforward and fully mechanizable comparison between different many-valued logics, and one that is applicable both to the challenging cases of logics $\mathcal{L}_x$ and $\mathcal{L}_y$ with $\sharp(\mathcal{V}_x) \neq \sharp(\mathcal{V}_y)$ and to the hereby illustrated more subtle cases in which $\sharp(\mathcal{V}_x) = \sharp(\mathcal{V}_y)$ yet $\sharp(\mathcal{D}_x) \neq \sharp(\mathcal{D}_y)$. The next section will show in detail, as an illustration of the application of general methods that will be identified further on, how one of our 4-valued logics may be recast in a classic-like fashion, using only two truth-values. The methods are fully general, and will capitalize in our specific illustrations on the primitive expressivity strength of the language $\mathcal{T}_\mathcal{B}$. The new semantics that will originate, as a matter of fact, is to be formulated in such a way that we will be able to show, in the succeeding section, that a classic-like tableau-theoretic presentation is available for that same logic. The chosen formalism will have the property of analyticity, making the associated decision procedures fully automatable.

3 Alternative Bivalent Semantics

The definition of p-entailment that we have been using to characterize the underlying inference relations of our logics in no way depends on the fact that the proposed collections of valuations provide truth-functional interpretations. Indeed, any such inference relation may be determined using only two 'logical values'. Classical Logic, in that case, has the best of both worlds, being truth-functional (and in fact functionally complete) over the collection $\mathcal{V}_2 = \{1, 0\}$, naturally partitioned into 'logical truth' ($\mathcal{D}_2 = \{1\}$) and 'logical falsity' ($\mathcal{U}_2 = \{0\}$). As the following exposition will make clear, however, it is not hard to realize that any other specific p-entailment relation can alternatively be characterized in a similar, classic-like way.

Let's call a 'bivaluation' any mapping of the form b : $\mathcal{T}_\mathcal{B} \longrightarrow \mathcal{V}_2$, and call 'bivalent' any collection Biv of bivaluations. The corresponding notion of p-entailment, that will here be denoted by $\models_{\text{Biv}}$, is defined just as before, but now substituting Biv for Hom. For each w $\in$ Hom and associated partition of the truth-values, a 'bivalent counterpart' b_w may immediately be defined with the help of the characteristic function r2 : $\mathcal{V} \longrightarrow \mathcal{V}_2$ that takes each designated value into logical truth and each undesignated valued into logical falsity: Just consider the 'bivalent reduction' b_w = r2 ∘ w. There is of course also a bivalent counterpart for the associated entailment relation $\models^p$: Take Biv = $\{b_w : w \in \text{Hom}\}$, and notice that $\Gamma \models_{\text{Biv}} \Delta$ iff $\Gamma \models^p \Delta$. A different question is whether there is anything as convenient, concise and useful as truth-tables that might be used to describe the bivaluation semantics obtained from such a bivalent reduction.

A constructive positive answer to the above question may be provided if the very language of our logic turns out to be expressive enough so as to distinguish between any pair of designated values, and any pair of undesignated values, even after having them 'flattened', in a sense, by the bivalent reduction. In the particular case of $\mathcal{V}_b$, for instance, what we are looking for is a way of distinguishing the three logically true values in $\mathcal{D}_b$ (and similarly, in the case of $\mathcal{V}_n$, for the three logically false values in $\mathcal{U}_n$). This will be possible and easy to describe in case one can find convenient one-variable 'separating formulas' $\circledS_{v_1 v_2}(p)$ to the effect that, given $\mathsf{w}_1, \mathsf{w}_2 \in \mathsf{Hom}$ with $v_1 = \mathsf{w}_1(p) \neq \mathsf{w}_2(p) = v_2$, then $\mathsf{b}_{\mathsf{w}_1}(\circledS_{v_1 v_2}(p)) \neq \mathsf{b}_{\mathsf{w}_2}(\circledS_{v_1 v_2}(p))$. If v_1 and v_2 come from different partitions, one may use p itself as a separating formula. For the case of the logics obtained from choices $[\mathcal{V}_b]$ and $[\mathcal{V}_n]$, in particular, the values t and f may be distinguished from the values $\top$ and $\bot$ through the separating formula $\neg_1 p$. Now, can $\top$ and $\bot$ also be distinguished from each other, using the primitive linguistic resources of $\mathcal{B}$? Yes, they can. Indeed, here is one way of doing it:

p	$\neg_1 p$	$\copyright_n(p) \triangleq p \wedge_1 \neg_2 p$	$\copyright_b(p) \triangleq \neg_1 \copyright_n(\neg_1 \neg_2(p))$	$\circledS(p) \triangleq \copyright_b(p \wedge_2 \copyright_n(p))$
t	f	t	t	t
$\top$	$\top$	f	t	f
$\bot$	$\bot$	f	t	t
f	t	f	f	f

Notice in particular how $\copyright_1$ and $\copyright_2$ might be regarded as characteristic functions of the sets of designated values $\mathcal{D}_n$ and $\mathcal{D}_b$.

REMARK 1. *For its expressive power, from this point on we will be working directly with the structure $\mathcal{B}_\circledS$, where the definable separating formula $\circledS$ is introduced among the primitive operators. The important thing to bear in mind is that the interpretation of the triple $\langle p, \neg_1 p, \circledS p \rangle$, after the bivalent reduction, gives a unique identification to each $v \in \mathcal{V}_x$, irrespective of the choice $[\mathcal{V}_x]$, so that each truth-value is given a unique 'binary print'.*

It is clear that choice $[\mathcal{V}_{e\ell}]$ is the most symmetric among the three ones we have chosen to distinguish. As a matter of fact, the logic behind this choice may be thought of as a combination of two 'De Morgan lattices', $\langle \mathcal{V}, \wedge_1, \vee_1, \neg_1 \rangle$ and $\langle \mathcal{V}, \wedge_2, \vee_2, \neg_2 \rangle$, and some classical results about *dualization* of logical operators do easily carry over from the components into the combined structure $\mathcal{B}$. To be more precise, consider the endomorphism ε on $\mathcal{T}_\mathcal{B}$ defined by setting:

$$\begin{aligned} p^\varepsilon &= p & &\text{where } p \text{ is an atom} \\ (\rtimes \alpha)^\varepsilon &= \rtimes(\alpha^\varepsilon) & &\text{where } \rtimes \in \{\neg_1, \neg_2, \circledS\} \\ (\alpha \oplus \beta)^\varepsilon &= (\alpha^\varepsilon \otimes \beta^\varepsilon) & &\text{where } \langle \oplus, \otimes \rangle \in \{\langle \wedge_i, \vee_i \rangle, \langle \vee_i, \wedge_i \rangle\} \text{ and } i \in \{1,2\} \end{aligned}$$

Then we have the following classic-like result:

THEOREM 2. *Given $\Gamma \cup \Delta \subseteq \mathcal{T}_{\mathcal{B}_\circledS}$, then $\Gamma \models_{e\ell} \Delta$ iff $\Delta^\varepsilon \models_{e\ell} \Gamma^\varepsilon$.*

In what follows we will see how such result may easily be proven, and also how it may be extended in order to reveal fascinating connections between our other two distinguished logics.

For all that matters, we will hereupon be concentrating on the choice $[\mathcal{V}_b]$: All our results and considerations will be easily adaptable and dualized for the choice $[\mathcal{V}_n]$. Accordingly, if we use $\neg_1$ and $\circledS$ to express at the object language level the difference between the three designated values behind choice $[\mathcal{V}_b]$, one might arrive to a set of axioms constraining the set of bivalent mappings having domain $\mathcal{T}_{\mathcal{B}_\circledS}$ and providing an alternative description of the underlying 4-valued logic, with its original entailment relation. In more precise terms, concerning the collection of bivaluations Biv^b described in the Appendix, the following may be proven:

THEOREM 3. Biv^b provides a sound and complete bivalent semantics for the paraconsistent logic behind choice $[\mathcal{V}_b]$.

This result is indeed a consequence of the following two lemmas, that can be checked in an entirely constructive fashion. Full details of the corresponding proofs will be exhibited for illustrative cases and subcases.

LEMMA 4. Given $\mathsf{w} \in \mathsf{Hom}$, define $\mathsf{b_w}$ by:

$$\mathsf{b_w}(\varphi) = \begin{cases} 0 & \text{if } \mathsf{w}(\varphi) = f, \\ 1 & \text{otherwise.} \end{cases}$$

Then, $\mathsf{b_w} \in \mathsf{Biv}^b$.

Proof. One must show that the bivaluation axioms $\mathsf{biv}[xy]\langle v \rangle$ are all respected by this definition. Consider for instance the case in which y is $\neg_2$.

Subcase $\mathsf{biv}[\neg_2]\langle 1 \rangle$. Assume $\mathsf{b_w}(\neg_2\varphi) = 1$. By the above definition, this means that $\mathsf{w}(\neg_2\varphi) \neq f$. From the truth-tables of $\mathcal{B}$ one may conclude that $\mathsf{w}(\varphi) \neq f$, and using again the above definition we have that $\mathsf{b_w}(\varphi) = 1$, as desired.

Subcase $\mathsf{biv}[\neg_2]\langle 0 \rangle$. Assume $\mathsf{b_w}(\neg_2\varphi) = 0$. By the definition of $\mathsf{b_w}$, $\mathsf{w}(\neg_2\varphi) = f$, and the 4-valued interpretation of $\neg_2$ tells us that $\mathsf{w}(\varphi) = f$. Here, from the definition of $\mathsf{b_w}$ we can say that $\mathsf{b_w}(\varphi) = 0$.

Subcase $\mathsf{biv}[\neg_1\neg_2]\langle 1 \rangle$. Assume $\mathsf{b_w}(\neg_1\neg_2\varphi) = 1$. The definition of $\mathsf{b_w}$ gives us $\mathsf{w}(\neg_1\neg_2\varphi) \neq f$. But the truth-tables of $\neg_1$ and of $\neg_2$ inform us that $\mathsf{w}(\neg_1\neg_2\varphi) \neq f$ iff $\mathsf{w}(\neg_1\varphi) \neq f$. So, using the definition of $\mathsf{b_w}$ again, $\mathsf{b_w}(\neg_1\varphi) = 1$.

Subcase $\mathsf{biv}[\neg_1\neg_2]\langle 0 \rangle$. Analogous to the previous subcase.

Subcase $\mathsf{biv}[\textcircled{S}\neg_2]\langle 1 \rangle$. Assume $\mathsf{b_w}(\textcircled{S}\neg_2\varphi) = 1$. From the definition of $\mathsf{b_w}$, we have that $\mathsf{w}(\textcircled{S}\neg_2\varphi) \neq f$. The truth-table of $\textcircled{S}$ tells us that $\mathsf{w}(\textcircled{S}\neg_2\varphi) = t$ and also that it must be the case that $\mathsf{w}(\neg_2\varphi) \in \{t, \bot\}$, and we should check next what the truth-table of $\neg_2$ has to tell us. The first option, $\mathsf{w}(\neg_2\varphi) = t$, is equivalent to writing that (i) $\mathsf{w}(\varphi) = f$; the second option, $\mathsf{w}(\neg_2\varphi) = \bot$, is equivalent to (ii) $\mathsf{w}(\varphi) = \top$. From the interpretation of $\textcircled{S}$ and (ii) it follows that (iii) $\mathsf{w}(\textcircled{S}\varphi) = f$. Now, the definition of $\mathsf{b_w}$ allows us to say, in case (i), that (iv) $\mathsf{b_w}(\varphi) = 0$, and to say, in case (ii)+(iii), that (v) $\mathsf{b_w}(\varphi) = 1$ and $\mathsf{b_w}(\textcircled{S}\varphi) = 0$. Conversely, (i) follows from (iv) and the definition of $\mathsf{b_w}$. Similarly from (v) and the definition of $\mathsf{b_w}$ we may conclude that both $w(\varphi) \neq f$ and $w(\textcircled{S}\varphi) = f$, and in this case the truth-table of $\textcircled{S}$ guarantees that we are talking about a situation in which (ii) holds good.

Subcase $\mathsf{biv}[\textcircled{S}\neg_2]\langle 0 \rangle$. Analogous to the previous subcase.

The proofs for the cases of the other connectives follow analogous patterns. The cases in which y is a binary connective, in fact, are quite similar to the latter two subcases. In simplifying the proofs of the corresponding converses it will often be useful to establish and use the following auxiliary facts:

$\mathsf{w}(\varphi) = \top$ iff $\mathsf{b_w}(\varphi) = 1$ and $\mathsf{b_w}(\textcircled{S}\varphi) = 0$
$\mathsf{w}(\varphi) = \bot$ iff $\mathsf{b_w}(\neg_1\varphi) = 1$ and $\mathsf{b_w}(\textcircled{S}\varphi) = 1$
$\mathsf{w}(\varphi) \in \{\top, \bot\}$ iff $\mathsf{b_w}(\varphi) = 1$ and $\mathsf{b_w}(\neg_1\varphi) = 1$

Finally, one must also verify the bivaluation axioms $\mathsf{biv}[mn]$. The case in which m is T is guaranteed to hold good, using the above definition, from the fact that each homomorphism w is total. Additionally, the cases in which m is C all reflect, from a bivalent perspective, semantic assignments that are unobtainable from the viewpoint of truth-tables. For $n = 0$, all one needs to guarantee is that each b_w is a function —and so it must be, as the characteristic mapping of the function w. Moreover, given that it is impossible, for instance, to attribute the values $w(\varphi) = f$ and $w(\neg_1\varphi) = f$, given the truth-table of $\neg_1$, the above definition tells us that $b_w(\varphi) = 0$ and $b_w(\neg_1\varphi) = 0$ cannot simultaneously obtain, proving thus biv$[C1]$. Similarly for $n \in \{2, 3\}$. ∎

LEMMA 5. Given $b \in \text{Biv}^b$, define w_b by:

$$\begin{array}{lll} w_b(\varphi) = t & \text{if} & b(\neg_1\varphi) = 0 \\ w_b(\varphi) = \top & \text{if} & b(\varphi) = 1 \text{ and } b(\text{⑤}\varphi) = 0 \\ w_b(\varphi) = \bot & \text{if} & b(\neg_1\varphi) = 1 \text{ and } b(\text{⑤}\varphi) = 1 \\ w_b(\varphi) = f & \text{if} & b(\varphi) = 0 \end{array}$$

Then, [A] $w_b(\varphi) \in \mathcal{D}$ iff $b(\varphi) = 1$. Moreover, [B] $w_b \in \text{Hom}$.

Proof. To check statement [A], the only non-obvious cases are those in which $w_b(\varphi) \in \{t, \bot\}$, that is, the cases in which either (a) $b(\neg_1\varphi) = 0$, or (b) both $b(\neg_1\varphi) = 1$ and $b(\text{⑤}\varphi) = 1$. On what concerns (a), biv$[C1]$ guarantees that $b(\varphi) = 0$ cannot be the case. As for (b), biv$[C2]$ and $b(\text{⑤}\varphi) = 1$ guarantee again that $b(\varphi) = 0$ is not the case. In both situations we must conclude from biv$[T0]$ that $b(\varphi) = 1$.

For statement [B] one has to check that w_b is well-defined as a 4-valued homomorphism, according to the corresponding truth-tables that interpret each operator from the language of $\mathcal{T}_{\mathcal{B}_\text{⑤}}$. Let's consider in detail the particular case in which φ has the form $\alpha \wedge_2 \beta$.

Subcase [$w_b(\varphi) = t$]. By the above definition of w_b, this is the same as asserting that $b(\neg_1(\alpha \wedge_2 \beta)) = 0$. But in this case, the bivaluation axiom biv$[\neg_1\wedge_2]\langle 0\rangle$ guarantees that this is a situation in which either (a) $b(\neg_1\alpha) = 0$ and $b(\neg_1\beta) = 0$, or (b) $b(\neg_1\alpha) = 0$ and $b(\beta) = 1$ and $b(\text{⑤}\beta) = 0$; or else (c) $b(\alpha) = 1$ and $b(\text{⑤}\alpha) = 0$ and $b(\neg_1\beta) = 0$. Using again the definition of w_b, we see that this corresponds to having either (aw) $w_b(\alpha) = t$ and $w_b(\beta) = t$, or (bw) $w_b(\alpha) = t$ and $w_b(\beta) = \top$; or else (cw) $w_b(\alpha) = \top$ and $w_b(\beta) = t$. But this is in accordance to what we desired, as it describes exactly the three pairs of inputs for which the truth-table of $\wedge_2$ outputs the value t.

Subcase [$w_b(\varphi) = \top$]. This time, by the definition of w_b, we know that both (a) $b(\alpha \wedge_2 \beta) = 1$ and (b) $b(\text{⑤}(\alpha \wedge_2 \beta)) = 0$. From (a) and biv$[\wedge_2]\langle 1\rangle$, we are left with three situations to consider. Combining them with what we obtain from (b) and biv$[\text{⑤}\wedge_2]\langle 0\rangle$, and in view of biv$[T0]$, we are left with but one situation, in which (c) $b(\alpha) = 1$ and $b(\text{⑤}\alpha) = 0$, and also (d) $b(\beta) = 1$ and $b(\text{⑤}\beta) = 0$. So, from the definition of w_b, we must conclude that $w_b(\alpha) = \top$ and $w_b(\beta) = \top$, again in accordance with the truth-table of $\wedge_2$.

Subcase [$w_b(\varphi) = \bot$]. Analogous to the previous case, but now using biv$[\neg_1\wedge_2]\langle 1\rangle$ and biv$[\text{⑤}\wedge_2]\langle 1\rangle$ to conclude that either $w_b(\alpha) = \bot$, or $w_b(\beta) = \bot$, or else $\langle w_b(\alpha), w_b(\beta)\rangle \in \{\langle t, f\rangle, \langle f, t\rangle\}$.

Subcase [$w_b(\varphi) = f$]. Use biv$[\wedge_2]\langle 0\rangle$, and reason as before.

The analysis follows the same pattern for the case of the other operators, using in each case the appropriate bivaluation axioms. One should still check in separate, however, the cases in which φ has the form $\bowtie p$, where $\bowtie p$ represents a separating formula applied to an atom, that is, $\bowtie \in \{\neg_1, \text{⑤}\}$.

Consider first the case in which φ has the form $\neg_1 p$.

Subcase $[w_b(\varphi) = t]$. Using the above definition of w_b, this is to say that $b(\neg_1\neg_1 p) = 0$. But in this case the bivaluation axiom $\text{biv}[\neg_1\neg_1]\langle 0\rangle$ says that we are exactly in a situation in which $b(p) = 0$. The definition of w_b guarantees that such is the case iff $w_b(p) = f$.

Subcase $[w_b(\varphi) = \top]$. In that case, the definition of w_b says that we are in a situation in which both (a) $b(\neg_1 p) = 1$ and (b) $b(\text{\textcircled{S}}\neg_1 p) = 0$. However, (b) informs us, given the bivaluation axiom $\text{biv}[\text{\textcircled{S}}\neg_1]\langle 0\rangle$, that either (c) $b(\neg_1 p) = 0$ or else (d) both $b(p) = 1$ and $b(\text{\textcircled{S}}p) = 0$. Now, we know that (a) and (c) are jointly untenable, given axiom $\text{biv}[T0]$. The only option left, (d), means, by the definition of w_b, that we have $w_b(p) = \top$.

Subcase $[w_b(\varphi) = \bot]$. Follows the same line as the two previous subcases, but now using $\text{biv}[\neg_1\neg_1]\langle 1\rangle$ and $\text{biv}[\text{\textcircled{S}}\neg_1]\langle 1\rangle$.

Subcase $[w_b(\varphi) = f]$. Immediate from the definition of w_b, as we have $w_b(\neg_1 p) = f$ iff $b(\neg_1 p) = 0$ iff $w_b(p) = t$.

Consider at last the case in which φ has the form $\text{\textcircled{S}}p$.

Subcase $[w_b(\varphi) = t]$. By the definition of w_b, this is exactly the situation in which $b(\neg_1 \text{\textcircled{S}}p) = 0$. Using the bivaluational axiom $\text{biv}[\neg_1 \text{\textcircled{S}}p]\langle 0\rangle$, this corresponds to (i) $b(\text{\textcircled{S}}p) = 1$. From the definition of w_b, this already guarantees that (ii) $w_b(p) \neq \top$. Now, from $\text{biv}[C2]$ and (i) one may also conclude that (iii) $b(p) \neq 0$, and from this $\text{biv}[T0]$ gives us (iv) $b(p) = 1$. Recalling statement [A], from (iv) we may conclude that $w_b(p) \in \{t, \top, \bot\}$. Taking (ii) into account, we're left with $w_b(p) \in \{t, \bot\}$.

Subcase $[w_b(\varphi) = \top]$. By the definition of w_b this would imply that both (i) $b(\text{\textcircled{S}}p) = 1$ and (ii) $b(\text{\textcircled{S}}\text{\textcircled{S}}p) = 0$. From (ii) and $\text{biv}[\text{\textcircled{S}}\text{\textcircled{S}}]\langle 0\rangle$ we conclude (iii) $b(\text{\textcircled{S}}p) = 0$, contradicting (i) in view of $\text{biv}[C0]$.

Subcase $[w_b(\varphi) = \bot]$. Again impossible, as the previous subcase. Use $\text{biv}[\text{\textcircled{S}}\text{\textcircled{S}}]\langle 1\rangle$.

Subcase $[w_b(\varphi) = f]$. Here the definition of w_b puts us in the situation in which (i) $b(\text{\textcircled{S}}p) = 0$ and also tells us, in such situation, that $w_b(p) \neq \bot$. In one direction, the bivaluation axiom $\text{biv}[C3]$ uses (i) to establish that (ii) $b(\neg_1 p) \neq 1$, from which $\text{biv}[T0]$ informs us that (iii) $b(\neg_1 p) = 1$. The latter rules out, from the definition of w_b, the possibility that $w_b(p) = t$. So, we're left with $w_b(p) \in \{\top, f\}$. In the other direction, from $\text{biv}[T0]$ we know that either (iv) $b(p) = 0$ or (v) $b(p) = 1$. The definition of w_b uses (iv) by itself to say that $w_b(p) = f$, and uses (i) and (v) together to say that $w_b(p) = \top$.

In both the latter cases and all their subcases we see that $w_b(\times p)$ behaves in accordance to the corresponding truth-tables. ∎

While the 4-valued truth-functional presentation of the logic behind choice $[\mathcal{V}_b]$ brings along an immediate associated decision procedure in terms of truth-tables, it might not be obvious at first glance at its apparently complicated bivalent presentation that an alternative such procedure is also available in such case. The next section will show though how that can be achieved, once we associate a very simple analytic classic-like proof system to the proposed bivalent semantics.

4 A Uniform Tableau-theoretic Framework

Given the specific format in which the bivaluation axioms for choice $[\mathcal{V}_b]$ were presented, it is easy to extract from them a two-signed tableau system that can be used to check the inferences of the corresponding logic. The restrictions on the bivaluations that appear in the Appendix may indeed be converted into tableau rules in the following manner:

(P0) each expression of the form $\mathsf{b}(\varphi) = v$ is rewritten as a signed formula $S_v{:}\varphi$, using one of the two signs $S_1 = T$ or $S_0 = F$;

(P1) a '$\Leftrightarrow$' is read as separating the head of a tableau rule (to the left) from its conclusions (to the right);

(P2) each ',' is understood as separating nodes (signed formulas) from the same branch;

(P3) an '|' at the right of a '$\Leftrightarrow$' demarcates bifurcations in the output of a given rule;

(P4) an expression of the form '$h_1, \ldots, h_n \Rightarrow *$' denotes a closure rule.

After such a conversion, for each bivaluation axiom $\mathsf{biv}[xy]\langle v \rangle$ there will correspond a tableau rule $\mathsf{tab}[xy]\langle S_v \rangle$, and for each axiom $\mathsf{biv}[Cn]$ there will be, in the tableau system, a corresponding closure rule $\mathsf{tab}[Cn]$. The standard decision procedure that will be presented below shows that the tableau rule that corresponds to $\mathsf{biv}[T0]$ (a kind of dual-cut rule for tableaux), while certainly admissible, does not need to be taken as primitive in the proof system. Now, while the closure rule $\mathsf{tab}[C0]$ is also to be found in two-signed tableaux for classical logic, rules $\mathsf{tab}[C1]$ to $\mathsf{tab}[C3]$ are all distinctive marks of the 4-valued choice $[\mathcal{V}_b]$. We will call Tab^b the tableau system thereby obtained, and use for tableaux here the usual terminology and definitions that are found in the standard literature of the area.

THEOREM 6. Tab^b provides a sound and complete proof system for the logic behind choice $[\mathcal{V}_b]$.

Using the bivalent semantics discussed in the previous section, one direction of this result is obvious, as the tableau rules directly translate bivaluation axioms. To check the converse direction, it will be useful to define a convenient 'complexity measure' $\mathsf{cm} : \mathcal{T}_{\mathcal{B}_\circledast} \longrightarrow \mathbb{N}$ over the structure of a formula φ, according to which:

$$\mathsf{cm}(\varphi) = \begin{cases} 0 & \text{if } \varphi \text{ is an atom } p, \\ \mathsf{cm}(\alpha) & \text{if } \varphi \text{ is } \ltimes\alpha \text{ for a separating connective } \ltimes, \\ 1 + \mathsf{cm}(\alpha) & \text{if } \varphi \text{ is } \rtimes\alpha \text{ for any other unary connective } \rtimes, \\ 1 + \mathsf{Max}(\mathsf{cm}(\alpha), \mathsf{cm}(\beta)) & \text{if } \varphi \text{ is } \alpha \bowtie \beta \text{ for some binary connective } \bowtie. \end{cases}$$

To extend the complexity measure for signed formulas, just ignore the signs. Now the following lemma suffices to complete the proof of the above result:

LEMMA 7. *For any collection of signed formulas it is always possible to produce an exhausted tableau in* Tab^b.

Proof. Just notice that every signed formula with non-null complexity, whichever form it has, is the head of a (uniquely determined) tableau rule, and that the body of such rule only contains formulas of lower complexity. ∎

As a corollary of the previous result we also immediately obtain a completely standard tableau-theoretic decision procedure for our logic. Indeed, if one wants to check whether $\{\gamma_1, \gamma_2, \ldots, \gamma_m\} \models \{\delta_n, \ldots, \delta_2, \delta_1\}$ is the case, one should construct a tableau for $\{T{:}\gamma_1, T{:}\gamma_2, \ldots, T{:}\gamma_m\} \cup \{F{:}\delta_n, \ldots, F{:}\delta_2, F{:}\delta_1\}$. If it closes at any moment, the inference is valid. Otherwise, when one arrives to an exhausted tableau and it still has open branches, counter-models may be extracted from such branches by collecting all the thereby occurring formulas of complexity 0.

As an illustration, an exhausted tableau to test the validity of the inference $\{\neg_2 p \wedge_1 \neg_2 q)\} \models \{\neg_1(p \vee_2 q)\}$ will use the tableau rules $\mathsf{tab}[\neg_2 \vee_2]\langle F \rangle$, $\mathsf{tab}[\wedge_1]\langle T \rangle$, $\mathsf{tab}[\neg_1 \neg_2]\langle F \rangle$, $\mathsf{tab}[\neg_2]\langle T \rangle$, $\mathsf{tab}[\mathsf{S} \neg_2]\langle T \rangle$, $\mathsf{tab}[\mathsf{S} \neg_2]\langle F \rangle$, plus the closure rules $\mathsf{tab}[C0]$, and $\mathsf{tab}[C3]$, not necessarily in this order, and produce some open branches. A first class of such open branches is the one that contains both the formula $F{:}\neg_1 p$ and the formula $F{:}\neg_1 q$. This is enough to tell us that they represent the 4-valued counter-models in which $\mathsf{w}(p) = t = \mathsf{w}(q)$. A second class of such open branches is the one that contains only one of the above formulas, say $F{:}\neg_1 p$, and on what concerns the other atom, q, it contains both $T{:}q$ and $T{:}\mathsf{S}q$ (*mutatis mutandis* if one exchanges the roles of p and q). All we can say in that case is that the 4-valued counter-models represented by the open tableau for the above inference allow for assignments in which $\mathsf{w}(p) = t$ and $\mathsf{w}(q) \in \{t, \bot\}$ (*mutatis mutandis*, the assignments in which $\mathsf{w}(q) = t$ and $\mathsf{w}(p) \in \{t, \bot\}$ also represent counter-models).

The logic-user might be satisfied with such procedure by itself in testing inferences of the logic underlying choice $[\mathcal{V}_b]$. But in this case there would hardly be a good technical reason to be found for not having stayed with the initial class of 4-valued models, which were even simpler and in many senses more well-behaved. One should insist here, though, that a much more attractive task, from the viewpoint of the logic-designer, is to *compare* different logics. There is a real advantage, in this case, in doing the bivalent reduction, from the semantical perspective, and working with the associated two-signed tableaux. For if that same reduction can be done for other logics, say the logics underlying the choices $[\mathcal{V}_{e\ell}]$ or $[\mathcal{V}_n]$, or, for all that matters, for any other finite-valued logics, for instance, checking whether a primitive or derived rule from the logic $\mathcal{L}_y$ is valid from the perspective of a logic $\mathcal{L}_x$, as long as such logics are written over comparable languages, costs just the same, in principle, as testing an inference of $\mathcal{L}_x$ inside $\mathcal{L}_x$.

For a practical illustration on how the present approach may also be used to produce easy proofs of some important meta-theoretical results concerning the comparison of the logics presented by way of bivalent semantics, consider the following extension of Theorem 2:

THEOREM 8. *Given* $\Gamma \cup \Delta \subseteq \mathcal{T}_{\mathcal{B}_\mathsf{S}}$, *then:*

$$\Gamma \models_b \Delta \text{ iff } \Delta^\varepsilon \models_n \Gamma^\varepsilon \quad \text{and} \quad \Gamma \models_n \Delta \text{ iff } \Delta^\varepsilon \models_b \Gamma^\varepsilon.$$

Proof. These may be verified by a quick inspection of the restrictions governing the classes of bivaluations Biv^b and Biv^n, in the Appendix — and in particular instructions **(DA)** and **(DB)**, that make the 'dual' relation between the bivaluation axioms quite explicit. ∎

Taking now the associated adequated tableau systems, Tab^b, $\mathsf{Tab}^{e\ell}$ and Tab^n, into account, an even more inclusive practical result that may immediately be established by 'classic-like dualization' is the following:

THEOREM 9. Let $\overline{T} \triangleq F$ and $\overline{F} \triangleq T$. Given $\{\varphi_1, \varphi_2, \ldots, \varphi_j\} \subseteq \mathcal{T}_{\mathcal{B}_\circledast}$ and given signs $S_i \in \{T, F\}$, for $1 \leq i \leq j$, then there is a closed tableau for $\{S_1{:}\varphi_1, S_2{:}\varphi_2, \ldots, S_j{:}\varphi_j\}$ using the rules of Tab^x iff there is a closed tableau for $\{\overline{S_1}{:}\varphi_1^\varepsilon, \overline{S_2}{:}\varphi_2^\varepsilon, \ldots, \overline{S_j}{:}\varphi_j^\varepsilon\}$ using the rules of Tab^y, where $\langle x, y \rangle \in \{\langle b, n \rangle, \langle e\ell, e\ell \rangle, \langle n, b \rangle\}$.

To sum up with, it is important to mention that general constructive reduction mechanisms such as the ones described above are indeed available, for any finite-valued logic, and the class of logics characterized by bivalent semantics that can be associated to adequate analytic tableaux goes in fact much beyond the class of finite-valued logics. In the next section we will finish by disclosing a bit more about the range of applicability of the above ideas and techniques.

5 Context, considerations, and future work

The algebraic structures now widely known as bilattices (cf. [Ginsberg, 1986]) were introduced in [Belnap, 1976] and frequently investigated in the literature since then (cf. [Belnap, 1977; Ginsberg, 1988; Fitting, 1990; Fitting, 1994; Arieli and Avron, 1998]) and even generalized (cf. [Wansing and Shramko, 2005]) for their potential applications to several areas of computer science. A useful source of information for developments on bilattices is [Arieli and Avron, 2000], a paper to whose title the title of present study explicitly refers. In a sense, there is no essential loss of generality here as we concentrate our efforts over a simple *four-valued* bilattice, as there is a representation theorem that supports it, and its structure stands among other bilattices in a similar way as the canonical structure underlying classical logic stands among other boolean algebras.

Some criticism has recently been raised to the whole idea that such logics could really serve as a foundation for computerized reasoning under inconsistent / incomplete information, arguing that their alleged significance is based on a "confusion between truth-values and information states" (cf. [Dubois, 2008]). This has almost immediately been counteracted by evidence suggesting that the critique is based on a "confusion of information states with belief states" (cf. [Wansing and Belnap, 2010]). The controversy does not affect our work in this paper, nonetheless, as all we do here is largely independent of how the two involved logical dimensions are explicated.

Still and all, we do have seen above how the *logic* represented by the bilattice is very sensible to how its underlying truth-values are grouped and interpreted. Having shown the difficulties that lie in effecting the comparison between the inferences sanctioned by such logics, from the viewpoint of the logic-designer, and the even more difficult task of comparing such logics with arbitrarily other many-valued logics, we have next illustrated a general method that is based in providing a uniform approach to these logics, starting with a bivalent reduction of their semantics. Even though the illustration has been presented in detail only for the choice that led to a paraconsistent logic based on the four-valued structure, the slightly extended language that we have used was chosen to be appropriate also for the application of the same techniques to the alternative paracomplete logic based on a different choice made over the same structure, and in several points we have commented on the adaptations that must be made on the statements and results of the former case so as to deal also with the latter case. In the currently detailed case study, as a matter of fact, what we have is a variant of the output of a general method that allows for the extraction of classic-like semantics for any finite-valued logic (check [Caleiro and Marcos, 2010] for a recent survey). The output of that method, as presented in the Appendix, has indeed been optimized here, by the use of equivalences stated in the classical metalanguage, in much the same way as formulas in DNF (namely, the sentences to the right of the '⇔' symbol, representing the body of the axioms / rules) can be manipulated into *reduced* DNF (the simplifications also take the closure rules into account).

Our two-valued formulation of the chosen illustration has also been carefully crafted so as to serve as input to another general method that allowed for the extraction of adequate proof-theoretical counterparts of our logic in terms of *analytic tableaux*. The method, in this case, is even more general than the previous one, and applies not only to logics whose classic-like semantic presentations result from the above mentioned reduction (cf. [Caleiro and Marcos, 2009]), but to many other logics that can be characterized by way of bivaluation axioms of a similar 'dyadic' format (cf. [Caleiro et al., 2005]). In that case, to facilitate the comparison between different logics it might be necessary to fiddle with structural rules in order to guarantee that the obtained proof systems are able to derive all the rules they have as admissible (cf. [Marcos and Mendonça, 2009])). In any case, for each logic obtained through the above methods, fully automated proof tactics are available (cf. [Marcos, 2010]). In particular, it should be noted that analyticity of the proof procedure was guaranteed in the present paper through a simpler strategy as the one used in our previous papers, where the introduction of a separating connective *by abbreviation* required a *proof strategy* to be associated, in general, to secure the *termination* of the construction of an arbitrary tableau proof. Here we simplified matters (and the definition of the *complexity measure*) by introducing the missing separating connective directly as part of the underlying propositional signature.

For sure, on what concerns the beautiful and well-developed theory of bilattices, the present approach provides a novel outlook, yet only starts scratching some of the many issues that are of interest for the logic-designer. For instance, even though we have tried to modify the initial language of the bilattice as little and harmlessly as possible, the introduction of a suitable implication into $\mathcal{B}$ can simplify many of the above tasks, in making the underlying language more expressive in a very helpful way (cf. [Arieli and Avron, 2000]). It rests to be shown, anyway, how much of the deeper interest behind the 4-valued approach can be thoroughly retained within the present classic-like bivalent / two-signed approach.

Finally, for a note on a completely different direction, we have given a few hints along this study on how the above four-valued logic could be seen as a sort of combination of simpler fragments, where interaction axioms have an important role to play. It should be interesting to investigate the underlying technique, for its own interest, as a new mechanism for the combination of truth-functional logics (contrast with [Coniglio and Fernández, 2005]), where 'coalescing' a logic into another would consist in a way of adding a new dimension to a given structure. In that sense, for instance, the inner structure of the four-valued formalism could be seen as a result from a natural combination of classical logic with itself.

Acknowledgments

The author is much obliged to partial funding received from CNPq and CAPES during the periods in which this paper was confected. Criticisms and stimuli received from Chris Fermüller and Carlos Caleiro were also fundamental. Finally, the presentation also owes a lot to the provoking audiences of the 31st Linz Seminar on Fuzzy Set Theory, the SQIG/IT, and the CLE/UNICAMP, where the ideas herein contained were presented at different stages.

Appendix

Here one can find the exhaustive bivalent description of all the above mentioned 4-valued logics.

In the metalinguistic notation below, a ',' replaces an *and*, a '|' replaces an *or*, a '⇒' replaces an *if-then* assertion, a '⇔' stands for an *iff* assertion, and a '⁕' represents the *absurd*. So, the first bivaluation axiom biv$[\wedge_2]\langle 1 \rangle$ below, for instance, should be read as restricting the class of bivaluations of interest to those in which $\mathsf{b}(\alpha \wedge_2 \beta) = 1$ is the case, for arbitrary formulas α and β of $\mathcal{T_B}$, if and only if one of three following situations occur: either $\mathsf{b}(\text{\textcircled{S}}\alpha) = 1$ is the case, or $\mathsf{b}(\text{\textcircled{S}}\beta) = 1$ is the case, or else both $\mathsf{b}(\alpha) = 1$ and $\mathsf{b}(\beta) = 1$ are simultaneously the case. Further, an axiom rule such as the first biv$[C2]$ below should be read as stating that $\mathsf{b}(\alpha) = 0$ and $\mathsf{b}(\text{\textcircled{S}}(\alpha)) = 1$ cannot simultaneously obtain, for any bivaluation b and formula α of $\mathcal{T_B}$.

The first bivalent semantics described below, Biv^b, corresponds to the logic underlying choice $[\mathcal{V}_b]$, whose completeness proof was presented in detail in the preceding text. It corresponds to the collection of all bivaluations that conform to the restrictions listed in what follows.

biv$[\wedge_1]\langle 1 \rangle$ $\mathsf{b}(\alpha \wedge_1 \beta) = 1 \Leftrightarrow$
                         $(\mathsf{b}(\neg_1\alpha) = 0 \,,\, \mathsf{b}(\beta) = 1)$
                        $|\ (\mathsf{b}(\alpha) = 1 \,,\, \mathsf{b}(\neg_1\beta) = 0)$
                        $|\ (\mathsf{b}(\text{\textcircled{S}}\alpha) = 1 \,,\, \mathsf{b}(\text{\textcircled{S}}\beta) = 1)$
                        $|\ (\mathsf{b}(\alpha) = 1 \,,\, \mathsf{b}(\text{\textcircled{S}}\alpha) = 0 \,,\, \mathsf{b}(\beta) = 1 \,,\, \mathsf{b}(\text{\textcircled{S}}\beta) = 0)$

biv$[\wedge_1]\langle 0 \rangle$ $\mathsf{b}(\alpha \wedge_1 \beta) = 0 \Leftrightarrow$
                         $(\mathsf{b}(\alpha) = 0)$
                        $|\ (\mathsf{b}(\beta) = 0)$
                        $|\ (\mathsf{b}(\neg_1\alpha) = 1 \,,\, \mathsf{b}(\text{\textcircled{S}}\alpha) = 1 \,,\, \mathsf{b}(\text{\textcircled{S}}\beta) = 0)$
                        $|\ (\mathsf{b}(\text{\textcircled{S}}\alpha) = 0 \,,\, \mathsf{b}(\neg_1\beta) = 1 \,,\, \mathsf{b}(\text{\textcircled{S}}\beta) = 1)$

biv$[\neg_1\wedge_1]\langle 1 \rangle$ $\mathsf{b}(\neg_1(\alpha \wedge_1 \beta)) = 1 \Leftrightarrow \mathsf{b}(\neg_1\alpha) = 1 \ |\ \mathsf{b}(\neg_1\beta) = 1$
biv$[\neg_1\wedge_1]\langle 0 \rangle$ $\mathsf{b}(\neg_1(\alpha \wedge_1 \beta)) = 0 \Leftrightarrow \mathsf{b}(\neg_1\alpha) = 0 \,,\, \mathsf{b}(\neg_1\beta) = 0$

biv$[\text{\textcircled{S}}\wedge_1]\langle 1 \rangle$ $\mathsf{b}(\text{\textcircled{S}}(\alpha \wedge_1 \beta)) = 1 \Leftrightarrow \mathsf{b}(\text{\textcircled{S}}\alpha) = 1 \,,\, \mathsf{b}(\text{\textcircled{S}}\beta) = 1$
biv$[\text{\textcircled{S}}\wedge_1]\langle 0 \rangle$ $\mathsf{b}(\text{\textcircled{S}}(\alpha \wedge_1 \beta)) = 0 \Leftrightarrow \mathsf{b}(\text{\textcircled{S}}\alpha) = 0 \ |\ \mathsf{b}(\text{\textcircled{S}}\beta) = 0$

biv$[\wedge_2]\langle 1\rangle$ $b(\alpha \wedge_2 \beta) = 1 \Leftrightarrow$
                         $(b(Ⓢ\alpha) = 1)$
                         $\mid (b(Ⓢ\beta) = 1)$
                         $\mid (b(\alpha) = 1, \; b(\beta) = 1)$

biv$[\wedge_2]\langle 0\rangle$ $b(\alpha \wedge_2 \beta) = 0 \Leftrightarrow$
                         $(b(\alpha) = 0, \; b(Ⓢ\beta) = 0)$
                         $\mid (b(Ⓢ\alpha) = 0, \; b(\beta) = 0)$

biv$[\neg_1\wedge_2]\langle 1\rangle$ $b(\neg_1(\alpha \wedge_2 \beta)) = 1 \Leftrightarrow$
                         $(b(\alpha) = 0)$
                         $\mid (b(\beta) = 0)$
                         $\mid (b(\neg_1\alpha) = 1, \; b(Ⓢ\alpha) = 1)$
                         $\mid (b(\neg_1\beta) = 1, \; b(Ⓢ\beta) = 1)$
                         $\mid (b(\neg_1\alpha) = 1, \; b(\neg_1\beta) = 1)$

biv$[\neg_1\wedge_2]\langle 0\rangle$ $b(\neg_1(\alpha \wedge_2 \beta)) = 0 \Leftrightarrow$
                         $(b(\neg_1\alpha) = 0, \; b(\neg_1\beta) = 0)$
                         $\mid (b(\neg_1\alpha) = 0, \; b(\beta) = 1, \; b(Ⓢ\beta) = 0)$
                         $\mid (b(\alpha) = 1, \; b(Ⓢ\alpha) = 0, \; b(\neg_1\beta) = 0)$

biv$[Ⓢ\wedge_2]\langle 1\rangle$ $b(Ⓢ(\alpha \wedge_2 \beta)) = 1 \;\;\Leftrightarrow\;\; b(Ⓢ\alpha) = 1 \mid b(Ⓢ\beta) = 1$
biv$[Ⓢ\wedge_2]\langle 0\rangle$ $b(Ⓢ(\alpha \wedge_2 \beta)) = 0 \;\;\Leftrightarrow\;\; b(Ⓢ\alpha) = 0, \; b(Ⓢ\beta) = 0$

. . . .

biv$[\vee_1]\langle 1\rangle$ $b(\alpha \vee_1 \beta) = 1 \;\;\Leftrightarrow\;\; b(\alpha) = 1 \mid b(\beta) = 1$
biv$[\vee_1]\langle 0\rangle$ $b(\alpha \vee_1 \beta) = 0 \;\;\Leftrightarrow\;\; b(\alpha) = 0, \; b(\beta) = 0$

biv$[\neg_1\vee_1]\langle 1\rangle$ $b(\neg_1(\alpha \vee_1 \beta)) = 1 \Leftrightarrow$
                         $(b(Ⓢ\alpha) = 0, \; b(Ⓢ\beta) = 0)$
                         $\mid (b(\alpha) = 0, \; b(\neg_1\beta) = 1)$
                         $\mid (b(\neg_1\alpha) = 1, \; b(\beta) = 0)$
                         $\mid (b(\neg_1\alpha) = 1, \; b(Ⓢ\alpha) = 1, \; b(\neg_1\beta) = 1, \; b(Ⓢ\beta) = 1)$

biv$[\neg_1\vee_1]\langle 0\rangle$ $b(\neg_1(\alpha \vee_1 \beta)) = 0 \Leftrightarrow$
                         $(b(\neg_1\alpha) = 0)$
                         $\mid (b(\neg_1\beta) = 0)$
                         $\mid (b(Ⓢ\alpha) = 1, \; b(\beta) = 1, \; b(Ⓢ\beta) = 0)$
                         $\mid (b(\alpha) = 1, \; b(Ⓢ\alpha) = 0, \; b(Ⓢ\beta) = 1)$

biv$[Ⓢ\vee_1]\langle 1\rangle$ $b(Ⓢ(\alpha \vee_1 \beta)) = 1 \;\;\Leftrightarrow\;\; b(Ⓢ\alpha) = 1 \mid b(Ⓢ\beta) = 1$
biv$[Ⓢ\vee_1]\langle 0\rangle$ $b(Ⓢ(\alpha \vee_1 \beta)) = 0 \;\;\Leftrightarrow\;\; b(Ⓢ\alpha) = 0, \; b(Ⓢ\beta) = 0$

. . . .

biv$[\vee_2]\langle 1\rangle$ $b(\alpha \vee_2 \beta) = 1 \Leftrightarrow$
                         $(b(\neg_1\alpha) = 0)$
                         $\mid (b(\neg_1\beta) = 0)$
                         $\mid (b(\alpha) = 1, \; b(Ⓢ\alpha) = 0)$
                         $\mid (b(\beta) = 1, \; b(Ⓢ\beta) = 0)$
                         $\mid (b(\alpha) = 1, \; b(\beta) = 1)$

biv$[\vee_2]\langle 0\rangle$ $b(\alpha \vee_2 \beta) = 0 \Leftrightarrow$
                         $(b(\alpha) = 0, \; b(\beta) = 0)$
                         $\mid (b(\alpha) = 0, \; b(\neg_1\beta) = 1, \; b(Ⓢ\beta) = 1)$
                         $\mid (b(\neg_1\alpha) = 1, \; b(Ⓢ\alpha) = 1, \; b(\beta) = 0)$

biv$[\neg_1\vee_2]\langle 1\rangle$ $b(\neg_1(\alpha \vee_2 \beta)) = 1 \Leftrightarrow$
                         $(b(Ⓢ\alpha) = 0)$
                         $\mid (b(Ⓢ\beta) = 0)$
                         $\mid (b(\neg_1\alpha) = 1, \; b(\neg_1\beta) = 1)$

biv$[\neg_1\vee_2]\langle 0\rangle$ $b(\neg_1(\alpha \vee_2 \beta)) = 0 \Leftrightarrow$
                         $(b(Ⓢ\alpha) = 1, \; b(\neg_1\beta) = 0)$
                         $\mid (b(\neg_1\alpha) = 0, \; b(Ⓢ\beta) = 1)$

biv[$\text{Ⓢ}\vee_2$]$\langle 1 \rangle$	$b(\text{Ⓢ}(\alpha \vee_2 \beta)) = 1$	$\Leftrightarrow$	$b(\text{Ⓢ}\alpha) = 1$, $b(\text{Ⓢ}\beta) = 1$
biv[$\text{Ⓢ}\vee_2$]$\langle 0 \rangle$	$b(\text{Ⓢ}(\alpha \vee_2 \beta)) = 0$	$\Leftrightarrow$	$b(\text{Ⓢ}\alpha) = 0 \mid b(\text{Ⓢ}\beta) = 0$

. . . .

biv[$\neg_1\neg_1$]$\langle 1 \rangle$	$b(\neg_1(\neg_1\alpha)) = 1$	$\Leftrightarrow$	$b(\alpha) = 1$
biv[$\neg_1\neg_1$]$\langle 0 \rangle$	$b(\neg_1(\neg_1\alpha)) = 0$	$\Leftrightarrow$	$b(\alpha) = 0$
biv[$\text{Ⓢ}\neg_1$]$\langle 1 \rangle$	$b(\text{Ⓢ}(\neg_1\alpha)) = 1$	$\Leftrightarrow$	$b(\alpha) = 0 \mid (b(\neg_1\alpha) = 1, b(\text{Ⓢ}\alpha) = 1)$
biv[$\text{Ⓢ}\neg_1$]$\langle 0 \rangle$	$b(\text{Ⓢ}(\neg_1\alpha)) = 0$	$\Leftrightarrow$	$b(\neg_1\alpha) = 0 \mid (b(\alpha) = 1, b(\text{Ⓢ}\alpha) = 0)$

. . . .

biv[$\neg_2$]$\langle 1 \rangle$	$b(\neg_2\alpha) = 1$	$\Leftrightarrow$	$b(\alpha) = 1$
biv[$\neg_2$]$\langle 0 \rangle$	$b(\neg_2\alpha) = 0$	$\Leftrightarrow$	$b(\alpha) = 0$
biv[$\neg_1\neg_2$]$\langle 1 \rangle$	$b(\neg_1(\neg_2\alpha)) = 1$	$\Leftrightarrow$	$b(\neg_1\alpha) = 1$
biv[$\neg_1\neg_2$]$\langle 0 \rangle$	$b(\neg_1(\neg_2\alpha)) = 0$	$\Leftrightarrow$	$b(\neg_1\alpha) = 0$
biv[$\text{Ⓢ}\neg_2$]$\langle 1 \rangle$	$b(\text{Ⓢ}(\neg_2\alpha)) = 1$	$\Leftrightarrow$	$b(\neg_1\alpha) = 0 \mid (b(\alpha) = 1, b(\text{Ⓢ}\alpha) = 0)$
biv[$\text{Ⓢ}\neg_2$]$\langle 0 \rangle$	$b(\text{Ⓢ}(\neg_2\alpha)) = 0$	$\Leftrightarrow$	$b(\alpha) = 0 \mid (b(\neg_1\alpha) = 1, b(\text{Ⓢ}\alpha) = 1)$

. . . .

biv[$\neg_1\text{Ⓢ}$]$\langle 1 \rangle$	$b(\neg_1(\text{Ⓢ}\alpha)) = 1$	$\Leftrightarrow$	$b(\text{Ⓢ}\alpha) = 0$
biv[$\neg_1\text{Ⓢ}$]$\langle 0 \rangle$	$b(\neg_1(\text{Ⓢ}\alpha)) = 0$	$\Leftrightarrow$	$b(\text{Ⓢ}\alpha) = 1$
biv[$\text{Ⓢ}\text{Ⓢ}$]$\langle 1 \rangle$	$b(\text{Ⓢ}(\text{Ⓢ}\alpha)) = 1$	$\Leftrightarrow$	$b(\text{Ⓢ}\alpha) = 1$
biv[$\text{Ⓢ}\text{Ⓢ}$]$\langle 0 \rangle$	$b(\text{Ⓢ}(\text{Ⓢ}\alpha)) = 0$	$\Leftrightarrow$	$b(\text{Ⓢ}\alpha) = 0$

. . . .

biv[$T0$]		$\Rightarrow$	$b(\alpha) = 0 \mid b(\alpha) = 1$
biv[$C0$]	$(b(\alpha) = 0, b(\alpha) = 1)$	$\Rightarrow$	$\ast$
biv[$C1$]	$(b(\alpha) = 0, b(\neg_1\alpha) = 0)$	$\Rightarrow$	$\ast$
biv[$C2$]	$(b(\alpha) = 0, b(\text{Ⓢ}\alpha) = 1)$	$\Rightarrow$	$\ast$
biv[$C3$]	$(b(\neg_1\alpha) = 0, b(\text{Ⓢ}\alpha) = 0)$	$\Rightarrow$	$\ast$

Using precisely the same technique illustrated in the preceding text, a bivalent semantics Biv^n corresponding to the logic underlying choice $[\mathcal{V}_n]$ may be produced. Its description is quite simple to present if we take into account the above exhaustive description of Biv^b. Indeed, all we have to do is:

(DA) rewrite each expression of the form $b(\varphi) = v$ as $b(\varphi) = 1 - v$

(DB) exchange each $\wedge$ for a $\vee$, and vice-versa

So, a bivaluation axiom for Biv^b such as:

biv[$\neg_1\wedge_2$]$\langle 0 \rangle$ $\quad b(\neg_1(\alpha \wedge_2 \beta)) = 0 \Leftrightarrow$
$\qquad\qquad\qquad (b(\neg_1\alpha) = 0, b(\neg_1\beta) = 0)$
$\qquad\qquad\qquad \mid (b(\neg_1\alpha) = 0, b(\beta) = 1, b(\text{Ⓢ}\beta) = 0)$
$\qquad\qquad\qquad \mid (b(\alpha) = 1, b(\text{Ⓢ}\alpha) = 0, b(\neg_1\beta) = 0)$

becomes in Biv^n the axiom:

biv[$\neg_1\vee_2$]$\langle 1 \rangle$ $\quad b(\neg_1(\alpha \vee_2 \beta)) = 1 \Leftrightarrow$
$\qquad\qquad\qquad (b(\neg_1\alpha) = 1, b(\neg_1\beta) = 1)$
$\qquad\qquad\qquad \mid (b(\neg_1\alpha) = 1, b(\beta) = 0, b(\text{Ⓢ}\beta) = 1)$
$\qquad\qquad\qquad \mid (b(\alpha) = 0, b(\text{Ⓢ}\alpha) = 1, b(\neg_1\beta) = 1)$

and a closure rule for Biv^b such as

biv[$C3$] $\quad (b(\neg_1\alpha) = 0, b(\text{Ⓢ}\alpha) = 0) \Rightarrow \ast$

becomes in Biv^n the closure rule:

biv[$C3$] $\quad (b(\neg_1\alpha) = 1, b(\text{Ⓢ}\alpha) = 1) \Rightarrow \ast$

Finally, a bivalent semantics $\mathsf{Biv}^{e\ell}$ corresponding to the logic underlying choice $[\mathcal{V}_{e\ell}]$ is even easier to describe than the previously described semantics, given that there is no real need to use for that effect more than one separating connective — and each one of $\neg_1$, $\neg_2$ and $\circledS$ will do the job equally well. As a matter of stipulation, choosing for that effect the latter connective, $\circledS$, the corresponding straightforward bivaluation axioms that we obtain are presented in what follows.

$$
\begin{array}{llll}
\mathsf{biv}[\neg_1]\langle 0\rangle & \mathsf{b}(\neg_1\alpha) = 0 & \Leftrightarrow & \mathsf{b}(\circledS\alpha) = 1 \\
\mathsf{biv}[\circledS\neg_1]\langle 0\rangle & \mathsf{b}(\circledS\neg_1\alpha) = 0 & \Leftrightarrow & \mathsf{b}(\alpha) = 1 \\[4pt]
\mathsf{biv}[\neg_2]\langle 0\rangle & \mathsf{b}(\neg_2\alpha) = 0 & \Leftrightarrow & \mathsf{b}(\circledS\alpha) = 0 \\
\mathsf{biv}[\circledS\neg_2]\langle 0\rangle & \mathsf{b}(\circledS\neg_2\alpha) = 0 & \Leftrightarrow & \mathsf{b}(\alpha) = 0 \\[4pt]
\mathsf{biv}[\circledS\circledS]\langle 0\rangle & \mathsf{b}(\circledS\circledS\alpha) = 0 & \Leftrightarrow & \mathsf{b}(\circledS\alpha) = 0 \\[4pt]
\mathsf{biv}[\wedge_1]\langle 0\rangle & \mathsf{b}(\alpha \wedge_1 \beta) = 0 & \Leftrightarrow & \mathsf{b}(\alpha) = 0 \mid \mathsf{b}(\beta) = 0 \\
\mathsf{biv}[\circledS\wedge_1]\langle 0\rangle & \mathsf{b}(\circledS(\alpha \wedge_1 \beta)) = 0 & \Leftrightarrow & \mathsf{b}(\circledS\alpha) = 0 \mid \mathsf{b}(\circledS\beta) = 0 \\[4pt]
\mathsf{biv}[\wedge_2]\langle 0\rangle & \mathsf{b}(\alpha \wedge_2 \beta) = 0 & \Leftrightarrow & \mathsf{b}(\alpha) = 0 \mid \mathsf{b}(\beta) = 0 \\
\mathsf{biv}[\circledS\wedge_2]\langle 0\rangle & \mathsf{b}(\circledS(\alpha \wedge_2 \beta)) = 0 & \Leftrightarrow & \mathsf{b}(\circledS\alpha) = 0,\ \mathsf{b}(\circledS\beta) = 0 \\[4pt]
\mathsf{biv}[\vee_1]\langle 0\rangle & \mathsf{b}(\alpha \vee_1 \beta) = 0 & \Leftrightarrow & \mathsf{b}(\alpha) = 0,\ \mathsf{b}(\beta) = 0 \\
\mathsf{biv}[\circledS\vee_1]\langle 0\rangle & \mathsf{b}(\circledS(\alpha \vee_1 \beta)) = 0 & \Leftrightarrow & \mathsf{b}(\circledS\alpha) = 0,\ \mathsf{b}(\circledS\beta) = 0 \\[4pt]
\mathsf{biv}[\vee_2]\langle 0\rangle & \mathsf{b}(\alpha \vee_2 \beta) = 0 & \Leftrightarrow & \mathsf{b}(\alpha) = 0,\ \mathsf{b}(\beta) = 0 \\
\mathsf{biv}[\circledS\vee_2]\langle 0\rangle & \mathsf{b}(\circledS(\alpha \vee_2 \beta)) = 0 & \Leftrightarrow & \mathsf{b}(\circledS\alpha) = 0 \mid \mathsf{b}(\circledS\beta) = 0
\end{array}
$$

The above listed bivaluation axioms for $\mathsf{Biv}^{e\ell}$ clearly tell only half of the story. However, to obtain the other half one simply has to follow instructions **(DA)** and **(DB)**. Axioms $\mathsf{biv}[T0]$ and $\mathsf{biv}[C0]$ must still be added to the above list, but no extra closure axioms are needed for a complete presentation, in the present case.

BIBLIOGRAPHY

[Arieli and Avron, 1998] Ofer Arieli and Arnon Avron. The value of the four values. *Artificial Intelligence*, 102:97–141, 1998.

[Arieli and Avron, 2000] Ofer Arieli and Arnon Avron. Bilattices and paraconsistency. In D. Batens, C. Mortensen, G. Priest, and J. P. Van Bendegem, editors, *Frontiers of Paraconsistent Logic*, Proceedings of the I World Congress on Paraconsistency, held in Ghent, BE, July 29–August 3, 1997, pages 11–28. Research Studies Press, Baldock, 2000.

[Belnap, 1976] Nuel D. Belnap. How a computer should think. In G. Ryle, editor, *Contemporary Aspects of Philosophy*, pages 30–55. Oriel Press, 1976.

[Belnap, 1977] Nuel D. Belnap. A useful four-valued logic. In J. M. Dunn and G. Epstein, editors, *Modern Uses of Multiple-Valued Logic*, pages 8–37. D. Reidel, 1977.

[Caleiro and Marcos, 2009] Carlos Caleiro and João Marcos. Classic-like analytic tableaux for finite-valued logics. In H. Ono, M. Kanazawa, and R. de Queiroz, editors, *Proceedings of the XVI Workshop on Logic, Language, Information and Computation* (WoLLIC 2009), held in Tokyo, JP, June 2009, volume 5514 of *Lecture Notes in Artificial Intelligence*, pages 268–280. Springer, 2009. Preprint available at:
http://sqig.math.ist.utl.pt/pub/CaleiroC/09-CM-ClATab4FVL.pdf.

[Caleiro and Marcos, 2010] Carlos Caleiro and João Marcos. Two many values: An algorithmic outlook on Suszko's Thesis. In *Proceedings of the XL International Symposium on Multiple-Valued Logic* (ISMVL 2010), pages 93–97. IEEE Computer Society, 2010. Preprint available at: http://sqig.math.ist.utl.pt/pub/CaleiroC/10-CM-ismvl.pdf.

[Caleiro et al., 2005] Carlos Caleiro, Walter Carnielli, Marcelo E. Coniglio, and João Marcos. Two's company: "The humbug of many logical values". In J.-Y. Béziau, editor, *Logica Universalis*, pages 169–189. Birkhäuser Verlag, Basel, Switzerland, 2005. Preprint available at: http://sqig.math.ist.utl.pt/pub/CaleiroC/05-CCCM-dyadic.pdf.

[Coniglio and Fernández, 2005] Marcelo E. Coniglio and Víctor L. Fernández. Plain fibring and direct union of logics with matrix semantics. In *Proceedings of the 2nd Indian International Conference on Artificial Intelligence*, pages 1648–1658. IICAI, 2005.

[Dubois, 2008] Didier Dubois. On ignorance and contradiction considered as truth-values. *Logic Journal of IGPL*, 16(2):195–216, 2008.

[Fitting, 1990] Melvin Fitting. Bilattices in logic programming. In *Proceedings of the XX International Symposium on Multiple-Valued Logic* (ISMVL 1990), pages 238–246. IEEE Computer Society, 1990.

[Fitting, 1994] Melvin Fitting. Kleene's three valued logics and their children. *Fundamenta Informaticae*, 20(1–3):113–131, 1994.

[Ginsberg, 1986] Matthew L. Ginsberg. Multi-valued logics. In *AAAI-86 Proceedings*, pages 243–249. AAAI, 1986.

[Ginsberg, 1988] Matthew L. Ginsberg. Multivalued logics: A uniform approach to inference in artificial intelligence. *Computational Intelligence*, 4:265–316, 1988.

[Marcos and Mendonça, 2009] João Marcos and Dalmo Mendonça. Towards fully automated axiom extraction for finite-valued logics. In W. Carnielli, M. E. Coniglio, and I. M. L. D'Ottaviano, editors, *The Many Sides of Logic*, Studies in Logic, pages 425–440. College Publications, London, 2009. Preprint available at:
http://sqig.math.ist.utl.pt/pub/MarcosJ/08-MM-towards.pdf.

[Marcos, 2010] João Marcos. Automatic generation of proof tactics for finite-valued logics. *Electronic Proceedings in Theoretical Computer Science*, 21:91–98, 2010. Available at: http://arxiv.org/abs/1003.4802v1.

[Wansing and Belnap, 2010] Heinrich Wansing and Nuel Belnap. Generalized truth values: A reply to Dubois. *Logic Journal of the IGPL*, 18(6):921–935, 2010.

[Wansing and Shramko, 2005] Heinrich Wansing and Yaroslav Shramko. Some useful 16-valued logics: How a computer network should think. *Journal of Philosophical Logic*, 34:121–153, 2005.

João Marcos
Department of Informatics and Applied Mathematics
Federal University of Rio Grande do Norte (UFRN)
Natal, Brazil
E-mail: jmarcos@dimap.ufrn.br

A Note On The Logic Of *Una Tantum* Truth

CLAUDIO PIZZI

ABSTRACT. The paper begins by arguing that even if Prior's attempt of deriving a logic of earlier-later from a logic of tenses cannot be considered successful, it is full of interesting suggestions which can be developed dropping the quantificational apparatus used by Prior. The central part of the paper aims at defining the notion of a *una tantum* true proposition, i.e. of a proposition true at only one instant of a history, and at exploring its properties in the light of a notion of temporal necessity (inspired by V. Smirnov) which is stronger than so-called Megaric necessity. Section §3 introduces by definition the notions of realization and of precedence (consequence) between positions described by *una tantum* truths. After introducing a system called CRS+IP, in section §4 a semantic proof is given of some intuitive properties of the chronological order between positions, including irriflexivity and asymmetry.

§1. In the same years in which W.V.O. Quine was engaged in launching an attack against modal logic and its metaphysical dangers, A.N. Prior developed his project of a *tense logic*, i.e. of a bimodal logic having two primitives P, F to be read "sometimes in the past" and "sometimes in the future"[1]. Prior was hostile to the idea that the tensed language should be eliminated, along the lines recommended by Quine, in favour of an atemporal language containing simply the before-after relation between events. Being strongly influenced by the study of ancient and Medieval philosophy, Prior was convinced (a) that the commonsensical ontology of persisting substances (Johnsons's *continuants*) needed a tensed language in order to express the fact that an object acquires and loses properties in time; (b) that there is no reason to claim that atemporal language has a logical priority over tensed language.

In order to give substance to his philosophical assumptions, in various papers Prior tried to show that in a suitable framework a logic of earlier-later could be constructed in terms of a tensed language, rather than *vice versa*. In Chapter XI of *Papers on Time and Tense* Prior outlined what he called "four grades of temporal involvement".

The four grades are represented as four different system named I, II, III, IV.

(I) The language of System I admits two sets of variables, $Var = \{p, q, r, \ldots\}$

---

[1]See A.N.Prior, *Time and Modality* (Clarendon Press, Oxford 1957); *Past Present and Future* (id., 1967); *Papers on Time and Tense* (id., 1968). For the conflict between *"tensers"* and *"detensers"* see J. Massey, *Tense Logic. Why bother?*, Nous, 3, 7-32 (1969).

(*propositional variables*) and $I = \{t, u, v, \ldots\}$ (*instant variables*) and has $\supset, \neg$, P, F, =, U, R, ∀ as primitive operators. The symbol U is read as "is earlier than" and R is an operator of realization such that Rtp is read "p is realized at t". ∧ and ∨ are defined as usual and G and H are defined as duals of F, P respectively.

Prior gives minimal axioms for U and R and introduces the following definitions:

(1a) RtFp $=_{Df}$ $\exists t'(Utt' \wedge Rt'p)$

(1b) RtPp $=_{Df}$ $\exists t'(Ut't \wedge Rt'p)$

In System I all the theorems of the minimal PF-logic K_t^2 preceded by the operator Rt are derived from the minimal UR logic.

II. The second system is an axiomatic extension of the first. Further postulates for R are introduced to take care of the interaction of R with quantifiers and of the iteration of R-operators. In System II it is allowed to substitute UR-formulas for $p, q, r \ldots$ inside tensed formulas (so to have, say, such formulas as F(Rtp)). By defining $\Box A$ as $\forall t R t A$ Prior shows that one can derive in K_t the whole modal system S5.

III. In grade III the instant variables $t, t', t'' \ldots$ are allowed to be substituted for $p, q, r \ldots$ Instants are thus treated as so-called "clock-propositions", i.e. special propositions describing univocally an instant of time. Among the wffs we find now such expression as Rtt', which, however, is put axiomatically equivalent to $t = t'$ thanks to the following axiom:

(R=) R$tt' \equiv t = t'$

Other new axioms are

T6. Rtt

and

T7. R$tp \supset (t \supset p)$

Thanks to this extension Prior reaches the equivalences U$tt' \equiv$ RtFt', U$t't \equiv$ RtPt', R$tp \equiv \Box(t \supset p)$ and also the two theorems $\Diamond t$ (i.e. $\neg \forall t R t \neg t$) and $\exists t t$.

Thanks to the preceding equivalences System III may be reformulated by taking □, G, H as primitives beyond ∀. The language again admits two sorts of variables: $t, t', t''' \ldots$ and $p, q, r \ldots$. The new axioms for III are the axioms for K_t extended with

1) S5-axioms for □
2) $\Box p \supset Gp$
3) $\Box p \supset Hp$
4) $\Diamond t$
5) $\exists tt$
6) $\Box(t \supset p) \vee \Box(t \supset \neg p)$

Thanks to the two definitions

---

[2]The system K_t is axiomatized by extending PC (standard propositional calculus) with 1.H$(p \supset q) \supset$ (H$p \supset$ Hq), 2. $p \supset$ H$\neg$G$\neg p$, the rule RH: $\vdash A \rightarrow \vdash HA$ and the mirror-images of the mentioned formulas with G in place of H and viceversa: 3. G$(p \supset q) \supset$ (G$p \supset$ Gq), 4.$p \supset$ G$\neg$H$\neg p$, RG: $\vdash A \rightarrow \vdash GA$.

(Def R) $Rtp =_{Df} \Box(t \supset p)$
(Def U) $Utt' =_{Df} \Box(t \supset Ft')$

the old theorems involving R and U turn out to be derivable.

IV. The last grade of involvement in Prior's view expresses the full subordination of the UR-logic to PF-logic. The key step is performed by eliminating $\Box$ in terms of G and H. The definition proposed by Prior is as follows:

(i) $\Box^0 p = p$
(ii) $\Box^{n+1} A = A \wedge G\Box^n A \wedge H\Box^n A$
(iii) $\Box A = \forall n \Box^n A$

The reason of this unusual definition is that Prior found that the traditional definition of necessity as $A \wedge GA \wedge HA$ or, equivalently, of $\Diamond A$ as to $A \vee FA \vee PA$ (*Megaric* definition) is insufficient to cover all the points of any non linear model. (In order to distinguish the Megaric notion of modality from other temporal notions of modalities, we will use henceforth the symbols $\Box^m$ and $\Diamond^m$). For instance, $\Diamond^m q$ cannot receive value 1 at the instant x of the following K_t temporal model (where every arrow from y' to y'' represents the relation "y' before y''" or — due to the conversion property — the relation "y'' after y' ").

Figure 1.

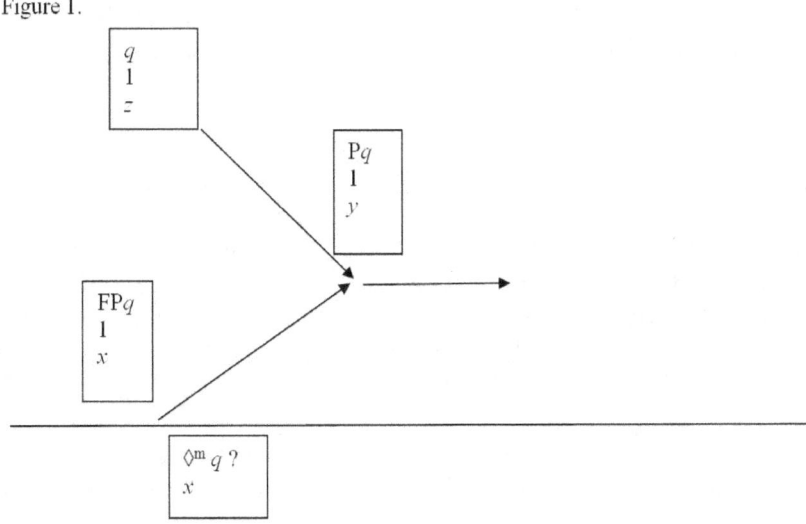

Two comments about Prior's four grades of temporal involvement are in order. *a)* As a general remark, the inversion obtained in system IV is reached in a framework which is not expressed by the tools of a propositional intensional language. In fact, Prior's language admits two sorts of variables and quantification over times. *b)* Furthermore, as remarked by Rescher e Urquhart[3], the recursive definition of $\Box$ introduced by Prior is an oddity, since it allows

---
[3] N. Rescher and A. Urquhart, *Temporal Logic*, Springer, NY, 1971, p.183.

eliminating indexed □-operators, but does not allow eliminating □ in a finite number of steps.

Even neglecting the preceding considerations, if we agree that in Prior's construction a derivation of a before-after logic inside a PF-logic is reached to some extent, in such a derivation an essential use is made of quantification and identity. One could reply that in grade III quantification is applied to instants treated as propositions, so that one could argue that we are not actually in front of first-order quantification. In many papers, as a matter of fact, Prior extends the expressivity of tensed language with propositional quantifiers and treats propositional quantification as a natural extension of propositional calculus. However, propositional quantification is after all a special case of second order quantification, and obviously cannot be treated without suitable axioms for propositional quantifiers[4].

§2. In 1977, after Prior's death, Kit Fine edited a collection of Prior's essays, *Worlds, Times and Selves*[5]. The selection of Prior's essays is followed by a long Postscript written by Fine (pp.115-168), in which the central topic is the relation between modal logic and quantificational logic. Here Fine gives special attention to the notion of a *world-proposition*, namely of a proposition which is true in one and only one world: as already seen, this notion in tense logic (where such propositions are called clock-propositions) had a central role in Prior's argument. The statement "p is a world-proposition" is defined as follows in terms of propositional quantifiers and modal operators:

(Def Q) $Qp =_{Df} \Diamond(p \land \forall q(q \supset \Box(p \supset q)))$.

$\Box A$ may then be rendered as $\forall p(Qp \supset \Box(q \supset Tr(A))$, where $Tr(A)$ is the translation of A. In this way we translate a modal logic into another one: $\Diamond r$ for instance is translated into the modal formula $\exists p(Qp \land \Box(p \supset r))$.

At p. 153 of his book Fine extends his analysis to the logic of tensed statements. The translation is double-sided: from tensed language to untensed language and *vice versa*. The translation in the second direction is commented in the following way. He writes (p. 134):

> "If $<$ is linear, then $\Box A$ is defined as $HA \land GA \land A$. If time is not linear, then a new definition or a new primitive may be required. That p is an instant-proposition — in symbols Ip — can now be defined as $\Diamond(p \land \forall q(q \supset \Box(p \supset q)))$. If time is linear, then Ip can be defined without quantifiers as
>
> (Def I) $Ip =_{Df} \Diamond(p \land G \neg p \land H \neg p)$.

---

[4] For a basic analysis of the interplay between propositional quantification and modality see K. Fine, *Propositional Quantifiers in Modal Logics*, Theoria, vol. 36, 1970, 336–346.

[5] A.N. Prior and K. Fine, *Worlds, Times and Selves*, Duckworth, Worcester and London, 1977.

However, this is not of much help since quantifiers will eventually need to be introduced. The reverse translation then is as follows: If p and q are instant propositions, R$'xyt$ is replaced with $\Box(p \supset \text{R}xy)$ and $t < u$ with $\Box(p \supset \text{F}q)$; $\forall t$ is replaced with the quantifier $\forall p$ over instant-propositions; and $\forall x$ is replaced with $\Box \forall x$." (p.154).

From the preceding quotation it turns out that also Fine cannot avoid using propositional quantifiers, even if he ambiguously suggests (see Def I) that we can avoid them in treating linear time.

Here we want to discuss the following problem suggested by Fine's remarks. Is it possible to reconstruct a logic of earlier — later inside a tense logic without using propositional quantifiers or standard quantifiers? If a full reconstruction is not possible, which are the properties of the earlier-later relation which it is possible to represent? As Fine himself suggests in the above reported quotation, in treating with logics for nonlinear time the first step will be to define temporal modalities in a way which is different from the one provided by Megaric definition.

The key-notion we will use here is the notion of what we will call "*una tantum* truth". The concept of a "*una tantum*" proposition, i.e. of a proposition being true at one and only one instant, is akin to the notion of a clock-proposition or instant- proposition, but does not depend on the existence of a clock or a metric system and in principle can be applied to every kind of topological logic[6].

Our starting point will be a system which is a simple extension of K_t which we will call CRS. CRS is K_t extended with the following axioms:

5. $Gp \supset GGp$
6. $Hp \supset HHp$
7. $Gp \supset Fp$
8. $Hp \supset Pp$

This system is equivalent to Cocchiarellas system CR (for causal relativistic time) extended with axioms 7 and 8 for seriality (this meaning that frames have no end-point in both directions). We will call this system CRS[7].

We take for granted that CRS is complete w.r. to the class of temporal frames $\mathfrak{F} = \langle T, >, < \rangle$ where $T = \{x, x, x'', \ldots, y, y', y'', \ldots\}$ is an non-empty set of objects named instants, $<$ and $>$ are converse relations defined over T which are transitive and serial[8]. In what follows completeness will allow proving results indifferently in a syntactical or a semantical way. A CRS-model is a

---

[6]See N.Rescher and J.Garson, *Topological Logic*, Journal of Symbolic Logic, 33, 537–548 (1968).

[7]This system is implicitly introduced by V.Smirnov in *The Definition of Modal Operators by Means of Tense Operators*, in I. Niiniluoto and E. Saarinen, *Intensional Logics and its Applications*, Acta Phil. Fennica, 35, 50–69 (1982), where the system K_t+5+6 is called K_c. CRS is K_c+ 7 +8.

[8]The proof will not be given here, but is easily obtained *via* analysis of the canonical frame.

CRS-frame endowed with an evaluation function V from atomic variables to $\{1,0\}$.

Notice that thanks to seriality we derive the following implications:

a) $(p \supset \mathrm{HF}p) \supset (p \supset \mathrm{PF}p)$
b) $(p \supset \mathrm{GP}p) \supset (p \supset \mathrm{FP}p)$

so, by a) and b), Modus Ponens and Axioms 3, 4 (see note 2)

c) $(p \supset \mathrm{PF}p)$
d) $(p \supset \mathrm{FP}p)$

and, by Uniform Substitution of $\mathrm{P}p$ for p,

e) $\mathrm{P}p \supset \mathrm{FPP}p$

so by Axiom 6, which is equivalent to $\mathrm{PP}p \supset \mathrm{P}p$,

f) $\mathrm{P}p \supset \mathrm{FP}p$

As temporal modalities are concerned, we begin by recalling that in 1982 Smirnov (see note 7) introduced a definition of modalities which is as follows:

(Def S□′) $\square' A =_{Df} \mathrm{HG}A$
(Def S◊′) $\lozenge' A =_{Df} \mathrm{PF}A$

A problem left open by Smirnov is, however, to understand the exact relation between PFA and FPA. Does FPA imply PFA (or *vice versa*) in the given system CRS? The preceding Fig. 1 says no. Suppose by Reductio that, at some instant x, FPq is 1 and PFq is 0[9]. This means that at all instants t such that $t < x$, Fq is 0 and at all instants w such that $w > t$, q is 0. So, at instant y, q is 0 but z is not an instant such that $z > t$. So no contradiction follows from the hypothesis. By a suitable reformulation of the same model, it is straighforward to see that also PFp does not imply FPp. So FPA is not in general equivalent to PFA, at least in this system. Smirnov's definition then creates an asymmetry between PFA and FPA which has no intuitive justification.

A simple correction of Smirnov's proposal is to introduce the definition

(Def g◊) $\lozenge A =_{Df} \mathrm{PF}A \vee \mathrm{FP}A$

or equivalently

(Def g□) $\square A =_{Df} \mathrm{GH}A \wedge \mathrm{HG}A$.

---

[9] In what follows we will use the expression "A is 1(0)" to mean that the value which the function V associates to A is 1(0).

This new kind of necessity (possibility) will be called here Global Necessity (Global Possibility).

A possible criticism of this definition is that, in the minimal system K_t, Fp does not imply PFp nor PFp, so Fp does not imply $\Diamond p$. Furthermore, in K_t, p implies HFp, but does not imply PFp, so p does not imply $\Diamond p$. But the criticism is unsound if referred to the present system CRS and to all its extensions. Suppose in fact Fp. Then Fp implies $HFFp$ by 2., $PFFp$ by 7. (seriality), and by 5. (transitivity), PFp and $PFp \vee FPp$. So we have $Fp \supset \Diamond p$ and also $Pp \supset p$. So, conversely, $\Box p$ implies both Gp and Hp. Even if CRS is not minimal, it embodies properties which are intuitive for any standard time series.

From the preceding remark it turns out that $\Box p$ implies $Gp \wedge Hp$. Since by c) above we have $p \supset \Diamond p$, so $\Box p \supset p$, $\Box p$ implies $Hp \wedge Gp \wedge p$: which means $\Box A \supset \Box^m A$, i.e. that global necessity implies Megaric necessity.

The converse implication from $\Box^m A$ to $\Box A$ is however invalid. If it were a theorem, we would have also $(FPA \vee PFA) \supset (PA \vee FA \vee A)$ for every A, so $PFA \supset (PA \vee FA \vee A)$. This formula is frequently used to express linearity (semiconnexivity) of the time series. In fact the non linear model in Fig.1 falsifies the formula: this is proved by showing that Pq, Fq and q might be all 0 at an instant x at which PFq is 1 (to see this, it is enough to add the information that q is 0 at every instant belonging to the straight line at the bottom of the graph).

We have to prove now that our definition of $\Box$ and $\Diamond$ satisfies some standard properties of necessity and possibility. The most outstanding properties are the following:

B) $p \supset \Box \Diamond p$

The proof is as follows:

1) $p \supset FPp$ $p \supset GPp$, $Gp \supset Fp$
2) $Pp \supset HFPPp$ $p \supset HFp$, Pp/p
3) $Pp \supset HFPp$ 2), $PPp \supset Pp$
4) $GPp \supset GHFPp$ 3), K_t
5) $p \supset GPp$ K_t
6) $p \supset GHFPp$ 4), 5)
7) $p \supset GH(FPp \vee PFp)$ 6), $GHp \supset GH(p \vee q)$

The argument for $p \supset HG(FPp \vee PFp)$ is specular and moves from $p \supset PFp$ at step 1). So we have $p \supset (GH(FPp \vee PFp) \wedge HG(FPp \vee PFp))$, which means $p \supset \Box \Diamond p$.

T) $\Box p \supset p$ (see line 1) of the preceding proof).

K) $\Box(p \supset q) \supset (\Box p \supset \Box q)$

K is proved starting from axiom 4. $H(p \supset q) \supset (Hp \supset Hq)$ and the rule $\vdash A \to \vdash GA$. The other steps are $GH(p \supset q) \supset G(Hp \supset Hq)$ and $HG(p \supset$

$q) \supset (HGp \supset HGq)$. So axiom K follows by Leibniz's *Theorema Praeclarum*.

The proof of the rule of Necessitation Nec $\vdash A \to \vdash \Box A$ i.e. $\vdash A \to \vdash GHA \land HGA$ is trivial. Obviously from T one has straightforwardly:

D) $\Box p \supset \Diamond p$

The problem of knowing which is exactly the fragment of CRS containing only $\Box$ and truth-functional connectives will not be discussed here. What is clear is that such fragment is at least as strong as the well-known modal system KTB (i.e. PC+K+T + B). So if $\mathfrak{F}$ is a frame $\langle W, \mathbf{R} \rangle$ verifying all and only the theorems of CRS, the relation $\mathbf{R}$ of $\mathfrak{F}$ is at least reflexive and symmetric.

The main question concerns of course the transitivity of $\mathbf{R}$. Since $<$ and $>$ have the properties of being transitive, may we conclude that $\mathbf{R}$ is also transitive? This means asking whether we have in CRS the theorem $(GHA \land HGA) \supset (GHGHA \land GHHGA \land HGHGA \land HGGA)$, which is the same as $\Box p \supset \Box\Box p$.

We may reply by considering the dual formula wff $\Diamond\Diamond p \supset \Diamond p$. Is

(*) $(PF(PFp \lor FPp) \lor FP(PFp \lor FPp)) \supset (PFp \lor FPp)$

a CSR-theorem? A negative answer comes by showing that there is a CSR-countermodel to $PFFPp \supset (PFp \lor FPp)$, a wff which is a consequence of (*) thanks to the distribution of PF over disjunction and to the so-called simplification of disjunctive antecedents. Let us look at figure 2 and suppose that, at x, $PFFPq$ is 1 and PFq is 0. If q is everywhere 0 in the bottom line, PFq and FPq are 0 at x, but no contradiction follows by the assignment 1 to $PFFPq$:

Figure 2

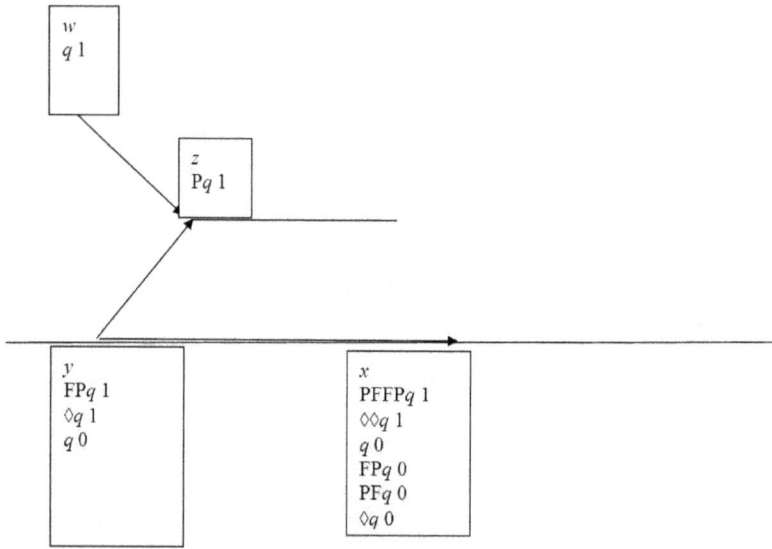

§3. Now let us introduce the notion of "Una tantum truth", which is a hybrid kind of necessity since, as a matter of fact, it has some features in common with impossibility. Let us call *history* a linear maximal order of elements $x, y, z \ldots$ such that $x < y < z \ldots$ or $x > y > z \ldots$. Let us call H_t the set of all histories "passing through" an instant t (i.e. such that t belongs to the intersection of the elements of H_t). We could call H_t the *chronological universe* (*c-universe*) of t. When we say that A is a *una tantum* truth at t we mean that A is true at t and false at all other instants of H_t. The rigorous definition is as follows:

(Def ▲) $\blacktriangle A =_{Df} A \wedge G\neg A \wedge H\neg A$

To say that $\blacktriangle A$ is true at x means that A is true at x but false at every other point of every history passing through x. The dual operator will be symbolized by ▼:

(Def ▼) $\blacktriangledown A =_{Df} \neg \blacktriangle \neg A$

From (Def ▼) we derive the equivalence

(*) $\blacktriangledown A \equiv A \vee F\neg A \vee P\neg A$

$\blacktriangledown A$ means "A is not *una tantum* false" (Note that the right side of (*) is equivalent to $(GA \wedge HA) \supset A$). Some apparently anomalous theorems concerning ▲ and ▼ are the following (where $\top$ is $p \vee \neg p$ and $\bot$ is $\neg \top$):

a) $\blacktriangledown \top$
b) $\blacktriangledown \bot$

REMARK. $\blacktriangledown \bot$ is $\bot \vee F\neg\bot \vee P\neg\bot$ and is equivalent to $F\top \vee P\top$, where $F\top$ and $P\top$ are the axioms for forward and backward seriality. So $\blacktriangledown \bot$ expresses the absence of an end-point in at least one of the two directions.

c) $(\blacktriangle A \wedge B) \supset \Box^m(A \supset B)$

(Proof: by the paradoxes of material implication and temporal implication B implies $A \supset B$, while $G\neg A \wedge H\neg A$ implies $G(A \supset B) \wedge H(A \supset B)$.)

Let us remark that, even if $G\neg A$ implies $G(A \supset B)$ and $H\neg A$ implies $H(A \supset B)$, $\blacktriangle A$ alone does not imply $\Box^m(A \supset B)$. In fact, if $\blacktriangle A$ were to imply $\Box^m A$, then $A \wedge G\neg A \wedge H\neg A$ would imply GA, so $\blacktriangle A$ would imply $GA \wedge G\neg A$, so $G\bot$, which is impossible since $\neg G\bot$ is a thesis of CRS.

The following are other theorems of CRS involving ▲ and ▼. It turns out that ▲ is a highly non standard modal operator, and its logic is not surely a normal modal logic, even if it exhibits a S4 behaviour:

1) $\blacktriangle p \supset p$
2) $\blacktriangle p \wedge p \equiv \blacktriangle p$ (from 1))

3) $p \supset \blacktriangledown p$
4) $\blacktriangle p \supset \blacktriangledown(p \wedge \neg q)$
5) $\Box \blacktriangledown \bot$
6) $\blacktriangle(p \supset q) \supset (\blacktriangle p \supset \blacktriangle r)$
(Proof: Suppose $\blacktriangle(p \supset q)$ and $\blacktriangle p$. Then it follows $G\neg(p \supset q)$ and $G\neg p$, which, however, implies $G(p \supset q)$. So $\blacktriangle(p \supset q)$ and $\blacktriangle p$ jointly imply $G\bot$ and $\bot$, or equivalently the simple r).
7) $\blacktriangle(p \supset q) \supset \neg \blacktriangle p$ (from 5))
8) $\blacktriangle(p \supset q) \supset (\blacktriangle p \supset \blacktriangle q)$
9) $\blacktriangle p \supset \blacktriangle \blacktriangle p$
(Proof: $G\neg p$ implies $G(\neg p \vee Fp \vee Pp)$, so $G\neg \blacktriangle p$.)
10) $\blacktriangledown \blacktriangledown p \supset \blacktriangledown p$
11) $(q \wedge \blacktriangle p) \supset \blacktriangle(p \wedge q)$

Some mixed theorems are the following:

12) $\blacktriangle p \supset F\neg p$
13) $\Box \blacktriangle p \supset \Box F \neg p$
14) $\Diamond \blacktriangle p \supset \Diamond F \neg p$
15) $\Box p \supset \blacktriangledown p$
16) $\blacktriangle p \supset GG\neg p$
17) $\blacktriangle p \supset HH\neg p$
18) $\blacktriangle p \supset \neg \Box p$
19) $\blacktriangle p \supset \neg \Box \blacktriangle p$
20) $\neg \Box \blacktriangle p$[10]
21) $\neg \Box \blacktriangle \top$
22) $\Box \blacktriangledown \top$
23) $\blacktriangle \bot \equiv \bot$
24) $\blacktriangle \top \equiv \blacktriangle \bot$
25) $\Box(p \supset q) \supset \Box(\blacktriangle p \supset q)$

Among the rules of inference the following are noteworthy:

$\vdash A \rightarrow \vdash \blacktriangledown A$
$\vdash A \equiv B \rightarrow \vdash \blacktriangle A \equiv \blacktriangle B$

Non-theorems are

26) $p \supset \blacktriangle p$
27) $\blacktriangle p \supset (p \wedge GHp)$
28) $\Diamond \blacktriangle p \supset \blacktriangle p$
29) $\blacktriangle p \supset \Box \blacktriangle p$
30) $(\blacktriangle p \wedge \blacktriangle q) \supset \blacktriangle(p \wedge q)$

---

[10] Suppose in fact by a contradiction $GH\blacktriangle p$, from which we have $GH(p \wedge H\neg p)$. So by distributivity $GHp \wedge GHHp$, so $GHH\bot$ and $\Box \bot$, which is inconsistent with $\Diamond \top$. The same with $HG\blacktriangle p$, so $\Box \bot$, which is negation of $\Diamond \top$.

31) $\blacktriangle(p \wedge q) \supset (\blacktriangle p \vee \blacktriangle q)$

Note that $\blacktriangle p \supset \Box p$ and $\blacktriangle p \supset \mathrm{GH}p$ are simply contradictions. The same holds for $\blacktriangle p \supset (p \supset \mathrm{FP}\neg p)$ and $\blacktriangle p \supset (p \supset \Diamond \neg p)$.

Now we may define two notions which correspond to the key-notions in Prior's construction, the notion of realization and the notion of precedence / consequence. In order to stress a difference with the elements of temporal models called instants, we will say that every proposition which is *una tantum* true describes a position. We stipulate that minor bold letters $\mathbf{a}, \mathbf{b}, \mathbf{c} \ldots$ are typographical variants of the capitals $A, B, C \ldots$ when they are used in association with the three symbols $\mathrm{R}, <, >$, which are introduced by definition.

(Def R) $\mathrm{R}\mathbf{b}A =_{Df} \Diamond \blacktriangle B \wedge \Box(\blacktriangle B \supset A)$.

From (Def R) we have consequently $\mathrm{R}\mathbf{b}A \supset \neg(\mathrm{R}\mathbf{b}\neg A)$. $\Diamond \blacktriangle B \wedge \Box(\blacktriangle B \supset A)$ in fact implies $\neg(\Diamond \blacktriangle B \wedge \Box(\blacktriangle B \supset \neg A))$. The proof rests on the fact that $\blacktriangle B$ implies both B and $\Diamond B$.

The converse formula $\neg(\mathrm{R}\mathbf{b}\neg A) \supset \mathrm{R}\mathbf{b}A$, i.e. $\mathrm{R}\mathbf{b}\neg A \vee \mathrm{R}\mathbf{b}A$, is not CSR-valid. The next figure shows that there is no contradiction in stating that the same wff A can be true at an instant x_1 accessible to k at which $\blacktriangle B$ is true and also false at a different instant x_2 accessible to k at which $\blacktriangle B$ is false.

If B is a proposition which is equivalent to the first argument of the operator R, we have $\mathrm{R}\mathbf{b}B \equiv \Diamond \blacktriangle B \wedge \Box(\blacktriangle B \supset B)$. Thus, being $\Box(\blacktriangle B \supset B)$ a thesis,

32) $\mathrm{R}\mathbf{b}B \equiv \Diamond \blacktriangle B$

The special case in which B is a tautology (recalling that $\top$ stands for $p \vee \neg p$) yields the equivalence $\mathrm{R}(p \vee \neg p)\top \equiv \Diamond \blacktriangle \top \wedge \Box(\blacktriangle \top \supset \top)$. But since we know that among the theorems we have $\Box(\blacktriangle \top \supset \top)$ and also $\neg \Diamond \blacktriangle \top$ (see 20) above), a theorem is then

33) $\neg \mathrm{R}(p \vee \neg p)\top$

The relation of *precedence* (*consequence*) between positions may be defined in this way:

(Def $<$) $\mathbf{b}' < \mathbf{b}'' =_{Df} \Diamond \blacktriangle B' \wedge \Diamond \blacktriangle B'' \wedge \Box(\blacktriangle B' \supset \mathrm{F}\blacktriangle B'')$
(Def $>$) $\mathbf{b}'' > \mathbf{b}' =_{Df} \Diamond \blacktriangle B' \wedge \Diamond \blacktriangle B'' \wedge \Box(\blacktriangle B'' \supset \mathrm{P}\blacktriangle B)$

Now an important theoretical question concerns the uniqueness of *una-tantum* sentences. In fact, the operators $\mathrm{G}\neg A$ and $\mathrm{H}\neg A$ state that A is false at all instants which belong to some history passing through the reference instant, not at all instants which have some **R**-relation with it[11]. In other words there is a class of models which in the most simple case have the following structure, where x_1 has a **R**-relation with x_2 but no relation of antecedence and consequence with x_2.

---

[11] For the present purposes it is enough to say that x has a **R**-relation with y if and only if either there is a z such that $x < z$ and $y < z$ or such that $x > z$ and $y > z$.

Figure 3

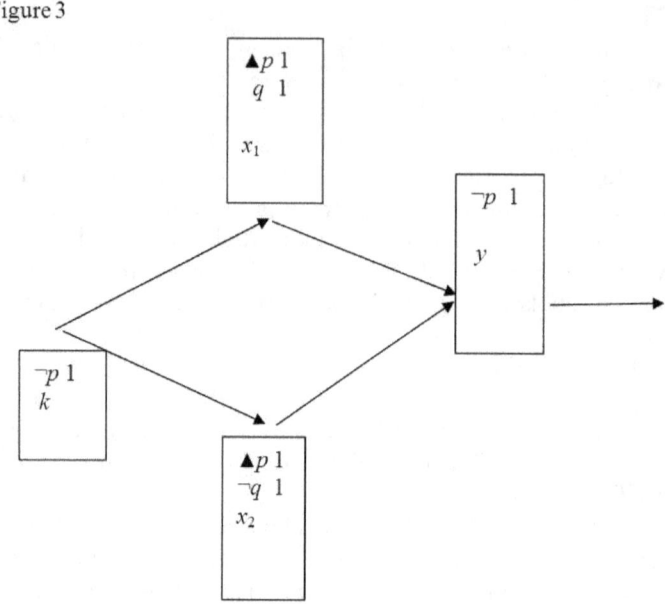

In front of such a situation it appears that we are faced with two different intuitions. On the one hand, it seems to make sense to say that there are distinct alternative "presents" having, say, the clock marking the same time but in different possible systems or, as we say, in different chronological universes[12].

On the other hand, it seems that if a position is possible with respect to another position, as x_1 and x_2 in Figure 3, and if both positions share a proposition which is an Una Tantum truth, the two positions should be considered to be identical. To be more clear, if an instant x is intended as an object which is described by all the propositions true in it and x contains ▲T, all instants containing ▲T in the same chronological universe should be considered to be identical instants. This "Identity Principle" for instants is expressible in this way:

(IP) $(\Diamond(▲T \wedge A)) \supset \Box(▲T \supset A)$.

Now (IP) is equivalent to

(IP+) $\Box(▲T \supset \neg A) \vee \Box(▲T \supset A)$

and since from (IP+) we have

---

[12]It is not excluded that some physical interpretation of branching models with alternative presents may be of some interest. In this connection it is enough to remind that in the '90es Nuel Belnap developed the idea of 'branching space-time' as a simple blend of relativity and indeterminism: see N. Belnap, *Branching Space-Time*, Synthese 92, 385–434, 1992.

(IP++) $\Diamond \blacktriangle T \supset ((\Diamond \blacktriangle T \wedge \Box(T \supset \blacktriangle \neg A)) \vee (\Diamond \blacktriangle T \wedge (\Box \blacktriangle T \supset A)))$

we reach the implication

(REM) $\Diamond \blacktriangle T \supset (RtA \vee Rt \neg A)$

from which we have $\Diamond \blacktriangle T \supset (RtA \equiv \neg(Rt \neg A))$

Under the condition that $\Diamond \blacktriangle p$ is true, a model as the one described in the last figure cannot exist. At instant k, in fact, both $\Diamond(\blacktriangle p \supset q)$ and $\Diamond(\blacktriangle p \supset \neg q)$ should be 1, which implies that $\Diamond \blacktriangle p$ is 1 at k. But (IP) entails that $\Box(\blacktriangle p \supset q)$ and $\Box(\blacktriangle p \supset \neg q)$ should also be 1 at k, which by standard modal logic implies $\neg \Diamond \blacktriangle p$: contradiction.

Many semantic conditions on temporal frames are sufficient to justify (IP): for instance the functionality of **R** (expressed by $\Diamond p \supset \Box p$). But if we have to take into consideration the occurrence of propositions expressing *Una Tantum* truths, a minimal semantic property of models validating (IP) is the following.

If V is any arbitrary evaluation function defined on CRS-models and A any CSR-wff:
(IPp) If, for some y and some y', $x\mathbf{R}y$, $x\mathbf{R}y'$, $V(\blacktriangle A, y) = 1$ and $V(\blacktriangle A, y') = 1$, then $y = y'$.

We may now prove what follows:

(IP) For any CRS+IP- model $\mathfrak{M}$, (IP) holds in $\mathfrak{M}$ if and only if (IPp) is a property of $\mathfrak{M}$.
Proof.
($\Rightarrow$) Let us suppose that (IPp) is a property of a model $\mathfrak{M}$ of CRS+IP and suppose by Reductio that, for some arbitrary x of $\mathfrak{M}$, $V(\Diamond(\blacktriangle T \wedge A), x) = 1$ and $V(\Box(\blacktriangle T \supset A), x) = 0$. Then at some y such that $x\mathbf{R}y$, $\blacktriangle T \wedge A$ is 1 and at some y' such that $x\mathbf{R}y'$, $\Box(\blacktriangle T \supset A)$ is 0. So at both y and y', $\blacktriangle T$ has value 1; by (IPp) this implies $y = y'$, but this is absurd since A should have value 0 in y' and value 1 in y.

($\Leftarrow$) Suppose conversely that (IPp) is not a property of $\mathfrak{M}$. Then, for some instant x of $\mathfrak{M}$, it may be that $\blacktriangle T$ is 1 at two instants y and y' such that $x\mathbf{R}y$ and $x\mathbf{R}y'$ but $y \neq y'$. It may then happen that, for at least a wff A, $V(\blacktriangle T \wedge A)$ is 1 at y and $V(\blacktriangle T \wedge \neg A)$ is 1 at y', so at x we have both $\Diamond(\blacktriangle T \wedge A)$ and $\Diamond(\blacktriangle T \wedge \neg A)$, what yields a counterexample to (IP.) Q.E.D.

The consequence of (IPp) (a condition on models) on the properties of the CRS+IP-frames is an open problem. The central question, which will not be treated in the present paper, is whether (IPp) is independent from the condition of linearity or has some logical relation with it.

§4. In what follows we show some results about formulas expressing the relations between positions in the system CRS + IP. We omit the trivial proof of the soundness of CRS + IP with respect to the outlined semantics, i.e. the proof that any wff A is a CRS+IP-thesis only if A is valid in all serial and transitive K_t- models enjoying the (IP)-property. In lack of a completeness proof what will be proved is that the wffs are valid for the mentioned semantics.

i) $(t > t' \land t' > t'') \supset (t > t'')$

Proof: We have to prove $(((\Diamond \blacktriangle T \land \Diamond \blacktriangle T' \land \Box(\blacktriangle T \supset F \blacktriangle T')) \land (\Diamond \blacktriangle T' \land \Diamond \blacktriangle T'' \land \Box(\blacktriangle T' \supset F \blacktriangle T''))) \supset (\Diamond \blacktriangle T \land \Diamond \blacktriangle T'' \land \Box(\blacktriangle T \supset F \blacktriangle T'')))$.

A simple syntactical proof of the formula is at our disposal. $\Box(\blacktriangle T' \supset F \blacktriangle T'')$ implies $\Box^m(\blacktriangle T' \supset F \blacktriangle T'')$, so $G(\blacktriangle T' \supset F \blacktriangle T'')$, so also $F \blacktriangle T' \supset FF \blacktriangle T''$; hence, by the theorem $FF \blacktriangle T'' \supset F \blacktriangle T''$, we obtain $F \blacktriangle T' \supset F \blacktriangle T''$. Then, from the premise $(\Diamond \blacktriangle T' \land \Diamond \blacktriangle T'' \land \Box(\blacktriangle T' \supset F \blacktriangle T''))$ (i.e. $t' > t''$), we obtain, by PC, $(\Diamond \blacktriangle T' \land \Diamond \blacktriangle T'' \land \Diamond \blacktriangle T \land \Box(\blacktriangle T \supset F \blacktriangle T'))$ and, by the transitivity of strict implication, $(\Diamond \blacktriangle T' \land \Diamond \blacktriangle T'' \land \Diamond \blacktriangle T \land \Box(\blacktriangle T \supset F \blacktriangle T''))$, so the conclusion $(\Diamond \blacktriangle T \land \Diamond \blacktriangle T'' \land \Box(\blacktriangle T \supset F \blacktriangle T''))$, i.e. $(t > t'')$. Being a theorem of CDR + IP, $(t > t' \land t' > t'') \supset (t > t'')$ is valid for the proposed semantics.

ii) $(t < t') \supset (t' > t)$

Proof. Let us suppose that $t < t'$ is 1 at a certain instant x of a certain CRS+IP-model $\mathfrak{M}$. Then we suppose that $\Diamond \blacktriangle T$ is 1 at x and also that $\Box(\blacktriangle T \supset F \blacktriangle T')$ is such. As already seen, this implies $\Diamond \blacktriangle T'$. Since $t' > t$ is the same as $\Diamond \blacktriangle T' \land \Diamond \blacktriangle T \land \Box(\blacktriangle T' \supset P \blacktriangle T)$, we have to prove simply $\Box(\blacktriangle T' \supset P \blacktriangle T)$ at x. Let us suppose by Reductio that $\blacktriangle T' \supset P \blacktriangle T$ is 0 at a certain $y \neq x$ such that $x R y$. So, at y, $\blacktriangle T'$ is 1 and $P \blacktriangle T$ is 0.

Since $\Diamond \blacktriangle T$ is 1 at x, this means that there is an instant w s.t. $x R w$ at which $\blacktriangle T$ is 1 and $\blacktriangle T \supset F \blacktriangle T'$ is 1. Then there is an instant z such that $z > w$, where $\blacktriangle T'$ is 1. Since we already established that, at y, $\blacktriangle T'$ is 1 and, by (IPp), $z = y$, so not only in y, but in z, $\blacktriangle T' \supset P \blacktriangle T$ is 0. Then $P \blacktriangle T$ is 0 in both z and y and T is 0 at all instants which precede z, including w. Contradiction. Q.E.D.

iii) $(t' > t) \supset (t < t')$

The argument is as for ii), *mutatis mutandis*.

iv) $\neg(t < t)$

iv) boils down simply to $\neg(\Diamond \blacktriangle T \land \Box(\blacktriangle T \supset F \blacktriangle T))$. Let us evaluate this wff by Reductio. Suppose 4. is 0 at some x of a certain CRS+IP- model $\mathfrak{M}$, so that $\Diamond \blacktriangle T$ is 1 at x and $\Box(\blacktriangle T \supset F \blacktriangle T)$ is also 1 at x. So by virtue of seriality there is some point y s.t. $x R y$, $\blacktriangle T$ is 1 and $\blacktriangle T \supset F \blacktriangle T$ is also 1. So $F \blacktriangle T$ and FT are also 1 at y. But this is impossible since, due to $\blacktriangle T$, $G \neg T$, i.e. $\neg FT$, is true at y. Q.E.D.

v) $(t < t') \supset \neg(t' < t)$

Suppose that both $\Diamond\blacktriangle T$, $\Diamond\blacktriangle T'$, $\Box(\blacktriangle T \supset F\blacktriangle T')$, $\Box(\blacktriangle T' \supset F\blacktriangle T)$ are all 1 in x. Let y be an accessible instant to x at which $\blacktriangle T$ is 1. So T is 1 at y and also $F\blacktriangle T'$ is such, so $\blacktriangle T'$ is 1 at some future instant z such that $y < z$ and $x < z$, while T is stably 0 at every $y'' > y$, due to the fact that $\blacktriangle T$ is 1 at y.

On the other hand, due to $\Diamond\blacktriangle T'$ being 1 at x, there is a y' s.t. $x\mathbf{R}y'$ where $\blacktriangle T'$ is 1 and $\blacktriangle T' \supset F\blacktriangle T$ is also 1. Since $\blacktriangle T'$ is also 1 at z, then $z = y'$ by (IPp).

Now, since $\blacktriangle T' \supset F\blacktriangle T$ is 1 at y' and $z = y'$, $\blacktriangle T'$ is 1 at y' ($= z$), $\blacktriangle T' \supset F\blacktriangle T$ is 1 at z and both $\blacktriangle T$ and T must be 1 at some instant $z' > z$: so, by the transitivity of $>$, T must be 1 at some $z' > y$. But this is impossible since we already established that T must be 0 at every instant y'' such that $y'' > y$. Q.E.D.

Up to know we were able to derive properties of the sequences of positions some of which are normally considered unexpressible in tensed language, such as irreflexivity and asymmetry. It is more difficult to treat with properties which in first order language need the use of (infinitary) existential quantifiers. An example is provided by seriality itself:

(S) $\forall x \exists y (x < y)$

However, it is possible to extend in some suitable way the language used to speak of the relation between positions As we already know, from the definition (Def $<$) we have the equivalence

$t' < t'' \equiv \Diamond\blacktriangle T' \wedge \Diamond\blacktriangle T'' \wedge \Box(\blacktriangle T' \supset F\blacktriangle T'')$.

Let us now define a relation called *counter-antecedence* in this way:

(CA) $t' < -t'' =_{Df} \Box(\blacktriangle T' \supset F\blacktriangle \neg T'')$

and a second one called *counter-consequence* in this way:

(CC) $t' > -t'' =_{Df} \Box(\blacktriangle T' \supset P\blacktriangle \neg T'')$

Forward Seriality and Backward Seriality may be expressed now by the following two wffs:

(FS) $t < -t$
(BS) $t > -t$

To understand the point, let us consider only FS, which means $\Box(\blacktriangle T \supset F\blacktriangle \neg T)$. The latter formula is a CRS-theorem thanks to the definition of $\blacktriangle T$ as $T \wedge G\neg T \wedge H\neg T$: in fact, $\blacktriangle T$ implies $G\neg T$ and $G\neg T$ implies $F\neg T$ thanks to the axiom expressing seriality (axiom 7 above). On the other hand axiom 7 entails $\Box(G\neg T \supset F\neg T)$ and also $\Box(\blacktriangle T \supset F\neg T)$ by *a fortiori*. So $t < -t$ is equivalent to $\Box(G\neg T \supset F\neg T)$ and then, via simple steps, to $\Box F(T \vee \neg T)$: but this is equivalent to $\Box F\top$, so to $F\top$ and to axiom 7.

In conclusion, a theorem involving simply **t** and the relation of counterantecedence and asserting, intuitively, that every position **t** is counterantecedent to itself is equivalent to an axiom expressing the inexistence of an end point in the future, i.e. forward seriality, as it was intended.

The same argument holds, *mutatis mutandis*, for BS[13].

Claudio Pizzi
Dipartimento di Filosofia e Scienze Sociali
Università di Siena
Siena, Italy
E-mail: pizzic@msn.com

---

[13] The author is grateful to Alberto Zanardo for his useful remarks to an earlier version of this paper.

Outline of a theoretical framework of Architecture of Information: a School of Brasilia proposal

MAMEDE LIMA–MARQUES

1 Introduction

The motivation for this paper came about with the perception that, from the beginning of 2000, there has been a need for theoretical fundaments for the Architecture of Information (AI) [Haverty, 2002; Dillon, 2002; Robins, 2002; Dale, 2002]. The challenge to be faced is the possibility of proposing a theoretical framework of Architecture of Information with the aim of lessening the conceptual gap currently presented.

The situation becomes more complex when it is noticed that all sciences and all institutions in our society deal with information. The human being is immersed in information. As a consequence a fundamental question arises: what is the organization of information underlying human understanding and interest?

The notion of information has become a crucial topic in several emerging scientific discipline [Doucette et al., 2007] and it shall lead a clear epistemology for the Science of Information. In line with various authors [Bates, 2005; Hofkirchner, 1999], the Science of Information is considered a wide field of human knowledge and so is its object of study: information. As a consequence, our vision aligns with this approach in the sense of creating a "Science of Information" (SI) [Doucette et al., 2007], and thus an "Architecture of Information".

Therefore it aligns itself with the initiative of 3rd International Conference on the Foundations of Information Science, Paris, July 2005, where a new expanded field was proposed: 'Science of Information'. This not to be confused with the older term 'Information Science', which sometimes is understood as advanced "library science", rather it is to take into consideration a newer and larger perspective encompassing many academic disciplines and new fields of interest.

2 On the basis

The search for adequate epistemological elements for engineering a solid foundation for the scientific explanations within SI and AI is crucial. Positioning oneself as regards matters of the core elements – data, information and knowledge – is very complex not only due to the high level of polysemy their usage comprises but also due to what is described by [Floridi, 2004] and revised by [Crnkovic and Hofkirchner, 2011] as unsolved problems. Such positioning is, however, fundamental for building coherent scientific theories and developing advanced applications.

A position about each of these fundamental elements is presented in what follows. These positions, far from solving unsolved problems, define epistemological frameworks from which is hoped to advance in the discussions.

2.1 On knowledge

The nature of knowledge is closely connected to the idea of AI. Phenomenology is suggested as the theoretical framework for understanding the phenomenon of knowledge. Therefore, the ideas of its most influential thinkers will be addressed, what will serve as basis for the discussions presented in this paper.

Phenomenology as a philosophy postulates that "all doing is, in essence, significant" began with the ideas of German philosopher Franz Clemens Brentano (1838-1917). Brentano defined two classes of phenomena: physical and mental. According to him, research about physical phenomena could be conducted through traditional positivist methods insofar as these phenomena become direct objects of sense perception; on the other hand, the positivist method would not the applied for mental phenomena due to the primary characteristic of such phenomena: 'intentionality', defined hereinafter [Hirschheim, 1985].

Edmund Husserl (1859-1938), philosopher of Israeli ascent born in Moravia (region in the Check Republic) and a follower of Brentano, is considered to be the founder of the phenomenological movement. His Phenomenology consists in a philosophical method proposing the description of lived experience from consciousness, the manifestations of which are purged from its real or empirical characteristics and considered against the plan of essential generality.

Phenomenon, from Husserl's point of view [Husserl, 1961; Husserl, 1963], does not mean "the simple appearance that opposes to the truth of a being or number", as it is in Plato and Kant, it is apparition rather than the appearance; it is the appearance of the object accessible through consciousness; the full manifestation of sense. And Philosophy needs find clarification of this sense.

A core point in Husserl's phenomenological concept is the 'intentional' character of consciousness, according to which 'consciousness is always consciousness about something'. Therefore, 'intentionality' consists of consciousness tending to an object and giving it a meaning. Kant's Phenomenology describes consciousness and experience, but abstains from considerations regarding its intentional content [Smith, 2011].

For Husserl, Phenomenology is fundamentally interested in the structure of various forms of experience: perception, thought, memory, imagination, emotion, will and volition to bodily awareness, embodied action, and social activity, including linguistic activity. The structure of these forms of experience comprises various intentionalities. This view, influenced by Brentano, determines the direction of experience for the objects in the world, meaning that viewing an object as data, as imaginary or as past is possible [Smith, 2011].

The phenomenon of knowledge, for Husserl, presents itself in its fundamental aspects. As reported by [Hessen, 1978], in knowledge the 'subject' and the 'object' face one another. Knowledge appears as a relation between these two elements, which remain eternally separate from each other. The subject-object dualism pertains to the essence of knowledge. The relation between these two elements is balanced – as a correlation, not equivalence – meaning that subject is subject and object is object. Both are only for one another. The function of the subject is to apprehend the object in terms of their properties, and function of the object is be apprehended by the subject. The subject is altered according to knowledge. In the subject arises an 'image' of the object, i.e. a set of object properties.

Therefore, knowledge is an image, a set of properties of the object apprehended by the subject. Knowledge is different from subject and from object. Knowledge appears as a third element that through correlation connects with those two elements thus forming a trinity.

2.2 On the Data

In the context of the theory of knowledge, supported by the phenomenology of Husserl, knowledge plays an important role in the sense of convergence of the intentionality of the subject, ready to deliver a unique experience: the set of object properties.

Intentionality refers to the notion that consciousness is always consciousness of something. Consciousness occurs as the simultaneity of a conscious act and its object. Intentionality is often summed up as 'aboutness'. Whether this something that consciousness is about is in direct perception or in fantasy is not consequential to the concept of intentionality itself; whatever consciousness is directed at, i.e. what consciousness is consciousness of. Therefore the object of consciousness doesn't have to be a physical object apprehended in perception. It can just as well be abstractness or an 'ideal object'. These 'structures' of consciousness, i.e., perception, memory, fantasy, are called intentionality.

Based on these conditions, taking intentionality as central pivot of the phenomenon, the nature of data can be its genesis related to the moment in which the apprehension occurs. It is proposed therefore that data is the state of the object properties to the instant immediately prior to his apprehension by the subject.

Unlike content found in literature that relates data and information, a direct relationship of data and knowledge is here. Both can be understood as different dimensions of the intentionality of the subject.

2.3 On the ontological status of information

The concept of information is diverse in its meanings, from daily to technical use [Crnkovic and Hofkirchner, 2011]. Generally, the concept of information is closely connected to the notions of restriction, communication, control, data, form, instruction, knowledge, mental stimuli, pattern, perception, representation, record, among others.

The Greek concept of *form* is represented by various words related particularly with *view*: the view or appearance of something. The ancient words are μορφή (*morphē*), εἶδος (*eidos*) and ἰδέα (*idia*), "the type, the idea, the form". "*Eidos*" was used by Plato – and, later on, by Aristotle – to indicate the ideal identity or essence of something (Theory of Forms). This word may also be associated with the concept of thought, proposition or even of concept. The words φαινόμενα (*phainomena*), "appearance" and φαινό (*phainō*) "of glow and light" still carry similar meaning.

The *Theory of Forms*, developed in Phaedo one of Plato's dialogues, realizes an understanding of various concepts and proceeds in his theory. For Plato the idea of concept, according to this new doctrine, is immutable, timeless, intellectually comprehensible and capable of a precise definition in a pure reasoning, because it is a real thing and it exists as an independent thing, an entity. Immortality of the soul is evidenced through our capacity of apprehending the concept of eternal, the object Plato called *form*.

In Physics, 'information' has a well-established meaning examples of which include quantum phenomena, and even the possibility of violating the second Law of thermodynamics by Maxwell's demon.

In information theory, entropy is a measure of the uncertainty associated with a random variable. Thus, the term usually refers to the Shannon entropy, which quantifies the expected value of the information contained in a message, usually in units such as bits. In this context, a 'message' means a specific realization of the random variable.

In statistical thermodynamics, Boltzmann's equation regards to probability related to the entropy S of an ideal gas to the quantity W (*Wahrscheinlichkeit*), which is the number of microstates corresponding to a given macrostate, or a configuration, or, yet, *complexions* (from Latin *complex* and the Greek suffix for diminutive *ions*):

(1) $\quad S = k \, log \, W$

In statistical mechanics, a microstate is a particular microscopic configuration of a thermodynamic system that the system may occupy with certain probability, in the course of its thermal fluctuations. As a counterpart, the macrostate of a system refers to its macroscopic properties, such as its temperature and pressure.

Macrostate is characterized by a distribution of probabilities of possible states in a particular statistical set of all microstates. This distribution describes the probability of finding the system in a determined microstate. In the thermodynamic limit, the microstate visited by a macroscopic system during its fluctuation possesses the same macroscopic properties.

The idea of information arises, then, from equation 1 and then we have:

(2) $\quad S = \quad k \, ln2 \, log_2 \, W$

where:
$$k = \text{Boltzman's constant} = 1,3806505 \cdot 10^{-23} \cdot J/K^{-1}$$
$$log_2 W = \text{information}$$
$$kln2 = 0.69k = \text{minimum of information}$$

The choice of a logarithmic base corresponds to the choice of a unit for measuring information. If the base 2 is used the resulting units may be called binary digits, or more briefly bits [Shannon, 1948].

Any experiment by which an information about a physical system is obtained corresponds in average to an increase of entropy in the system or in its surroundings. This average increase is always larger than (or equal to) the amount of information obtained [Brillouin, 1953].

Used in the more specific sense of information theory, information is a quantity that can be measured in bits [Lloyd, 2008].

The idea of information measures is a more general definition of information than either Shannon information or algorithmic information content. Information measures allow for the identification of effective complexity, a measure of the amount of information required to describe a system's regularities or rule-governed behavior.

Due to the real configuration of atoms and molecules of a gas in a specific space being unknown, *entropy* is associated to the *information we do not have*, in such way that when information is obtained, entropy is reduced. Entropy measures ignorance [Gell-Mann and Lloyd, 1996]. Therefore, the nature of *information* is related to *change in entropy*. In microscopic states, *information* is *complexions* of a gas, of a system or of an object. Further study of these microscopic states allows inference of some properties of the information.

Despite the reluctance of Claude Shannon (1916–2001) to extend the scope of his Mathematical Theory of Communication to other areas of knowledge such as physics, it is possible to conceive an integration, or at least a correlation between the entropy as defined in Shannon'n theory, and Heisenberg's Uncertainty Principle, through the concepts of uncertainty inherent to these two theories [de Carvalho Pineda, 2006].

Matter in subatomic levels, as evidenced experimentally, is presented in a superposition state, may take simultaneously more than one physical microstate within a set of possibilities. These quantum states of subatomic particles, as provided by the Uncertainty Principle formulated in 1927 by Werner Heisenberg (1901–1976), had an uncertainty inherent in probabilistic models, and are only persisted in the observer at the moment of his apprehension (in this case, it is the scientific instruments used in the experiments that allow observation of subatomic particles behavior, such as extension of human senses of perception).

In a quantum sense, a position from a perspective of information as to how the representation of primary properties of matter would occur becomes necessary. Following this line of thought, it is possible to define data as a snapshot of this information, produced by the process of decoherence at the time of his apprehension and when occurs the decaying of superposition state to a single state persisted. Information on a quantum level would thus have an inherent uncertainty and be correlated to the set of possibilities of different quantum states, which could be assumed by subatomic particles.

Based on an analogy between the messages of Shannon's Mathematical Theory of Communication and a perspective of quantum information, as described above, it is possible to arrive at this correlation. Assuming that there is a finite set of possible quantum states, each of these quantum states with a certain probability of occurrence, it is clear that the uncertainty inherent in the perspective of quantum information will be reduced by the occurrence of the decaying of superposition state to a single state at the time of his apprehension. The amount of uncertainty reduced by the process of decoherence is related to the probability of occurrence of the quantum state apprehended at the moment of observation, in the same way that the receipt of a most probable message reduces the uncertainty less than the receipt of a message less probable.

This approach seems to adhere to both the theory of Shannon's and the concept of perspective Quantum information: "Information is a reduction of uncertainty given when you get an answer to a question".

In this approach, information represents the primary properties of the object independently of the subject and therefore strictly ontological, while this information would be data persisted at the exact moment of his apprehension by the subject, a snapshot. For this approach, there would be a fundamental distinction between data and information, making it necessary to agree to a proper terminology for this the model. From the arguments presented one can conclude that information is 'thing', i.e. information belongs to ontological level, and that data is the condition of the object properties on the instant immediately prior to its apprehension by the subject.

3 Elements of AI

3.1 On space

Spencer Brown, in his book *Laws of Form*, [Brown, 1969], introduces the idea of form as a 'distinction' in a space, proposes a logical system and overcomes a few boundaries between mathematics and philosophy. The idea of distinction and the idea of indication, and that we cannot make an indication without drawing a distinction, are taken as a given. We take, therefore, the form of distinction for the form. By definition, distinction is perfect continence. A distinction is drawn by arranging a boundary with separate sides so that a point on one side cannot reach the other side without crossing the boundary. For example, in a plane space a circle draws a distinction. So, it is possible to postulate that there is no space without distinction. Once a distinction is drawn, the spaces, states, or contents on each side of the boundary, being distinct, can be indicated.

3.2 On state

Distinguished space has a state. Time is related to state. Distinguished space has content. Content is composed by things. Things have properties. Therefore, we assume that 'space of information' is the set of distinguished information in a distinguished space.

DEFINITION 1. A state $\mathcal{E}$ is an unique configuration of information in an interval of time Δ_t, denoted as $\mathcal{E}_{\Delta_t}$.

DEFINITION 2. A dynamic $\mathcal{D}$, of the spaces of information e is defined as:

$$\mathcal{D} = \{\mathcal{E}_{\Delta_{t_1}}, \mathcal{E}_{\Delta_{t_2}}, \mathcal{E}_{\Delta_{t_3}}, \cdots, \mathcal{E}_{\Delta_{t_n}}\}$$

3.3 On AI

It is shown that the concept of AI can be applied in any information space. Examples of information spaces can be characterized generally as any set of things. We may think of a DNA structure as an information space, hence an AI, or, the solar system, or my office desk with its objects over, or even any object as particulars information spaces, i.e. it is possible consider all these meanings from the point of view of an AI.

The concern in organizing and structuring knowledge accompanies human history for centuries. The phenomenon of information explosion took even greater from the World Wide Web, and caused a growing concern with the systematization and access to knowledge. The concept of AI is to be inserted in this context, despite its origin dating from ancient times.

The term 'information architecture', as recorded in the literature, was first used by architect Richard Saul Wurman in 1976, who described it as "science and art of creating instructions for organized spaces". Wurman viewed the problem of searching, organizing and presenting information as analogous to the problems of the architecture of buildings, which will provide for the needs of its residents, because the architect needs to identify these needs, organize them into a coherent pattern that determines its nature and their interactions, and design a building that satisfies them.

The publications: Information Anxiety, [Wurman, 1989] and Information Anxiety 2, [Wurman, 2001] show an overview of the fundamental principles that motivated the author in his previous work, highlighting how dramatic is the explosion of information.

In view of Wurman, the assemblage, the organization and presentation of information served the purpose of the tasks characteristic of Architecture. The Information Architecture would be an expansion of the profession of architecture, but applied to spaces of information. And information structures interactions influence the world in the same way that the building structures encourage or limit social interactions. In 1976, Wurman organized the National Conference of the American Institute of Architects (AIA) and chose "The Architecture of Information" as the theme of the conference, coincidentally 100 years after the first meeting of the American Library Association. Today we have worked with a much broader idea of AI, particularly in the proposal of the School of Brasilia.

The first concept of AI arises as a result of above subsection, where for a given information space we consider the configuration of information, or the information being.

DEFINITION 3. AI is the states configuration of the constituent elements of the thing itself and its properties, characterized by space-temporality of distinguished information.

It is shown that this concept of AI can be applied in any information space. AI is inherent to any information space, in any domain. As a consequence of the definitions, one can say that there is no space of information without AI.

From an extension of the first concept of AI, we get the second concept, characterized by the need of changing of state. Better saying, considering a time interval Δ_t for an information space, an initial state $\mathcal{E}_{\Delta_{t_1}}$ of a configuration of properties of their constituent elements (snapshot information), a changing to future state $\mathcal{E}_{\Delta_{t_{n+1}}}$ is performed by a transformation. Evidently the intermediate states are characterized by a dynamic $\mathcal{D}$ belonging to this context.

DEFINITION 4. A *transformation* is a set of events, applied to a particular state, in order to provoke changes to future states.

This perspective of AI can be applied in any situation. For example, one can show that there is no information system without AI. In fact, when designing an organization's information, and information systems are patterns of organization of information, you can not do so without regard to AI, or AI is inherent to any information system.

4 AI as Social Science

The third concept is related to the perspective of applying a *transformation*. This perspective is related to the performance of a *subject*. As a result, we are in the field of Applied Social Sciences, or on the application of transformations performed by an individual or by a *subject*.

Philosophy of Language originated the *Theory of Speech Acts*, at the beginning of the sixties having been appropriated by Pragmatics later on. John Langshaw Austin (1911-1960), philosopher of the Oxford Analitic School, followed by John Searle and others, understood language as a form of action: "all speech is an action". The various types of human action realized through language were reflected upon: *the speech acts*. The Theory of Speech Acts was published posthumously, in 1962, in the book *How to do Things with Words* [Austin, 1962, page23]. For Austin, speech is not only a way of passing on information, but also – and foremost – it is a way of acting on the interlocutor and on the world.

Until then, linguists and philosophers, in general, thought that the claims serve only to describe a state of things, and thus were true or false. Austin calls into question this view of descriptive language, showing that certain statements do not serve to describe anything, but to take action.

The *School of Brasilia* proposes an extension of Austin's Speech Act in the sense that "all doing is an act" and "all act is a transformation".

DEFINITION 5. *Transformation acts* are sets of events, applied to a particular state, by subject, in order to provoke changes to future states.

In the literature there are dozens of books whose titles we see the term 'information architecture'. With the theoretical framework presented, examples of AI in the literature become particular cases or examples, of the general concept. Some of the titles available are: *Enterprise Information Architecture, Information Assurance Architecture, e-Gov Information Architecture, Federal Enterprise Architecture, Strategic Information Architecture, Supply Chain Information Architecture, Web Information Architecture, Information Security Architecture*, etc.

Enterprise Information Architecture is the most recognized by industry. In the proposal of this approach, we can consider an enterprise as an information space. If we take a snapshot at some time interval, we obtain the configuration information of the enterprise. This is the initial situation. We may analyze this situation from the perspectives of the current paradigms for organizing information in Enterprises. From a future perspective of a possible desired state of the current situation, it is possible to design a future state. Acts of transformation will occur to make it possible to achieve the desired future state. This same approach can be applied to any space, where it is necessary to consider any aspect of the "information life cycle", from a perspective of SI, and consequently AI.

BIBLIOGRAPHY

[Austin, 1962] John L. Austin. *How to Do Things With Words*. Oxford University Press, Oxford, 1962.

[Bates, 2005] M.J. Bates. Information and knowledge: an evolutionary framework for information science. *Information research*, 10(4):10–4, 2005.

[Brillouin, 1953] L. Brillouin. The negentropy principle of information. *Journal of Applied Physics*, 24(9), April 1953.

[Brown, 1969] George Specer Brown. *Laws of form*. George Allen & Unwin London, 1969.

[Crnkovic and Hofkirchner, 2011] G.D. Crnkovic and W. Hofkirchner. Floridi's "open problems in philosophy of information", ten years later. *Information*, 2(2):327–359, 2011.

[Dale, 2002] Adrian Dale. Letter 12: Information architecture: the next professional battleground? *Journal of Information Science*, 28(6):523–525, 2002.

[de Carvalho Pineda, 2006] José Octávio de Carvalho Pineda. A entropia segundo Claude Shannon: desenvolvimento do conceito fundamental da teoria da informa cão. Master's thesis, PUC-SP, São Paulo, 2006.

[Dillon, 2002] A. Dillon. Information architecture in jasist: just where did we come from? *Journal of the American Society for Information Science and Technology*, 53(10):821–823, 2002.

[Doucette et al., 2007] D. Doucette, R. Bichler, W. Hofkirchner, and C. Raffl. Toward a new science of information. *Data Science Journal*, 6(0):198–205, 2007.

[Floridi, 2004] Luciano Floridi. Open problems in the philosophy of information. *Metaphilosophy*, Volume 35(Number 4):pp. 554–582, July 2004.

[Gell-Mann and Lloyd, 1996] M. Gell-Mann and S. Lloyd. Information measures, effective complexity, and total information. *Complexity*, 2(1):44–52, 1996.

[Haverty, 2002] Marsha Haverty. Information architecture without internal theory: an inductive design process. *Journal of the American Society for Information Science and Technology*, 53(10):839–845, 2002.

[Hessen, 1978] Johannes Hessen. *Teoria do conhecimento*. A. Amado, Coimbra, 7 edition, 1978.

[Hirschheim, 1985] R. Hirschheim. Information systems epistemology: An historical perspective. *Research methods in information systems*, pages 13–35, 1985.

[Hofkirchner, 1999] W. Hofkirchner. *The quest for a unified theory of information: proceedings of the Second International Conference on the Foundations of Information Science*. Routledge, 1999.

[Husserl, 1961] Edmund Husserl. *Recherches logiques*. Presses universitaires de France, France, 1961.

[Husserl, 1963] Edmund Husserl. *Idées directrices pour une phénoménologie*. Gallimard, France, 1963.

[Lloyd, 2008] Seth Lloyd. Quantum information science. USA, 2008.

[Robins, 2002] David Robins. Information archtecture in library and information sicence curricula. *Bulletin of the American Society for Information Science and Technology*, 28(2):20–22, 2002.

[Shannon, 1948] Claude Elwood Shannon. *A Mathematical Theory of Communication*, volume 27. Bell System Technical Journal, Octuber 1948.

[Smith, 2011] David Woodruff Smith. Phenomenology. In Edward N. Zalta, editor, *The Stanford Encyclopedia of Philosophy*. The Metaphysics Research Lab, Stanford University, fall 2011 edition, 2011.

[Wurman, 1989] Richard Saul Wurman. *Information anxiety*. Doubleday, New York, 1989.

[Wurman, 2001] Richard Saul Wurman. *Information Anxiety 2*. Hayden/Que, Indianapolis, USA, 2001.

Mamede Lima-Marques
Centre for Research on Architecture of Information – FACE/UnB and
Faculty of Information Science
University of Brasília
Brasília, Brazil
E-mail: mamede@unb.br

Non-deterministic combination of connectives

A. SERNADAS, C. SERNADAS, J. RASGA, P. MATEUS

ABSTRACT. Combined connectives arise when combining logics [Sernadas *et al.*, in print] and are also useful for analyzing the common properties of two connectives within a given logic [Sernadas *et al.*, 2011]. A non-deterministic semantics and a Hilbert calculus are proposed for the meet-combination of connectives (and other language constructors) of any matrix logic endowed with a Hilbert calculus. The logic enriched with such combined connectives is shown to be a conservative extension of the original logic. It is also proved that both soundness and completeness are preserved. Illustrations are provided for classical propositional logic.

1 Introduction

When combining logics one frequently wants to impose some interaction between connectives.[1] For instance, in the logic $\mathcal{L}$ resulting from the fibring [Gabbay, 1996] of any two given logics $\mathcal{L}_1$ and $\mathcal{L}_2$ (where one finds all the inference rules from those two logics), if one imposes the sharing of two constructors c_1 and c_2 with the same arity, the resulting *shared constructor* $\langle c_1 c_2 \rangle$ in $\mathcal{L}$ enjoys the logical properties inherited from c_1 together with those inherited from c_2. Such a sharing may easily lead to inconsistency.

One may wonder if it would not be better instead to endow the combined constructor only with the *common* logical properties of the component constructors. As we conjectured in [Sernadas *et al.*, 2011], one should expect to be led to a different way of combining logics, at least with respect to the behavior of the combined constructors, hopefully avoiding inconsistency in more situations. This expectancy was fully vindicated in [Sernadas *et al.*, in print].

Herein, we go back to the problem addressed in [Sernadas *et al.*, 2011]: provide the means for reasoning about the common properties of any two connectives in a given logic. Given a logic $\mathcal{L}$, an enriched logic $\mathcal{L}^\times$ was proposed in [Sernadas *et al.*, 2011] for deriving such common properties by dealing with ordered pairs of connectives as language constructors. For instance, for reasoning about the common properties of conjunction and disjunction one can analyze in $\mathcal{L}^\times$ the properties either of the combined connective $\lceil \wedge \vee \rceil$ or of $\lceil \vee \wedge \rceil$. Although these two combined connectives were shown to be equivalent to some extent in $\mathcal{L}^\times$, it would be nice to avoid the product semantics given in [Sernadas *et al.*, 2011] to such combined connectives.

---
[1] Connectives are to be taken here in a general sense, including, besides the propositional connectives, other language constructors like modal operators.

By adopting instead a non-deterministic semantics for the combination of connectives, we manage herein to collapse $\lceil \wedge \vee \rceil$ and $\lceil \vee \wedge \rceil$, as envisaged. A calculus is also proposed. The resulting logic $\mathcal{L}^\sqcap$ is shown to be sound and complete, as long as the original logic $\mathcal{L}$ is sound and complete. It is also proved that, as envisaged, $\mathcal{L}^\sqcap$ is a conservative extension of $\mathcal{L}$.

For the sake of simplicity, we assume that the given logic $\mathcal{L}$ is of a propositional nature.[2] Furthermore, we assume that $\mathcal{L}$ is endowed with a Hilbert calculus and a matrix semantics. It should be stressed that the proposed logic $\mathcal{L}^\sqcap$ is presented with a Hilbert calculus, but its non-deterministic semantics is outside the universe of matrix semantics.

In Section 2, after recalling the adopted notion of matrix logic with Hilbert calculus, we show how to enrich any such given logic with non-deterministic combinations of constructors. The main results on the enriched logic are established in Section 3. Illustrations are provided for classical propositional logic in Section 4, including a comparison with the results in [Sernadas et al., 2011]. Finally, in Section 5 we assess what was achieved and what still lies ahead.

2 Meet-combining constructors

For the purpose of this paper, by a *suitable logic* we mean a triple $\mathcal{L} = (\Sigma, \Delta, \mathcal{M})$ where

- The *signature* Σ is a family $\{\Sigma_n\}_{n \in \mathbb{N}}$ with each Σ_n being a finite set of n-ary language *constructors*. Formulas are built as usual with these constructors. We use L for denoting the set of *formulas*.

- The Hilbert *calculus* Δ is a set of finitary rules of the form

$$\frac{\alpha_1, \ldots, \alpha_m}{\beta}$$

  where formulas $\alpha_1, \ldots, \alpha_m$ are said to be the *premises* of the rule and formula β is said to be its *conclusion*. As expected, the conclusion of a rule without premises is said to be an *axiom*. Derivations are defined as usual for Hilbert calculi. We write

$$\Gamma \vdash \varphi$$

  for stating that there is a derivation of formula φ from set Γ of hypotheses.

- The matrix *semantics* $\mathcal{M}$ is a class of matrices over Σ. Recall that a matrix over Σ is a pair $M = (\mathfrak{A}, D)$ where

$$\mathfrak{A} = (A, \{\underline{c} : A^n \to A \mid c \in \Sigma_n\}_{n \in \mathbb{N}})$$

  is an algebra over Σ and $D \subseteq A$. The elements of A are known as *truth values* and those of D are the *distinguished* or *designated* ones. Denotation, satisfaction and entailment are as expected for matrix semantics. We write

$$[\![\varphi]\!]_M$$

---

[2]That is, we address only logics without variables and binding operators.

for the denotation of formula φ given by the algebra of matrix M, and we say that M *satisfies* φ, written

$$M \Vdash \varphi,$$

if $[\![\varphi]\!]_M \in D$. As usual, given a set Φ of formulas, we write $M \Vdash \Phi$ if $M \Vdash \varphi$ for every $\varphi \in \Phi$. Set Γ of formulas *entails* formula φ, written

$$\Gamma \vDash \varphi,$$

if $M \Vdash \varphi$ whenever $M \Vdash \Gamma$ for every $M \in \mathcal{M}$.

In short, by a suitable logic we mean a logic of propositional nature, endowed with a Hilbert calculus and a matrix semantics.

Given such a suitable logic $\mathcal{L} = (\Sigma, \Delta, \mathcal{M})$, the objective now is to define a logic

$$\mathcal{L}^{\sqcap} = (\Sigma^{\sqcap}, \Delta^{\sqcap}, \mathcal{M}^{\sqcap})$$

where, without disturbing the original properties of the constructors in $\mathcal{L}$, one can also reason with and about the envisaged meet-combinations of constructors in the original logic $\mathcal{L}$ by adopting a non-deterministic semantics for those constructors.

Language

The signature $\Sigma^{\sqcap}$ is composed of all possible meet-combinations of constructors in Σ. More concretely,

$$\Sigma^{\sqcap} = \{\Sigma_n^{\sqcap}\}_{n \in \mathbb{N}}$$

with

$$\Sigma_n^{\sqcap} = \{\lceil c_1 c_2 \rceil \mid c_1, c_2 \in \Sigma_n\}.$$

For any constructors $c_1, c_2 \in \Sigma$ of the same arity, $\lceil c_1 c_2 \rceil$ is said to be their meet-combination. Moreover, c_1 and c_2 are said to be the *components* of $\lceil c_1 c_2 \rceil$. Since this paper addresses only such meet-combined constructors, from now on we refer to them simply as combined constructors. As expected, we use $L^{\sqcap}$ for denoting the set of formulas of $\mathcal{L}^{\sqcap}$. Examples are delayed until Section 4.

We look at $\Sigma^{\sqcap}$ as an enrichment of Σ via the embedding

$$\eta : c \mapsto \lceil cc \rceil.$$

Furthermore, in the context of $\Sigma^{\sqcap}$ we may write c for $\lceil cc \rceil$. In due course, this notational shortcut will be fully vindicated. Accordingly, we take that $L \subset L^{\sqcap}$.

Given a formula φ over Σ and a formula ψ over $\Sigma^{\sqcap}$, we say that φ is a *determination* of ψ if $\varphi \sqsubseteq \psi$ where the binary relation $\sqsubseteq \, \subseteq L \times L^{\sqcap}$ is inductively defined as follows:

- $c_k \sqsubseteq \lceil c_1 c_2 \rceil$, for each $k = 1, 2$ and $c_1, c_2 \in \Sigma_0$;

- $c_k(\varphi_1, \ldots, \varphi_n) \sqsubseteq \lceil c_1 c_2 \rceil(\psi_1, \ldots, \psi_n)$ whenever $\varphi_1 \sqsubseteq \psi_1, \ldots, \varphi_n \sqsubseteq \psi_n$, for each $k = 1, 2$, $n \in \mathbb{N}^+$ and $c_1, c_2 \in \Sigma_n$.

In short, $\varphi \sqsubseteq \psi$ if φ is obtained by choosing one component of each occurrence of a combined connective in ψ. It should be stressed that, when a combined connective occurs more than once, those choices need not be the same. Observe that $c \sqsubseteq c$ since $c \sqsubseteq \lceil cc \rceil$ which justifies our notation for determination.

Clearly, the determination relation extends the componentship relation to formulas. This extension is needed for setting up in due course the calculus $\Delta^{\sqcap}$.

For defining the envisaged non-deterministic semantics, it becomes handy to denote by $|\psi|$ the *number of constructors* in a given formula $\psi \in L^{\sqcap}$. The map $\psi \mapsto |\psi|$ is inductively defined as expected:

- $|\lceil c_1 c_2 \rceil| = 1$, for each $c_1, c_2 \in \Sigma_0$;
- $|\lceil c_1 c_2 \rceil(\psi_1, \ldots, \psi_n)| = 1 + |\psi_1| + \cdots + |\psi_n|$, for each $n \in \mathbb{N}^+$, $c_1, c_2 \in \Sigma_n$ and $\psi_1, \ldots, \psi_n \in L^{\sqcap}$.

This notation is extended to finite sequences of formulas as follows:

$$|\psi_1 \ldots \psi_n| = |\psi_1| + \cdots + |\psi_n|.$$

Non-deterministic semantics

The non-deterministic semantics $\mathcal{M}^{\sqcap}$ is not matricial. Each interpretation in $\mathcal{M}^{\sqcap}$ is a matrix in $\mathcal{M}$ enriched with a choice stream. More concretely, $\mathcal{M}^{\sqcap}$ is the following class of interpretations

$$\{(M, \omega) \mid M \in \mathcal{M}, \omega \in \{1, 2\}^{\mathbb{N}}\}.$$

In each interpretation, the matrix M provides the meaning of the original constructors in Σ and the *choice sequence* ω determines the meaning of the combined connectives.

To this end, we need some notation concerning segments of infinite sequences. Given $\omega \in \{1, 2\}^{\mathbb{N}}$, for each $k \in \mathbb{N}$, let

$$\omega_{k:\infty}$$

denote the infinite sequence $\omega_k \omega_{k+1} \ldots$ (the tail of ω starting at element k). Later on, given $k_1 \leq k_2$, we shall also need to write

$$\omega_{k_1:k_2}$$

for denoting the finite sequence $\omega_{k_1} \ldots \omega_{k_2}$ (the segment of ω from element k_1 to element k_2 inclusive).

The *denotation* of a formula given by (M, ω) is inductively defined as follows:

- $[\![\lceil c_1 c_2 \rceil]\!]_{M\omega}^{\sqcap} = c_{\omega_0}$;
- $[\![\lceil c_1 c_2 \rceil(\psi_1, \ldots, \psi_n)]\!]_{M\omega}^{\sqcap} = c_{\omega_0}([\![\psi_1]\!]_{M\omega_{1:\infty}}^{\sqcap}, \ldots, [\![\psi_n]\!]_{M\omega_{1+|\psi_1 \ldots \psi_{n-1}|:\infty}}^{\sqcap})$.

Clearly, the semantics of $\mathcal{L}^{\sqcap}$ is not homomorphic, since the denotation of a combined connective may vary if used more than once in a formula.

Satisfaction is defined as expected. We say that (M,ω) *satisfies* ψ, written

$$M\omega \Vdash^{\sqcap} \psi,$$

if $[\![\psi]\!]^{\sqcap}_{M\omega}$ is a distinguished value of M. Furthermore, we say that M *satisfies* ψ, written

$$M \Vdash^{\sqcap} \psi,$$

if $M\omega \Vdash^{\sqcap} \psi$ for every ω. As usual, we may write $M \Vdash^{\sqcap} \Psi$ for asserting that $M \Vdash^{\sqcap} \psi$ for every $\psi \in \Psi$.

Entailment is made to be global over choices as follows. We say that Θ entails ψ, written

$$\Theta \vDash^{\sqcap} \psi,$$

if, for every M,

$$M \Vdash^{\sqcap} \psi \text{ whenever } M \Vdash^{\sqcap} \Theta.$$

Examples are delayed until Section 4. In due course (in Section 3), we prove that, as envisaged, $\vDash^{\sqcap}$ extends $\vDash$.

Calculus

The Hilbert calculus $\Delta^{\sqcap}$ is an enrichment of Δ (via the embedding $c \mapsto \lceil cc \rceil$), including, besides the rules inherited from Δ, the following rules for dealing with the combined constructors.

For each formula $\psi \in L^{\sqcap}$, the *lifting rule* (in short LFT)

$$\frac{\varphi \quad \text{for each } \varphi \sqsubseteq \psi}{\psi}.$$

This rule is motivated by the idea that each $\lceil c_1 c_2 \rceil$ inherits at least the common properties of c_1 and c_2. Accordingly, ψ should hold whenever all of its determinations hold.

For each formula $\psi \in L^{\sqcap}$, the *co-lifting rule* (in short cLFT)

$$\frac{\psi}{\varphi} \quad \text{provided that } \varphi \sqsubseteq \psi.$$

This rule is motivated by the idea that each $\lceil c_1 c_2 \rceil$ should enjoy at most the common properties of c_1 and c_2. In fact, this rule guarantees more. It guarantees that $\lceil c_1 c_2 \rceil$ enjoys at most the common *original* properties of c_1 and c_2 because, in due course, we show that $\mathcal{L}^{\sqcap}$ is a conservative extension of $\mathcal{L}$.

In the sequel, we write

$$\Theta \vdash^{\sqcap} \psi$$

for stating that ψ is derivable from Θ within $\mathcal{L}^{\sqcap}$. Examples are delayed until Section 4. The following result shows that, as intended, $\vdash^{\sqcap}$ extends $\vdash$.

PROPOSITION 1. *Let* $\Gamma \cup \{\varphi\} \subset L$. *Then* $\Gamma \vdash^{\sqcap} \varphi$ *whenever* $\Gamma \vdash \varphi$.

Proof. Taking into account that $\mathcal{L}^{\sqcap}$ contains all the rules of $\mathcal{L}$, the proof is a trivial induction on the length of the derivation of φ from Γ within $\mathcal{L}$. ∎

3 Main results

First, assuming that the suitable logic $\mathcal{L}$ is (strongly) sound, we establish the (strong) soundness of $\mathcal{L}^{\sqcap}$. To this end, we need to introduce some notation and prove the relevant technical lemmas, including the fact that $\vDash^{\sqcap}$ extends $\vDash$.

Given a formula ψ of $\mathcal{L}^{\sqcap}$ and a choice sequence ω, we denote by

$$\psi^{\omega}$$

the formula of $\mathcal{L}$ resulting from determining ψ according to ω. This determination map $\psi, \omega \mapsto \psi^{\omega}$ is inductively defined as follows:

- $\lceil c_1 c_2 \rceil^{\omega} = c_{\omega_0}$, for each $c_1, c_2 \in \Sigma_0$;
- $\lceil c_1 c_2 \rceil (\psi_1, \ldots, \psi_n)^{\omega} = c_{\omega_0}(\psi_1^{\omega_{1:\infty}}, \ldots, \psi_n^{\omega_{1+|\psi_1 \ldots \psi_{n-1}|:\infty}})$, for each $n \in \mathbb{N}^+$, $c_1, c_2 \in \Sigma_n$ and $\psi_1, \ldots, \psi_n \in L^{\sqcap}$.

The dual of this notion provides for each given $\varphi \sqsubseteq \psi$ the set

$$\Omega_{\varphi}^{\psi}$$

composed of the choice sequences specified by φ in ψ. This specification map $\varphi, \psi \mapsto \Omega_{\varphi}^{\psi}$ is inductively defined as follows:

- $\Omega_{c_k}^{\lceil c_1 c_2 \rceil} = \{k\omega \mid \omega \in \{1,2\}^{\mathbb{N}}\}$, for each $c_1, c_2 \in \Sigma_0$ and $k = 1, 2$;
- $\Omega_{c_k(\varphi_1, \ldots, \varphi_n)}^{\lceil c_1 c_2 \rceil (\psi_1, \ldots, \psi_n)}$ is the set

$$\{k\omega_{0:|\varphi_1|-1}^1 \ldots \omega_{0:|\varphi_n|-1}^n \omega \mid \omega^j \in \Omega_{\varphi_j}^{\psi_j}, j = 1, \ldots, n \text{ and } \omega \in \{1,2\}^{\mathbb{N}}\},$$

for each
$n \in \mathbb{N}^+$, $c_1, c_2 \in \Sigma_n$, $\psi_1, \ldots, \psi_n \in L^{\sqcap}$, $k = 1, 2$ and $\varphi_1 \sqsubseteq \psi_1, \ldots, \varphi_n \sqsubseteq \psi_n$.

PROPOSITION 2. *Let* $\psi \in L^{\sqcap}$, $\varphi \in L$ *and* $\omega \in \{1,2\}^{\mathbb{N}}$. *Then*

$$\psi^{\omega} \sqsubseteq \psi \quad \text{and} \quad \varphi^{\omega} = \varphi.$$

Proof. The theses are established by straightforward inductions. ∎

PROPOSITION 3. *Let* $\psi \in L^{\sqcap}$, $\omega \in \{1,2\}^{\mathbb{N}}$ *and* $M \in \mathcal{M}$. *Then*

$$[\![\psi^{\omega}]\!]_M = [\![\psi]\!]^{\sqcap}_{M\omega}.$$

Moreover,

$$M \Vdash \psi^{\omega} \text{ iff } M\omega \Vdash^{\sqcap} \psi.$$

Proof. The proof of the first assertion follows by induction on the structure of ψ.

(1) ψ is $\lceil c_1 c_2 \rceil$. Then

$$[\![\lceil c_1 c_2 \rceil^{\omega}]\!]_M = [\![c_{\omega_0}]\!]_M = \underline{c_{\omega_0}} = [\![\lceil c_1 c_2 \rceil]\!]^{\sqcap}_{M\omega}.$$

(2) ψ is $\lceil c_1 c_2 \rceil(\psi_1, \ldots, \psi_n)$. Then

$$\begin{aligned}
[\![\lceil c_1 c_2 \rceil(\psi_1, \ldots, \psi_n)^\omega]\!]_M &= [\![c_{\omega_0}(\psi_1^{\omega_{1:\infty}}, \ldots, \psi_n^{\omega_{1+|\psi_1\ldots\psi_{n-1}|:\infty}})]\!]_M \\
&= \underline{c_{\omega_0}}([\![\psi_1^{\omega_{1:\infty}}]\!]_M, \ldots, [\![\psi_n^{\omega_{1+|\psi_1\ldots\psi_{n-1}|:\infty}}]\!]_M) \\
&= \underline{c_{\omega_0}}([\![\psi_1]\!]^\sqcap_{M\omega_{1:\infty}}, \ldots, [\![\psi_n]\!]^\sqcap_{M\omega_{1+|\psi_1\ldots\psi_{n-1}|:\infty}}) \\
&= [\![\lceil c_1 c_2 \rceil(\psi_1, \ldots, \psi_n)]\!]^\sqcap_{M\omega}.
\end{aligned}$$

With respect to the second assertion,

$$M \Vdash \psi^\omega \text{ iff } [\![\psi^\omega]\!]_M \in D \text{ iff } [\![\psi]\!]^\sqcap_{M\omega} \in D \text{ iff } M\omega \Vdash^\sqcap \psi.$$

∎

PROPOSITION 4. *Let* $\Gamma \cup \{\varphi\} \subset L$. *Then* $\Gamma \models^\sqcap \varphi$ *whenever* $\Gamma \models \varphi$.

Proof. Let $M \in \mathcal{M}$ such that $M \Vdash^\sqcap \Gamma$. We have to show that $M \Vdash^\sqcap \varphi$, that is, $M\omega \Vdash^\sqcap \varphi$ for every choice sequence ω. Indeed, for an arbitrary choice sequence ω, $M\omega \Vdash^\sqcap \Gamma$ and, so, by Proposition 3, $M \Vdash \{\gamma^\omega \mid \gamma \in \Gamma\}$. Thus, by Proposition 2, $M \Vdash \Gamma$ and so, since $\Gamma \models \varphi$, $M \Vdash \varphi$. Therefore, by Proposition 2, $M \Vdash \varphi^\omega$ and so, by Proposition 3, $M\omega \Vdash^\sqcap \varphi$. ∎

PROPOSITION 5. *A sound rule in* $\mathcal{L}$ *is sound in* $\mathcal{L}^\sqcap$.

Proof. Consider a sound rule in Δ with premises $\alpha_1, \ldots, \alpha_m$ and conclusion β. Then, $\alpha_1, \ldots, \alpha_m \models \beta$ and, so, by Propostion 4, $\alpha_1, \ldots, \alpha_m \models^\sqcap \beta$. ∎

PROPOSITION 6. *The lifting rule is sound in* $\mathcal{L}^\sqcap$.

Proof. We must show that

$$\{\varphi \mid \varphi \sqsubseteq \psi\} \models^\sqcap \psi.$$

Let $M \in \mathcal{M}$. Assume that $M \Vdash^\sqcap \{\varphi \mid \varphi \sqsubseteq \psi\}$. Let $\omega \in \{1,2\}^\mathbb{N}$ be a choice sequence. Then $M \Vdash^\sqcap \psi^\omega$ since $\psi^\omega \sqsubseteq \psi$ by Proposition 2. Hence, $M\omega' \Vdash^\sqcap \psi^\omega$ for every $\omega' \in \{1,2\}^\mathbb{N}$. Therefore, by Proposition 3, $M \Vdash (\psi^\omega)^{\omega'}$ for every $\omega' \in \{1,2\}^\mathbb{N}$. Thus, by Proposition 2, $M \Vdash \psi^\omega$. So, by Proposition 3, $M\omega \Vdash^\sqcap \psi$. ∎

PROPOSITION 7. *Let* $\psi \in L^\sqcap$, $\varphi \in L$ *and* $\omega, \omega' \in \{1,2\}^\mathbb{N}$. *Assume that* $\varphi \sqsubseteq \psi$, $\omega \in \Omega^\psi_\varphi$ *and*

$$\omega_{0:|\varphi|-1} = \omega'_{0:|\varphi|-1}.$$

Then $\omega' \in \Omega^\psi_\varphi$.

Proof. The proof follows by case analysis.

(1) ψ is $\lceil c_1 c_2 \rceil$. Then $|\psi| = 1$. Since $\omega \in \Omega^\psi_\varphi$ we conclude that $\varphi = c_{\omega_0}$. On the other hand, $\omega'_0 = \omega_0$ and so $\varphi = c_{\omega'_0}$. Thus, $\omega' \in \Omega^\psi_\varphi$.

(2) ψ is $\lceil c_1 c_2 \rceil(\psi_1, \ldots, \psi_n)$. Then $|\lceil c_1 c_2 \rceil(\psi_1, \ldots, \psi_n)| = 1 + |\psi_1| + \cdots + |\psi_n|$. Assume that φ is $c_k(\varphi_1, \ldots, \varphi_n)$ where $\varphi_j \sqsubseteq \psi_j$ for $j = 1, \ldots, n$. Since $\omega \in \Omega^\psi_\varphi$ we conclude that ω is of the form $k\omega^1_{0:|\varphi_1|-1} \cdots \omega^n_{0:|\varphi_n|-1}\omega''$ where $\omega^j \in \Omega^{\psi_j}_{\varphi_j}$ for $j = 1, \ldots, n$ and $\omega'' \in \{1,2\}^\mathbb{N}$. Therefore $\omega'_{0:|\varphi|-1} = k\omega^1_{0:|\varphi_1|-1} \cdots \omega^n_{0:|\varphi_n|-1}$ and so $\omega' \in \Omega^\psi_\varphi$. ∎

PROPOSITION 8. *Let $\psi \in L^\sqcap$ and $\varphi \in L$ be such that $\varphi \sqsubseteq \psi$. Then*

$$\psi^\omega = \varphi$$

for every $\omega \in \Omega_\varphi^\psi$.

Proof. The proof follows by induction on ψ.

(1) ψ is $\lceil c_1 c_2 \rceil$. Then φ is c_k for some $k = 1, 2$. Let $\omega \in \Omega_\varphi^\psi$. Hence $\omega_0 = k$. Moreover, by definition, $\psi^\omega = c_{\omega_0}$. Thus, $\psi^\omega = c_k = \varphi$.

(2) ψ is $\lceil c_1 c_2 \rceil (\psi_1, \ldots, \psi_n)$. Let φ be $c_k(\varphi_1, \ldots, \varphi_n)$ where $k \in \{1, 2\}$ and $\varphi_j \sqsubseteq \psi_j$ for $j = 1, \ldots, n$. Let $\omega \in \Omega_\varphi^\psi$. Then ω is of the form $k\omega_{0:|\varphi_1|-1}^1 \cdots \omega_{0:|\varphi_n|-1}^n \omega'$ where $\omega^j \in \Omega_{\varphi_j}^{\psi_j}$ for $j = 1, \ldots, n$ and $\omega' \in \{1, 2\}^\mathbb{N}$. Therefore, by the induction hypothesis, $\psi_j^{\omega^j} = \varphi_j$ for $j = 1, \ldots, n$. The thesis follows since ψ^ω is

$$c_k(\psi_1^{\omega_{0:|\varphi_1|-1}^1 \cdots \omega_{0:|\varphi_n|-1}^n \omega'}, \ldots, \psi_n^{\omega_{0:|\varphi_n|-1}^n \omega'})$$

and $\psi_j^{\omega_{0:|\varphi_j|-1}^j \cdots \omega_{0:|\varphi_n|-1}^n \omega'}$ is $\psi_j^{\omega^j} = \varphi_j$ for $j = 1, \ldots, n$. ∎

PROPOSITION 9. *The co-lifting rule is sound in $\mathcal{L}^\sqcap$.*

Proof. We must prove that

$$\psi \vDash^\sqcap \varphi$$

for every $\varphi \sqsubseteq \psi$. Let $M \in \mathcal{M}$. Assume that $M \Vdash^\sqcap \psi$. Let φ be such that $\varphi \sqsubseteq \psi$ and $\omega \in \{1, 2\}^\mathbb{N}$. Consider $\omega' \in \Omega_\varphi^\psi$. Then, $M\omega' \Vdash^\sqcap \psi$ since $M \Vdash^\sqcap \psi$. So, by Proposition 3, $M \Vdash \psi^{\omega'}$. Hence, by Proposition 8, $M \Vdash \varphi$. Using Proposition 2, $M \Vdash \varphi^\omega$ and so, again by Proposition 3, $M\omega \Vdash^\sqcap \varphi$. ∎

THEOREM 10 (Soundness).
If $\mathcal{L}$ is sound then $\mathcal{L}^\sqcap$ is sound.

Proof. Assume that $\mathcal{L}$ is sound. Then, in particular, the rules in Δ are sound in $\mathcal{L}$ and so, by Proposition 5, all the rules in Δ are sound in $\mathcal{L}^\sqcap$. Moreover, the rules LFT and cLFT are sound thanks to Proposition 6 and Proposition 9, respectively. ∎

With the technical lemmas established towards the soundness result, it is also possible to show that, as envisaged, the combined connectives $\lceil c_1 c_2 \rceil$ and $\lceil c_2 c_1 \rceil$ are not distinguished at all by the proposed non-deterministic semantics, a major improvement over the limited interchangeability result obtained in [Sernadas et al., 2011]. It is in addition possible to prove that, as required, $\mathcal{L}^\sqcap$ is a conservative extension of $\mathcal{L}$.

THEOREM 11 (Model-theoretic interchangeability).
Let $\psi, \psi' \in L^\sqcap$ be such that ψ' is obtained from ψ by replacing zero or more occurrences of any combined connective $\lceil c_1 c_2 \rceil$ by $\lceil c_2 c_1 \rceil$. Then

$$\psi \vDash^\sqcap \psi'.$$

Proof. Let $M \in \mathcal{M}$. Assume that $M \Vdash^\sqcap \psi$. Let $\omega \in \{1,2\}^\mathbb{N}$ and $\omega' \in \Omega^\psi_{\psi'\omega}$. Then $M\omega' \Vdash^\sqcap \psi$ and so, by Proposition 3, $M \Vdash \psi^{\omega'}$. Hence, by Proposition 8, $M \Vdash \psi'^\omega$. Again, by Proposition 3, $M\omega \Vdash^\sqcap \psi'$. ∎

THEOREM 12 (Model-theoretic conservativeness).
For every $\Gamma \cup \{\varphi\} \subset L$,

$$\Gamma \vDash \varphi \text{ whenever } \Gamma \vDash^\sqcap \varphi.$$

Proof. Assume that $\Gamma \vDash^\sqcap \varphi$ and let $M \in \mathcal{M}$ such that $M \Vdash \Gamma$. Let $\omega \in \{1,2\}^\mathbb{N}$. Then, by Proposition 2, $M \Vdash \{\gamma^\omega \mid \gamma \in \Gamma\}$. Therefore, by Proposition 3, $M\omega \Vdash^\sqcap \Gamma$ and so $M \Vdash^\sqcap \Gamma$. Thus, by hypothesis, $M \Vdash^\sqcap \varphi$. Let $\omega' \in \{1,2\}^\mathbb{N}$. Then $M\omega' \Vdash^\sqcap \varphi$ and, by Proposition 3, $M \Vdash \varphi^{\omega'}$. The thesis follows by Proposition 2. ∎

As an immediate corollary of the conservativeness of the proposed enrichment of $\mathcal{L}$, we obtain the following result on the consistency of $\mathcal{L}^\sqcap$.

THEOREM 13 (Model-theoretic consistency).
If there is $\varphi \in L$ *such that* $\nvDash \varphi$, *then there is* $\psi \in L^\sqcap$ *such that* $\nvDash^\sqcap \psi$.

Proof. Thanks to Theorem 12, just take ψ to be such φ. ∎

Observe that if $\mathcal{L}$ is sound and complete, the preservation of soundness allows us to carry over the conservativeness and the consistency results to the proof-theoretic level.

THEOREM 14 (Proof-theoretic conservativeness).
Assume that $\mathcal{L}$ *is sound and complete. Then, for every* $\Gamma \cup \{\varphi\} \subset L$,

$$\Gamma \vdash \varphi \text{ whenever } \Gamma \vdash^\sqcap \varphi.$$

THEOREM 15 (Proof-theoretic consistency).
Assuming that $\mathcal{L}$ *is sound and complete, if there is* $\varphi \in L$ *such that* $\nvdash \varphi$, *then there is* $\psi \in L^\sqcap$ *such that* $\nvdash^\sqcap \psi$.

Finally, concerning completeness, we show that if $\mathcal{L}$ is (strongly) complete then so is $\mathcal{L}^\sqcap$. To this end, we need some technical lemmas.

PROPOSITION 16. *Assume that* $\mathcal{L}$ *is complete. Then,*

$$\Gamma \vdash^\sqcap \varphi \text{ whenever } \Gamma \vDash^\sqcap \varphi \quad \text{for every } \Gamma \cup \{\varphi\} \subset L.$$

Proof. Assume that $\Gamma \vDash^\sqcap \varphi$. We start by showing that $\Gamma \vDash \varphi$. Let $M \in \mathcal{M}$. Assume that $M \Vdash \Gamma$. By Proposition 2,

$$\Gamma = \{\gamma^\omega \mid \gamma \in \Gamma, \omega \in \{1,2\}^\mathbb{N}\}.$$

Hence, $M \Vdash \gamma^\omega$ for $\gamma \in \Gamma$ and $\omega \in \{1,2\}^\mathbb{N}$. Thus, by Proposition 3,

$$M\omega \Vdash^\sqcap \Gamma$$

for every $\omega \in \{1,2\}^\mathbb{N}$. Therefore, by hypothesis, $M\omega \Vdash^\sqcap \varphi$. Again by Proposition 3, $M \Vdash \varphi^\omega$ and so, by Proposition 2, $M \Vdash \varphi$. Then, by completeness of $\mathcal{L}$, $\Gamma \vdash \varphi$. Thus, by Proposition 1 $\Gamma \vdash^\sqcap \varphi$. ∎

PROPOSITION 17. *Assume that*

$$\Gamma' \vdash^\sqcap \varphi' \text{ whenever } \Gamma' \vDash^\sqcap \varphi' \quad \text{for every } \Gamma' \cup \{\varphi'\} \subset L.$$

Then,

$$\Gamma \vdash^\sqcap \psi \text{ whenever } \Gamma \vDash^\sqcap \psi \quad \text{for every } \Gamma \subset L \text{ and } \psi \in L^\sqcap.$$

Proof. Assume that $\Gamma \not\vdash^\sqcap \psi$. Then, taking into account the lifting rule, $\Gamma \not\vdash^\sqcap \delta$ for some $\delta \sqsubseteq \psi$. Observe that $\Gamma \cup \{\delta\} \subset L$ and so, by the hypothesis, $\Gamma \not\vDash^\sqcap \delta$. Let $M \in \mathcal{M}$ be such that

$$M \Vdash^\sqcap \Gamma \text{ and } M\omega \not\Vdash^\sqcap \delta \text{ for some } \omega \in \{1,2\}^{\mathbb{N}}.$$

Then, by Proposition 3, $M \not\Vdash \delta^\omega$ and so $M \not\Vdash \delta$ by Proposition 2. Let $\omega' \in \Omega_\delta^\psi$. Then, by Proposition 8, $\psi^{\omega'} = \delta$. Therefore, $M \not\Vdash \psi^{\omega'}$ and so, by Proposition 3, $M\omega' \not\Vdash^\sqcap \psi$. Thus, $M \not\Vdash^\sqcap \psi$. ∎

PROPOSITION 18. *Assume that*

$$\Gamma \vdash^\sqcap \psi \text{ whenever } \Gamma \vDash^\sqcap \psi \quad \text{for every } \Gamma \subset L \text{ and } \psi \in L^\sqcap.$$

Then,

$$\Theta \vdash^\sqcap \psi \text{ whenever } \Theta \vDash^\sqcap \psi \quad \text{for every } \Theta \cup \{\psi\} \in L^\sqcap.$$

Proof. Assume that $\Theta \cup \{\psi\} \in L^\sqcap$ and $\Theta \not\vdash^\sqcap \psi$. Then, by the co-lifting rule,

$$\{\gamma \mid \gamma \sqsubseteq \theta, \theta \in \Theta\} \not\vdash^\sqcap \psi.$$

Therefore, by hypothesis,

$$\{\gamma \mid \gamma \sqsubseteq \theta, \theta \in \Theta\} \not\vDash^\sqcap \psi.$$

Let $M \in \mathcal{M}$ be such that $M \Vdash^\sqcap \gamma$ for each $\gamma \sqsubseteq \theta$, $\theta \in \Theta$ and $M \not\Vdash^\sqcap \psi$. By soundness of the LFT rule, $M \Vdash^\sqcap \Theta$ and so $\Theta \not\vDash^\sqcap \psi$. ∎

THEOREM 19 (Completeness).
If $\mathcal{L}$ is complete then $\mathcal{L}^\sqcap$ is complete.

Proof. Assume that $\mathcal{L}$ is complete. Then, by Proposition 16,

$$\text{if } \Gamma \vDash^\sqcap \varphi \text{ then } \Gamma \vdash^\sqcap \varphi$$

for every $\Gamma \cup \{\varphi\} \subset L$. Hence, for every $\Gamma \subset L$ and $\psi \in L^\sqcap$,

$$\text{if } \Gamma \vDash^\sqcap \psi \text{ then } \Gamma \vdash^\sqcap \psi$$

using Proposition 17. Thus, thanks to Proposition 18,

$$\text{if } \Theta \vDash^\sqcap \psi \text{ then } \Theta \vdash^\sqcap \psi$$

for every $\Theta \cup \{\psi\} \in L^\sqcap$. ∎

As an immediate corollary of the preservation of completeness, we are able to carry the interchangeability result (Theorem 11) to the proof-theoretic level when enriching a complete logic.

THEOREM 20 (Proof-theoretic interchangeability).
Let $\psi, \psi' \in L^{\sqcap}$ be such that ψ' is obtained from ψ by replacing zero or more occurrences of any combined connective $\lceil c_1 c_2 \rceil$ by $\lceil c_2 c_1 \rceil$. Then, assuming that $\mathcal{L}$ is complete,
$$\psi \vdash^{\sqcap} \psi'.$$

4 The case of classical propositional logic

For illustrating the proposed calculus and non-deterministic semantics for meet-combined constructors, we choose classical propositional logic (CPL). In fact, $\mathsf{CPL}^{\sqcap}$ is sufficiently rich for providing interesting examples, as well as for showing significant differences between its non-deterministic combination of constructors and the product combination in $\mathsf{CPL}^{\times}$ (as defined in [Sernadas et al., 2011]).

We assume that the CPL signature contains the propositional symbols (q_i for $i \in \mathbb{N}$), negation, conjunction, disjunction, implication and equivalence. Moreover, we assume that the CPL calculus includes the tautologies as axioms plus modus ponens (MP). Finally, we assume that the CPL semantics is composed of the matrices induced by valuations. Recall that each valuation $v : \{q_i \mid i \in \mathbb{N}\} \to \{0,1\}$ canonically induces a matrix M_v with $A_v = \{0,1\}$, providing the denotation of the propositional symbols imposed by v. Clearly, M_v satisfies precisely the same formulas as v.

Observe that $\mathsf{CPL}^{\sqcap}$ is sound and complete, thanks to the results of the previous section, given the soundness and completeness of CPL as presented above. Furthermore, $\mathsf{CPL}^{\sqcap}$ is a conservative extension of CPL.

We start by looking at the combination of conjunction and disjunction, providing an example of a common property of those connectives and showing that, as required, the property is carried over to their non-deterministic combination in $\mathcal{L}^{\times}$. We also compare it with their product combination. Afterward, we examine the impact of non-determinism on implication and equivalence, concluding that only a mitigated form of the metatheorem of substitution of equivalents holds in $\mathsf{CPL}^{\sqcap}$, contrarily to what happens in $\mathsf{CPL}^{\times}$ where this metatheorem holds in full form. Finally, we establish that the metatheorem of deduction still holds in $\mathsf{CPL}^{\sqcap}$ with provisos similar to those in $\mathsf{CPL}^{\times}$.

Combining conjunction and disjunction

Commutativity, for instance, is a common property of conjunction and disjunction. Thus, we should be able to derive it for $\lceil \wedge \vee \rceil$ and, so, thanks to interchangeability and completeness, also for $\lceil \vee \wedge \rceil$. Indeed, we can build the derivation in Figure 1 for
$$q_1 \lceil \wedge \vee \rceil q_2 \vdash^{\sqcap}_{\mathsf{CPL}} q_2 \lceil \wedge \vee \rceil q_1.$$

In fact, it is straightforward to provide a derivation for
$$\psi_1 \lceil \wedge \vee \rceil \psi_2 \vdash^{\sqcap}_{\mathsf{CPL}} \psi_2 \lceil \wedge \vee \rceil \psi_1$$

1	$q_1 \lceil \wedge \vee \rceil q_2$	HYP
2	$q_1 \wedge q_2$	cLFT : 1
3	$q_1 \vee q_2$	cLFT : 1
4	$(q_1 \wedge q_2) \supset (q_2 \wedge q_1)$	TAUT
5	$(q_1 \vee q_2) \supset (q_2 \vee q_1)$	TAUT
6	$q_2 \wedge q_1$	MP : 2, 4
7	$q_2 \vee q_1$	MP : 3, 5
8	$q_2 \lceil \wedge \vee \rceil q_1$	LFT : 6, 7

Figure 1. Derivation of commutativity of $\lceil \wedge \vee \rceil$ in $\mathsf{CPL}^\sqcap$.

given arbitrary $\psi_1, \psi_2 \in L^\sqcap_{\mathsf{CPL}}$. For an example of a derivation of this type, involving determinations of arbitrary formulas, see Figure 2.

This desideratum was already achieved for the product combination of conjunction and disjunction in [Sernadas *et al.*, 2011]. But therein we had two different but related combinations ($\lceil \wedge \vee \rceil$ and $\lceil \vee \wedge \rceil$), while herein the non-deterministic semantics collapses them, thanks to Theorem 11:

$$\begin{cases} q_1 \lceil \wedge \vee \rceil q_2 \vdash^\sqcap_{\mathsf{CPL}} q_1 \lceil \vee \wedge \rceil q_2 \\ q_1 \lceil \vee \wedge \rceil q_2 \vdash^\sqcap_{\mathsf{CPL}} q_1 \lceil \wedge \vee \rceil q_2. \end{cases}$$

At this point, the reader will ask whether we are able to establish

$$\vdash^\sqcap_{\mathsf{CPL}} (q_1 \lceil \wedge \vee \rceil q_2) \equiv (q_1 \lceil \vee \wedge \rceil q_2).$$

The answer is negative, not because of any limitation on the collapsing of the two combined connectives, but for a deeper reason: implication does not behave as might be expected in the presence of non-determinism, as we proceed to analyze.

Impact of non-determinism

Observe that $\psi \supset \psi$ is not always valid in $\mathsf{CPL}^\sqcap$. Clearly, if $\varphi \in L_{\mathsf{CPL}}$, then $\varphi \supset \varphi$ is valid in $\mathsf{CPL}^\sqcap$. However, if non-deterministic constructors are present in the formula ψ, then $\psi \supset \psi$ is not necessarily valid. For instance, the formula

$$(\mathsf{ff} \lceil \wedge \vee \rceil \mathsf{tt}) \supset (\mathsf{ff} \lceil \wedge \vee \rceil \mathsf{tt})$$

is not valid. Indeed, this formula is not satisfied by an interpretation (M, ω) such that $\omega_1 = 2$ and $\omega_4 = 1$, since such a choice sequence ensures that the $\lceil \wedge \vee \rceil$ on the left is evaluated as $\vee$ and the $\lceil \wedge \vee \rceil$ on the right as $\wedge$.

Therefore, the presence of non-deterministic constructors in a formula also disturbs the closure for instantiation, even for validity. In particular, the result of instantiating a tautology with non-deterministic formulas is not necessarily valid. For instance,

$$q_1 \supset q_1$$

1	ψ_1	HYP	
2	$\psi_1 \supset \psi_2$	HYP	
3_{φ_1}	φ_1	cLFT : 1	for each $\varphi_1 \sqsubseteq \psi_1$
$4_{\varphi_1 \varphi_2}$	$\varphi_1 \supset \varphi_2$	cLFT : 2	for each $\varphi_1 \sqsubseteq \psi_1, \varphi_2 \sqsubseteq \psi_2$
5_{φ_2}	φ_2	MP : $3_{\varphi_1}, 4_{\varphi_1\varphi_2}$	for each $\varphi_2 \sqsubseteq \psi_2$
6	ψ_2	LFT : 5	

Figure 2. Derivation of MP in $\mathsf{CPL}^{\sqcap}$.

is valid in $\mathsf{CPL}^{\sqcap}$ while some of its instances (like the one above) are not. As another example, the instance

$$\lceil \mathsf{ttff} \rceil \supset (q_2 \supset \lceil \mathsf{ttff} \rceil)$$

of $q_1 \supset (q_2 \supset q_1)$ is not valid. On the other hand, for example, every (even non-deterministic) instance of

$$q_1 \supset \mathsf{tt}$$

is valid in $\mathsf{CPL}^{\sqcap}$.

However, the presence of non-deterministic constructors does not affect modus ponens. Indeed, as depicted in Figure 2, MP still holds in general within $\mathsf{CPL}^{\sqcap}$:

$$\psi_1, \psi_1 \supset \psi_2 \vdash^{\sqcap}_{\mathsf{CPL}} \psi_2.$$

Given the impact of non-determinism on implication and, thus, also on equivalence, one should not expect to be able to prove the full version of the metatheorem of of substitution of equivalents in $\mathsf{CPL}^{\sqcap}$. In fact, we are only able to prove the following weak version (substitution only up to interderivability):

THEOREM 21 (Metatheorem of substitution of equivalents in $\mathsf{CPL}^{\sqcap}$).
Let $\psi, \psi', \theta, \theta' \in L^{\sqcap}_{\mathsf{CPL}}$ be such that:

- $\vdash^{\sqcap}_{\mathsf{CPL}} \psi \equiv \psi'$;

- θ' is obtained from θ by replacing zero or more occurrences of ψ by ψ'.

Then:
$$\theta \vdash^{\sqcap}_{\mathsf{CPL}} \theta'.$$

Proof. Assume that $\vdash^{\sqcap}_{\mathsf{CPL}} \psi \equiv \psi'$. The proof follows by induction on the structure of θ.

Base: θ is a propositional symbol. Consider two cases:
(1) θ is ψ. Consider two subcases: (a) θ' is ψ'. Then $\vdash^{\sqcap}_{\mathsf{CPL}} \theta \equiv \theta'$ by hypothesis. Hence $\theta \vdash^{\sqcap}_{\mathsf{CPL}} \theta'$ by MP; (b) θ' is θ. Then $\theta \vdash^{\sqcap}_{\mathsf{CPL}} \theta'$.
(2) θ is not ψ. Then ψ does not occur in θ and, so, θ' is θ. Thus, $\theta \vdash^{\sqcap}_{\mathsf{CPL}} \theta'$.

Step: Let θ be $\lceil c_1 c_2 \rceil(\theta_1, \ldots, \theta_n)$. Consider two cases:

(1) θ is ψ. Consider two subcases: (a) θ' is ψ'. Then $\vdash_{\mathsf{CPL}}^{\sqcap} \theta \equiv \theta'$ by hypothesis. Hence $\theta \vdash_{\mathsf{CPL}}^{\sqcap} \theta'$ by MP; (b) θ' is θ. Then $\theta \vdash_{\mathsf{CPL}}^{\sqcap} \theta'$.

(2) θ is not ψ. Then θ' is $\lceil c_1 c_2 \rceil(\theta_1', \ldots, \theta_n')$ where, for $j = 1, \ldots, n$, the formula θ_j' is obtained from θ_j by replacing zero or more occurrences of ψ by ψ'. Assume without loss of generality that n is 1. Observe that,

$$\vdash_{\mathsf{CPL}}^{\sqcap} \theta_1 \equiv \theta_1'$$

by the induction hypothesis. Therefore, by co-lifting,

$$\vdash_{\mathsf{CPL}}^{\sqcap} \varphi_1 \equiv \varphi_1' \quad \text{for every } \varphi_1 \sqsubseteq \theta_1 \text{ and } \varphi_1' \sqsubseteq \theta_1'.$$

Hence, by the conservativeness of $\mathsf{CPL}^{\sqcap}$, see Theorem 12,

$$\vdash_{\mathsf{CPL}} \varphi_1 \equiv \varphi_1' \quad \text{for every } \varphi_1 \sqsubseteq \theta_1 \text{ and } \varphi_1' \sqsubseteq \theta_1'.$$

Thus,

$$\vdash_{\mathsf{CPL}} c_k(\varphi_1) \equiv c_k(\varphi_1') \quad \text{for every } \varphi_1 \sqsubseteq \theta_1, \varphi_1' \sqsubseteq \theta_1' \text{ and } k = 1, 2$$

since the metatheorem of substitution of equivalents holds in CPL. Moreover,

$$\vdash_{\mathsf{CPL}}^{\sqcap} c_k(\varphi_1) \equiv c_k(\varphi_1') \quad \text{for every } \varphi_1 \sqsubseteq \theta_1, \varphi_1' \sqsubseteq \theta_1' \text{ and } k = 1, 2$$

since $\mathsf{CPL}^{\sqcap}$ is an extension of CPL. Hence,

$$c_k(\varphi_1) \vdash_{\mathsf{CPL}}^{\sqcap} c_k(\varphi_1') \quad \text{for every } \varphi_1 \sqsubseteq \theta_1, \varphi_1' \sqsubseteq \theta_1' \text{ and } k = 1, 2$$

by MP, and so

$$\{c_k(\varphi_1) \mid \varphi_1 \sqsubseteq \theta_1, k = 1, 2\} \vdash_{\mathsf{CPL}}^{\sqcap} \{c_k(\varphi_1') \mid \varphi_1' \sqsubseteq \theta_1', k = 1, 2\}.$$

Therefore, $\theta \vdash_{\mathsf{CPL}}^{\sqcap} \theta'$. Indeed, θ is $\lceil c_1 c_2 \rceil(\theta_1)$ and, by co-lifting,

$$\lceil c_1 c_2 \rceil(\theta_1) \vdash_{\mathsf{CPL}}^{\sqcap} \{c_k(\varphi_1) \mid \varphi_1 \sqsubseteq \theta_1, k = 1, 2\}.$$

Furthermore, by lifting,

$$\{c_k(\varphi_1') \mid \varphi_1' \sqsubseteq \theta_1', k = 1, 2\} \vdash_{\mathsf{CPL}}^{\sqcap} \lceil c_1 c_2 \rceil(\theta_1')$$

and, $\lceil c_1 c_2 \rceil(\theta_1')$ is θ'. ∎

Note that, under the assumptions of Theorem 21,

$$\vdash_{\mathsf{CPL}}^{\sqcap} \theta \equiv \theta'$$

does not hold in general. For instance we have

$$\begin{cases} \vdash_{\mathsf{CPL}}^{\sqcap} \mathrm{ff} \equiv \mathrm{ff} \\ \vdash_{\mathsf{CPL}}^{\sqcap} \mathrm{tt} \equiv \mathrm{tt} \end{cases}$$

but
$$\nvdash_{\mathsf{CPL}}^{\sqcap} (\mathsf{ff}\lceil \wedge \vee \rceil \mathsf{tt}) \equiv (\mathsf{ff}\lceil \wedge \vee \rceil \mathsf{tt})$$
because
$$\nvdash_{\mathsf{CPL}}^{\sqcap} (\mathsf{ff}\lceil \wedge \vee \rceil \mathsf{tt}) \supset (\mathsf{ff}\lceil \wedge \vee \rceil \mathsf{tt})$$
as we saw above at the semantic level.

Thus, concerning substitution of equivalents, $\mathsf{CPL}^{\times}$ and $\mathsf{CPL}^{\sqcap}$ are quite different. On the other hand, concerning the metatheorem of deduction, the two logics are alike, as we proceed to establish.

Metatheorem of deduction

Towards the metatheorem of deduction in $\mathsf{CPL}^{\sqcap}$ we need the following auxiliary result and to borrow from [Sernadas et al., 2011] the notion of essential co-lifting.

PROPOSITION 22. *Let* $\psi \in L_{\mathsf{CPL}}^{\sqcap}$. *Then*
$$\vdash_{\mathsf{CPL}}^{\sqcap} (\bigwedge_{\varphi \sqsubseteq \psi} \varphi) \supset \psi.$$

Proof. We prove the result semantically making use of the soundness and completeness of CPL and $\mathsf{CPL}^{\sqcap}$. Assume that
$$M_v\, \omega \Vdash^{\sqcap} \bigwedge_{\varphi \sqsubseteq \psi} \varphi.$$

Then, by Proposition 3,
$$M_v \Vdash \varphi^{\omega} \quad \text{for each } \varphi \sqsubseteq \psi.$$

Hence, by Proposition 2,
$$M_v \Vdash \varphi \quad \text{for each } \varphi \sqsubseteq \psi,$$

and so, by the same proposition,
$$M_v \Vdash \varphi^{\omega'} \quad \text{for each } \omega' \in \{1,2\}^{\mathbb{N}} \text{ and } \varphi \sqsubseteq \psi.$$

Thus, by Proposition 3,
$$M_v\, \omega' \Vdash^{\sqcap} \varphi \quad \text{for each } \omega' \in \{1,2\}^{\mathbb{N}} \text{ and } \varphi \sqsubseteq \psi,$$

that is,
$$M_v \Vdash^{\sqcap} \varphi \quad \text{for each } \varphi \sqsubseteq \psi.$$

Then
$$M_v \Vdash^{\sqcap} \psi$$
by Proposition 6, and, so, in particular $M_v\, \omega \Vdash^{\sqcap} \psi$. ∎

Let $\psi_1 \ldots \psi_n$ be a derivation for $\Theta \vdash_{\mathsf{CPL}}^{\sqcap} \psi$. Formula ψ_i *depends* on $\theta \in \Theta$ in this derivation if

- either ψ_i is θ;

- or ψ_i is obtained using a rule (either MP or LFT or cLFT) with at least one of the premises depending on θ.

Formula ψ_i is an *essential co-lifting* over a dependent of θ in the derivation $\psi_1 \ldots \psi_n$ if

- ψ_i is obtained using cLFT from ψ_j, that is, $\psi_i \sqsubseteq \psi_j$;
- ψ_j depends on θ;
- θ is in $L_{\mathsf{CPL}}^{\square} \setminus L_{\mathsf{CPL}}$.

THEOREM 23 (Metatheorem of deduction in $\mathsf{CPL}^{\square}$).
Let $\Theta \cup \{\eta, \psi\} \subseteq L_{\mathsf{CPL}}^{\square}$ and $\psi_1 \ldots \psi_n$ be a derivation for

$$\Theta, \eta \vdash_{\mathsf{CPL}}^{\square} \psi$$

without essential co-liftings over dependents of η. Then

$$\Theta \vdash_{\mathsf{CPL}}^{\square} \eta \supset \psi.$$

Proof. The proof is carried out by induction on the length of the given derivation.

Base. There are three cases to consider:
(a) ψ is an axiom of $\mathsf{CPL}^{\square}$. Then ψ is a tautology of CPL. Hence $\vdash_{\mathsf{CPL}} \psi$. Observe that $\vdash_{\mathsf{CPL}} \psi \supset (\delta \supset \psi)$, in particular, for each $\delta \sqsubseteq \eta$. Thus, for each $\delta \sqsubseteq \eta$, $\vdash_{\mathsf{CPL}} \delta \supset \psi$, and, so, $\Theta \vdash_{\mathsf{CPL}}^{\square} \delta \supset \psi$. Therefore, by LFT, $\Theta \vdash_{\mathsf{CPL}}^{\square} \eta \supset \psi$.
(b) ψ is in Θ. Then $\Theta \vdash_{\mathsf{CPL}}^{\square} \psi$ and so, by cLFT, $\Theta \vdash_{\mathsf{CPL}}^{\square} \varphi$ for each $\varphi \sqsubseteq \psi$. Recall that $\Theta \vdash_{\mathsf{CPL}}^{\square} \varphi \supset (\delta \supset \varphi)$ for each $\delta \sqsubseteq \eta$ and $\varphi \sqsubseteq \psi$. Thus, for each $\delta \sqsubseteq \eta$ and $\varphi \sqsubseteq \psi$, $\Theta \vdash_{\mathsf{CPL}}^{\square} \delta \supset \varphi$. Therefore, by LFT, $\Theta \vdash_{\mathsf{CPL}}^{\square} \eta \supset \psi$.
(c) ψ is η. Observe that, for each $\delta \sqsubseteq \eta$, $\vdash_{\mathsf{CPL}} \delta \supset \delta$ and, so, $\Theta \vdash_{\mathsf{CPL}}^{\square} \delta \supset \delta$. Hence, by LFT, $\Theta \vdash_{\mathsf{CPL}}^{\square} \eta \supset \psi$.

Step. There are three additional cases to consider:
(a) ψ is obtained by MP from ψ_{i_1} and ψ_{i_2} with ψ_{i_2} being $\psi_{i_1} \supset \psi$ and $\psi_{i_1}, \psi \in L_{\mathsf{CPL}}$. By the induction hypothesis, $\Theta \vdash_{\mathsf{CPL}}^{\square} \eta \supset \psi_{i_1}$ and $\Theta \vdash_{\mathsf{CPL}}^{\square} \eta \supset (\psi_{i_1} \supset \psi)$. Therefore, by cLFT, $\Theta \vdash_{\mathsf{CPL}}^{\square} \delta \supset \psi_{i_1}$ and $\Theta \vdash_{\mathsf{CPL}}^{\square} \delta \supset (\psi_{i_1} \supset \psi)$ for every $\delta \sqsubseteq \eta$. Thus,

$$\Theta \vdash_{\mathsf{CPL}}^{\square} \delta \supset \psi, \text{ for every } \delta \sqsubseteq \eta.$$

using a tautology and MP. Finally, by LFT, $\Theta \vdash_{\mathsf{CPL}}^{\square} \eta \supset \psi$.
(b) ψ is obtained by LFT from $\psi_{i_1}, \ldots, \psi_{i_m}$. Then $\{\psi_{i_j} \mid j = 1, \ldots, m\} = \{\varphi \mid \varphi \sqsubseteq \psi\}$ and so, for every $\varphi \sqsubseteq \psi$, there is a derivation with length less than n and without essential co-liftings over dependents of η, for $\Theta, \eta \vdash_{\mathsf{CPL}}^{\square} \varphi$. Hence, by the induction hypothesis, $\Theta \vdash_{\mathsf{CPL}}^{\square} \eta \supset \varphi$ for every $\varphi \sqsubseteq \psi$. Therefore, by cLFT, $\Theta \vdash_{\mathsf{CPL}}^{\square} \delta \supset \varphi$ for every $\delta \sqsubseteq \eta$ and $\varphi \sqsubseteq \psi$. Thus,

$$\Theta \vdash_{\mathsf{CPL}}^{\square} \delta \supset (\bigwedge_{\varphi \sqsubseteq \psi} \varphi), \text{ for every } \delta \sqsubseteq \eta.$$

Hence, taking into account Proposition 22,

$$\Theta \vdash_{\mathsf{CPL}}^{\sqcap} \delta \supset \psi, \text{ for every } \delta \sqsubseteq \eta.$$

Finally, by cLFT and LFT, $\Theta \vdash_{\mathsf{CPL}}^{\sqcap} \eta \supset \psi$.
(c) ψ is obtained by cLFT from ψ_j. Then $\psi \sqsubseteq \psi_j$ and there is a derivation with length less than n and without essential co-liftings over dependents of η, for $\Theta, \eta \vdash^n \psi_j$. Since the co-lifting is non essential there are two cases to consider:
(i) ψ does not depend on η. Then $\Theta \vdash_{\mathsf{CPL}}^{\sqcap} \psi$. Hence, by cLFT, $\Theta \vdash_{\mathsf{CPL}}^{\sqcap} \varphi$ for every $\varphi \sqsubseteq \psi$. Thus, $\Theta \vdash_{\mathsf{CPL}}^{\sqcap} \varphi \supset (\delta \supset \varphi)$ for every $\delta \sqsubseteq \eta$ and $\varphi \sqsubseteq \psi$. So, by MP, $\Theta \vdash_{\mathsf{CPL}}^{\sqcap} \delta \supset \varphi$ for every $\delta \sqsubseteq \eta$ and $\varphi \sqsubseteq \psi$. Therefore, by LFT, $\Theta \vdash_{\mathsf{CPL}}^{\sqcap} \eta \supset \psi$.
(ii) η is in L_{CPL}. Observe that, by the induction hypothesis, $\Theta \vdash_{\mathsf{CPL}}^{\sqcap} \eta \supset \psi_j$. Then $\Theta \vdash_{\mathsf{CPL}}^{\sqcap} \eta \supset \psi$ by cLFT. ∎

In short, in both $\mathsf{CPL}^{\sqcap}$ (with non-deterministic combination of connectives) and $\mathsf{CPL}^{\times}$ (with product combination of connectives) the metatheorem of deduction holds with provisos on the use of the cLFT rule.

5 Outlook

The non-deterministic semantics proposed herein for combining connectives (and other constructors) was shown to have some advantages over the original semantics that we proposed in [Sernadas et al., 2011], namely the collapse of $\lceil c_1 c_2 \rceil$ and $\lceil c_2 c_1 \rceil$, while still endowing them only with the common properties of their components.

However, a price had to be paid, given the complicated nature of non-determinism. Although the calculus is surprisingly simple, the use of the LFT rule requires much more work since a high number of premises has to be established before. This number grows exponentially with the number of non-deterministic constructors in the conclusion.

Furthermore, non-determinism has a deep impact on the logic. For instance, $\psi \supset \psi$ is not any more a valid formula in general.

It is worthwhile to point out that one can easily conceive other ways of introducing non-determinism in the semantics of combined constructors. For instance, we considered the alternative of working with non-deterministic algebras, but the path towards a sound and complete calculus was not so clear.

Actually, the solution adopted herein of making a choice for each use of a non-deterministic constructor in a formula seems much more interesting for applications, namely for reasoning about non-deterministic circuits [Wolf, 1994]. Indeed, each non-deterministic gate is expected to behave independently of the other similar gates in the circuit.

For a survey of other approaches to non-determinism in logic see [Avron and Zamansky, In print], including those motivated by paraconsistency [Carnielli, 1990; Carnielli and Coniglio, 2005; Avron, 2007] and by reasoning about computation errors [Avron and Konikowska, 2009].

Our results on non-deterministically combined constructors endowed with the common properties of their components generalize the pioneering work in [Rautenberg, 1989b; Rautenberg, 1989a] on a calculus for the common properties of conjunction and disjunction.

Concerning future research, by replacing the choice sequences by stochastic processes and quantum processes, the results presented herein should be carried over to logics with probabilistic combination of constructors and to logics with quantum superposition of constructors.

Such developments will open the way to much more exciting applications in the field of logics for the verification of faulty hardware [Kleitman et al., 1997] and quantum programs [Nielsen and Chuang, 2000].

Acknowledgments

This work was partially supported by FCT and EU FEDER, namely via project AMDSC UTAustin/MAT/0057/2008 and under the MCL (Meet-Combination of Logics) initiative of SQIG at IT.

BIBLIOGRAPHY

[Avron and Konikowska, 2009] A. Avron and B. Konikowska. Proof systems for reasoning about computation errors. *Studia Logica*, 91(2):273–293, 2009.

[Avron and Zamansky, In print] A. Avron and A. Zamansky. Non-deterministic semantics for logic systems. In D. Gabbay and F. Guenthner, editors, *Handbook of Philosophical Logic, 2nd Edition*. Springer, In print.

[Avron, 2007] A. Avron. Non-deterministic semantics for logics with a consistency operator. *International Journal of Approximate Reasoning*, 45(2):271–287, 2007.

[Carnielli and Coniglio, 2005] W. Carnielli and M. Coniglio. Splitting logics. In S. Artemov, H. Barringer, A. S. d'Avila Garcez, L. C. Lamb, and J. Woods, editors, *We Will Show Them: Essays in Honour of Dov Gabbay, Volume One*, pages 389–414. King's College Publications, 2005.

[Carnielli, 1990] W. Carnielli. Many-valued logics and plausible reasoning. In *Proceedings of the XX International Congress on Many-Valued Logics*, pages 328–335. IEEE Computer Society, 1990.

[Gabbay, 1996] D. M. Gabbay. Fibred semantics and the weaving of logics. I. Modal and intuitionistic logics. *The Journal of Symbolic Logic*, 61(4):1057–1120, 1996.

[Kleitman et al., 1997] D. Kleitman, T. Leighton, and Y. Ma. On the design of reliable Boolean circuits that contain partially unreliable gates. *Journal of Computer and System Sciences*, 55(3):385–401, 1997.

[Nielsen and Chuang, 2000] M. A. Nielsen and I. L. Chuang. *Quantum Computation and Quantum Information*. Cambridge University Press, 2000.

[Rautenberg, 1989a] W. Rautenberg. Axiomatization of semigroup consequences. *Archive for Mathematical Logic*, 29(2):111–123, 1989.

[Rautenberg, 1989b] W. Rautenberg. A calculus for the common rules of $\wedge$ and $\vee$. *Studia Logica*, 48(4):531–537, 1989.

[Sernadas et al., 2011] A. Sernadas, C. Sernadas, and J. Rasga. On combined connectives. *Logica Universalis*, 5(2):205–224, 2011.

[Sernadas et al., in print] A. Sernadas, C. Sernadas, and J. Rasga. On meet-combination of logics. *Journal of Logic and Computation*, in print. DOI:10.1093/logcom/exr035.

[Wolf, 1994] J. M. Wolf. Nondeterministic circuits, space complexity and quasigroups. *Theoretical Computer Science*, 125(2):295–313, 1994.

A. Sernadas, C. Sernadas, J. Rasga, P. Mateus
Instituto Superior Técnico
Universidade Técnica de Lisboa Lisboa, Portugal
and
SQIG - Instituto de Telecomunicações
Lisboa, Portugal
E-mails: {acs,css,jfr,pmat}@math.ist.utl.pt

Metavaluations, Naive Set Theory and Inconsistency[1]

Ross T. Brady

ABSTRACT. This paper proves the non-triviality of inconsistent naive set theory using metavaluations. The proof is based on a 4-valued logic rather than the 3-valued logics previously used, and metavaluations are, for the first time, used without a completeness theorem, thus, disassociating themselves somewhat from the proof theory. Further, it is shown that some axioms and a rule can be added to the logic, which could not be included in earlier model-theoretic approaches.

1 Introduction.

Metavaluations are essentially valuations that express the provability of formulae, and can thus be thought of as capturing the inductive features of proof systems. They are particularly useful in determining systemic properties of formulae that are explicitly about negation, disjunction, conjunction and the quantifiers, and not explicitly about the inference '$\rightarrow$'. They were introduced for quantified positive logics by Meyer in [1976] primarily to prove the Priming Property: if $\vdash A \vee B$ then $\vdash A$ or $\vdash B$, and the corresponding Existential Property: if $\vdash \exists x A$ then $\vdash A^t/_x$, for some term t. Slaney in [1984] continued with this aim when adding negation, whilst restricting the logics to the relevant logics, TW, EW and RW. In doing so, he introduced a second *-metavaluation in order to express the provability of negated formulae. Interestingly, he connects the metavaluations for the logics EW and RW with the Lukasiewicz 3-valued logic, with an adjustment in the $\rightarrow$-matrix for the logic TW, and finishes, on p.167, with the quote: 'The questions of what other algebraic structures yield metavaluations and of what purposes these might serve must await a future investigation'. This paper expands the 3-valued matrix logics to corresponding 4-valued ones and employs their metavaluations to show that an inconsistent naive set theory is non-trivial, for the first time applying metavaluations to model inconsistency.

---

[1] This project was initiated by comments made by Walter Carnielli during discussion of my paper "The Simple Consistency of Naive Set Theory using Metavaluations" when it was presented to the Richard Routley Memorial Conference, held at the University of Melbourne, during July, 2011. He raised the question of using other matrices to drive further metavaluations, and in particular 4-valued ones. And, it is fitting that this paper concerns inconsistency as this has driven the large part of Walter's research. In the audience, Greg Restall also raised the point about the starting conditions at $v_{0,0}$ and $v_{0,0}{}^*$, and changes to these have been quite instrumental in achieving the results of this paper. Further, Zach Weber had earlier offered encouragement for this enterprise and he is part of the reason that I have continued to pursue this course.

In his [1987], Slaney introduced two types of metacomplete logics, called M1 and M2, as ones that support corresponding metavaluations for logics with negation. The logic TW, mentioned above, is M1-metacomplete, whilst the logics EW and RW are M2-metacomplete, the key differentiation being that M1-logics have no negated entailment theorems, whilst M2-logics contain the rule, $A \Rightarrow \sim(A \rightarrow \sim A)$, which is deductively equivalent to: $A, \sim B \Rightarrow \sim(A \rightarrow B)$. M2-metacomplete logics are then shown to have the converse property: if $\vdash \sim(A \rightarrow B)$ then $\vdash A \ \& \sim B$. The M1-metacomplete logics also include Brady's logic MC of meaning containment. (See the earlier logic DJ^d in his [1996] and [2006], and the latest logic MC in Brady and Meinander [forthcoming].)

Brady in [2008] expanded the scope of use of metavaluations by using them to analyse negation. Subsequently, he has used them to show the simple consistency of arithmetic in [2010] and naive set theory in [2011]. This paper will further expand their use in proving the non-triviality of inconsistent naive set theory.

To show that a logic is metacomplete, one requires two theorems, a soundness theorem and a completeness theorem, the soundness theorem establishing 'If $\vdash A$ then $v(A) = T$' and the completeness theorem 'If $v(A) = T$ then $\vdash A$', where v is a metavaluation assigning T or F to each formula A. If one carefully examines the proofs in the above literature, one can see that only the soundness half is used to prove simple consistency, and similarly it is only the soundness that would be needed for a proof of non-triviality. Completeness is needed for the Priming Property and the Existential Property. However, completeness is strangely needed to show soundness for the RW axiom, $A \rightarrow .A \rightarrow B \rightarrow B$, and the axiom $A \rightarrow .B \rightarrow A$ of RWK. Nevertheless, soundness of the axioms and rules of the logic TW do not require the completeness theorem. This will be very useful for our present purpose, as completeness will not be shown for the inconsistent naive set theory. This will also be the first time that metavaluations will be used without completeness, this having the effect of separating metavaluations from their corresponding proof systems. These metavaluations will result in a kind of model that is nevertheless driven by proof. This separation will occur at the set-theoretic level, though completeness is still provable at the level of the predicate logic, with the use of sentential and predicate variables.

Being similar in approach to that of the simply consistent naive set theory of Brady [2011], we will closely follow the methodology of that paper, which in turn closely follows the simple consistency proof in Brady [1983]. We will focus on the points of deviation with [2011], the key deviations being as follows:

(i) One such deviation is the Law of Excluded Middle ($A \lor \sim A$), this not being included in the simply consistent theory as it yields inconsistency. Here, we will add this Law as far as we are able, no longer being constrained by the Priming Property. In order to extend the Law to include $\rightarrow$-formulae for M2-logics, we alter all the *-metavaluations for $A \rightarrow B$ by conjoining $\vdash A \rightarrow B$. However, this extension to $\rightarrow$-formulae does not apply to M1-logics.

(ii) Nevertheless, for the M1-logics, as in [2011], the metavaluations will reduce to a single transfinite sequence, thus expanding the range of logics from §6.3 of [2006] that enable a proof of non-triviality to go through. However, without completeness, this facility is somewhat reduced, with only certain rule- and entailment-forms of $A \to .A \to B \to B$ and $A \to .B \to A$ included in this paper, which are not valid in the modelling used in §6.3 of [2006].

(iii) In the soundness proof for the Extensionality Rule and elsewhere, the two metavaluations v and v^* together function as a 4-valued logic, as they can independently assign a formula A to be T or F. These four values reduce to three values when the consistency result, if $v(A) = T$ then $v^*(A) = T$, is shown or the negation-completeness result, if $v^*(A) = T$ then $v(A) = T$, is shown. The former result will not hold here and the latter result will only be partially shown, yielding a partial Law of Excluded Middle, leaving the logic as essentially 4-valued. [The metavaluations v and v^* will be introduced in their fullness in §3, and their associated 4-valued matrix logics in §6.]

Thus, this paper extends and complements earlier model-theoretic proofs of the non-triviality of inconsistent naive set theories in Brady and Routley [1989], in Brady [1989] and in [2006]. Initially, non-triviality was shown in Brady and Routley [1989] for extensional naive set theory, i.e. without '$\to$' included in the formula A of the Comprehension Axiom, used to generate a set. The logic here was the 3-valued R-mingle logic, RM3. Brady in [1989] extended this result by removing this restriction on '$\to$', although weakening the logic to that of DK, which is MC (below in §2) plus the distribution axioms and the Law of Excluded Middle. Further, [2006] strengthens this full non-triviality result to include the logic TK (below), which essentially adds Hypothetical Syllogism to DK.

2 The Proof Systems.

We start by setting out some axioms, rules and meta-rules, in order to eventually gather together some appropriate M1- and M2-metacomplete logics. We will generally call such a logic LQ.

Primitives:

$\sim$, &, $\vee$, $\to$, $\forall$, $\exists$.

$w, x, y, z, \ldots$ bound variables.

$a, b, c, \ldots$ free variables.

$A, B, C, \ldots$ meta-variables over formulae.

Formation Rules.

1. If A and B are formulae then $\sim A$, $A \& B$, $A \vee B$ and $A \to B$ are formulae.

2. If a is a free variable, x is a bound variable and A is a formula then $\forall x A^x/_a$ and $\exists x A^x/_a$ are formulae.

Axioms.

1. $A \to A$.
2. $A \,\&\, B \to A$.
3. $A \,\&\, B \to B$.
4. $(A \to B) \,\&\, (A \to C) \to .A \to B \,\&\, C$.
5. $A \to A \vee B$.
6. $B \to A \vee B$.
7. $(A \to C) \,\&\, (B \to C) \to .A \vee B \to C$.
8. $A \,\&\, (B \vee C) \to (A \,\&\, B) \vee (A \,\&\, C)$.
9. $\sim\sim A \to A$.
10. $A \to \sim B \to .B \to \sim A$.
11. $(A \to B) \,\&\, (B \to C) \to .A \to C$.
12. $A \to B \to .B \to C \to .A \to C$.
13. $A \to B \to .C \to A \to .C \to B$.
14. $(A \to B \vee C) \,\&\, (A \,\&\, B \to C) \to .A \to C$.
15. $A \to .A \to B \to B$.
16. $A \to .B \to A$.
17. $A \vee \sim A$. [Set-theoretic restrictions on A17 will be determined in §5.]

Rules.

1. $A, A \to B \Rightarrow B$.
2. $A, B \Rightarrow A \,\&\, B$.
3. $A \to B, C \to D \Rightarrow B \to C \to .A \to D$.
4. $A \Rightarrow \sim(A \to \sim A)$.
5. $A \Rightarrow A \to B \to B$.
6. $A \Rightarrow (A \to .B \to C) \to .B \to C$.
7. $A \to B \Rightarrow C \to .A \to B$.

Meta-Rule.

1. If $A, B \Rightarrow C$ then $D \vee A, D \vee B \Rightarrow D \vee C$.

Quantifier Axioms.

1. $\forall x A \to A^a/_x$.
2. $\forall x(A \to B) \to .A \to \forall x B$.
3. $A^a/_x \to \exists x A$.
4. $\forall x(A \to B) \to .\exists x A \to B$.
5. $\forall x(A \vee B) \to A \vee \forall x B$.
6. $A \,\&\, \exists x B \to \exists x(A \,\&\, B)$.

Quantifier Rules.

1. $A^a/_x \Rightarrow \forall x A$, where a does not occur in A.

Quantifier Meta-Rule.

1. If $A, B^a/_x \Rightarrow C^a/_x$ then $A, \exists x B \Rightarrow \exists x C$, provided, in the proof of $C^a/_x$ from A and $B^a/_x$, QR1 does not generalize on a free variable in either of the premises. This proviso also applies to free variables in the premises of the rule $A, B \Rightarrow C$ of MR11.

Logics.

$\quad$ MCQ = A1-7,9-11 + R1-3 + MR1 + QA1-4 + QR1 + QMR1.

$\quad$ TW^{d}Q = A1-10,12-13 + R1-2 + MR1 + QA1-6 + QR1 + QMR1.

$\quad$ TJ^{d}Q = TW^{d}Q + A11.

$\quad$ TK^{d}Q = TJ^{d}Q + A17.

$\quad$ EW^{d}Q = TW^{d}Q + R5.

$\quad$ RW^{d}Q = EW^{d}Q $-$ R5 + A15.

$\quad$ RWK^{d}Q = RW^{d}Q + A16.

[M1- and M2-logics will be differentiated in §3 using these systems.]

We now add the naive set-theoretic axiomatization, including the Comprehension Axiom (CA), and the Extensionality Rule (ER), added to logics LQ determined from the above axioms, rules and meta-rules, with help from the Soundness Theorems in §5 and §6 and their Corollaries in §7.

Primitives:

$\quad$ {, }, : (set-forming symbols)

$\quad$ $\in$ (membership relation)

$\quad$ 1, 0 (sentential constants)

The free and bound variables become free and bound set variables, respectively, and the meta-variables over formulae apply to the formulae introduced below.

Terms:

1. If x is a bound variable and A is a formula then $\{x : A\}$ is a term.

2. A free variable is a term.

   $s, t, u, v, \ldots$ meta-variables over terms.

Formulae:

If s and t are terms then $s \in t$ is an atomic formula.

1 and 0 are atomic formulae.

The formation of formulae using logical connectives and quantifiers follows that for LQ. We replace the free variable a in QA1 and QA3 by a term t. As in Brady's previous work in [1983] and [2006], we also introduce the concept of an <u>initial formula</u> as one which is either an atomic formula or one of the form $A \to B$.

Axioms and Rule:

1. $t \in \{x : A\} \leftrightarrow A^t/_x$. (CA)

2. $\forall z(z \in s \leftrightarrow z \in t) \Rightarrow \forall w(s \in w \leftrightarrow t \in w)$. (ER)

3. 1.

4. ~ 0.

We can define identity s=t between sets as $\forall z(z \in s \leftrightarrow z \in t)$. For M1-logics (see later in §3), we will also use the Extensionality Axiom:

$\forall z(z \in s \leftrightarrow z \in t) \to \forall w(s \in w \leftrightarrow t \in w)$. (EA)

The following Substitution of Identity Rule is derivable, as in Brady [2011].

$s = t \Rightarrow A(s) \leftrightarrow A(t)$, where t is substituted for a single argument place of s in the formula A. (SI)

3 The Sequences of Metavaluations.

As in Brady [2011], we introduce the transfinite sequences for both the v and $v^\star$ metavaluations:

$$
\begin{array}{cccc}
v_{0,0}, & v_{0,1}, & \ldots, & v_{0,\lambda_0}. \\
v_{1,0}, & v_{1,1}, & \ldots, & v_{1,\lambda_1}. \\
\ldots & \ldots & \ldots & \ldots \\
v_{\tau,0}, & v_{\tau,1}, & \ldots, & v_{\tau,\lambda_\tau}. \\
\end{array}
$$

$\ldots \quad \ldots \quad \ldots \quad \ldots$

$$
\begin{array}{cccc}
v_{0,0}^\star, & v_{0,1}^\star, & \ldots, & v_{0,\lambda_0}^\star. \\
v_{1,0}^\star, & v_{1,1}^\star, & \ldots, & v_{1,\lambda_1}^\star. \\
\ldots & \ldots & \ldots & \ldots \\
v_{\tau,0}^\star, & v_{\tau,1}^\star, & \ldots, & v_{\tau,\lambda_\tau}^\star. \\
\ldots & \ldots & \ldots & \ldots \\
\end{array}
$$

The metavaluations $v_{\tau,v}$ and $v_{\tau,v}^\star$ in these sequences are each paired off, to produce a combined metavaluation. The metavaluations, $v_{0,\lambda_0}, v_{1,\lambda_1}, \ldots, v_{\tau,\lambda_\tau}$, $\ldots$ and $v_{0,\lambda_0}^\star, v_{1,\lambda_1}^\star, \ldots, v_{\tau,\lambda_\tau}^\star, \ldots$ of these transfinite sequences are all fixed points of those sequences, established as such by Fixed Point Lemma 1 in §4. They themselves have respective fixed points, $v_{\kappa,\lambda_\kappa}$ and $v_{\kappa,\lambda_\kappa}^\star$, due to Fixed Point Lemma 2, also in §4.

We let v and $v^\star$ be any such metavaluation $v_{\tau,v}$ and *-metavaluation $v_{\tau,v}^\star$, respectively, and assign the following:

For the atomic formulae, 1 and 0:

$$v(1) = T, \ v^\star(1) = T, \ v(0) = F, \text{ and } v^\star(0) = F.$$

For atomic formulae of form $t \in a$, for the free variable a:

$$v(t \in a) = F \text{ and } v^\star(t \in a) = T.$$

For the connectives (except '$\to$') and quantifiers, following Meyer [1976], and Slaney [1984] and [1987]:

$v(A \ \& \ B) = T$ iff $v(A) = T$ and $v(B) = T$.

$v^\star(A \ \& \ B) = T$ iff $v^\star(A) = T$ and $v^\star(B) = T$.

$v(A \lor B) = T$ iff $v(A) = T$ or $v(B) = T$.

$v^\star(A \lor B) = T$ iff $v^\star(A) = T$ or $v^\star(B) = T$.

$v(\sim A) = T$ iff $v^\star(A) = F$.

$v^\star(\sim A) = T$ iff $v(A) = F$.

$v(\forall x A) = T$ iff $v(A^t/_x) = T$, for all terms t.

$v^\star(\forall x A) = T$ iff $v^\star(A^t/_x) = T$, for all terms t.

$v(\exists x A) = T$ iff $v(A^t/_x) = T$, for some term t.

$v^\star(\exists x A) = T$ iff $v^\star(A^t/_x) = T$, for some term t.

We set out below the remaining metavaluations and *-metavaluations for formulae of the form $t \in \{x : A\}$ and for $\to$-formulae, for each member of the above sequences. For the v^* metavaluation of $\to$-formulae, we need to distinguish M1- and M2-logics. As predicate logics, TJ^dQ and its weaker logics are M1-metacomplete, whilst RWK^dQ and its weaker logics, provided they include the rule $A \Rightarrow \sim(A \to \sim A)$, are M2-metacomplete (see Slaney [1987]), but these need to be checked for the set-theoretic system against the metavaluations in this section. We will sort this out in the proof of the Soundness Theorem in §5 and also in the Corollaries in §7.

This process is carried out by double transfinite induction, firstly within each of the τ-sequences, $v_{\tau,0}, v_{\tau,1}, \ldots, v_{\tau,\lambda_\tau}$, and $v_{\tau,0}{}^*, v_{\tau,1}{}^*, \ldots, v_{\tau,\lambda_\tau}{}^*$, and then over both of these sequences as a whole, for all ordinals τ. We simply divide this induction into the four cases:

1. $v_{0,0}$ and $v_{0,0}{}^*$;

2. $v_{\tau,0}$ and $v_{\tau,0}{}^*$, for $\tau > 0$;

3. $v_{\tau,v}$ and $v_{\tau,v}{}^*$, for v a successor ordinal;

4. $v_{\tau,v}$ and $v_{\tau,v}{}^*$, for v a limit ordinal.

Note the differences with Brady [2011] in Case 1 for $t \in \{x : A\}$, which is set up to be understood as inconsistent, and in the M2-*-metavaluations for the $\to$-formulae in all Cases, which are set up so as to prove negation completeness, i.e. if $v^*(A \to B) = T$ then $v(A \to B) = T$. We also need to adjust the quantifiers for $t \in \{x : A\}$ in Case 4 to align with the form of Persistence Theorem 1.

Case 1. For $v_{0,0}$ and $v_{0,0}{}^*$:

$v_{0,0}(t \in \{x : A\}) = T$; $v_{0,0}{}^*(t \in \{x : A\}) = F$.

$v_{0,0}(A \to B) = T$ iff $\vdash A \to B$.

$v_{0,0}{}^*(A \to B) = T$, for M1-logics.

$v_{0,0}{}^*(A \to B) = T$ iff $\vdash A \to B$, for M2-logics.

Case 2. For $v_{\tau,0}$ and $v_{\tau,0}{}^*$, where $\tau > 0$:

$v_{\tau,0}(t \in \{x : A\}) = T$; $v_{\tau,0}{}^*(t \in \{x : A\}) = F$.

$v_{\tau,0}(A \to B) = T$ iff $\vdash A \to B$ and, if $v_{\rho,\lambda_\rho}(A) = T$ then $v_{\rho,\lambda_\rho}(B) = T$, and if $v_{\rho,\lambda_\rho}{}^*(A) = T$ then $v_{\rho,\lambda_\rho}{}^*(B) = T$, for all $\rho < \tau$.

$v_{\tau,0}{}^*(A \to B) = T$, for M1-logics.

$v_{\tau,0}{}^*(A \to B) = T$ iff, $\vdash A \to B$ and, if $v_{\rho,\lambda_\rho}(A) = T$ then $v_{\rho,\lambda_\rho}{}^*(B) = T$, for all $\rho < \tau$, for M2-logics.

Case 3. For $v_{\tau,v}$ and $v_{\tau,v}{}^*$, where v is a successor ordinal:

$v_{\tau,v}(t \in \{x : A\}) = T$ iff $v_{\tau,v-1}(A^t/x) = T$.

$v_{\tau,v}{}^*(t \in \{x : A\}) = T$ iff $v_{\tau,v-1}{}^*(A^t/x) = T$.

$v_{\tau,v}(A \to B) = T$ iff $v_{\tau,0}(A \to B) = T$.

$v_{\tau,v}{}^*(A \to B) = T$ iff $v_{\tau,0}{}^*(A \to B) = T$, for M1- and M2-logics.

<u>Case 4.</u> For $v_{\tau,v}$ and $v_{\tau,v}{}^*$, where v is a limit ordinal:

$v_{\tau,v}(t \in \{x : A\}) = T$ iff $v_{\tau,\rho}(t \in \{x : A\}) = T$, for all $\rho < v$.

$v_{\tau,v}{}^*(t \in \{x : A\}) = T$ iff $v_{\tau,\rho}{}^*(t \in \{x : A\}) = T$, for some $\rho < v$.

$v_{\tau,v}(A \to B) = T$ iff $v_{\tau,0}(A \to B) = T$.

$v_{\tau,v}{}^*(A \to B) = T$ iff $v_{\tau,0}{}^*(A \to B) = T$, for M1- and M2-logics.

We need the following Persistence Theorem 1 to establish the fixed points v_{τ,λ_τ} and $v_{\tau,\lambda_\tau}{}^*$ for the τ-sequences, and Persistence Theorem 2 to establish the further fixed points $v_{\kappa,\lambda_\kappa}$ and $v_{\kappa,\lambda_\kappa}{}^*$.

4 The Persistence Theorems.

THEOREM 1 (Persistence Theorem 1). *For ordinals τ, ρ and v, and a formula A, let $\rho < v$. Then, if $v_{\tau,\rho}(A) = F$ then $v_{\tau,v}(A) = F$, and if $v_{\tau,\rho}{}^*(A) = T$ then $v_{\tau,v}{}^*(A) = T$.*

[Note that F persists for v and T persists for v^*, the opposite of that for the simple consistency proof in Brady [2011].]

<u>Proof.</u> As in Brady [2011], we prove both components together by double induction on the ordinal $v > 0$ and on formula construction.

(i) Let $v = 1$ and hence $\rho = 0$. Then, $v_{\tau,0}(t \in \{x : A\}) = T$ and $v_{\tau,0}{}^*(t \in \{x : A\}) = F$, satisfying the theorem.

As in [2011], we apply induction on the remaining formulae here, the same argument then applying to all ρ and v with $\rho < v$. The theorem is satisfied for 1, 0 and $t \in a$ as their values do not change.

If $v_{\tau,\rho}(\sim A) = F$ then $v_{\tau,\rho}{}^*(A) = T$, and, by induction hypothesis, $v_{\tau,v}{}^*(A) = T$ and $v_{\tau,v}(\sim A) = F$. Similarly, if $v_{\tau,\rho}{}^*(\sim A) = T$ then $v_{\tau,v}{}^*(\sim A) = T$.

A similar argument applies to $A \& B$, $A \lor B$, $\forall x A$ and $\exists x A$, and the argument for $A \to B$ is trivial.

(ii) Let v be a successor ordinal. Assume the theorem for ordinals up to $v-1$. Let ρ be 0. The theorem holds as for $v = 1$ above.

Let ρ be a successor ordinal. Then, if $v_{\tau,\rho}(t \in \{x : A\}) = F$ then $v_{\tau,\rho-1}(A^t/x) = F$, $v_{\tau,v-1}(A^t/x) = F$, and $v_{\tau,v}(t \in \{x : A\}) = F$. Similarly, if $v_{\tau,\rho}{}^*(t \in \{x : A\}) = T$ then $v_{\tau,v}{}^*(t \in \{x : A\}) = T$.

Let ρ be a limit ordinal. Then, if $v_{\tau,\rho}(t \in \{x : A\}) = F$ then $v_{\tau,\mu}(t \in \{x : A\}) = F$, for some $\mu < \rho$, and $v_{\tau,\eta}(A^t/x) = F$, for some $\eta < \mu$. So, by induction hypothesis, $v_{\tau,v-1}(A^t/x) = F$ and $v_{\tau,v}(t \in \{x : A\}) = F$. Similarly, if $v_{\tau,\rho}^*(t \in \{x : A\}) = T$ then $v_{\tau,v}^*(t \in \{x : A\}) = T$.

(iii) Let v be a limit ordinal. Then, if $v_{\tau,\rho}(t \in \{x : A\}) = F$ then $v_{\tau,v}(t \in \{x : A\}) = F$, since $\rho < v$. Similarly, for $v_{\tau,v}^*(t \in \{x : A\}) = T$. □

LEMMA 2 (Fixed Point Lemma 1). *For any ordinal τ, the τ-sequences, $v_{\tau,0}$, $v_{\tau,1}$, ..., v_{τ,λ_τ}, and $v_{\tau,0}^*$, $v_{\tau,1}^*$, ..., v_{τ,λ_τ}^*, have respective fixed points v_{τ,λ_τ} and v_{τ,λ_τ}^*, such that $v_{\tau,\lambda_\tau}(A) = v_{\tau,\lambda_\tau+1}(A)$ and $v_{\tau,\lambda_\tau}^*(A) = v_{\tau,\lambda_\tau+1}^*(A)$.*

Proof. Since there are only denumerably many formulae, there can only be denumerably many ordinals v where either of $v_{\tau,v}(A) \neq v_{\tau,v+1}(A)$ or $v_{\tau,v}^*(A) \neq v_{\tau,v+1}^*(A)$ occur. Thus, the fixed points v_{τ,λ_τ} and v_{τ,λ_τ}^* exist. □

THEOREM 3 (Persistence Theorem 2). *For ordinals μ and τ, and a formula $A \to B$, let $\mu < \tau$. Then, if $v_{\mu,0}(A \to B) = F$ then $v_{\tau,0}(A \to B) = F$, and if $v_{\mu,0}^*(A \to B) = F$ then $v_{\tau,0}^*(A \to B) = F$.*

Proof. As in Brady [2011], the result follows from the quantifiers used in the above metavaluations. We consider the first conditional, letting $v_{\mu,0}(A \to B) = F$. If $\mu = 0$ then not-⊢$A \to B$ and hence $v_{\tau,0}(A \to B) = F$. So, let $\mu > 0$. Then, not-⊢$A \to B$ or, for some $\rho < \mu$, $[v_{\rho,\lambda_\rho}(A) = T$ and $v_{\rho,\lambda_\rho}(B) = F$, or $v_{\rho,\lambda_\rho}^*(A) = T$ and $v_{\rho,\lambda_\rho}^*(B) = F]$. Since $\mu < \tau$, $\rho < \tau$, and the condition yielding $v_{\tau,0}(A \to B) = F$ is satisfied.

Now, let $v_{\mu,0}^*(A \to B) = F$. For M1-logics, $v_{\mu,0}^*(A \to B) = T$ and the conditional holds. For M2-logics, let $\mu = 0$. Then, not-⊢$A \to B$, in which case, $v_{\tau,0}^*(A \to B) = F$. So, let $\mu > 0$. Then, not-⊢$A \to B$ or, for some $\rho < \mu$, $[v_{\rho,\lambda_\rho}(A) = T$ and $v_{\rho,\lambda_\rho}^*(B) = F]$. Again, since $\mu < \tau$, $\rho < \tau$ and $v_{\tau,0}^*(A \to B) = F$. □

LEMMA 4 (Fixed Point Lemma 2). *The transfinite sequences of fixed points, v_{0,λ_0}, v_{1,λ_1}, ..., v_{τ,λ_τ}, ... and v_{0,λ_0}^*, v_{1,λ_1}^*, ..., v_{τ,λ_τ}^*, ..., of all of the τ-sequences have themselves fixed points, $v_{\kappa,\lambda_\kappa}$ and $v_{\kappa,\lambda_\kappa}^*$, such that $v_{\kappa,\lambda_\kappa}(A) = v_{\kappa+1,\lambda_{\kappa+1}}(A)$ and $v_{\kappa,\lambda_\kappa}^*(A) = v_{\kappa+1,\lambda_{\kappa+1}}^*(A)$.*

Proof. Again, as in [2011], there are only denumerably many formulae of form $A \to B$, and each τ-sequence is differentiated only by its metavaluations of $A \to B$ formulae. So, there can only be denumerably many ordinals τ such that either $v_{\tau,\lambda_\tau}(A \to B) \neq v_{\tau+1,\lambda_{\tau+1}}(A \to B)$ or $v_{\tau,\lambda_\tau}^*(A \to B) \neq v_{\tau+1,\lambda_{\tau+1}}^*(A \to B)$. So, the fixed points $v_{\kappa,\lambda_\kappa}$ and $v_{\kappa,\lambda_\kappa}^*$ exist. □

5 The Meta-Theorems.

In the absence of a completeness theorem, we prove a Negation Completeness Theorem in order to determine the scope of application of the Law of Excluded Middle, for the purpose of applying axiom A17 in §2. We then prove a Soundness Theorem to ensure that the metavaluations $v_{\kappa,\lambda_\kappa}$ and $v_{\kappa,\lambda_\kappa}^*$ provide a kind of model for the naive set theory, thus enabling non-triviality to be shown.

In order to define a range of formulae for which the Negation Completeness Theorem holds, we inductively introduce the Atomic Set S_B for a formula B.

Let B_0 be the set of all atomic formulae of a formula B.

Let B_{i+1} be the set of all atomic formulae occurring in B_i after each atomic formula of the form $t \in \{x : A\}$ is replaced by $A^t/_x$.

Then, S_B is the union of all such B_i, for all natural numbers i.

Thus, S_B contains all atoms that can be reached through successive formulae occurring inside curly brackets and also includes circularities by only mentioning repetitions once. Nevertheless, the use of S_B will not exhaust all possible formulae B for which the Theorem will hold, but it will contain the interesting cases, including the self-membership of the Russell Set R, $\{x : \sim x \in x\}$, in which case $B_0 = \{R \in R\}$ and $B_1 = \{R \in R\}$, and finally $S_{R \in R} = \{R \in R\}$.

THEOREM 5 (The Negation Completeness Theorem). *Let a formula B be such that S_B does not contain any atomic formula of the form $s \in a$ (where a is a free variable). M1-logics have the additional restriction that neither does B contain an '$\rightarrow$' nor does S_B contain any atomic formula of the form $t \in \{x : A\}$ where A contains an '$\rightarrow$'. Then, for all ordinals τ and v, if $v_{\tau,v}{}^*(B) = T$ then $v_{\tau,v}(B) = T$.*

[Note that if such a term t contains an '$\rightarrow$' inside curly brackets, then, if this is significant for the proof below, the '$\rightarrow$' will still occur in S_B as a result of a subsequent replacement step.]

<u>Proof.</u> We prove 'If $v_{\tau,v}{}^*(B) = T$ then $v_{\tau,v}(B) = T$', for all ordinals τ and v, by double transfinite induction over the sequence of sequences of metavaluations, whilst internally applying induction over such formulae B. We do this by examining each of the four Cases, used to specify metavaluations for $t \in \{x : A\}$ and $C \rightarrow D$.

(i) Case 1 holds trivially for $t \in \{x : A\}$, since $v_{0,0}{}^*(t \in \{x : A\}) = F$ and $v_{0,0}(t \in \{x : A\}) = T$. Also, $v_{0,0}{}^*(C \rightarrow D) = v_{0,0}(C \rightarrow D)$, for M2-logics. Further, for all metavaluations: $v^*(1) = v(1)$ and $v^*(0) = v(0)$. The theorem also holds for the remaining connectives and quantifiers: $\sim C$, $C \& D$, $C \vee D$, $\forall x C$ and $\exists x C$, by induction on formulae, again for all ordinals τ and v and thus for all four Cases.

(ii) For Case 2, the theorem applies for $t \in \{x : A\}$, as in Case 1. Also, for M2- logics, 'if $v_{\tau,0}{}^*(C \rightarrow D) = T$ then $v_{\tau,0}(C \rightarrow D) = T$' holds, by applying the induction hypothesis to D and to C, for all $\rho < \tau$. This also extends to Case 3 and Case 4.

(iii) Returning to Case 3, for a successor ordinal v, if $v_{\tau,v}{}^*(t \subset \{x : A\}) = T$ then $v_{\tau,v-1}{}^*(A^t/_x) = T$, and, by induction hypothesis, $v_{\tau,v-1}(A^t/_x) = T$ (since $A^t/_x$ also satisfies the restrictions), and hence $v_{\tau,v}(t \in \{x : A\}) = T$.

(iv) For Case 4, for a limit ordinal v, if $v_{\tau,v}{}^*(t \in \{x : A\}) = T$ then $v_{\tau,\rho}{}^*(t \in \{x : A\}) = T$, for some $\rho < v$. For reductio, also let $v_{\tau,v}(t \in \{x : A\}) = F$, in which case, $v_{\tau,\mu}(t \in \{x : A\}) = F$, for some $\mu < v$. Take the maximum of ρ and μ and call it η. Then, since $\rho \leq \eta$ and $\mu \leq \eta$, by Persistence Theorem 1, $v_{\tau,\eta}{}^*(t \in \{x : A\}) = T$ and $v_{\tau,\eta}(t \in \{x : A\}) = F$, which contradicts the induction hypothesis because $\eta < v$. □

For example, for M1- and M2-logics, this theorem applies to any atomic formulae of the form $t \in \{x : x \notin x\}$, where t is a constant term, enabling the Law of Excluded Middle to apply, in particular, to $R \in R$ and thus ensuring that the contradiction $R \in R \,\&\, {\sim}R \in R$ is derivable and our naive set theory is inconsistent.

A similar point can be made about the Curry set C, $\{x : x \in x \to p\}$, for constant p, except that the specifically M1-logics are excluded. But, of course, the derivability of p is prevented by the absence of $A \to .A \to B \Rightarrow A \to B$ in the logic, and indeed non-triviality will be shown in that the constant 0 is not derivable.

Nevertheless, free variables are allowed in atoms of the form $t \in \{x : A\}$, provided in $A^t/_x$ that they occur on the left-hand side of any '$\in$' and not on the right-hand side, in any of the stages of construction of $A^t/_x$ used in defining $S_A{}^t/_x$. The point here is that $s \in a$ does not satisfy the Negation Completeness Theorem, as its metavaluation v takes F and its metavaluation v^* takes T, for all ordinals τ and υ.

For the following Soundness Theorem, we can take provability as for a rather flexible logic LQ with the results in §5 and §7 pinning down the possible M1- and M2-logics.

THEOREM 6 (The Soundness Theorem). *If* $\vdash A$ *then* $v_{\kappa,\lambda_\kappa}(A) = T$, *and hence if* $\vdash {\sim}A$ *then* $v_{\kappa,\lambda_\kappa}{}^*(A) = F$.

<u>Proof</u>. As usual, we prove '$\vdash A$ then $v_{\kappa,\lambda_\kappa}(A) = T$', by induction on proof steps, by showing $v_{\kappa,\lambda_\kappa}(A) = T$ for each axiom A, and by showing that each rule preserves metavaluations $v_{\kappa,\lambda_\kappa}$ with value T.

As in Brady [2011], we leave the Extensionality Rule and Axiom for §6 below, and proceed with a small sample of key axioms and rules of our possible logics LQ. We then follow with the Comprehension Axiom (CA), given that 1 and ${\sim}0$ are clearly sound.

One important difference, however, with [2011] is the addition of $\vdash A \to B$ to the valuation $v_{\tau,0}{}^*(A \to B) = T$ for M2-logics. This will arise in axioms and rules with second-degree shapes, i.e. A4,7,10-14, QA2,4 and R3. On each such occasion $v_{\tau,\lambda_\tau}{}^*(A \to B) = T$ runs in tandem with $v_{\tau,\lambda_\tau}(A \to B) = T$, as far as the assumption and/or proof of $\vdash A \to B$ is concerned. This lets all these soundness results through, together of course with all the other zero and first-degree shapes, thus accounting for axioms A1-14 and QA1-6, rules R1-3 and QR1, and meta-rules MR1 and QMR1. (A11 and A14 are special and are discussed below.) Further, in the case of R4, for M2-logics only, not-$\vdash A \to {\sim}A$ appears as a disjunct which does not need to be satisfied, and thus R4 preserves T.

Another difference is the absence of the Completeness Theorem (if $v_{\tau,\upsilon}(A) = T$ then $\vdash A$), which will negatively affect A15,16. Their rule-forms R5-7, however, are not so affected, but have other problems, as stated below. Further, the absence of the Consistency Theorem (if $v_{\tau,\upsilon}(A) = T$ then $v_{\tau,\upsilon}{}^*(A) = T$) will negatively affect R7. Again, look below.

<u>A11</u>. $(A \to B) \,\&\, (B \to C) \to .A \to C$. (for M1-logics)

To prove $v_{\kappa,\lambda_\kappa}((A \to B) \,\&\, (B \to C) \to .A \to C)$, we just need to show that, for all $\tau < \kappa$, if $v_{\tau,\lambda_\tau}((A \to B) \,\&\, (B \to C)) = T$ then $v_{\tau,\lambda_\tau}(A \to C) = T$, since, for M1-logics, $v_{\tau,\lambda_\tau}{}^*(A \to C) = T$. So, we let $v_{\tau,\lambda_\tau}((A \to B) \,\&\, (B \to C)) = T$. Hence, $\vdash (A \to B) \,\&\, (B \to C)$ and $\vdash A \to C$. So, for $\tau = 0$, $v_{\tau,\lambda_\tau}(A \to C) = T$ holds, and we let $\tau > 0$.

By metavaluations, for all $\rho < \tau$,

if $v_{\rho,\lambda_\rho}(A) = T$ then $v_{\rho,\lambda_\rho}(B) = T$, and if $v_{\rho,\lambda_\rho}{}^*(A) = T$ then $v_{\rho,\lambda_\rho}{}^*(B) = T$, and

if $v_{\rho,\lambda_\rho}(B) = T$ then $v_{\rho,\lambda_\rho}(C) = T$, and if $v_{\rho,\lambda_\rho}{}^*(B) = T$ then $v_{\rho,\lambda_\rho}{}^*(C) = T$.

So, if $v_{\rho,\lambda_\rho}(A) = T$ then $v_{\rho,\lambda_\rho}(C) = T$, and if $v_{\rho,\lambda_\rho}{}^*(A) = T$ then $v_{\rho,\lambda_\rho}{}^*(C) = T$, and hence $v_{\tau,\lambda_\tau}(A \to C) = T$.

However, for M2-logics, by letting $v_{\tau,\lambda_\tau}{}^*((A \to B) \,\&\, (B \to C)) = T$, then, for all $\rho < \tau$, if $v_{\rho,\lambda_\rho}(A) = T$ then $v_{\rho,\lambda_\rho}{}^*(B) = T$, and if $v_{\rho,\lambda_\rho}(B) = T$ then $v_{\rho,\lambda_\rho}{}^*(C) = T$, in which case we require: if $v_{\rho,\lambda_\rho}{}^*(B) = T$ then $v_{\rho,\lambda_\rho}(B) = T$. But this would follow from the Negation Completeness Theorem, provided B satisfies the property that S_B does not contain any atomic formula of the form $s \in a$ (where a is a free variable). So, here, we can include A11 in the M2-logics for a wide range of its instantiations.

<u>A14</u>. $(A \to B \vee C) \,\&\, (A \,\&\, B \to C) \to .A \to C$. (for M1-logics)

As in Brady [2011], this remains sound for M1-logics, but, as for A11, the proviso 'if $v_{\rho,\lambda_\rho}{}^*(B) = T$ then $v_{\rho,\lambda_\rho}(B) = T$' will enable soundness to be shown for M2-logics whenever this proviso holds.

Soundness holds for A17 at least in the cases covered by the Negation Completeness Theorem. The rules, R5 and R6 also fail for both M1- and M2- logics at this stage, and so does their axiom-form A15. However, R7 holds for M1-logics, but its axiom-form A16 fails. Also, note that $A \to B \to .C \to .A \to B$ is sound, but for M1-logics only. Further, Corollary 3 in §7 will enable one to add two more sound axioms and rules for M1-logics.

We now proceed with the Comprehension Axiom, as in Brady [2011].

<u>CA</u>. $t \in \{x : A\} \leftrightarrow A^t/_x$.

For $\tau \leq \kappa$, if $v_{\tau,\lambda_\tau}(t \in \{x : A\}) = T$ then $v_{\tau,\lambda_\tau+1}(t \in \{x : A\}) = T$, by Fixed Point Lemma 1, and hence $v_{\tau,\lambda_\tau}(A^t/_x) = T$. Similarly, for the *-metavaluation. For the converse, if $v_{\tau,\lambda_\tau}(A^t/_x) = T$ then $v_{\tau,\lambda_\tau+1}(t \in \{x : A\}) = T$, and, by Fixed Point Lemma 1, $v_{\tau,\lambda_\tau}(t \in \{x : A\}) = T$. Again, the *-metavaluation is similar. $\square$

This leaves just the Extensionality Rule and Axiom to check, which we do in §6 below.

6 The Soundness of the Extensionality Rule and Axiom.

For metavaluational purposes, we replace the quantifiers in (ER) by terms u and v, as follows:

If $v_{\kappa,\lambda_\kappa}(u \in s \leftrightarrow u \in t) = T$, for all terms u, then $v_{\kappa,\lambda_\kappa}(s \in v \leftrightarrow t \in v) = T$, for all terms v.

We use both M1- and M2-logics for this result, but then show that the axiom-form is sound for M1-logics.

We follow the method used in the previous works: Brady and Routley [1989], Brady [1983], [1989], [2006] (with [1983] being the most accessible), but there are important differences. The main difference with [1983] is the use of formulae rather than closed formulae or sentences. This also impacts on terms, as we use general terms rather than the constant terms of earlier work. We also closely follow the simply consistent approach using metavaluations in [2011], but making allowance for inconsistencies.

In conjunction with this, as in [2011], one should note that $v(t \in a) = F$ and $v^*(t \in a) = T$, for free variables a, where v represents an arbitrary metavaluation. This will mean that $v(s \in v \leftrightarrow t \in v) = T$ whenever v is a free variable, subject of course to $\vdash s \in v \leftrightarrow t \in v$ holding. So, we do not need to consider the case of this term w being a free variable in our soundness proof, which leaves us with w taking the form $\{x{:}E\}$. Further, one should note that the Law of Excluded Middle is not sound for expressions of the form $t \in a$, leaving what amounts to a truth-value gap in the system. This ensures the use of what amounts to four values in our logic, which we now set out.

Similar to that in [2011], we will link the current method with the method in [1983] by introducing four "values" for a metavaluation v:

$v(A) = T$ and $v^*(A) = T$,

$v(A) = T$ and $v^*(A) = F$,

$v(A) = F$ and $v^*(A) = T$, and

$v(A) = F$ and $v^*(A) = F$.

These four "values" correspond to the values t, b, n and f, respectively, as used in Brady [1982], and this will make reference to the earlier proofs much easier. Note that in [2011], essentially these values were also used, except that the value b was absent, due to the simple consistency of the system.

In the previous proofs, there was a focus on the values 1 and 0, representing t and f above and, since there was just a third value $1/2$, this value could be determined by default. In our case,

1 and 0 will correspond to $v^*(A) = T$ and $v(A) = F$, respectively,

and these two values will enable the determination of the four "values", according as the values 1 and 0 hold or not. That is:

The "value" t holds iff 1 holds and 0 fails to hold,
b holds iff both 1 and 0 fail,
n holds iff both 1 and 0 hold, and
f holds iff 0 holds and 1 fails.

This will in turn enable the determination of the equivalence $s \in v \leftrightarrow t \in v$, once the equivalence of the respective 1's and 0's are established. So, to make reference to earlier work easier, we will use the values 1 and 0, as above, dispensing with the asterisk $*$.

We can establish a 4-valued matrix system for M1- and M2-logics, using these four "values", as in Brady [2008, pp. 350–1]. For M2-logics, we obtain the following matrices:

$\sim$	
t	f
b	b
n	n
f	t

&	t	b	n	f
t	t	b	n	f
b	b	b	f	f
n	n	f	n	f
f	f	f	f	f

$\vee$	t	b	n	f
t	t	t	t	t
b	t	b	t	b
n	t	t	n	n
f	t	b	n	f

$\rightarrow$	t	b	n	f
t	t	f	n	f
b	t	b	n	f
n	t	n	t	n
f	t	t	t	t

The designated values are t and b, i.e. where $v(A) = T$, and validity is defined as usual in terms of these. The matrices can be recognized as those for the logic BN4, introduced and axiomatized in Brady [1982, pp. 10, 21–3]. However, for M1-logics, we obtain the following:

$\sim$	
t	f
b	b
n	n
f	t

&	t	b	n	f
t	t	b	n	f
b	b	b	f	f
n	n	f	n	f
f	f	f	f	f

$\vee$	t	b	n	f
t	t	t	t	t
b	t	b	t	b
n	t	t	n	n
f	t	b	n	f

$\rightarrow$	t	b	n	f
t	t	n	n	n
b	t	t	n	n
n	t	n	t	n
f	t	t	t	t

As above, the designated values are t and b, with validity defined as usual. As in Brady [2008], we call this matrix logic BN4-1, as it models the M1-logics in a similar manner to the modelling of M2-logics by BN4. [These two $\rightarrow$-matrices are, of course, subject to $\vdash A \rightarrow B$ holding in the cases of $v(A \rightarrow B) = t$ or n (i.e. $= 1$) and $v(A \rightarrow B) = t$ or b (i.e. $\neq 0$).]

Getting back to our proof, given the assumption $v_{\kappa,\lambda_\kappa}(u \in s \leftrightarrow u \in t) = T$, for all terms u, and for given terms s and t, we need to establish, for an arbitrary term of the form $\{x : E\}$, for any $\tau < \kappa$,

$$v_{\tau,\lambda_\tau}(s \in \{x : E\}) = T \text{ iff } v_{\tau,\lambda_\tau}(t \in \{x : E\}) = T, \text{ and}$$

$$v_{\tau,\lambda_\tau}{}^*(s \in \{x : E\}) = T \text{ iff } v_{\tau,\lambda_\tau}{}^*(t \in \{x : E\}) = T.$$

With reference to the above four combinations of our numerical values 1 and 0, this can be recast as follows, without the asterisk:

If, for all $\tau < \kappa$, $v_{\tau,\lambda_\tau}(u \in s) = v_{\tau,\lambda_\tau}(u \in t)$, for all terms u, then, for any $\tau < \kappa$, $v_{\tau,\lambda_\tau}(s \in \{x : E\}) = v_{\tau,\lambda_\tau}(t \in \{x : E\})$, for any term of form $\{x : E\}$.

Given the antecedent, we prove the consequent by transfinite induction with induction hypothesis:

For any $\rho < \tau$, $v_{\rho,\lambda_\rho}(s \in \{x : A\}) = v_{\rho,\lambda_\rho}(t \in \{x : A\})$, for any term of form $\{x : A\}$, where $\tau > 0$.

As in Brady and Routley [1989] and Brady [1983] and [1989], we undertake the required analysis of how $v_{\tau,\lambda_\tau}(s \in \{x : E\}) = 1$ or 0 within the τ-sequence, so as to be able to show $v_{\tau,\lambda_\tau}(t \in \{x : E\}) = 1$ or 0, respectively, by replacing the 's' by 't' in the sequence. In making this replacement, we will utilize our assumption: $v_{\tau,\lambda_\tau}(u \in s) = v_{\tau,\lambda_\tau}(u \in t)$, which holds for any $\tau < \kappa$ and for any term u.

However, with reference to Brady [2011], 1 was used there to represent t, whilst 0 represented f. Here, 1 represents t or n, whilst 0 represents f or n. So, we need to be aware of these differences on the way through.

As in Brady [2011], we need the following 7 definitions.

(1) Let $v_{\tau,\lambda_\tau}(A) = 1$ or 0. Then, define $\nu(\tau, A)$ to be the least ordinal σ such that $v_{\tau,\sigma}(A) = 1$ or 0.

(2) Define $I(A)$ as the set of all maximal initial subformulae of A, where an initial formula is either atomic or of the form $A \to B$. However, we do temporarily allow bound variables to occur in these initial formulae. The maximality is with respect to subformula containment. So, A can be constructed from members of $I(A)$ by using $\sim$, &, $\vee$, $\forall$ and $\exists$, without variable replacement.

(3) Furthermore, define $S(A)$ as the set of substitution instances of initial formulae in $I(A)$, obtained by substituting terms for each of the bound variables. Distinct from earlier work, these terms can contain free variables and free variables are not substituted upon in this process.

(4) Define the set $D(A)$, the determining set of A, of all those members B of $S(A)$ which satisfy $v_{\tau,\nu(\tau,A)}(B) = 1$ or 0. Unlike in Brady [2011], these members B of $D(A)$ can include atoms of the form $u \in a$, for a free variable a, since $v(u \in a)$ equals both 1 and 0, for all metavaluations v. Hence, for such atoms $u \in a$, $\nu(\tau, u \in a)$ would equal 0.

(5) We introduce the set of designated occurrences of the term s in A as those occurrences traced back from the starting point '$s \in \{x : E\}$'. We use the symbolisation, $A(s)$, for the formula A with the set of designated occurrences of the term s. If the term t is substituted for these and only these designated occurrences of s, we represent this by $A(^t/_s)$. This tracing back of designated occurrences of s in $A(s)$ into $D(A(s))$ can be seen as follows:

> Let $v_{\tau,\lambda_\tau}(A) = 1$ or 0. Then, for any initial formula $B(s)$ of $D(A(s))$, the designated occurrences of s in $B(s)$ are those occurrences of s in the maximal initial subformula of $I(A(s))$ of which $B(s)$ is a substitution instance, and which are designated occurrences of $A(s)$.

(6) Let $A(s)$ be of the form $u \in \{x : B\}(s)$, $\{x : B\}$ not being a designated occurrence of s. Define the corresponding formula of $A(s)$, $C(A(s))$, as

$B^u/_x$. Note that $v_{\tau,\nu}(A(s)) = v_{\tau,\nu-1}(C(A(s)))$, for ν a successor ordinal, and hence if $v_{\tau,\lambda_\tau}(A(s)) = 1$ or 0 then $v_{\tau,\lambda_\tau}(C(A(s))) = 1$ or 0.

(7) Let $v_{\tau,\lambda_\tau}(A(s)) = 1$ or 0. Define the general determining set $G(A(s))$ of $A(s)$ by transfinite recursion on $\nu(\tau, A(s))$ as follows:

$$G(A(a)) = (D(A(s)) - D'(A(s))) \sqcup \bigcup_{B(s) \in D'(A(s))} G(C(B(s))),$$

where $D'(A(s)) = \{B(s) \in D(A(s)) : B(s)$ contains a designated occurrence of s and is of form $u \in \{x : C\}(s)$, $\{x : C\}$ not being a designated occurrence of $s\}$. Note that all members of $G(A(s))$ are initial formulae, as can be shown by induction on $\nu(\tau, A(s))$.

As in Brady [2011], we now follow the sequence of 8 lemmas in Brady [1983], but with an additional one to cater for initial formulae of the form $u \in a$. We make the necessary terminological changes regarding terms, formulae and values in each case. In comparison with [2011], we also need to make the metavaluational adjustments for 1 and 0. The proofs go through using the same techniques as in [1983] (or as referenced in Brady and Routley [1989]), and these are just indicated for each lemma.

The first lemma gives the basic property of determining sets.

LEMMA 7 (Extensionality Lemma 1). *Let $A(s)$ be a formula with a set of designated occurrences of s, such that $v_{\tau,\lambda_\tau}(A(s)) = 1$ or 0. Then, if, for all $B(s) \in D(A(s))$, $v_{\tau,\lambda_\tau}(B(^t/_s)) = v_{\tau,\lambda_\tau}(B(s))$, then $v_{\tau,\lambda_\tau}(A(^t/_s)) = v_{\tau,\lambda_\tau}(A(s))$.*

Proof. Here, we follow Brady and Routley [1989], instead of Brady [1983]. Proof is by induction on the valuation procedure for the above formulae $A(s)$, starting with the initial formulae in $S(A(s))$ and building up to $A(s)$ using $\sim$, &, $\vee$, $\forall$ and $\exists$. We note that the metavaluations standing for 1 and 0 continue to work for this inductive procedure, in a similar manner to that in Brady [2011]. □

In order to show that $G(A(s))$ is well-defined, we need to prove the following lemma. We return to Brady [1983], setting out the lemmas in the style of Brady [2011].

LEMMA 8 (Extensionality Lemma 2). *Let $A(s)$ be a formula with a set of designated occurrences of s, such that $v_{\tau,\lambda_\tau}(A(s)) = 1$ or 0. Let $B(s) \in D'(A(s))$. Then $v_{\tau,\lambda_\tau}(C(B(s))) = 1$ or 0 and $\nu(\tau, C(B(s))) < \nu(\tau, A(s))$.*

Proof. By the metavaluation procedure for $B(s)$ in the form $u \in \{x : C\}(s)$, $v_{\tau,\lambda_\tau}(C(B(s))) = v_{\tau,\lambda_\tau}(B(s))$ and $\nu(\tau, C(B(s))) = \nu(\tau, B(s)) - 1$. □

We follow with a property needed for Extensionality Lemma 8.

LEMMA 9 (Extensionality Lemma 3). *Let $A(s)$ be a formula with a set of designated occurrences of s, such that $v_{\tau,\lambda_\tau}(A(s)) = 1$ or 0. Then, for all $C(s) \in G(A(s))$, $v_{\tau,\lambda_\tau}(C(s)) = 1$ or 0 and $\nu(\tau, C(s)) \leq \nu(\tau, A(s))$.*

Proof. By transfinite induction on the ordinals $\nu(\tau, A(s))$, making use of Extensionality Lemma 2. □

The following lemma gives the basic property of general determining sets.

LEMMA 10 (Extensionality Lemma 4). *Let $A(s)$ be a formula with a set of designated occurrences of s, such that $v_{\tau,\lambda_\tau}(A(s)) = 1$ or 0. Then, if, for all $B(s) \in G(A(s))$, $v_{\tau,\lambda_\tau}(B(^t/_s)) = v_{\tau,\lambda_\tau}(B(s))$, then $v_{\tau,\lambda_\tau}(A(^t/_s)) = v_{\tau,\lambda_\tau}(A(s))$.*

Proof. By transfinite induction on $\nu(\tau, A(s))$, using Extensionality Lemmas 1 and 2. □

The following lemma gives the analysis of the initial formulae of general determining sets.

LEMMA 11 (Extensionality Lemma 5). *Let $A(s)$ be a formula with a set of designated occurrences of s, such that $v_{\tau,\lambda_\tau}(A(s)) = 1$ or 0. Then, all the initial formulae of $G(A(s))$ that contain a designated occurrence of s are of one of the three types:*

(I) $(u \in s)(s)$, where at least the displayed occurrence of s is designated and s takes the form $\{x : E\}$,

(II) $(B \to C)(s)$, where a designated occurrence of s occurs in either B or C, or

(III) $(u \in a)(s)$, where a designated occurrence of s occurs in the term u or as the displayed free variable a.

Proof. By transfinite induction on $\nu(\tau, A(s))$, using the definition of $D'(A(s))$ and Extensionality Lemma 2. □

The next lemma deals with substitution of t for s into formulae of type (III), needed for Extensionality Lemma 8.

LEMMA 12 (Extensionality Lemma 6). *Let $(u \in a)(s)$, where a designated occurrence of s occurs in the term u or as the displayed free variable a, such that $v_{\tau,\lambda_\tau}((u \in a)(s)) = 1$ or 0. Also let $v_{\tau,\lambda_\tau}(v \in s) = v_{\tau,\lambda_\tau}(v \in t)$, for all terms v. Then, $v_{\tau,\lambda_\tau}((u \in s)(^t/_s)) = v_{\tau,\lambda_\tau}((u \in s)(s))$.*

Proof. Indeed, $v_{\tau,\lambda_\tau}((u \in a)(s)) = 1$ and 0, since a is a free variable. If the free variable a is not a designated occurrence of s, then $v_{\tau,\lambda_\tau}((u \in a)(^t/_s)) = 1$ and 0, as well. If a is a designated occurrence of s then, since, for all terms v, $v_{\tau,\lambda_\tau}(v \in a) = v_{\tau,\lambda_\tau}(v \in t)$, both $v_{\tau,\lambda_\tau}((u(s) \in a))$ and $v_{\tau,\lambda_\tau}(u(^t/_s) \in t) = 1$ and $= 0$. □

The next lemma deals with substitution of t for s into formulae of type (II), also needed for Extensionality Lemma 8.

LEMMA 13 (Extensionality Lemma 7). *Let $(B \to C)(s)$ be a formula with a set of designated occurrences of s, such that $v_{\tau,\lambda_\tau}((B \to C)(s)) = 1$ or 0. If $\tau > 0$, let $v_{\rho,\lambda_\rho}(s \in \{x : A\}) = v_{\rho,\lambda_\rho}(t \in \{x : A\})$, for any term of form $\{x:A\}$, for all $\rho < \tau$. Then, $v_{\tau,\lambda_\tau}((B \to C)(^t/_s)) = v_{\tau,\lambda_\tau}(B(s) \to C(s))$.*

Proof. Similarly to that in Brady [2011], we need to especially consider the case where $v_{\tau,\lambda_\tau}((B \to C)(s)) = 1$, since then $\vdash (B \to C)(s)$ also holds. Here, as a result of its *-metavaluation being T, we only need consider M2-logics. From the associated proof, we have $\vdash \forall w(s \in w \leftrightarrow t \in w)$, given $\vdash \forall z(z \in s \leftrightarrow z \in t)$, and, as in the derivation of the rule (SI), we can substitute $\{x : (B \to C)(x)\}$ for w, yielding $\vdash (B \to C)(^t/_s)$. Thus:

(⊢1) if $\vdash (B \to C)(s)$ then $\vdash (B \to C)(^t/_s)$.

We then use this to establish $v_{\tau,\lambda_\tau}((B \to C)(^t/_s)) = 1$, in conjunction with the given proof of Lemma 11 in Brady [1983]. However, we also need to consider $v_{\tau,\lambda_\tau}((B \to C)(s)) = 0$ for both M1- and M2-logics, where, due to its metavaluation F, we will need to show:

(⊢2) if not-$\vdash (B \to C)(s)$ then not-$\vdash (B \to C)(^t/_s)$.

So, we let $\vdash (B \to C)(^t/_s)$. As above, $\vdash (B \to C)(s)$ then follows.

The proof proceeds by transfinite induction on the ordinals τ. For M1-logics, $v_{\tau,\lambda_\tau}((B \to C)(s)) = 1$ and also $v_{\tau,\lambda_\tau}((B \to C)(^t/_s)) = 1$, for any τ. For the remaining cases, we first let $\tau = 0$. Then, for M1- and M2-logics, if $v_{\tau,\lambda_\tau}((B \to C)(s)) = 0$ then not-$\vdash (B \to C)(s)$ and, by (⊢2), not-$\vdash (B \to C)(^t/_s)$ and $v_{\tau,\lambda_\tau}((B \to C)(^t/_s)) = 0$. For M2-logics, if $v_{\tau,\lambda_\tau}((B \to C)(s)) = 1$ then $\vdash (B \to C)(s)$ and, by (⊢1), $\vdash (B \to C)(^t/_s)$ and $v_{\tau,\lambda_\tau}((B \to C)(^t/_s)) = 1$. For $\tau > 0$, we first establish the following, as in Brady [1983]:

(+) $v_{\rho,\lambda_\rho}(B(^t/_s)) = v_{\rho,\lambda_\rho}(B(s))$, for all $\rho < \tau$.

We now proceed in some detail with the cases where $v_{\tau,\lambda_\tau}((B \to C)(s)) = 1$ and $= 0$, as they are a little different from [1983].

(i) Let $v_{\tau,\lambda_\tau}((B \to C)(s)) = 1$. We only need consider M2-logics. Then, $\vdash (B \to C)(s)$ and, for all $\rho < \tau$, if $v_{\rho,\lambda_\rho}(B(s)) \neq 0$ then $v_{\rho,\lambda_\rho}(C(s)) = 1$, where either B or C or both have a designated occurence of s. By (+), if $v_{\rho,\lambda_\rho}(B(^t/_s)) \neq 0$ then $v_{\rho,\lambda_\rho}(C(^t/_s)) = 1$, for all $\rho < \tau$, and, together with (⊢1), $v_{\tau,\lambda_\tau}((B \to C)(^t/_s)) = 1$.

(ii) For M1- and M2-logics, let $v_{\tau,\lambda_\tau}((B \to C)(s)) = 0$. Then, not-$\vdash (B \to C)(s)$ or, for all $\rho < \tau$, $v_{\rho,\lambda_\rho}(B(s)) \neq 0$ and $v_{\rho,\lambda_\rho}(C(s)) = 0$, or, $v_{\rho,\lambda_\rho}(B(s)) = 1$ and $v_{\rho,\lambda_\rho}(C(s)) \neq 1$. Then, by (⊢2) and (+), $v_{\tau,\lambda_\tau}((B \to C)(^t/_s)) = 0$. □

We now proceed with the substitution of t for s into formulae of type (I).

LEMMA 14 (Extensionality Lemma 8). *Let $(u \in s)(s))$ be a formula with a set of designated occurrences of s including the displayed 's', where s is of the form $\{x : E\}$ and such that $v_{\tau,\lambda_\tau}((u \in s)(s)) = 1$ or 0. If $\tau > 0$, let $v_{\rho,\lambda_\rho}(s \in \{x : A\}) = v_{\rho,\lambda_\rho}(t \in \{x : A\})$, for any term of form $\{x : A\}$, for all $\rho < \tau$. Also let $v_{\tau,\lambda_\tau}(v \in s) = v_{\tau,\lambda_\tau}(v \in t)$, for all terms v. Then,*

$$v_{\tau,\lambda_\tau}((u \in s)(^t/_s)) = v_{\tau,\lambda_\tau}((u \in s)(s)).$$

Proof. By transfinite induction on $\nu(\tau, (u \in s)(s))$, and here $\nu(\tau, (u \in s)(s))$ must be a successor ordinal. The proof follows Lemma 12 of Brady [1983] and use is made of Extensionality Lemmas 2-7. □

This lemma sets out the requirements for the Extensionality Soundness Theorem.

LEMMA 15 (Extensionality Lemma 9). *If $v_{\tau,\lambda_\tau}(v \in s) = v_{\tau,\lambda_\tau}(v \in t)$, for all terms v, for all $\tau < \kappa$, then $v_{\tau,\lambda_\tau}(s \in \{x : E\}) = v_{\tau,\lambda_\tau}(t \in \{x : E\})$, for any term of form $\{x : E\}$, for any $\tau < \kappa$.*

Proof. By transfinite induction on τ. In considering $v_{\tau,\lambda_\tau}(s \in \{x : E\}) = 1$ or 0, we take the set of designated occurrences of s as just that occurrence on the left of the '$\in$' in $s \in \{x : E\}$, and similarly for t in considering $v_{\tau,\lambda_\tau}(t \in \{x : E\}) = 1$ or 0. We show, using Extensionality Lemmas 3-8 much as in Brady [1983], that if $v_{\tau,\lambda_\tau}(s \in \{x : E\}) = 1$ or 0 then $v_{\tau,\lambda_\tau}(t \in \{x : E\}) = v_{\tau,\lambda_\tau}(s \in \{x : E\})$, and if $v_{\tau,\lambda_\tau}(t \in \{x : E\}) = 1$ or 0 then $v_{\tau,\lambda_\tau}(s \in \{x : E\}) = v_{\tau,\lambda_\tau}(t \in \{x : E\})$. From this, it can be seen that $v_{\tau,\lambda_\tau}(s \in \{x : E\}) = v_{\tau,\lambda_\tau}(t \in \{x : E\})$, as argued earlier. □

We now revert to the previous T and F values for the metavaluations v and $v^\star$.

THEOREM 16 (Extensionality Soundness Theorem). *The Extensionality Rule (ER) is sound, i.e. if $v_{\kappa,\lambda_\kappa}(\forall z(z \in s \leftrightarrow z \in t)) = T$ then $v_{\kappa,\lambda_\kappa}(\forall w(s \in w \leftrightarrow t \in w)) = T$.*

For M1-logics, the Extensionality Axiom (EA) is sound, i.e. $v_{\kappa,\lambda_\kappa}(\forall z(z \in s \leftrightarrow z \in t) \to .\forall w(s \in w \leftrightarrow t \in w)) = T$.

Proof. Let $v_{\kappa,\lambda_\kappa}(\forall z(z \in s \leftrightarrow z \in t)) = T$. Then, for all terms u, $v_{\kappa,\lambda_\kappa}(u \in s \leftrightarrow u \in t) = T$, and hence $\vdash u \in s \leftrightarrow u \in t$ and, for all $\tau < \kappa$, $v_{\tau,\lambda_\tau}(u \in s) = T$ iff $v_{\tau,\lambda_\tau}(u \in t) = T$, and $v_{\tau,\lambda_\tau}{}^\star(u \in s) = T$ iff $v_{\tau,\lambda_\tau}{}^\star(u \in t) = T$. By Extensionality Lemma 9, $v_{\tau,\lambda_\tau}(s \in \{x : E\}) = T$ iff $v_{\tau,\lambda_\tau}(t \in \{x : E\}) = T$, and $v_{\tau,\lambda_\tau}{}^\star(s \in \{x : E\}) = T$ iff $v_{\tau,\lambda_\tau}{}^\star(t \in \{x : E\}) = T$, for all $\tau < \kappa$, for all terms of form $\{x:E\}$. Thus, since $\vdash s \in \{x : E\} \leftrightarrow t \in \{x : E\}$, for all $\{x : E\}$, $v_{\kappa,\lambda_\kappa}(\forall w(s \in w \leftrightarrow t \in w)) = T$ follows, given our earlier remark concerning $s \in a$, where a is a free variable.

For M1-logics, let $v_{\kappa,\lambda_\kappa}{}^\star(\forall z(z \in s \leftrightarrow z \in t)) = T$. Then, $v_{\kappa,\lambda_\kappa}{}^\star(\forall w(s \in w \leftrightarrow t \in w)) = T$, because $v_{\kappa,\lambda_\kappa}{}^\star(s \in v \leftrightarrow t \in v) = T$, for all terms v. Thus, given $\vdash \forall z(z \in s \leftrightarrow z \in t) \to .\forall w(s \in w \leftrightarrow t \in w)$, $v_{\kappa,\lambda_\kappa}(\forall z(z \in s \leftrightarrow z \in t) \to .\forall w(s \in w \leftrightarrow t \in w)) = T$ follows. □

7 The Soundness Theorem and its Corollaries.

Thus, we can reiterate:

THEOREM 17 (The Soundness Theorem). *If $\vdash A$ then $v_{\kappa,\lambda_\kappa}(A) = T$, and hence if $\vdash \sim A$ then $v_{\kappa,\lambda_\kappa}{}^\star(A) = F$.*

Proof. From the previous Soundness and Extensionality Soundness Theorems. □

COROLLARY 18 (Corollary 1. (Non-Triviality)). *There is a non-theorem of the system, viz. 0. Thus, our naive set theory is non-trivial, as well as being inconsistent.*

Proof. Since $v(0) = F$, 0 is not provable, by the Soundness Theorem. Inconsistency can be shown in the usual way, since, as indicated earlier, the Russell set $\{x : \sim x \in x\}$ satisfies the restrictions on the Negation Completeness Theorem. □

COROLLARY 19 (Corollary 2). *For any logic for which the Soundness Theorem holds for whatever ordinal κ might be, for all $0 \leq \tau \leq \kappa$, $v_{\tau,\lambda_\tau}(A \to B) = T$ iff $\vdash A \to B$.*

Proof. $v_{0,\lambda_0}(A \to B) = T$ iff $\vdash A \to B$, and, by the Soundness Theorem, $v_{\kappa,\lambda_\kappa}(A \to B) = T$ iff $\vdash A \to B$. By Persistence Theorem 2, for any τ such that $0 \leq \tau \leq \kappa$, if $v_{\kappa,\lambda_\kappa}(A \to B) = T$ then $v_{\tau,\lambda_\tau}(A \to B) = T$, and further $v_{0,\lambda_0}(A \to B) = T$. Thus, $v_{\tau,\lambda_\tau}(A \to B) = T$ iff $\vdash A \to B$, for any τ. □

COROLLARY 20 (Corollary 3). *For all M1-logics for which the Soundness Theorem holds, $\kappa = 0$, meaning that the single pair of transfinite sequences, $v_{0,0}, v_{0,1}, \ldots, v_{0,\lambda_0}$, and $v_{0,0}{}^\star, v_{0,1}{}^\star, \ldots, v_{0,\lambda_0}{}^\star$, will suffice to show non-triviality.*

Proof. Since, for all metavaluations v, $v(A \to B) = T$ iff $\vdash A \to B$, by Corollary 2, and also $v^\star(A \to B) = T$, both $v_{0,\lambda_0}(A \to B) = v_{1,\lambda_1}(A \to B)$ and $v_{0,\lambda_0}{}^\star(A \to B) = v_{1,\lambda_1}{}^\star(A \to B)$ hold. Thus, $\kappa = 0$, as the 0-sequence and 1-sequence are only differentiated by the valuations to $A \to B$. □

If the M2-logics were to satisfy Corollary 3, then $A \,\&\, (A \to B) \to B$ would become sound, at least for constant instances, since, by Negation Completeness, $v_{0,\lambda_0}{}^\star(A) = T$ and $v_{0,\lambda_0}{}^\star(A \to B) = T$ would then imply $v_{0,\lambda_0}{}^\star(B) = T$. So, M2-logics cannot satisfy Corollary 3, since $A \,\&\, (A \to B) \to B$ leads to triviality for constant instances, contradicting the non-provability of 0, established by Corollary 1.

COROLLARY 21 (Corollary 4). *The additional axioms and rule for M1-logics,*

$$A \to B \to .(A \to B \to .C \to D) \to .C \to D,$$

$$(A \to .B \to C \to .D \to E) \to .B \to C \to .A \to .D \to E \text{ and}$$

$$A \Rightarrow (A \to .B \to C) \to .B \to C$$

are also sound

Proof. We make use of the single pair of metavaluations, v_{0,λ_0} and $v_{0,\lambda_0}{}^\star$ of Corollary 3. □

In conclusion, we make the following two points. Firstly, the axioms and rules, given in Corollary 4, together with $A \to B \to .C \to .A \to B$ (noted in §5), were not able to be included in our previous model-theoretic method for the non-triviality of inconsistent naive set theory in Brady [1989] and in [2006], upon which ideas used in setting up our current method are based. Secondly, as shown in §5, conjunctive syllogism (A11) and strong distribution (A14), both of which are normally restricted to being in M1-logics without being in M2-logics (see for example Brady [2011]), can here be so included in M2-logics, subject to the Negation Completeness Theorem holding for B, for each of these axioms.

BIBLIOGRAPHY

[Brady and Meinander, forthcoming] R. T. Brady and A. Meinander. Distribution in the Logic of Meaning Containment and in Quantum Mechanics. In K. Tanaka, F. Berto, E. Mares, and F Paoli, editors, *Paraconsistency: Logic and Applications*. Springer Publishing, Dordrecht, forthcoming.

[Brady and Routley, 1989] R. T. Brady and R. Routley. The Non-Triviality of Extensional Dialectical Set Theory. In G. Priest, R. Routley, and J. Norman, editors, *Paraconsistent Logic, Essays on the Inconsistent*. Philosophia Verlag, Munich, 1989.

[Brady, 1982] R. T. Brady. Completeness Proofs for the Systems RM3 and BN4. *Logique et Analyse*, 25:9–32, 1982.

[Brady, 1983] R. T. Brady. The Simple Consistency of a Set Theory Based on the Logic CSQ. *Notre Dame Journal of Formal Logic*, 24:431–449, 1983.

[Brady, 1989] R. T. Brady. The Non-Triviality of Dialectical Set Theory. In G. Priest, R. Routley, and J. Norman, editors, *Paraconsistent Logic, Essays on the Inconsistent*. Philosophia Verlag, Munich, 1989.

[Brady, 1996] R. T. Brady. Relevant Implication and the Case for a Weaker Logic. *Journal of Philosophical Logic*, 25:151–183, 1996.

[Brady, 2006] R. T. Brady. *Universal Logic*. CSLI Publications, Stanford, 2006.

[Brady, 2008] R. T. Brady. Negation in Metacomplete Relevant Logics. *Logique et Analyse*, 51:331–354, 2008.

[Brady, 2010] R. T. Brady. The Simple Consistency of Arithmetic, for Metacomplete Logics without a Quantified Form of Distribution. Presented at the World Congress in Universal Logic, Lisbon, Portugal, 2010.

[Brady, 2011] R. T. Brady. The Simple Consistency of Naive Set Theory using Metavaluations. Presented at the Richard Routley Memorial Conference, held at the University of Melbourne, July 2011.

[Meyer, 1976] R. K. Meyer. Metacompleteness. *Notre Dame Journal of Formal Logic*, 17:501–516, 1976.

[Slaney, 1984] J. K. Slaney. A Metacompleteness Theorem for Contraction-free Relevant Logics. *Studia Logica*, 43:159–168, 1984.

[Slaney, 1987] J. K Slaney. Reduced Models for Relevant Logics Without WI. *Notre Dame Journal of Formal Logic*, 28:395–407, 1987.

Ross T. Brady
Philosophy Program
La Trobe University
Victoria, Australia
Email: Ross.Brady@latrobe.edu.au

A Deductive Language for Querying Inconsistent Databases

SANDRA DE AMO AND MÔNICA SAKURAY PAIS

ABSTRACT. The presence of inconsistent information is a very common problem faced by database management systems users and database researchers. A typical scenario where inconsistent information naturally appears is when multiple different data sources are integrated in a unique repository. In the last two decade, a lot of research work has been undertaken by the database community proposing solutions to the information integration problem. In recent years, with the emerging of ubiquitous computing technologies, another interesting research scenario has been attracting the interest of database researchers, namely the incorporation of preference facilities in database query languages, in order to personalize query answering according to the user's specific preferences. Very often, users are not consistent when expressing their preferences and depending on the moment they can prefer A to B and vice-versa. In this paper, we focus on the problem of querying inconsistent data and present the P-Datalog query language, a deductive language for querying databases containing inconsistent information. We use the paraconsistent logic (**LFI1**) that we have introduced in a previous work, as the underlying logic of P-Datalog. We present a declarative semantics which captures the desired meaning of a recursive query executed over a database containing inconsistent facts and whose rules allow inferring information from inconsistent premises. Besides, we introduce a bottom-up evaluation method for P-Datalog programs based on an alternating fixpoint operator. A prototype of the P-Datalog (PDLog) is also presented in order to validate our approach.

1 Introduction

In many real-world database applications we are faced with inconsistent information and very often we cannot simply get rid off them since it may constitute relevant information. Many research topics in database involve dealing with inconsistent information. Two topics of research are the most cited in the literature: data integration and preference handling. When integrating data coming from multiple different sources we are faced with the possibility of inconsistency in databases. Indeed, with the huge amount of data available in the Internet, it is likely to find some conflicting information when gathering data from different sites. Also, in situations requiring user feedback one are often faced with contradicting information. For instance, when the preferences of multiple users are collected or when a single user describes his preferences through a series of interactions with a system, one are likely to have inconsistent information.

1.1 Inconsistency originating from data integration

The treatment of inconsistencies arising from the integration of multiple sources has been a topic increasingly studied in the past years and has become an important field of research in databases. Roughly, there are two approaches to handle the inconsistency problem in knowledge bases: belief revision ([Kifer and Lozinskii, 1992; Subrahmanian, 1994]) and paraconsistent logic ([Blair and Subrahmanian, 1987]). The goal of the first approach is to make an inconsistent theory consistent, either by revising it or by representing it by a consistent semantics. The main concern of this approach is to avoid contradictions. On the other hand, the paraconsistent approach allows reasoning in the presence of inconsistency, and contradictory information can be derived or introduced without trivialization. Besides these two approaches, there are "hybrid" approaches based on formalisms which associates degrees of belief, reliability or uncertainty to the source knowledge bases ([Cholvy, 1998; Benferhat et al., 1995]).

1.2 Inconsistency originating from user feedback

In most AI applications involving the ability of making decisions, users are required to compare different alternatives and must be able to choose those which better conform to their needs or personal preferences. Thus, such AI applications must support the ability of automate the preference elicitation process and also provide mechanisms to infer user preferences from some previous feedback obtained from the user. Several formalisms and frameworks have been proposed so far for preference modeling and reasoning. The approach of CP-Nets [Boutilier et al., 2004] is a very elegant, natural and intuitive formalism based on a graphical model which captures users qualitative *conditional preference* over tuples, under a *ceteris paribus* semantics (the "all else being equal" assumption). The presence of cycles in a CP-Net occurs often in real life situations. In fact, a cycle in the CP-Net may denote inconsistent or contradicting information, which may occur when the preferences of multiple users are collected or when a single user describes his preferences through a series of interactions with a system. In this case, contradicting information is easily generated: at different stages of the process he may affirm that he prefers A to B as well as B to A.

Besides the simple CP-Net graphical model, two logical formalisms are used for preference representation: Qualitative Choice Logic (QCL) and Possibilistic Logic. QCL [Brewka et al., 2004] is a non-monotonic propositional logic for representing alternatives, or ranking options for problem solutions. The logic is different from usual non-monotonic logics: it does not employ non-standard inference rules or modal operators expressing consistency or belief. It simply incorporate a new connective $\times$, called *ordered disjunction* with the following semantics: *if it is possible to have A, great ! If not, I may accept B instead.* Possibilistic Logic [Dubois and Prade, 2004] manipulates pairs of classical logic formulas associated with priority levels. *Priority levels* are incorporated into the language by means of specific logic symbols. This has several advantages such as enabling the simultaneous handling of generic preferences and case-based preferences as well as positive and negative preferences. Negative preferences refer to what is rejected and should be avoided. Positive preferences point out what is desirable or satisfactory [Dubois and Prade, 2008].

At the best of our knowledge there is no formalism for preference modeling and reasoning based on a paraconsistent logic. In Section 8 we briefly present our ongoing research on preference modeling using a stochastic method and discuss how the ideas presented in this paper could be further explored in order to produce a more flexible formalism for reasoning with preferences.

1.3 The objective of this paper

In this paper we present P-Datalog, a deductive language for querying databases containing inconsistencies. The theoretical foundations of the language are presented in Sections 3 to 6. These sections constitute an slight extended version of [de Amo and Pais, 2007]. In Section 7 we discuss the implementation of PDLog, an automate prover based on P-Datalog. In Section 8 we discuss some ideas we are developing to extend P-Datalog to other scenarios where inconsistent information is frequently present.

P-Datalog is based on a paraconsistent approach, that is, inconsistencies are kept and not simply rejected. Following the arguments presented in [Gabbay and Hunter, 1991], our choice is motivated by the assumption that, in most situations, inconsistent information can be useful, unavoidable and even desirable. Thus, in most situations, the goal of retaining all available information is quite legitimate since discarding inconsistent information would imply *losing* information.

P-Datalog is a language which allows inferring facts from a database $\mathcal{K}$ obtained by integrating *local consistent* sources which may be *globally inconsistent*: they may be contradictory with respect to each other. In this paper, we focus our attention in the integrated database, which possibly contains inconsistent information. We are not interested in the integration process, that is, which information is kept in the source databases or how the integrated database is built. These issues have been treated in a previous paper [de Amo et al., 2002], where we proposed a system based on a deductive proof mechanism to integrate information coming from multiple databases. The intuitive idea of the semantics of facts stored in the integrated database $\mathcal{K}$ is the following: a fact $A \in \mathcal{K}$ is **true** (resp. **false**) if it is true (resp. false) in *all* source databases. It is **inconsistent** if it is true in *some* local source and false in *some* other. The language P-Datalog allows to infer new facts from the facts stored in the integrated database $\mathcal{K}$. These new facts inferred are related to the facts which *would be inferred* in each individual consistent source. If an inferred fact A is **true** in the global database (the integrated one), then it would be locally inferred as true in *all* individual sources: it is a safe information and it is surely true. If it is globally **inconsistent** then it would be locally inferred as true in *some* individual sources and as false in others: there is some evidence of A in $\mathcal{K}$. If it is globally **false**, then it would be locally inferred as false in *all* individual sources.

The syntax of P-Datalog slightly differs from Datalog¬ syntax ([Abiteboul et al., 1995]). As in Datalog¬, P-Datalog programs are normal logic programs ([Lloyd, 1993]) (the same as the *general logic programs* of [Van Gelder et al.,

1991]): a set of rules where negation may appear in the body but not in the head of rules. P-Datalog programs may also include rules with the truth-value **i** (inconsistent) in their bodies. In fact, the main difference between P-Datalog and Datalog¬ concerns their semantics. In the classical context of Datalog¬, the rules are first-order formulas (Horn clauses with (possibly) negated literals in the body). The answers to a Datalog¬ query constitute a set of facts where each fact has an associated truth-value **t** (true), **f** (false) or **u** (unknown). In our approach, the rules of a P-Datalog program are Horn clauses like in Datalog¬, but their *semantics* is related to the *paraconsistent logic* **LFI1** which was originally introduced in [de Amo et al., 2000; de Amo et al., 2002] as a logical framework to model database integration[1]. An answer to a P-Datalog query is a set of facts, where each fact has an associated truth-value which can be **t** (true), **f** (false), **u** (unknown) or **i** (inconsistent).

In order to define the 4-valued well-founded semantics of a P-Datalog query, we take advantage of the natural 3-valued semantics of the paraconsistent logic **LFI1** (where the truth-values are **t**, **f** and **i**). We adapt the ideas of [Przymusinski, 1990; Van Gelder, 1989] originally presented in the context of Datalog¬ programs. In this classical setting, 2-valued first-order logic models are augmented with a third truth-value **u** (unknown). In our setting, 3-valued **LFI1** models are augmented with the truth-value **u** (unknown) as well. Thus, the 4-valued semantics for P-Datalog programs we propose is a natural extension of the 3-valued well-founded semantics of Datalog¬ programs.

Differently from some approaches treating paraconsistent query languages ([Pereira and Alferes, 1992; Sakama, 1992; Blair and Subrahmanian, 1987; Subrahmanian, 1994]), our well-founded semantics is a natural extension of the well-founded semantics for Datalog¬ programs proposed by [Przymusinski, 1990][2]. In this paper, we also present a bottom-up evaluation procedure for computing the well-founded semantics based on the alternating fixpoint computation introduced in [Van Gelder, 1989].

This paper is a full version of the conference paper [de Amo and Pais, 2005].

1.4 Motivating Example

The following example illustrates our approach for querying inconsistent data:

EXAMPLE 1 (Motivation). Suppose we have the following rule in a dishonest public contest for hiring civil servants: "if there is *some evidence* that the candidate is supported by an influential person which is not a civil servant himself and if the candidate has no debts towards the income tax services, then there is some evidence that this candidate will get the job." The intuitive meaning behind the expression *there is some evidence* is that this information is supported by *at least* one source, even though some sources may affirm the contrary. This information is *controversial*, that is, it comes from the integra-

---

[1] In fact, **LFI1** belongs to a family of paraconsistent logics called *Logics of Formal Inconsistency* (**LFI**). An extensive and comprehensive presentation of **LFI1** and the family of paraconsistent logics **LFI** can be found in [de Amo et al., 2000] and [Carnielli and Marcos, 2002] respectively.

[2] Other approaches for associating a meaning to a program by means of a Herbrand model verifying some special properties have been proposed in the literature ([Van Gelder et al., 1991; Bidoit and Legay, 1990]).

tion of different sources $K_1, ..., K_n$. In some of these sources, the information is true and in some other it is false. We can translate the story above in the following P-Datalog program P_{job}:

$$job(x) \leftarrow \sim owe(x), supportedby(x,y), \sim job(y)$$

Notice that the unique rule of P_{job} follows the syntax of Datalog¬. However, its semantics is evaluated according to the semantic laws of the paraconsistent logic **LFI1**. In this logic, an atomic formula $R(\vec{x})$ is *verified* if its truth-value is **t** or **i** (in a paraconsistent approach, inconsistencies are not rejected). Thus in the P_{job} program, literals *supportedby(x,y)* (in the body) and *job(x)* (in the head) represent information that are true or inconsistent. On the other hand, the literals $\sim owe(x)$ and $\sim job(y)$ represent negative information. Intuitively, this means that all information sources affirm the fact that x has no records in the income tax services files concerning debts and that y is not a civil servant. Let us suppose that we have the following facts stored in the integrated database:

$\mathcal{K}$ = {∘ *supportedby(charles,joseph)*, ∘ *supportedby(joseph,charles)*,
∘ *supportedby(paul,james)*, • *supportedby(john,kevin)*,
∘ *supportedby(james,kevin)*, • *owe(james)*}

The symbols ∘ and • attached to each fact in the database mean that the fact is *sure* and *controversial*, respectively. We notice that the facts stored in the database must be explicitly declared as sure or controversial (by attaching these symbols ∘ and •). Following the closed-world assumption, facts that are not in the database are considered false. In order to better understand the semantics of each fact stored in the database $\mathcal{K}$, we can view it as the integration of several local consistent sources $\mathcal{K}_1, \ldots, \mathcal{K}_n$. In each $\mathcal{K}_i$, each fact *supportedby(a,b)* and *owe(c)* has truth-value **t** or **f**. The fact *supportedby(charles,joseph)* is **true** in every $\mathcal{K}_i$ $(i = 1, \ldots, n)$ and *owe(john)* is **false** in every $\mathcal{K}_i$ $(i = 1, \ldots, n)$. Whereas *owe(james)* is **true** in some $\mathcal{K}_j$ $(j = 1, ..., n)$ and **false** in some $\mathcal{K}_p$ $(p = 1, ..., n)$, there is a $\mathcal{K}_i$ $(i = 1, \ldots, n)$ where *owe(james)* is **true**, and there is a $\mathcal{K}_j$ $(j = 1, \ldots, n)$ where *owe(james)* is **true**.

We now show a 4-valued model $\mathcal{J}$ of P_{job} which includes the facts of the database $\mathcal{K}$, i.e., $\mathcal{J}$ agrees with $\mathcal{K}$ on the values of *owe* and *supportedby* atoms. This 4-valued model $\mathcal{J}$ contains the facts *job(x)* which corresponds to the answer to the query "For which people is there some evidence that they will get the job?" As we will show later (see Example 23), this model $\mathcal{J}$ is the P-Datalog well-founded semantics of P_{job} on input $\mathcal{K}$. The values of the *job* atoms in the derived database $\mathcal{J}$ are the following:

true (t): *job(paul)*
false (f): *job(kevin), job(james)*
inconsistent (i): *job(john)*
unknown (u): *job(charles), job(joseph)*

This model asserts that *James* surely does not get the job: there is some evidence that he owes to the taxation office. From this we can infer that *Paul* surely get the job. Indeed, *Paul* does not owe any tax return and he is supported by *James* who is not a civil servant. It also can be deduced that *Kevin* does not succeed in getting the job because nobody supports him. In *John*'s case, he does not owe the taxation office but it is controversial that he is supported by *Kevin*, who is not a public servant himself. Thus it is controversial that *John* gets the job. On the other hand, it is unknown that *Charles* and *Joseph* succeed in the public contest. They fulfill almost all the requirements: they do not have debts, they have the support of an influential person but they depend on each other: *Charles* supports *Joseph* and *Joseph* supports *Charles*. The only chance for Charles getting the job is if Joseph (his only support) does not get it. And vice-versa, the only chance for Joseph getting the job is if Charles (his only support) does not get it. Therefore it is not possible to infer which one will get the job: either Charles or Joseph. This means that this information is *unknown*: we *are not able to infer* the existence or nonexistence of any source supporting it.

The answer to our query : *"For which people is there some evidence that they will get the job?"* is Paul and John. Besides, we know that Paul surely gets the job, but in John's case, we only can affirm that it is controversial that he gets the job. This is derived from an inconsistent fact (•*supportedby(john, kevin)*) in the integrated database $\mathcal{K}$. That means: (1) ¿From the point of view of local sources $\mathcal{K}_i$ affirming that John *is* supported by Kevin, it is inferred that John gets the job, (2) From the point of view of sources in $\mathcal{K}_j$ affirming that John *is not* supported by Kevin, it is inferred that John does not get the job (we remind that in all local sources the information *owe(john)* is false and *job(kevin)* is *derived* as false, so ∼*owe(john)* and ∼*job(kevin)* are true). Therefore, in the integrated database $\mathcal{K}$, as expected, the information *job(john)* is *derived* as inconsistent.

1.5 Paper organization

The paper is organized as follows: In Section 2 we briefly describe the basic notions of the logic **LFI1**. In Section 3, we introduce P-Datalog programs and generalize the notion of database instance to allow the storage of inconsistent information in databases. In Section 4 we describe the well-founded semantics of a P-Datalog program. In Section 5, we present a bottom-up method for evaluating P-Datalog programs based on an alternating fixpoint operator and briefly discuss its implementation. In Section 6, we discuss some related work. In Section 7, we briefly discuss the implementation of the method we proposed for evaluating P-Datalog queries. Finally, in Section 8 we present our perspectives for further work as well as our ongoing research.

2 LFI1 : A 3-valued Paraconsistent Logic

In this section we briefly describe the syntax and semantics of **LFI1**. A detailed presentation can be found in [Carnielli and Marcos, 2002; de Amo et al., 2000]. The semantics of a P-Datalog program is based on the semantics of **LFI1**. Even if P-Datalog programs constitute a small fragment of the set of **LFI1** formulas (we only consider Prolog-like Horn clauses), inference in P-Datalog is based on the paraconsistent framework of **LFI1**.

Let **R** be a finite signature without functional symbols and **Var** a set of

	¬	•
t	f	f
i	i	t
f	t	f

∨	t	i	f
t	t	t	t
i	t	i	i
f	t	i	f

∧	t	i	f
t	t	i	f
i	i	i	f
f	f	f	f

→	t	i	f
t	t	i	f
i	t	i	f
f	t	t	t

Figure 1. LFI1 3-valued connective matrices

variables symbols. We assume that formulas of **LFI1** are defined in the usual way, as in the classical first-order logic setting, with the addition of a new symbol • (read "it is inconsistent"). A formula of **LFI1** is defined inductively by the following statements (and only by them) :
1. If R is a predicate symbol of arity k and $x_1, ..., x_k$ are constants or variables, then $R(x_1, ..., x_k)$ and $x_1 = x_2$ are atomic formulas or atoms. The former is called a *relational* atom and the later an *equality* atom.
2. If F, G are formulas and x is a variable then $F \vee G$, $\neg F$, $\forall x F$, $\exists x F$ and $\bullet F$ are formulas.

A *sentence* is a formula without free variables. A *fact* is a relational atom without free variables. We denote by $\mathcal{F}$ the set of facts.

In the remainder, we often will use the symbols • and ○ in two different contexts: a syntactic one and a semantic one. In **LFI1**, the symbols • and ○ are new modalities which allow extending the syntax of first-order formulas. So, in **LFI1**, these symbols are used in a syntactic context. In the Example 1, these symbols have been used in a semantic context, in order to identify the truth-value of facts stored in the database. We think that these two forms of using the symbols ○ and • will not cause confusion, because the P-Datalog syntax do not include them in its specification. So, in this paper, the use of ○ and • will be restricted to the semantic context.

DEFINITION 2. Let **R** be a finite signature and $\mathcal{F}$ the set of facts defined over **R**. An *interpretation* over **R** is an application $\delta : \mathcal{F} \to \{\mathbf{f}$ (false), $\mathbf{t}$ (true), $\mathbf{i}$ (inconsistent)$\}$.

An interpretation of facts can be extended to the propositional sentences in a natural way by using the connective matrices described in Figure 1. Notice that the major difference between **LFI1** and Kleene's three-valued logic concerns the way both logics interpret the implication $\mathbf{i} \to v$, where $v \in \{\mathbf{t}, \mathbf{i}, \mathbf{f}\}$[3]. This can be explained by the fact that in Kleene's Logic the truth-value $\mathbf{i}$ characterizes *undefined* information, whereas in **LFI1**, it stands for *overdefined* (*inconsistent*) information.

The connective $\wedge$ is derived from of $\vee, \neg$: $A \wedge B = \neg(\neg A \vee \neg B)$ and the connective $\to$ is derived from $\neg, \vee, \bullet$: $A \to B = \neg(A \vee \bullet A) \vee B$. In the next section, we will use a derived connective $\sim$, called *negation by default*, defined as $\sim A = \neg A \wedge \neg \bullet A$.

The extension of δ to the quantified sentences is not treated here. The reason is that the formulas we will deal with in the next sections are Horn clauses, which are interpreted over a finite Herbrand Universe. So, the universal

---
[3] In Kleene's Logic, the truth-value of $\mathbf{i} \to v$ is $\mathbf{i}$, for all $v \in \{\mathbf{t}, \mathbf{i}, \mathbf{f}\}$.

quantifiers appearing in the clauses can be viewed as a bounded conjunction. The details of the semantics of quantifiers can be found in [de Amo et al., 2000].

obtained by means of the concept of *distribution quantifiers*, introduced in [Carnielli, 1987]. Basically, this concept translates our basic intuition that an universal quantifier should work as a kind of unbounded conjunction and an existential quantifier as an unbounded disjunction. Due to lack of space, we do not present this extension here. For more details, see [de Amo et al., 2000]. The formulas We extend δ to the quantified sentences as follows :

- $\delta(\forall x A(x)) = 1$ iff for all valuations v we have $\delta(A[v(x)/x]) = 1$,
- $\delta(\forall x A(x)) = 0$ iff there exists a valuation v such that $\delta(A[v(x)/x]) = 0$,
- $\delta(\forall x A(x)) = \frac{1}{2}$ iff for all valuations v we have $\delta(A[v(x)/x]) = 1$ or $\frac{1}{2}$, and there exists a valuation v' such that $\delta(A[v'(x)/x]) = \frac{1}{2}$
- $\delta(\exists x A(x)) = 1$ iff there exists a valuation v such that $\delta(A[v(x)/x]) = 1$,
- $\delta(\exists x A(x)) = 0$ iff for all valuations v we have $\delta(A[v(x)/x]) = 0$,
- $\delta(\exists x A(x)) = \frac{1}{2}$ iff for all valuations v we have $\delta(A[v(x)/x]) = 0$ or $\frac{1}{2}$, and there exists a valuation v' such that $\delta(A[v'(x)/x]) = \frac{1}{2}$.

It is easy to see that $\delta(\forall x A(x)) = \delta(\neg \exists x \neg A(x))$ as usual in classical first-order logic. We denote by **Dom** the Herbrand Universe of **R** (the constant symbols of **R**). In fact, we are supposing that the universe domain of any interpretation δ is **Dom** (thus, δ interprets the constant symbols by themselves). A *valuation* is an application $v : $ **Var** $\rightarrow$ **Dom**.

DEFINITION 3. Let $F(x_1, ..., x_n)$ be a formula of **LFI1** with free variables $x_1, ..., x_n$, v a valuation and δ an interpretation. We say that (δ, v) *satisfies* $F(x_1, ..., x_n)$ (denoted by $(\delta, v) \models F(x_1, ..., x_n)$) iff $\delta(F[v(x_1), ..., v(x_n)/x_1, ..., x_n])$ is **t** or **i**. If $(\delta, v) \models F$ for each valuation v, we say that δ is a *model* of F (denoted $\delta \models F$). We also say that F is *verified* or *satisfied* by δ.

Remark. We notice that the notion of *equivalence* between formulas in **LFI1** is different from the one we are used in classical logic. We say that two formulas F and G are *equivalent* if for all interpretations δ, δ satisfies F if and only if δ satisfies G. In fact, this is the same definition used in classical logic, except that in the latter, *being equivalent* is the same as *having the same truth-value*. This is not the case for **LFI1**. For instance, the formulas $\sim\sim A$ and A are equivalent in **LFI1**, but they don't have the same truth-value under some interpretations. If δ is an interpretation such that $\delta(A) = $ **i**, we have $\delta(\sim\sim A) = $ **t**. So, the two formulas have not the same truth-values, but both are satisfied by δ. Notice that $\neg A$ is *equivalent* to $\sim A \vee \bullet A$ in **LFI1**.

3 The P-Datalog Query Language

In this section we use the logical formalism **LFI1** to generalize the notion of database instance to allow the storage of inconsistent information in our databases. We also introduce the P-Datalog query language which is designed to query databases containing inconsistent information. We assume that the reader is familiar with traditional database terminology ([Abiteboul et al., 1995]). In what follows, we denote by $R(\vec{u})$ the formula $R(u_1, ..., u_k)$, where $u_1, ..., u_k$ are variables and we denote by $R(\vec{a})$ the ground atom (or *fact*) $R(a_1, ..., a_k)$, where $a_1, ..., a_k$ are constants (k is called the *arity* of R).

DEFINITION 4 (Paraconsistent Databases). Let **R** be a database schema (or signature), i.e., a set of relation names (or predicate names) and a set **Dom** of constants (the Herbrand Universe of its instances). A *3-valued instance* over **R** (or a *paraconsistent database*) is an interpretation **I** such that for each R ∈ **R** the set $\mathbf{I}_R = \{\vec{a} : \mathbf{I}(R(\vec{a})) = \mathbf{t} \text{ or } \mathbf{I}(R(\vec{a})) = \mathbf{i}\}$ is finite. So, an instance over **R** can be viewed as a finite set of facts over **R**, having truth-values **t** or **i**. The facts which are not in the instance **I** have truth-value **f**. A fact $R(\vec{a})$ such that $\mathbf{I}(R(\vec{a})) = \mathbf{i}$ is intended to be *controversial*. On the other hand, if $\mathbf{I}(R(\vec{a})) = \mathbf{t}$, $R(\vec{a})$ is intended to be a safe information.

Notation. In what follows, we will denote by $\circ R(\vec{a})$ the fact that $\mathbf{I}(R(\vec{a})) = \mathbf{t}$ and by $\bullet R(\vec{a})$ the fact that $\mathbf{I}(R(\vec{a})) = \mathbf{i}$. Here, we use the symbols ∘ and • in a semantic context.

P-Datalog is an extension of Datalog¬ ([Abiteboul et al., 1995]). This well-known deductive query language uses the classical first order logic as its underlying logic, and a Datalog¬ query applies over a classical database instance, i.e. a finite first-order interpretation. Rather than classical first-order logic, P-Datalog uses the paraconsistent logic **LFI1** as its underlying logic, and P-Datalog queries apply over paraconsistent databases. P-Datalog programs are first-order Horn clauses as in Datalog¬ programs, i.e. first-order clauses (Horn clauses) with positive and negative literals in their bodies. Negation in P-Datalog (as well as in Datalog¬) is understood as the *negation by default* ∼. The negation ¬ used in **LFI1** is called *weak negation*.

The intuitive meaning behind default and weak negations is the following: (1) the ground formula $\sim R(\vec{a})$ is verified by a paraconsistent database **I** if the fact $R(\vec{a})$ is not in **I**; (2) the ground formula $\neg R(\vec{a})$ is verified by **I** if the fact $R(\vec{a})$ is in **I** as controversial or if it is not in **I**[4].

DEFINITION 5 (P-Datalog Programs). A P-Datalog *program* is a finite set of rules $A \leftarrow L_1, ..., L_n$, where A is an atom of the form $R(\vec{u})$, and L_i are literals of the form: $R(\vec{u})$ or $\sim R(\vec{u})$. R is a relation name and $\vec{u}$ is a free tuple of appropriate arity. The atom A is called the *head* of the rule. The literals $L_1, ..., L_n$ constitute the *body* of the rule. One requires also that each variable occurring in the head of the rule must occur in at least one of the free tuples in the body.

Let P be a P-Datalog program and **I** a paraconsistent database (a 3-valued instance). We denote by $P_\mathbf{I}$ the program obtained from P by adding to P unit clauses $A \leftarrow$ for each A such that $\mathbf{I}(A) = \mathbf{t}$ or **i**. From now on, we suppose that our programs include these unit clauses corresponding to the input facts of **I**. We denote by $sch(P)$ the set of relations (predicates) appearing in P, by $adom(P)$ the set of constants appearing in P and by $\mathbf{B}(P)$ all facts of the form $R(\vec{a})$ where $R \in sch(P)$ and $\vec{a}$ is a tuple of constants in $adom(P)$ (the Herbrand Base of P). The set of relations which appears in the head of rules are called the *intensional relations* and is denoted by $idb(P)$. The set of those appearing

---

[4] In [de Amo et al., 2000] the default negation ∼ is referred to as *strong negation*, since ∼ enjoys most of the *semantic* properties of classical negation. For instance: "for all interpretations I (in the Logic LFI1), I satisfies A if and only if I does not satisfy $\sim A$". Notice that weak negation ¬ does not verify this property.

only in the body of rules are called *extensional relations* and is denoted by $edb(P)$.

DEFINITION 6 (P-Datalog Query). A P-Datalog *query* is a pair $(P,Q(u_1,...,u_n))$ where P is a P-Datalog program, $Q \in idb(P)$ and $u_1, ..., u_n$ are variables or constants in $adom(P)$ (n is the *arity* of the relation Q).

EXAMPLE 7 (Running Example). Let us consider the same situation presented in Example 1. The rule P_{job} and the 3-valued instance **I** described in that example constitute a P-Datalog program P where $sch(P) = \{supportedby, job, owe\}$, $adom(P) = \{charles, john, james, joseph, paul, kevin\}$ and $\mathbf{B}(P) = \{owe(charles), owe(joseph), supportedby(charles,joseph), supportedby(joseph, charles), ...\}$. The intensional and extensional schemas are $edb(P) = \{supportedby, owe\}$, $idb(P) = \{job\}$. The pair $(P_{job}, job(x))$ is a P-Datalog query ("For which people is there some evidence that they will get the job?"). The pair $(P_{job}, job(Kevin))$ corresponds to the boolean query "Is there some evidence that Kevin may get the job?"

4 Answering P-Datalog Queries

In this section we introduce the well-founded semantics for P-Datalog programs. The well-founded semantics of a P-Datalog program P is designed to capture the natural semantics of queries $(P, Q(u_1, ..., u_n))$ where $Q \in idb(P)$, that is, what their answers are expected to be. Our approach is a natural extension of the well-founded semantics for Datalog¬ ([Przymusinski, 1990]). Our definition of a P-Datalog query makes use of *4-valued instances*, in which facts may assume one of the four truth-values in the set ***Val*** = $\{true(\mathbf{t}), false(\mathbf{f}), inconsistent(\mathbf{i}) \ unknown(\mathbf{u})\}$. In what follows, we assume that the reader is familiar with the notions of lattices, lattice operators, monotonicity and continuity, fixpoints, etc. For details, see [Lloyd, 1993].

4.1 4-valued Models

Let us consider the complete lattice (***Val***, $\leqslant$), where $\mathbf{f} \leqslant \mathbf{u} \leqslant \mathbf{i} \leqslant \mathbf{t}$. We notice that the 4-valued logic introduced by N. Belnap in [Belnap, 1977] can also deal with both inconsistent (**i**) and incomplete (**u**) information. The main difference between Belnap's logic and the underlying 4-valued logic of P-Datalog programs we will introduce in this section concerns the fact that in the latter, the lattice (***Val***, $\leqslant$) constitutes a *total* order, whereas in Belnap's approach the lattices considered are partial orders. As we will see, a total ordering over the set of truth-values ***Val*** is convenient for generalizing, in a natural way, the 3-valued well-founded semantics of Datalog¬ programs to a 4-valued well-founded semantics for P-Datalog programs.

DEFINITION 8. Let P be a P-Datalog program. A *4-valued instance* **I** over $sch(P)$ is an application $\mathbf{I} : \mathbf{B}(P) \longrightarrow \{\mathbf{t, f, u, i}\}$. If $\mathbf{I}(A) \neq \mathbf{u}$ for all $A \in \mathbf{B}(P)$, we say that **I** is *total*.

The answer of a P-Datalog program P is a special 4-valued instance which corresponds to the well-founded semantics of P. The main goal of this section is to define this particular instance.

~	
t	f
i	f
u	u
f	t

∨	t	i	u	f
t	t	t	t	t
i	t	i	i	i
u	t	i	u	u
f	t	i	u	f

∧	t	i	u	f
t	t	i	u	f
i	i	i	u	f
u	u	u	u	f
f	f	f	f	f

→	t	i	u	f
t	t	f	f	f
i	t	i	f	f
u	t	t	t	f
f	t	t	t	t

Figure 2. P-Datalog 4-valued connective matrices

There is a natural ordering $\preccurlyeq$ among 4-valued instances over $sch(P)$, defined by: $\mathbf{I} \preccurlyeq \mathbf{J}$ iff for each $A \in \mathbf{B}(P)$, $\mathbf{I}(A) \leqslant \mathbf{J}(A)$.

The set of 4-valued instances of a P-Datalog program P is denoted by 4-$Inst_P$. It is easy to verify that (4-$Inst_P, \preccurlyeq$) constitutes a complete lattice. We denote by $\top$ the *maximal* 4-valued instance (where all facts have truth-value $\mathbf{t}$) and by $\bot$ the *minimal* 4-valued instance (where all facts have truth-value $\mathbf{f}$). We also represent a 4-valued instance by listing the positive, inconsistent and negative facts, and omitting the unknown ones.

EXAMPLE 9 (4-valued instance). Let $\mathbf{J}$ be a 4-valued instance, where $\mathbf{J}(p){=}\mathbf{t}$, $\mathbf{J}(q){=}\mathbf{t}$, $\mathbf{J}(r){=}\mathbf{u}$ and $\mathbf{J}(s){=}\mathbf{f}$. $\mathbf{J}$ can be written as $\mathbf{J} = \{\circ p, \circ q, \sim s\}$. Let $\mathbf{J}' = \{\circ p, \circ q, \bullet s\}$. Then $\mathbf{J} \preccurlyeq \mathbf{J}'$.

We extend the 3-valued connective matrices of **LFI1** (see Figure 1), to the 4-valued matrices showed in Figure 2.

If F is the body of a P-Datalog rule and $\mathbf{J}$ is a 4-valued instance, we denote by $\mathbf{J}(F)$ the truth-value associated to F according to the matrices for the $\wedge$ and $\sim$ connectives given in Figure 2.

We notice that the matrix corresponding to the $\rightarrow$ connective in Figure 2 is not a straight extension of its counterpart in Figure 1: in P-Datalog, the truth-value of $\mathbf{t} \rightarrow \mathbf{i}$ is $\mathbf{f}$ and in **LFI1** this truth-value is $\mathbf{i}$. In fact, in the P-Datalog context an inconsistent fact cannot be derived from a set of consistent true facts. This kind of inference is accepted in **LFI1** (the truth-value of the implication being $\mathbf{i}$, and so, accepted in the paraconsistent logic)

Let $\mathbf{I}$ be a 4-valued instance and $A \leftarrow G$ be a P-Datalog rule (here, G denotes a conjunction of literals of the form B or $\sim B$). Then

$$\mathbf{I}(A \leftarrow G) = \begin{cases} \mathbf{i} & \text{if } \mathbf{I}(A)=\mathbf{I}(G)=\mathbf{i}, \\ \mathbf{t} & \text{if } \mathbf{I}(A) > \mathbf{I}(G) \text{ or } \mathbf{I}(A) = \mathbf{I}(G) \neq \mathbf{i} \\ \mathbf{f} & \text{otherwise} \end{cases}$$

DEFINITION 10. Let P be a P-Datalog program. An *instantiated rule* of P is a rule where all variables are replaced by constants in $adom(P)$. We denote by $ground(P)$ the set of instantiated rules of P. A 4-valued instance $\mathbf{J}$ over $sch(P)$ *satisfies* a boolean combination α of atoms in $\mathbf{B}(P)$ iff $\mathbf{J}(\alpha) \in \{\mathbf{t}, \mathbf{i}\}$. A *4-valued model* of P is a 4-valued instance $\mathbf{J}$ over $sch(P)$ satisfying each rule in $ground(P)$, i.e., the truth-value of each rule in $ground(P)$ is $\mathbf{t}$ or $\mathbf{i}$. A 4-valued model M is *minimal* iff for all $M' \subset M$, M' is not a model.

As we claimed in the introduction, **LFI1** is the underlying logic of P-Datalog. This is justified by the following proposition:

PROPOSITION 11. Let $\mathbf{J}$ be a paraconsistent instance. Then:

1. If **J** satisfies a ground P-Datalog rule $A \leftarrow L_1, \ldots, L_n$ in the P-Datalog semantics (using the matrices of Figure 2) then **J** satisfies the rule in the **LFI1** semantics (using the matrices of Figure 1).

2. If **J** satisfies a P-Datalog program P then **J** satisfies the **LFI1** formula corresponding to the finite conjunction of all instantiated rules of P.

Proof.
P-Datalog 4-valued connective matrices are an extended version of **LFI1** matrices with one particular difference in the $\leftarrow$ connective: the case **i**$\leftarrow$**t**, which is **f** in P-Datalog and **i** in **LFI1**.
(1) Suppose the instance **J** satisfies the rule $A \leftarrow G$ in P-Datalog, where A is an atom and G is a conjunction of literals of the form B or $\sim B$. By definition 10, $\mathbf{J}(A \leftarrow G) \in \{\mathbf{i}, \mathbf{t}\}$. By the definition of the semantics of the $\leftarrow$ connective, $\mathbf{J}(A) \geqslant \mathbf{J}(G)$:
(a) $\mathbf{J}(A)=\mathbf{t}$ and $\mathbf{J}(G) \in \{\mathbf{t,i,f}\} \Rightarrow$ **J** satisfies $A \leftarrow G$ in **LFI1**, because $\mathbf{J}(A \leftarrow G) = \mathbf{t}$ in **LFI1**.
(b) $\mathbf{J}(A)=\mathbf{i}$ and $\mathbf{J}(G) \in \{\mathbf{i, f}\} \Rightarrow$ **J** satisfies $A \leftarrow G$ in **LFI1**, because $\mathbf{J}(A \leftarrow G) \in \{\mathbf{t,i}\}$ in **LFI1**.
(c) $\mathbf{J}(A)=\mathbf{f}$ and $\mathbf{J}(G)=\mathbf{f} \Rightarrow$ **J** satisfies $A \leftarrow G$ in **LFI1**, because $\mathbf{J}(A \leftarrow G) = \mathbf{t}$ in **LFI1**.
(2) Follows from the fact that a P-Datalog program is a conjunction of rules and each rule is the conjunction of its ground instances[5]. ∎

4.2 Extended P-Datalog programs

The well-founded semantics of P-Datalog programs is based on the notion of stable models ([Gelfond and Lifschitz, 1988]). Stable models are usually defined as fixpoint of an immediate consequence operator. Following the same idea underlying the definition of 3-stable models in [Przymusinski, 1990], we introduce the notion of *extended P-Datalog programs*. We will see that for such programs we can define an immediate consequence operator which is monotonic and has a unique least fixpoint.

DEFINITION 12. An *extended P-Datalog program* is a P-Datalog program where (1) negative facts $\sim A$ do not appear in the body of rules; (2) truth-values **t, f, u** and **i** may occur as literals in the body of rules.

Next we define the *immediate consequence operator $4\text{-}T_P$ associated to an extended program P*.

DEFINITION 13.
Let P be an extended P-Datalog program. The *immediate consequence operator $4\text{-}T_P$ associated to P* is a mapping $4\text{-}T_P : 4\text{-}Inst_P \to 4\text{-}Inst_P$ defined as follows. Let **J** be a 4-valued instance and $A \in \mathbf{B}(P)$, then

---
[5] A rule is a closed universally quantified sentence in **LFI1**. Because the Herbrand Universe is finite, the rule reduces to the conjunction of its ground instances.

$$4\text{-}T_P(\mathbf{J})(A) = \begin{cases} \max\{\mathbf{J}(F_k)\} & \text{for all rules } A \leftarrow F_k \text{ in } ground(P), \\ & 0 \leqslant k \leqslant n. \\ \mathbf{f} & \text{otherwise} \end{cases}$$

We define the sequence $\{4\text{-}T_P^i(\bot)\}_{i\geqslant 0} = \{\mathbf{I}_0, \mathbf{I}_1, \mathbf{I}_2, \ldots\}$ of 4-valued instances as follows: $\mathbf{I}_0 = \bot$, $\mathbf{I}_1 = 4\text{-}T_P(\mathbf{I}_0)$, $\mathbf{I}_2 = 4\text{-}T_P(\mathbf{I}_1)$, ...

The following lemma says that the immediate consequence operator for an extended program P has a least fixpoint which coincides with the unique minimal 4-valued model of P.

LEMMA 14. *Let P be an extended P-Datalog program.*

1. *The operator $4\text{-}T_P$ is monotonic;*

2. *The 4-valued instance M is a 4-valued model of P iff $4\text{-}T_P(M) \preccurlyeq M$;*

3. *The sequence $\{4\text{-}T_P^i(\bot)\}_{i\geqslant 0}$ is increasing and converges to the least fixpoint of $4\text{-}T_P$;*

4. *P has a unique 4-valued minimal model (denoted by $P(\bot)$) that equals the least fixpoint of $4\text{-}T_P$.*

Proof. (1) Suppose that $A \in \mathbf{B}(P)$, $\mathbf{I}(A) \leqslant \mathbf{J}(A)$. We have to show that $4\text{-}T_P(\mathbf{I})(A) \leqslant 4\text{-}T_P(\mathbf{J})(A)$. We have the following possibilities: (a) There are rules of the form $A \leftarrow F$ in $ground(P)$, where F corresponds to a conjunction of atoms with truth-values $\mathbf{t}, \mathbf{i}, \mathbf{u}, \mathbf{f}$. Then $4\text{-}T_P(\mathbf{I})(A) = max\{\mathbf{I}(F)\}$ and $4\text{-}T_P(\mathbf{J})(A) = max\{\mathbf{J}(F)\}$. By the fact that $\mathbf{I}(A) \leqslant \mathbf{J}(A)$, $max\{\mathbf{I}(F)\} \leqslant max\{\mathbf{J}(F)\}$. So $4\text{-}T_P(\mathbf{I})(A) \leqslant 4\text{-}T_P(\mathbf{J})(A)$. (b) There are no rules in $ground(P)$ with A in the head. In this case, $4\text{-}T_P(\mathbf{I})(A) = 4\text{-}T_P(\mathbf{J})(A) = \mathbf{f}$.
($2 \rightarrow$): Let M a 4-valued model of P and let $A \in \mathbf{B}(P)$. If there are no rules in $ground(P)$ with A in the head, then $4\text{-}T_P(M)(A) = \mathbf{f} \preccurlyeq M(A)$. Otherwise, for all rules r_k of the form $A \leftarrow F_k$ in $ground(P)$, $M(A \leftarrow F_k) \in \{\mathbf{t}, \mathbf{i}\}$, and so, $max\{M(F_k)\} \leqslant M(A)$. Then, $4\text{-}T_P(M) \preccurlyeq M$. ($2 \leftarrow$): Let $4\text{-}T_P(M) \preccurlyeq M$. We show that M is a 4-model of P. Let $r_k, 0 \leq k \leq l$ the rules in $ground(P)$ of the form $A \leftarrow F_k$. Then, $max\{M(F_k) : 0 \leq k \leq l\} = 4\text{-}T_P(M)(A) \leqslant M(A)$. So, $M(F_k) \leqslant M(A)$, for all k. Thus, M satisfies the rules r_k.
(3) Because $4\text{-}Inst_P$ is a finite set, the sequence $\{4\text{-}T_P^i(\bot)\}_{i\geqslant 0}$ reaches a fixpoint after a finite number N of steps. Because $4\text{-}T_P$ is monotonic, $\mathbf{I}_0 \preccurlyeq \mathbf{I}_1 \preccurlyeq \mathbf{I}_2 \ldots \preccurlyeq \mathbf{I}_N$. So the sequence $\{4\text{-}T_P^i(\bot)\}_{i\geqslant 0}$ is increasing and converges to the least fixpoint of $4\text{-}T_P$ ($\mathbf{I}_N$).
(4) M is a minimal 4-valued model of $P \Longleftrightarrow M = inf\{M|M$ is a 4-valued model of $P\} \Longleftrightarrow M = inf\{M| 4\text{-}T_P(M) \preccurlyeq M\}$, by (2) $\Longleftrightarrow M = \mathbf{I}_N$ = least fixpoint of $4\text{-}T_P$ (by monotonicity of $4\text{-}T_P$, and by the fact that $\mathbf{I}_0 \preccurlyeq M$). ∎

4.3 4-stable Models

According to [Przymusinski, 1990], the semantics of a Datalog¬ program P is an *appropriate* 3-valued model of P. We extend this idea to P-Datalog programs and introduce the 4-stable models, a class of *special* models. The semantics of a P-Datalog query will be the intersection of all 4-stable models.

Let P be a P-Datalog program and $\mathbf{I}$ a paraconsistent database instance (a 3-valued instance). We denote by $P_\mathbf{I}$ the program obtained from P by adding to P unit clauses $A \leftarrow$ for each A such that $\mathbf{I}(A) = \mathbf{t}$, and clauses $A \leftarrow \mathbf{i}$ for each A such that $\mathbf{I}(A) = \mathbf{i}$. From now on, we suppose that our programs include these clauses corresponding to the facts of $\mathbf{I}$.

Let $\mathbf{I}$ be a 4-valued instance over $sch(P)$. The *positivised ground version* of P according to $\mathbf{I}$ (denoted $pg(P, \mathbf{I})$), is the P-Datalog program obtained from $ground(P)$ by replacing each negative literal $\sim A$ by $\mathbf{I}(\sim A)$ (i.e, by its respective truth-value: $\mathbf{t}, \mathbf{f}, \mathbf{u}, \mathbf{i}$). So, $pg(P, \mathbf{I})$ is an extended P-Datalog program, i.e., a program without negation. By lemma 14, the least fixpoint $pg(P, \mathbf{I})(\bot)$ of its immediate consequence operator exists. It contains all facts that are inferred from P and $\mathbf{I}$, by assuming the values for the negative premises as given by $\mathbf{I}$. We denote $pg(P, \mathbf{I})(\bot)$ by $conseq_P(\mathbf{I})$, i.e. $conseq_P(\mathbf{I})$ is the least fixpoint of the extended P-Datalog program $pg(P, \mathbf{I})$.

DEFINITION 15. Let P be a P-Datalog program. A 4-valued instance $\mathbf{I}$ over $sch(P)$ is a *4-stable model* of P iff $conseq_P(\mathbf{I}) = \mathbf{I}$.

The following example illustrates the notion of 4-stable model:

EXAMPLE 16 (4-stable model). Consider the P-Datalog program P_{job} and the input instance $\mathbf{J}$ given in the example 1. Let us check that $\mathbf{J}$ is a 4-stable model of P_{job}. For this, we have to compute $conseq(\mathbf{J})$ and show that $conseq(\mathbf{J}) = \mathbf{J}$. The program $P' = pg(P, \mathbf{J})$ is:

$$job(charles) \leftarrow \mathbf{t}, supportedby(charles, joseph), \mathbf{u}$$
$$job(joseph) \leftarrow \mathbf{t}, supportedby(joseph, charles), \mathbf{u}$$
$$\ldots$$
$$supportedby(paul, james) \leftarrow$$
$$supportedby(charles, joseph) \leftarrow$$
$$supportedby(john, kevin) \leftarrow \mathbf{i}$$
$$\ldots$$

The minimal 4-valued model of P' is obtained by iterating 4-$T_P(\bot)$ up to a fixpoint. The first execution of 4-T_P yields 4-$T_{P'}^1(\bot) = \{\sim job(charles), \sim job(joseph), \sim job(paul), \sim job(john), \sim job(james), \sim job(kevin)\}$. We can verify that 4-$T_{P'}^2(\bot) =$ 4-$T_{P'}^3(\bot) = \{\circ\ job(paul), \bullet\ job(john), \sim job(james), \sim job(kevin)\}$. Thus $conseq_P(\mathbf{J}) = \mathbf{J}$ and $\mathbf{J}$ is a *4-stable model* of P. The instance $\mathbf{J}$ coincides with $\mathbf{I}$ (given in the example 1) for the atoms *supportedby* and *owe*.

4.4 Well-founded Semantics

P-Datalog programs generally may have several 4-stable models, and each P-Datalog program has at least one 4-stable model (see theorem 22). Then it is reasonable to say that the desired answer to a P-Datalog query consists of the positive, *inconsistent* and negative facts belonging to *all* 4-stable models of the program.

DEFINITION 17. Let P be a P-Datalog program. The *well-founded semantics* of P is a 4-valued instance consisting of the positive, inconsistent and negative facts belonging to all 4-stable models of P. This semantics is denoted by P^{4wf}.

5 Bottom-up Evaluation of P-Datalog Queries

The previous description of the well-founded semantics, although effective, is inefficient. It involves checking all possible 4-valued instances of a program, determining which are 4-stable models, and then taking their intersection.

A much simpler method is based on an *alternating fixpoint* computation ([Van Gelder, 1989]), that converges to the well-founded semantics. The idea of the method is as follows. We define an alternating sequence $\{\mathbf{I}_i\}_{i \geq 0}$ of 4-valued instances that are underestimates and overestimates of the facts known in every 4-stable model of P. The alternating sequence $\{\mathbf{I}_i\}_{i \geq 0}$ is defined as follows: $\mathbf{I}_0 = \bot$ and $\mathbf{I}_{i+1} = conseq_P(\mathbf{I}_i)$, for $i \geq 0$.

We notice that each $\mathbf{I}_i$ is constructed starting from the total instance $\bot$ by repeated applications of $conseq_P$. Thus each $\mathbf{I}_i$ is a total instance, i.e., has no undefined truth-values.

THEOREM 18. The operator $conseq_P$ is antimonotonic. That is, if $\mathbf{I} \preccurlyeq \mathbf{J}$ then $conseq_P(\mathbf{J}) \preccurlyeq conseq_P(\mathbf{I})$.

Proof. Let us suppose that $\mathbf{I} \preccurlyeq \mathbf{J}$ and let $A \in \mathbf{B}(P)$. We will show that $conseq_P(\mathbf{J})(A) \leq conseq_P(\mathbf{I})(A)$. We remind that $conseq_P(\mathbf{I})$ is the least fixpoint of the extended P-Datalog program $pg(P, \mathbf{I})$, denoted by $pg(P, \mathbf{I})(\bot)$. We have the following cases:

(1) $conseq_P(\mathbf{J})(A) = \mathbf{t}$.
We affirm that for all $n \geq 1$, if n is such that $4\text{-}T^n_{pg(P,\mathbf{J})}(\bot)(A) = \mathbf{t}$, and $\forall j$, $j \geq n$, $4\text{-}T^j_{pg(P,\mathbf{J})}(\bot)(A) = \mathbf{t}$, then $conseq_P(\mathbf{I})(A) = \mathbf{t}$. This assertion is proved by induction on n.
• Induction Base: n=1. By our hypothesis, in $ground(P)$ there is a rule of the form: $A \leftarrow\sim B_1, \ldots, \sim B_n$, where B_k are atoms, $\mathbf{J}(B_k) = \mathbf{f}$, $\forall k$, $1 \leq k \leq n$.
By $\mathbf{I} \preccurlyeq \mathbf{J}$, $\mathbf{I}(B_k) = \mathbf{f}$ $\forall k$, $1 \leq k \leq n$.
Thus $conseq_P(\mathbf{I})(A) = \mathbf{t}$ and the induction base is proved.
• Induction step: by our hypothesis, in $ground(P)$ there exists a rule of the form: $A \leftarrow\sim B_1, \ldots, \sim B_n, D_1, \ldots, D_m$, where B_k, D_g are atoms, $\mathbf{J}(B_k) = \mathbf{f}$ $\forall k$, $1 \leq k \leq n$ e $4\text{-}T^n_{pg(P,\mathbf{J})}(\bot)(D_g) = \mathbf{t}$, $\forall g$, $1 \leq g \leq m$.
By $\mathbf{I} \preccurlyeq \mathbf{J}$, $\mathbf{I}(B_k) = \mathbf{f}$ $\forall k$, $1 \leq k \leq n$.
By the induction step hypothesis, $conseq_P(\mathbf{I})(D_g) = \mathbf{t}$, $\forall g$, $1 \leq g \leq m$.
Thus $conseq_P(\mathbf{I})(A) = \mathbf{t}$.
The induction is proved and $conseq_P(\mathbf{I})(A) = \mathbf{t}$.
(2) $conseq_P(\mathbf{J})(A) = \mathbf{i}$.
We affirm that for all $n \geq 1$, if n is such that $4\text{-}T^n_{pg(P,\mathbf{J})}(\bot)(A) = \mathbf{i}$, and $\forall j$, $j \geq n$, $4\text{-}T^j_{pg(P,\mathbf{J})}(\bot)(A) = \mathbf{i}$, then $conseq_P(\mathbf{I})(A) \geq \mathbf{i}$. This assertion is proved by induction on n.
• Induction base: n=1. By our hypothesis, in $ground(P)$ there is a rule of the

form: $A \longleftarrow \sim B_1, \ldots, \sim B_n, c_1, \ldots, c_p$, where B_k are atoms and c_g are truth-values, $\mathbf{J}(B_k) = \mathbf{f}$ $\forall k$, $1 \leq k \leq n$, $c_g \geq \mathbf{i}$ $\forall g$, $1 \leq g \leq p$, there is a w, $c_w = \mathbf{i}$, $1 \leq w \leq p$.
By $\mathbf{I} \preccurlyeq \mathbf{J}$, $\mathbf{I}(B_k) = \mathbf{f}$ $\forall k$, $1 \leq k \leq n$.
Thus $conseq_P(\mathbf{I})(A) \geq \mathbf{i}$ and the induction base is proved.
- Induction step: By our hypothesis, in $ground(P)$ there is a rule of the form: $A \longleftarrow \sim B_1, \ldots, \sim B_n, D_1, \ldots, D_m$, where B_k, D_g are atoms, $\mathbf{J}(B_k) = \mathbf{f}$ $\forall k$, $1 \leq k \leq n$ and $4\text{-}T^n_{pg(P,\mathbf{J})}(\perp)(D_g) \geq \mathbf{i}$, $\forall g$, $1 \leq g \leq m$.
By $\mathbf{I} \preccurlyeq \mathbf{J}$, $\mathbf{I}(B_k) = \mathbf{f}$ $\forall k$, $1 \leq k \leq n$.
By the induction step hypothesis, $conseq_P(\mathbf{I})(D_g) \geq \mathbf{i}$, $\forall g$, $1 \leq g \leq m$.
Thus $conseq_P(\mathbf{I})(A) = \mathbf{i}$.
The induction is proved and we can conclude that $conseq_P(\mathbf{I})(A) \geq \mathbf{i}$.

(3) $conseq_P(\mathbf{J})(A) = \mathbf{u}$.
We affirm that for all $n \geq 1$, if n is such that $4\text{-}T^n_{pg(P,\mathbf{J})}(\perp)(A) = \mathbf{u}$, and $\forall j$, $j \geq n$, $4\text{-}T^j_{pg(P,\mathbf{J})}(\perp)(A) = \mathbf{u}$, then $conseq_P(\mathbf{I})(A) \geq \mathbf{u}$. This assertion is proved by induction on n.
- Induction base: n=1. By our hypothesis, in $ground(P)$ there is a rule of the form: $A \longleftarrow \sim B_1, \ldots, \sim B_n, c_1, \ldots, c_p$, where B_k are atoms and c_g are truth-values, $\mathbf{J}(B_k) \in \{\mathbf{f}, \mathbf{u}\}$ $\forall k$, $1 \leq k \leq n$, $c_g \geq \mathbf{u}$ $\forall g$, $1 \leq g \leq p$, and there is a w, $1 \leq w \leq p$, $c_w = \mathbf{u}$ or there is a y, $1 \leq y \leq n$, $\mathbf{J}(B_y) = \mathbf{u}$.
By $\mathbf{I} \preccurlyeq \mathbf{J}$, $\mathbf{I}(B_k) \in \{\mathbf{f}, \mathbf{u}\}$ $\forall k$, $1 \leq k \leq n$.
Thus $conseq_P(\mathbf{I})(A) \geq \mathbf{u}$ and the induction base is proved.
- Induction step: by our hypothesis, in $ground(P)$ there is a rule of the form: $A \longleftarrow \sim B_1, \ldots, \sim B_n, D_1, \ldots, D_m$, where B_k, D_g are atoms, $\mathbf{J}(B_k) \in \{\mathbf{f}, \mathbf{u}\}$ $\forall k$, $1 \leq k \leq n$ and $4\text{-}T^n_{pg(P,\mathbf{J})}(\perp)(D_g) \geq \mathbf{u}$, $\forall g$, $1 \leq g \leq m$.
By $\mathbf{I} \preccurlyeq \mathbf{J}$, $\mathbf{I}(B_k) \in \{\mathbf{f}, \mathbf{u}\}$ $\forall k$, $1 \leq k \leq n$.
By the induction step hypothesis, $conseq_P(\mathbf{I})(D_g) \geq \mathbf{u}$, $\forall g$, $1 \leq g \leq m$.
Thus $conseq_P(\mathbf{I})(A) \geq \mathbf{u}$.
The induction is proved and we can conclude that $conseq_P(\mathbf{I})(A) \geq \mathbf{u}$.

(4) $conseq_P(\mathbf{J})(A) = \mathbf{f}$.
As $conseq_P(\mathbf{I})(A) \geq \mathbf{f}$, we conclude that $conseq_P(\mathbf{I})(A) \geq conseq_P(\mathbf{J})(A)$. ∎

From theorem 18, we can easily see that in the alternating sequence $\{\mathbf{I}_i\}_{i \geq 0}$ we have:

$$\mathbf{I}_0 \preccurlyeq \mathbf{I}_2 \preccurlyeq \cdots \preccurlyeq \mathbf{I}_{2i} \preccurlyeq \mathbf{I}_{2i+2} \preccurlyeq \cdots \preccurlyeq \mathbf{I}_{2i+1} \preccurlyeq \mathbf{I}_{2i-1} \preccurlyeq \cdots \preccurlyeq \mathbf{I}_3 \preccurlyeq \mathbf{I}_1 \quad (1)$$

Thus the even subsequence is increasing and the odd one is decreasing. Because there are finitely many 4-valued instances relatively to a given program P, each of these sequences becomes constant at some point: $\mathbf{I}_{2k_0} = \mathbf{I}_{2k_0+2} = \cdots \mathbf{I}_{2k_0+4} = $ and $\mathbf{I}_{2j_0+1} = \mathbf{I}_{2j_0+3} = \mathbf{I}_{2j_0+5} = \ldots$, for some $k_0 \geq 0$ and some $j_0 \geq 0$.

Let $\mathbf{I}_*$ be the *least upper bound* of the increasing sequence: $\mathbf{I}_* = lub\{\mathbf{I}_{2i}\}_{i \geq 0}$, and let $\mathbf{I}^*$ be the *greatest lower bound* of the decreasing sequence: $\mathbf{I}^* = glb\{\mathbf{I}_{2i+1}\}_{i \geq 0}$. From (1), it follows that $\mathbf{I}_* \preccurlyeq \mathbf{I}^*$.

LEMMA 19. Let $\mathbf{I}$ be a 4-valued instance of a P-Datalog program. Then $conseq_P(\mathbf{I}_*) = \mathbf{I}^*$ and $conseq_P(\mathbf{I}^*) = \mathbf{I}_*$.

Proof. Let $\{\mathbf{I}_i\}_{i \geq 0}$ be an alternating sequence of 4-valued instances, where $\mathbf{I}_* = \mathbf{I}_{2k} = \mathbf{I}_{2k+2}$ e $\mathbf{I}^* = \mathbf{I}_{2k+1} = \mathbf{I}_{2k+3}$, $k \geq 0$. We need to show that $conseq_P(\mathbf{I}^*) = \mathbf{I}_*$ and $conseq_P(\mathbf{I}_*) = \mathbf{I}^*$. In $\{\mathbf{I}_i\}_{i \geq 0}$ we have $\mathbf{I}_{i+1} = conseq_P(\mathbf{I}_i)$: $\mathbf{I}_{2k} = conseq_P(\mathbf{I}_{2k-1})$, $\mathbf{I}_{2k+1} = conseq_P(\mathbf{I}_{2k})$, $\mathbf{I}_{2k+2} = conseq_P(\mathbf{I}_{2k+1})$, $\mathbf{I}_{2k+3} = conseq_P(\mathbf{I}_{2k+2})$. Thus, $\mathbf{I}_* = conseq_P(\mathbf{I}^*)$ and $\mathbf{I}^* = conseq_P(\mathbf{I}_*)$. ∎

From the 4-valued instances $\mathbf{I}_*$ and $\mathbf{I}^*$ we can define the 4-valued instance $\mathbf{I}_*^*$ which coincides with the well-founded semantics of a P-Datalog program, as we will see in Theorem 22.

DEFINITION 20. Let $\mathbf{I}_*^*$ be a 4-valued instance of a P-Datalog program P, consisting of the facts known in both $\mathbf{I}_*$ and $\mathbf{I}^*$, that is:

$$\mathbf{I}_*^*(A) = \begin{cases} \mathbf{t} & \text{if } \mathbf{I}_*(A) = \mathbf{I}^*(A) = \mathbf{t} \\ \mathbf{i} & \text{if } \mathbf{I}_*(A) = \mathbf{I}^*(A) = \mathbf{i} \\ \mathbf{f} & \text{if } \mathbf{I}_*(A) = \mathbf{I}^*(A) = \mathbf{f} \\ \mathbf{u} & \text{otherwise} \end{cases}$$

THEOREM 21. Let $\mathbf{I}$ be a 4-valued instance of a P-Datalog program P. Then $\mathbf{I}_* \preccurlyeq \mathbf{I}_*^* \preccurlyeq \mathbf{I}^*$.

Proof. By the definition of $\mathbf{I}_*^*$, $\mathbf{I}_* \preccurlyeq \mathbf{I}_*^* \preccurlyeq \mathbf{I}^*$ is verified for all cases except when $\mathbf{I}_*(A) = \mathbf{i}$ and $\mathbf{I}^*(A) = \mathbf{t}$. We show that this cannot happen. Let us suppose that $\mathbf{I}_*(A) = \mathbf{i}$ and $\mathbf{I}^*(A) = \mathbf{t}$. By the fact that $conseq_P(\mathbf{I}_*) = \mathbf{I}^*$, $conseq_P(\mathbf{I}^*) = \mathbf{I}_*$, it follows that $conseq_P(\mathbf{I}^*)(A) = \mathbf{i}$ and $conseq_P(\mathbf{I}_*)(A) = \mathbf{t}$. We can prove by induction on n that : for all $n \geq 1$, if n is such that 4-$T^n_{pg(P,\mathbf{I}_*)}(\bot)(A) = \mathbf{t}$, and $\forall j, j \geq n$, 4-$T^j_{pg(P,\mathbf{I}_*)}(\bot)(A) = \mathbf{t}$, then $conseq_P(\mathbf{I}^*)(A) \in \{\mathbf{f},\mathbf{t}\}$. So, we can conclude that $conseq_P(\mathbf{I}^*)(A) \in \{\mathbf{f},\mathbf{t}\}$. Contradiction. ∎

The fixpoint construction yields the well-founded semantics for P-Datalog programs. The following theorem is the main result of this paper. It shows that each P-Datalog program has at least one 4-stable model ($\mathbf{I}_*^*$) and that the well-founded semantics coincides with $\mathbf{I}_*^*$.

THEOREM 22. For each P-Datalog program P, (a) $\mathbf{I}_*^*$ is a 4-stable model of P and (b) $\mathbf{I}_*^*$ equals the well-founded semantics of P (P^{4wf}).

Proof. For statement (a), we need to show that $conseq_P(\mathbf{I}_*^*) = \mathbf{I}_*^*$. ¿From Theorems 18 and 21, we can easily conclude that:

$$\mathbf{I}_* \preccurlyeq conseq_P(\mathbf{I}_*^*) \preccurlyeq \mathbf{I}^* \tag{2}$$

We affirm that $conseq_P(\mathbf{I}_*^*)(A) = \mathbf{I}_*^*(A)$, for all $A \in \mathbf{B}(P)$. Indeed: (1) If $\mathbf{I}_*^*(A) \in \{\mathbf{f}, \mathbf{t}, \mathbf{i}\}$, by the definition of $\mathbf{I}_*^*$ and from (2), we can conclude that $conseq_P(\mathbf{I}_*^*)(A) = \mathbf{I}_*^*(A)$.
(2) Let $\mathbf{I}_*^*(A) = \mathbf{u}$. By the definition of $\mathbf{I}_*^*$ we have the following possibilities:
(2a) $\mathbf{I}_*(A) = \mathbf{i}$ and $\mathbf{I}^*(A) = \mathbf{t}$. By Theorem 21, this case does not happen.
(2b) $\mathbf{I}_*(A) = \mathbf{f}$ and $\mathbf{I}^*(A) \in \{\mathbf{i},\mathbf{t}\}$. This can be written as $conseq_P(\mathbf{I}^*)(A) = \mathbf{f}$ and $conseq_P(\mathbf{I}_*)(A) \geq \mathbf{i}$. By definition, $conseq_P(\mathbf{I}_*)$ is the least fixpoint of $pg(P,\mathbf{I}_*)$

($\perp$). We affirm that for all $n \geq 1$, if n is such that $4\text{-}T^n_{pg(P,\mathbf{I}_*)}(\perp)(A) \geqslant \mathbf{i}$, and $\forall j,\ j \geqslant n$, $4\text{-}T^j_{pg(P,\mathbf{I}_*)}(\perp)(A) = 4\text{-}T^n_{pg(P,\mathbf{I}_*)}(\perp)(A)$, then $conseq_P(\mathbf{I}^*_*)(A) = \mathbf{u}$. This assertion can be proved by induction on n. So, we can conclude that $conseq_P(\mathbf{I}^*_*)(A) = \mathbf{I}_*(A) = \mathbf{u}$.

For statement (b), we will show that $P^{4wf} = \mathbf{I}^*_*$. Let $A \in \mathbf{B}(P)$.

• Let $P^{4wf}(A) = \mathbf{t}(\mathbf{i},\mathbf{f})$. By definition, P^{4wf} is the 4-valued instance consisting of facts that are true (resp. false, inconsistent) in all 4-stable models of P. From statement (1), we can affirm that $\mathbf{I}^*_*(A)$ is a 4-stable model. So $\mathbf{I}^*_*(A) = P^{4wf}(A)$.

• Let $P^{4wf}(A) = \mathbf{u}$. So, there exist 4-stable models M_1 and M_2 such that $M_1(A) \neq M_2(A)$. Let us suppose that $\mathbf{I}^*_*(A) = \mathbf{t}$ ($\mathbf{i}$, $\mathbf{f}$). We affirm that for all 4-stable model M of program P, and for all $i \geqslant 0$, we have:

$$\mathbf{I}_{2i} \preccurlyeq M \preccurlyeq \mathbf{I}_{2i+1}. \tag{3}$$

This assertion can be proved by induction on i and its proof does not present any difficulty.

By (3), $\mathbf{I}_* \preccurlyeq M_j \preccurlyeq \mathbf{I}^*$, for $j = 1, 2$.

• If $\mathbf{I}^*_*(A) = \mathbf{t}$, then $\mathbf{I}_*(A) = \mathbf{t}$, and by $\mathbf{I}_*(A) \leqslant M_j(A)$, we conclude that $M_1(A) = M_2(A) = \mathbf{t}$.

• If $\mathbf{I}^*_*(A) = \mathbf{i}$, then $\mathbf{I}_*(A) = \mathbf{I}^*(A) = \mathbf{i}$, and by $\mathbf{I}_*(A) \leqslant M_j(A) \leqslant \mathbf{I}^*(A)$, we conclude that $M_1(A) = M_2(A) = \mathbf{i}$.

• If $\mathbf{I}^*_*(A) = \mathbf{f}$, then $\mathbf{I}^*(A) = \mathbf{f}$, and by $\mathbf{I}^*(A) \geqslant M_j(A)$, we conclude that $M_1(A) = M_2(A) = \mathbf{f}$.

Contradiction. So, $\mathbf{I}^*_*(A) = \mathbf{u}$. ∎

We illustrate this computation in our running example:

EXAMPLE 23 ($\mathbf{I}^*_*$ computation). Consider again the program P_{job} and the database instance $\mathbf{I}$ of the running example 1. Note that for $\mathbf{I}_0$ the value of all facts is $\mathbf{f}$, and for each $j \geqslant 1$, $\mathbf{I}_j$ agrees with the input $\mathbf{I}$ on the predicates *supportedby* and *owe*. Therefore we only show the inferred *job*-facts:

$\mathbf{I}_0 = \{\sim job(charles), \sim job(james), \sim job(john), \sim job(joseph), \sim job(kevin), \sim job(paul)\}$.

$\mathbf{I}_1 = \{\circ\ job(charles), \circ\ job(james), \bullet\ job(john), \circ\ job(joseph), \sim job(kevin), \circ\ job(paul)\}$.

$\mathbf{I}_2 = \{\sim job(charles), \sim job(james), \bullet\ job(john), \sim job(joseph), \sim job(kevin), \sim job(paul)\}$.

$\mathbf{I}_3 = \{\circ\ job(charles), \sim job(james), \bullet\ job(john), \circ\ job(joseph), \sim job(kevin), \circ\ job(paul)\}$

$\mathbf{I}_4 = \{\sim job(charles), \sim job(james), \bullet\ job(john), \sim job(joseph), \sim job(kevin), \circ\ job(paul)\}$.

$\mathbf{I}_5 = \mathbf{I}_3$ and $\mathbf{I}_6 = \mathbf{I}_4$. So, $\mathbf{I}_* = \mathbf{I}_6$ and $\mathbf{I}^* = \mathbf{I}_5$. Thus $\mathbf{I}^*_* = \{\sim job(james), \bullet\ job(john), \sim job(kevin), \circ\ job(paul)\}$. This is exactly the natural answer for P_{job} we have informally discussed in example 1.

6 Related Work

The problem of inconsistent information management arising from the integration of heterogeneous sources has been widely studied in recent research. We can distinguish three main approaches to handle the inconsistency problem in knowledge bases: a consistency-based approach (where inconsistency is eliminated), a paraconsistent logic approach (where inconsistency is not rejected and inference methods can draw plausible conclusions from it) and a hybrid approach (formalisms which do not reject any information but instead associate degrees of belief, reliability or uncertainty to each source knowledge base).

In what follows, we discuss some recent papers which deal with the subject of integrating multi-source information and querying the resulting integrated database which maybe contains inconsistencies. The methods are grouped following the three main approaches mentioned above. We begin the discussion by summarizing the main issues treated in this paper.

Our approach. The present paper follows the paraconsistent logic approach. We assume that data integration has already been achieved following the method we proposed in [de Amo et al., 2002]. This method follows a paraconsistent approach for data integration and is based on a tableau proof system of **LFI1** ([Carnielli and Marcos, 2001]). It supposes the existence of n source databases $\mathcal{K}_1, \mathcal{K}_2, ..., \mathcal{K}_n$, where each database $\mathcal{K}_i$ satisfies a set of integrity constraitns C_i (properties expressed by first-order formulas). The method builds an integrated database $\mathcal{K}$ which satisfies the set $\{C_1, C_2, ..., C_n\}$ of integrity constraints, where the notion of satisfiability is taken from the paraconsistent logic **LFI1**. The integrated database $\mathcal{K}$ is a paraconsistent database: each fact stored in it can be either sure (positive or negative) or inconsistent. In the present paper, we propose a deductive language P-Datalog which allows to query this integrated database containing inconsistencies. The querying process semantics follows the paraconsistent semantics of the logic **LFI1**.

Consistent-based approaches. In [Arenas et al., 2000; Arenas et al., 2003] a logic framework based on annotated predicate logic ([Kifer and Lozinskii, 1992]) is proposed for obtaining consistent answers when querying a database with inconsistent information with respect to a set of integrity constraints. A consistent answer to a given query q is the one that every database *repair* would give in response to q. A procedure for building consistent answers without actually repairing the inconsistent database is described. In [Arenas et al., 1999a], a method based on the classical tableau system proof for first-order logic is introduced for producing consistent answers in a relational database that may violate given integrity constraints. Roughly speaking, these answers are those obtained in all repaired versions of the database (a repaired version is obtained by inserting or deleting information from the original database in order to make it *consistent* with the integrity constraints). In [Bravo and Bertossi, 2003], a method for specifying the database repairs of a mediated integration system under the paradigm "Local-as-View" for data integration has been proposed. The repairs are specified by means of a disjunctive logic program. The consistent answers to queries posed to such a system are computed by running a disjunctive logic program together with the specification of database repairs. Following this same line, in [Eiter et al., 2003], a generic logic programming framework for computing consistent answers to queries posed to a data integration system (in which inconsistency possibly arises) is proposed. This approach is also based on the specification of repairs and queries by means of a logic program. The main objective is optimizing the evaluation of queries expressed as logic programs.

In [Barceló and Bertossi, 2003], one proposes a method to specify database repairs using simple classical normal disjunctive logic programs with a stable model semantics. The database predicates in these programs contain annotations as extra arguments. The annotations come from the classical annotated predicated logic, contrarily to annotated or paraconsistent logic programs ([Blair and Subrahmanian, 1987; Subrahmanian, 1994]). Even though their method produces all the possible minimal sets of changes required to restore the consistency of a relational database, the main goal of this method is to obtain the consistent answers to a first order query. Our language P-Datalog is designed to query *paraconsistent* databases, i.e., where inconsistent information is kept inside the database with a special truth-value which distinguishes it from safe information. When using P-Datalog to query a paraconsistent database, one can specify the type of answers we are interested in: consistent or inconsistent, or both. P-Datalog classifies each fact in a query answer as true, false, inconsistent or unknown. In [Arieli, 2002], the Belnap 4-valued logic ([Belnap, 1977]) is used to represent different degrees of contradiction and partial information. The lattice of truth-values they use is different from ours. The intuition behind their semantics is that contradictory data corresponds to inadequate information about the real world, and therefore it should be minimized. In [Arieli et al., 2004], an abductive method for coherent integration of independent datasources is proposed. The method is based on SLDNF-Resolution. The general idea consists in computing a list of data-facts that should be inserted to the concatenated database or deleted from it in order to restore its consistency.

Hybrid approaches. In [Cholvy and Garion, 2001] a majority merging operator is used in the definition of a logic that allows us to reason with data coming from different sources. The possible inconsistencies that would come from the integration process are solved according to that majority operator. In our approach, we have assumed that our data have already been integrated and in this integrated database, inconsistent information has been detected and marked. In [Benferhat et al., 1995], several methods allowing inference of information coming from different sources are studied. The methods proposed can be grouped in two categories: those following a coherent-based approach (where inconsistencies are discarded) and those where inconsistent information is kept. The later ones are closer to our approach since they do not rely on restoring consistency in the integrated database. The authors suppose the knowledge bases are sets of formulas which are stratified, that is, there exists a total ordering between the sources based on a notion of *reliability*: source $\mathcal{K}_i$ is more reliable then source $\mathcal{K}_j$ if $j > i$. In our approach, the information kept in the integrated database is not stratified, since in most situations, there is no means of attaching a degree of reliability to information coming from different sources (for instance, simple information like *client addresses* may have different values in two different databases, and there is no means of saying that the first database is more reliable then the second one). Besides, the problem we proposed is different: our integrated database is a set of (possibly inconsistent) *facts* and we propose a language to query this set of facts, producing a new set of facts, where each element is clearly identified as positive sure information, negative sure information or inconsistent information.

Paraconsistent Approach. The approach proposed in [Sakama, 1992] is close to the approach we introduced in this paper. Among all the methods we mentioned so far for querying databases containing inconsistent information, this is the only one based on a paraconsistent logic programming framework, like ours. This work proposes a deductive query language and a declarative and fixpoint-based well-founded semantics for this language. having a syntax similar to Datalog¬ and P-Datalog syntax, except that negation appearing in the body of clauses is *explicit negation* rather than *default negation*. They also propose a declarative and fixpoint-based well-founded semantics for the language. The underlying logic is a paraconsistent logic with 7 truth-values: **true, inconsistent, false, indifferent, true by default, false by default,** and **unknown**. An interpretation satisfies a formula if the truth-value it associates to the formula is **true** or **inconsistent**. The main differences between this approach and ours are the following: (1) The underlying logic is far more complex than **LFI1** since it involves 7 truth-values (2) The body of the clauses contains explicit negation and rather than default negation, and the head may contain a negative atom. Thus the language is not a simple extension of Datalog¬. P-Datalog has been designed to extend to a paraconsistent context the classical Datalog¬ query language used for querying consistent databases. This has been achieved by augmenting the three-value semantics of Datalog¬ with only one new value (**inconsistent**).

7 Implementation Issues

We have implemented a first prototype of PDLog, an automated P-Datalog prover using the OCaml programming language. PDLog has been developed as a separate system and further it will be integrated in a relational database system. For the time being it is a helpful tool for validating the well-founded semantics of P-Datalog programs. It is based on the *"Match"* (pattern matching) and the *"Forward-chaining rule-based"* algorithms presented in [Winston and Horn, 1989].

The Objective Caml (OCaml) [Leroy, 2002] programming language used to implement the Automated P-Datalog Prover is a functional programming language from the *Metalanguages* (ML) programming languages family. OCaml is a free open source project managed and maintained by INRIA[6] (*Institut National de Recherche en Informatique et en Automatique*) and it is derived from the classic ML language designed by Robin Milner to allow users to write theorem-proving tactics. OCaml has common characteristics with ML languages and it also supports imperative and object-oriented programming styles.

The data structures used in the implementation are the following: (1) Basic elements: tuples of the form ([*String List*], *Value*), where *String list* corresponds to a list of atoms with true-value *Value*. For instance, (["*job*"; "*james*"], t) represents the list of literals ○ *job*, ○ *james*.

---
[6] http://caml.inria.fr

(2) Association List : List of tuples of the form (*Variable*; *Constant*).

The P-Datalog prover has three main modules:

1. The Association List Generator;
2. The Immediate Consequence Operator;
3. The Main Engine.

The input of PDLog is constituted by two files $\mathcal{P}$ and $\mathcal{F}$, the first one ($\mathcal{P}$) containing the set of rules of a P-Datalog program and the second one ($\mathcal{F}$) containing the starting database instance. The output of PDLog is the well-founded semantics of $\mathcal{P}$, that is, a set of true, false, inconsistent and undefined facts inferred from the program $\mathcal{P}$. The file $\mathcal{P}$ has the following main features: (1) each line contains a rule of the P-Datalog program $\mathcal{P}$, ended by the symbol ".``"; the literals inside the rules are separated by ";" ; (2) The variable are identified by strings starting by a capital letter and the constants identifiers by strings starting by a non-capital letters; (3) The symbols "o", "*", "~", ":-" denotes respectively the logic symbols "o", "•", "~", "←". The file $\mathcal{F}$ has the following main features: (1) each line contains a database fact; (2) before each fact there is a symbol "o" or "*" representing safe and inconsistent facts respectively. The symbol "?" before a fact in the output of PDLog represents the *undefined* symbol.

EXAMPLE 24. Let us consider the P-Datalog program P_{job} and the starting instance **I** of Example 1.

Arquivo texto com o programa:

```
job(X):-~ owe(X); supportedby(X,Y); ~ job(Y).
```

The input file $\mathcal{F}$ is the following:

```
o supportedby(charles,joseph)
o supportedby(joseph,charles)
o supportedby(paul,james)
* supportedby(john,kevin)
o supportedby(james,kevin)
o owe(james)
```

The output is presented below:
~ cargo(james), * cargo(john), ~ cargo(kevin), o cargo(paul),
? cargo(charles), ? cargo(joseph),
~ owe(charles), o owe(james), ~ owe(john), ~ owe(joseph),
~ owe(kevin), ~ owe(paul), ~ supportedby(charles,charles),
~ supportedby(charles,james), ~ supportedby(charles,john),
o supportedby(charles,joseph), ~ supportedby(charles,kevin),
~ supportedby(charles,paul), ~ supportedby(james,charles),
~ supportedby(james,james), ~ supportedby(james,john),
~ supportedby(james,joseph), o supportedby(james,kevin),
~ supportedby(james,paul), ~ supportedby(john,charles),

~ supportedby(john,james), ~ supportedby(john,john),
~ supportedby(john,joseph), * supportedby(john,kevin),
~ supportedby(john,paul), o supportedby(joseph,charles),
~ supportedby(joseph,james), ~ supportedby(joseph,john),
~ supportedby(joseph,joseph), ~ supportedby(joseph,kevin),
~ supportedby(joseph,paul), ~ supportedby(kevin,charles),
~ supportedby(kevin,james), ~ supportedby(kevin,john),
~ supportedby(kevin,joseph), ~ supportedby(kevin,kevin),
~ supportedby(kevin,paul), ~ supportedby(paul,charles),
o supportedby(paul,james), ~ supportedby(paul,john),
~ supportedby(paul,joseph), ~ supportedby(paul,kevin),
~ supportedby(paul,paul)

8 Conclusion

In this paper we have introduced P-Datalog, a deductive query language for querying databases obtained by the concatenation of several sources, which together can contain inconsistent information. We have provided a declarative well-founded semantics which can be evaluated by means of an alternating fixpoint. We have decided to implement the P-Datalog prover as a separate system and further to integrate it in a relational database system. For now, the P-Datalog prover is a helpful tool for validating the well-founded semantics we have proposed for P-Datalog programs. It has been implemented in Objective Caml ([Leroy, 2002]). The OCaml compiler generates code whose executing time is comparable to a C/C^{++} code, and it includes libraries for several platforms. Those characteristics and also the functional programming qualities allowed us to focus on the difficulties of our application and to develop a preliminary succinct solution.

Ongoing Research. Customizing database queries by considering user preferences is a research topic that has been raising a lot of interest within the database community in recent years. Such preferences are used for sorting and selecting the best tuples, those which most fulfill the user wishes. A first topic of interest within this context is the *elicitation of preferences*, consisting of methods to enable the user to inform his choice on pairs of objects belonging to a database. Depending on the size of the database, this task may require a great effort from the user, and consequently may discourage him to use the system. So, data mining techniques are been investigated nowadays in order to automate this task. As mentioned before in this paper, user's preferences are prone to inconsistency and uncertainty. In [de Amo and Silva, 2010] we proposed the method *CPrefMiner* to mining user's preferences from a sample of user's choices. Our learning method has focused on the uncertainty character of user's preferences and the preference model produced by the method is what we called a *Bayesian Preference Network*. In this work we suppose that the user's preferences are consistent. We intend to use the **LFI1** logic as a formalism to express general preference rules, including the inconsistent ones. A second topic of interest is the incorporation of preferences in standard query languages in order to customize the query answer for a specific user. A lot of work has already been done in this area so far [Kießling and Köstler, 2002;

Chomicki, 2003; de Amo and Pereira, 2010]. In [de Amo and Pereira, 2010] we have proposed CPrefSQL, an extension of the standard SQL query language that incorporates two new preference operators, enabling to filter the answers according to the user's preferences. CPrefSQL has been implemented as an extension of the PostGreSQL DBMS (database management system)[7]. CPrefSQL has been designed to treat only consistent preference rules. Before executing a CPrefSQL query, the query processor invokes a test which decides if the set of preference rules representing the use's preference (the preference theory) is consistent or not. If it is not consistent, the query is rejected by the query processor and an error message is issued. We are working on the development of a P-Datalog-like extension of CPrefSQL in order to deal with inconsistent preferences.

Future Research. Several work remains to be done in the P-Datalog context. We intend to extend the syntax of P-Datalog in order to allow literals of the form $\circ A$, $\bullet A$ and $\neg A$ in the body of the rules. The $\circ$, $\bullet$ and $\neg$ operators are non-monotonic and that will imply in modifications on the definition of extended P-Datalog program. Another direction for further research consists in relating the well-founded semantics of Datalog$^\neg$ queries posed to the source databases with the well-founded semantics of P-Datalog we have proposed in this paper. How the answers of Datalog$^\neg$ queries submitted to the (consistent) source databases $D_1, D_2, ..., D_n$ are related to the answer of a P-Datalog query submitted to the integrated database $D = D_1 \cup ... \cup D_n$. More precisely, given a P-Datalog query Q over D, there exists Datalog$^\neg$ queries $Q_1, Q_2, ..., Q_n$ over $D_1, ..., D_n$ respectively, such that the integration of their answers (using the integration method introduced in [de Amo et al., 2002]) coincides with the answer of the P-Datalog Q ? A complete investigation of how Datalog$^\neg$ queries over the consistent sources relate to P-Datalog queries over the concatenated database would formalize the intuitive idea we gave of P-Datalog query answering in Example 1. Finally, we intend to present a more efficient implementation of P-Datalog, based on a proof system for the **LFI1** paraconsistent logic. In [Carnielli and Marcos, 2001], a sound and complete tableau proof system for **LFI1** has been developed and in [de Amo et al., 2002], a database integration method based on this proof system has been introduced. We intend to use this proof system or an equivalent resolution proof system for evaluating the well-founded semantics of P-Datalog queries.

BIBLIOGRAPHY

[Abiteboul et al., 1995] S. Abiteboul, V. Vianu, and R. Hull. *Foundations of Databases*. Addison-Wesley, 1995.

[Alferes et al., 1996] J. J. Alferes, L. M. Pereira, and T. C. Przymusinski. Strong and explicit negation in non-monotonic reasoning and logic programming. In *Logics in Artificial Intelligence, European Workshop, JELIA, Évora, Portugal, September 30 - October 3, Proceedings*, volume 1126 of *Lecture Notes in Computer Science*, pages 143–163. Springer, 1996.

---

[7]The source code of CPrefSQL can be downloaded from http://www.lsi.ufu.br/cprefsql

[Arenas et al., 1999a] M. Arenas, L. Bertossi, and J. Chomicki. Consistent query answers in inconsistent databases. In *18th ACM SIGACT-SIGMOD-SIGART Symposium on Principles of Database Systems (PODS'99)*, pages 68–79, 1999.

[Arenas et al., 1999b] M. Arenas, L. Bertossi, and J. Chomicki. Consistent query answers in inconsistent databases. In *Proceedings of ACM Symposium on Principles of Database Systems-ACM PODS'99, Philadelphia*, pages 68–79, 1999.

[Arenas et al., 2000] M. Arenas, L. Bertossi, and M. Kifer. Applications of annotated predicate calculus to querying inconsistent databases. In *Proceedings of Computational Logic*, pages 926–941, 2000.

[Arenas et al., 2003] M. Arenas, L. Bertossi, and J. Chomicki. Answer sets for consistent query answering in inconsistent databases. In *Theory and Practice of Logic Programming*, volume 3, pages 393–424, 2003.

[Arieli et al., 2004] O. Arieli, M. Bruynooghe, M. Denecker, and B. Van Nuffelen. Coherent integration of databases by abductive logic programming. *Journal of Artificial Intelligence Research (JAIR)*, 21:245–286, 2004.

[Arieli, 1998] O. Arieli. Four-Valued Logics for Reasoning with Uncertainty in Prioritized Data. In *Proc. of the 7th Conference on Information Processing and Management of Uncertainty in Knowledge-Based Systems (IPMU'98)*, pages 503–510, 1998.

[Arieli, 2002] O. Arieli. Paraconsistent declarative semantics for extended logic programs. In *Annals of Mathematics and Artificial Intelligence.*, pages 36(4):381–417, 2002.

[Barceló and Bertossi, 2003] P. Barceló and L. Bertossi. Logic programs for querying inconsistent databases. In *Practical Aspects of Declarative Languages (PADL), 5th International Symposium*, pages 208–222, 2003.

[Belnap, 1977] N. D. Belnap. A useful four-valued logic. In G. Epstein and J.M. Dunn, editors, *Modern Uses of Many-valued Logic*, pages 30–56. Reidel, 1977.

[Benferhat et al., 1995] S. Benferhat, D. Dubois, and H. Prade. How to Infer from Inconsistent Beliefs without Revising? In *Proceedings of the Foureenth International Joint Conferences on Artificial Intelligence*, pages 1449–1455, 1995.

[Bertossi and Schwind, 2003] L. Bertossi and C. Schwind. Database repairs and analytic tableaux. *Annals of Mathematics in Artificial Intelligence journal*, 2003.

[Bidoit and Legay, 1990] N. Bidoit and P. Legay. Well!: An evaluation procedure for all logic programs. In S. Abiteboul and P. C. Kanellakis, editors, *ICDT'90, Third International Conference on Database Theory, Paris, France*, volume 470 of *Lecture Notes in Computer Science*, pages 335–348. Springer, 1990.

[Blair and Subrahmanian, 1987] H. A. Blair and V. S. Subrahmanian. Paraconsistent logic programming. In *Proceedings of the seventh conference on Foundations of software technology and theoretical computer science*, pages 340–360. Springer-Verlag, 1987.

[Boutilier et al., 2004] C. Boutilier, R. Brafman, H. Hoos, and D.Poole. Cp-nets: A tool for representing and reasoning about conditional ceteris paribus preference statements. *Journal of Artificial Intelligence Research*, 21:135–191, 2004.

[Bravo and Bertossi, 2003] L. Bravo and L. Bertossi. Logic programs for consistently querying data integration systems. *IJCAI*, pages 10–15, 2003.

[Brewka et al., 2004] G. Brewka, S. Benferhat, and D. Le Berre. Qualitative choice logi. *Artificial Intelligence*, 157:203–237, 2004.

[Carnielli and Lima–Marques, 1992] W. A. Carnielli and M. Lima–Marques. Reasoning under inconsistent knowledge. *Journal of Applied Non-classical Logics, Vol. 2 (1)*, pages 49–79, 1992.

[Carnielli and Marcos, 2001] W. A. Carnielli and J. Marcos. Tableau systems for logics of formal inconsistency. In *Proc. of the 2001 International Conference on Artificial Intelligence - IC AI*, pages 848–852, 2001.

[Carnielli and Marcos, 2002] W. A. Carnielli and J. Marcos. A taxonomy of C-systems. In *Paraconsistency - the Logical Way to the Inconsistent*, volume 228 of *Lecture Notes in Pure and Applied Mathematics*, pages 1–94, New York, 2002.

[Carnielli, 1987] W. A. Carnielli. Systematization of the finite many-valued through the method of tableaux. *The Journal of Symbolic Logic*, 52:473–493, 1987.

[Carnielli, 1990] W. A. Carnielli. Many-valued logics and plausible reasoning. *Proc. of International Symp. on Multiple-valued Logic, Charlotte, U.S.A. IEEE Computer Society Press*, pages 328–335, 1990.

[Ceri et al., 1990] S. Ceri, G. Gottlob, and L. Tanca. Logic programming and databases (surveys in computer science). *Springer-Verlag*, 1990.

[Cholvy and Garion, 2001] L. Cholvy and C. Garion. A logic to reason on contradictory beliefs with a majority approach. *Workshop IJCAI "Inconsistency in Data and Knowledge", Seattle*, August 2001.

[Cholvy, 1998] L. Cholvy. A general framework for reasoning about contradictory information and some of its applications. *ECAI Workshop "Conflicts among agents" Brighton*, August 1998.

[Chomicki, 2003] J. Chomicki. Preference formulas in relational queries. *ACM Transactions on Database Systems*, pages 427–466, 2003.

[Codd, 1970] E. F. Codd. A relational model of data for large shared data banks. *Communications of the ACM 13(6)*, pages 377–387, 1970.

[Cortes, 2003] A. Cortes. Consistent query answers from information source integration system. Master's thesis, Pontificia Universidad Catolica de Chile, Departamento de Ciencia de Computacion, Santiago, Chile, 2003.

[Costa and Subrahmanian, 1989] N. C. A. Costa and V. S. Subrahmanian. Paraconsistent logics as a formalism for reasoning about inconsistent knowledge bases. *Artificial Intelligence in Medicine 1*, pages 167–174, 1989.

[de Amo and Pais, 2005] S. de Amo and M. S. Pais. A paraconsistent logic programming approach for querying inconsistent knowledge bases. In *Proceedings of The 18th International FLAIRS Conference*, pages 747–752, 2005.

[de Amo and Pais, 2007] S. de Amo and M. Pais. A paraconsistent logic programming approach for querying inconsistent databases. *International Journal of Approximate Reasoning*, 46(2):366–386, 2007.

[de Amo and Pereira, 2010] Sandra de Amo and Fabiola Pereira. Evaluation of conditional preference queries. *Proceedings of the 25th Brazilian Symposium on Databases, October 2010, Belo Horizonte, Brazil. Journal of Information and Data Management (JIDM).*, 1(3):521–536, 2010.

[de Amo and Silva, 2010] S. de Amo and N.F. Silva. Cprefminer: A bayesian miner of conditional preferences. *Journal of Information and Data Management (JIDM) , Vol. 2 (1), 2011, pages 35-42.*, 2(1):35–42, 2010.

[de Amo et al., 2000] S. de Amo, W. A. Carnielli, and J. Marcos. Formal inconsistency and evolutionary databases. *Logic and Logical Philosophy*, 8:115–152, 2000.

[de Amo et al., 2002] S. de Amo, W. A. Carnielli, and J. Marcos. A logical framework for integrating inconsistent information in multiple databases. In *2nd Symposium on Foundations of Information and Knowledge Systems (FOIKS), , Salzau Castle, Germany*, volume 2284, pages 67–84. LNCS, 2002.

[Dubois and Prade, 2004] D. Dubois and H. Prade. Possibilistic logic: a retrospective and prospective view. *Fuzzy Sets and Systems*, 144:3–23, 2004.

[Dubois and Prade, 2008] D. Dubois and H. Prade. Handling bipolar queries in fuzzy information processing. in: Fuzzy information processing in databases. josé galindo (ed.). *Information Science Reference*, 1:97–114, 2008.

[Eiter et al., 2003] T. Eiter, M. Fink, G. Greco, and D. Lembo. Efficient evaluation of logic programs for querying data integration systems. In *Proc. 19th International Conference on Logic Programming (ICLP)*, pages 163–177, 2003.

[Fagin et al., 1983] R. Fagin, J. D. Ullman, and M. Vardi. On the semantics of updates in databases. 2nd ACM SIGACT-SIGMOD Symposium on Principles of Database Systems, pages 352–365, 1983.

[Gabbay and Hunter, 1991] D. Gabbay and A. Hunter. Making inconsistency respectable: A logical framework for inconsistency in reasoning. In P. Jorrand and J. Kelemen, editors, *Proceedings of Fundamentals of Artifical Intelligence Research (FAIR'91)*, pages 19–32. Springer-Verlag, 1991.

[Gärdenfors, 1988] P. Gärdenfors. Knowledge in flux - modeling the dynamics of epistemic states. *MIT Press*, 1988.

[Gelfond and Lifschitz, 1988] M. Gelfond and V. Lifschitz. The stable model semantics for logic programming. In *Proc. of the Fifth International Conference on Logic Programming*, pages 1070–1080, 1988.

[Greco and Zumpano, 2000] S. Greco and E. Zumpano. Querying inconsistent databases. In *Proc. Int. Conf. on Logic Programming and Automated Reasoning (LPAR2000), No. 1955 in LNAI*, page 308325. Springer-Verlag, 2000.

[Kießling and Köstler, 2002] W. Kießling and G. Köstler. Preference sql - design, implementation, experiences. In *VLDB*, pages 990–1001, 2002.

[Kifer and Lozinskii, 1992] M. Kifer and E. L Lozinskii. A logic for reasoning with inconsistency. *Journal of Automated Reasoning*, pages 179–215, 1992.

[Kleene, 1952] S. C. Kleene. *Introduction to Metamathematics*. Van Nostrand, New York, 1952.

[Leone et al., 1997] N. Leone, P. Rullo, and F. Scarcello. Disjunctive stable models: Unfounded sets, fixpoint semantics, and computation. *Information and Computation*, 135(2):69–112, 1997.

[Leroy, 2002] X. Leroy. The objective caml system - release 3.06. Documentation and users manual, August 2002.

[Lloyd, 1993] J. W. Lloyd. *Foundations of Logic Programming*. Springer-Verlag, 1993.

[Pereira and Alferes, 1992] L. M. Pereira and J. J. Alferes. Well founded semantics for logic programs with explicit negation. In *European Conference on Artificial Intelligence*, pages 102–106, 1992.

[Przymusinski, 1990] T. C. Przymusinski. Well-founded semantics coincides with three-valued stable semantics. *Fundamentae Informaticae, XIII*, pages 445–463, 1990.

[Przymusinski, 1992] T. C. Przymusinski. Two simple characterizations of well-founded semantics. In *Mathematical Foundations of Computer Science*, pages 451–462, 1992.

[Ramakrishnan and Ullman, 1995] R. Ramakrishnan and J. D. Ullman. A survey of deductive database systems. In *J. Logic Programming*, pages 125–149, 1995.

[Sakama, 1992] C. Sakama. Extended well-founded semantics for paraconsistent logic programs. In *Proceedings of the International Conference on Fifth Generation Computer Systems*, pages 592–599. ACM, 1992.

[Subrahmanian, 1994] V. S. Subrahmanian. Amalgamating knowledge bases. *ACM Trans. Database Syst.*, 19(2):291–331, 1994.

[Traverso, 2000] A. C. Traverso. Implementing consistent query answers in inconsistent databases. Master's thesis, Pontificia Universidad Catolica de Chile, Departamento de Ciencia de Computacion, Santiago, Chile, 2000.

[Van Gelder et al., 1991] A. Van Gelder, K. Ross, and J. S. Schlipf. The well-founded semantics for general logic programs. *Journal of the ACM*, 38(3):620–650, 1991.

[Van Gelder, 1989] A. Van Gelder. The alternating fixpoint of logic programs with negation. In *Proceedings of the eighth ACM Symposium on Principles of Database Systems-ACM PODS'89*, pages 1–10, 1989.

[Winston and Horn, 1989] P. H. Winston and B. K. P. Horn. *Lisp*. Addison-Wesley, 1989.

[You et al., 1995] J. H. You, S. Ghosh, L-Y Yuan, and R. Goebel. An introspective framework for paraconsistent logic programs. *Proceedings of the 1995 International Symposium on Logic Programming. 384-398*, pages 384–398, 1995.

Sandra de Amo
Faculdade de Computação
Universidade Federal de Uberlândia
Uberlândia, MG, Brazil
E-mail: deamo@ufu.br

Mônica Sakuray Pais
Instituto Federal Goiano
Urutaí, GO, Brazil
E-mail: monicaspais@gmail.com

Connexive Modal Logic Based on Positive S4

NORIHIRO KAMIDE AND HEINRICH WANSING

*Dedicated to Professor Walter Alexandre Carnielli
on the occasion of his 60th birthday*

ABSTRACT. A new connexive modal logic called CS4, which is based on the positive normal modal logic S4, is introduced as a Gentzen-type sequent calculus. The Kripke-completeness and cut-elimination theorems for CS4 are shown, and CS4 is shown to be embeddable into positive S4 and to be decidable. Moreover, it is shown that the basic constructive connexive logic C can be faithfully embedded into CS4 and into a subsystem of CS4 lacking syntactic duality between necessity and possibility.

1 Introduction

Connexive logics were originally introduced by Angell [Angell, 1962] and studied by McCall [McCall, 1966] in some philosophical motivations. A distinctive feature of connexive logics is that they validate the so-called Boethius' Theses:

$$(\alpha \to \beta) \to \sim(\alpha \to \sim\beta), \quad (\alpha \to \sim\beta) \to \sim(\alpha \to \beta).$$

A survey on connexive logic may be found in [Wansing, 2010]. A constructive *connexive modal logic* CK, which is a constructive connexive analogue of the smallest normal modal logic K, was introduced in [Wansing, 2005] by extending a certain *basic constructive connexive logic* C, which is a variant of *Nelson's constructive four-valued logic* [Almukdad and Nelson, 1984]. In CK also the converses of the Boethius' Theses are valid. The logic CK is a connexive version of the constructive logic FSK^d, which was introduced in [Odintsov and Wansing, 2004]. CK was shown to be embeddable into the positive fragment of Fischer Servi's *intuitionistic modal logic* FS in [Wansing, 2005].

In this paper, a new connexive modal logic called CS4, which is based on the positive fragment of the normal modal logic S4, is introduced as a Gentzen-type sequent calculus. The Kripke-completeness and cut-elimination theorems for CS4 are shown, and CS4 is shown to be embeddable into positive S4 and to be decidable. The logic CS4 is intended to give a natural modal analogue of the basic connexive logic C in a similar sense that S4 is regarded as a modal analogue of intuitionistic propositional logic in view of the Gödel translation from intuitionistic logic into S4 [Gödel, 1933]. It is also observed that C can be faithfully embedded into a subsystem of CS4 lacking syntactic duality between necessity and possibility.

The contents of this paper are then summarized as follows. In Section 2, the logic CS4 is introduced as a generalization of both a sequent calculus for positive S4 and a sequent calculus for C. In Section 3, the cut-elimination and decidability theorems for CS4 are proved using a translation of CS4 into positive S4. A similar translation has been used by Vorob'ev [Vorob'ev, 1952], Gurevich [Gurevich, 1977] and Rautenberg [Rautenberg, 1979] to embed Nelson's three-valued constructive logic [Almukdad and Nelson, 1984; Nelson, 1949] into intuitionistic logic. In Section 4, the Kripke-completeness theorem for CS4 is proved also using the translation of CS4 into positive S4. In Section 5, a theorem for embedding C into CS4 is proved using an analogue of the Gödel translation. Finally, in Section 6 it is observed that the embedding holds also for a subsystem $CS4^{d-}$ of CS4.

2 Sequent calculus

Formulas are constructed from propositional variables, $\to$ (implication), $\wedge$ (conjunction), $\vee$ (disjunction), $\sim$ (negation), $\Box$ (necessity) and $\Diamond$ (possibility). Lower-case letters $p, q, ...$ are used to denote propositional variables, Greek lower-case letters $\alpha, \beta, ...$ are used to denote formulas, and Greek capital letters $\Gamma, \Delta, ...$ are used to represent finite (possibly empty) multisets of formulas. An expression $\sharp\Gamma$ ($\sharp \in \{\sim, \Box, \Diamond\}$) means the multiset $\{\sharp\gamma \mid \gamma \in \Gamma\}$. A *sequent* is an expression of the form $\Gamma \Rightarrow \Delta$. An expression $\alpha \Leftrightarrow \beta$ means the two sequents $\alpha \Rightarrow \beta$ and $\beta \Rightarrow \alpha$, and we write $A \equiv B$ to indicate the syntactical identity between A and B. An expression of the form $L \vdash \Gamma \Rightarrow \Delta$ means that the sequent $\Gamma \Rightarrow \Delta$ is provable in a sequent calculus L. We will sometimes omit L in this expression. A rule R of inference is said to be *admissible* in a sequent calculus L if the following condition is satisfied: for any instance

$$\frac{S_1 \cdots S_n}{S}$$

of R, if $L \vdash S_i$ for all i, then $L \vdash S$.

We now define a sequent calculus CS4, cf. also [Kamide and Wansing, 2010].

DEFINITION 1 (CS4). The initial sequents of CS4 are of the form: for any propositional variable p,

$$p \Rightarrow p \qquad \sim p \Rightarrow \sim p.$$

The inference rules of CS4 are of the form:

$$\frac{\Delta \Rightarrow \Pi, \alpha \quad \alpha, \Sigma \Rightarrow \Gamma}{\Delta, \Sigma \Rightarrow \Pi, \Gamma} \text{ (cut)}$$

$$\frac{\alpha, \alpha, \Gamma \Rightarrow \Delta}{\alpha, \Gamma \Rightarrow \Delta} \text{ (co-left)} \qquad \frac{\Gamma \Rightarrow \Delta, \alpha, \alpha}{\Gamma \Rightarrow \Delta, \alpha} \text{ (co-right)}$$

$$\frac{\Gamma \Rightarrow \Delta}{\alpha, \Gamma \Rightarrow \Delta} \text{ (we-left)} \qquad \frac{\Gamma \Rightarrow \Delta}{\Gamma \Rightarrow \Delta, \alpha} \text{ (we-right)}$$

$$\frac{\Gamma \Rightarrow \Delta, \alpha \quad \beta, \Sigma \Rightarrow \Pi}{\alpha \to \beta, \Gamma, \Sigma \Rightarrow \Delta, \Pi} (\to\text{left}) \qquad \frac{\alpha, \Gamma \Rightarrow \Delta, \beta}{\Gamma \Rightarrow \Delta, \alpha \to \beta} (\to\text{right})$$

$$\frac{\alpha, \Gamma \Rightarrow \Delta}{\alpha \wedge \beta, \Gamma \Rightarrow \Delta} (\wedge\text{left1}) \qquad \frac{\beta, \Gamma \Rightarrow \Delta}{\alpha \wedge \beta, \Gamma \Rightarrow \Delta} (\wedge\text{left2})$$

$$\frac{\Gamma \Rightarrow \Delta, \alpha \quad \Gamma \Rightarrow \Delta, \beta}{\Gamma \Rightarrow \Delta, \alpha \wedge \beta} \; (\wedge\text{right}) \qquad \frac{\alpha, \Gamma \Rightarrow \Delta \quad \beta, \Gamma \Rightarrow \Delta}{\alpha \vee \beta, \Gamma \Rightarrow \Delta} \; (\vee\text{left})$$

$$\frac{\Gamma \Rightarrow \Delta, \alpha}{\Gamma \Rightarrow \Delta, \alpha \vee \beta} \; (\vee\text{right1}) \qquad \frac{\Gamma \Rightarrow \Delta, \beta}{\Gamma \Rightarrow \Delta, \alpha \vee \beta} \; (\vee\text{right2})$$

$$\frac{\alpha, \Gamma \Rightarrow \Delta}{\Box\alpha, \Gamma \Rightarrow \Delta} \; (\Box\text{left}) \qquad \frac{\Box\Gamma, {\sim}\Diamond\Sigma \Rightarrow \alpha}{\Box\Gamma, {\sim}\Diamond\Sigma \Rightarrow \Box\alpha} \; (\Box\text{right}^*)$$

$$\frac{\alpha \Rightarrow \Diamond\Gamma, {\sim}\Box\Sigma}{\Diamond\alpha \Rightarrow \Diamond\Gamma, {\sim}\Box\Sigma} \; (\Diamond\text{left}^*) \qquad \frac{\Gamma \Rightarrow \Delta, \alpha}{\Gamma \Rightarrow \Delta, \Diamond\alpha} \; (\Diamond\text{right})$$

$$\frac{\alpha, \Gamma \Rightarrow \Delta}{{\sim}{\sim}\alpha, \Gamma \Rightarrow \Delta} \; ({\sim}\text{left}) \qquad \frac{\Gamma \Rightarrow \Delta, \alpha}{\Gamma \Rightarrow \Delta, {\sim}{\sim}\alpha} \; ({\sim}\text{right})$$

$$\frac{\Gamma \Rightarrow \Delta, \alpha \quad {\sim}\beta, \Sigma \Rightarrow \Pi}{{\sim}(\alpha \rightarrow \beta), \Gamma, \Sigma \Rightarrow \Delta, \Pi} \; ({\sim}{\rightarrow}\text{left}) \qquad \frac{\alpha, \Gamma \Rightarrow \Delta, {\sim}\beta}{\Gamma \Rightarrow \Delta, {\sim}(\alpha \rightarrow \beta)} \; ({\sim}{\rightarrow}\text{right})$$

$$\frac{{\sim}\alpha, \Gamma \Rightarrow \Delta \quad {\sim}\beta, \Gamma \Rightarrow \Delta}{{\sim}(\alpha \wedge \beta), \Gamma \Rightarrow \Delta} \; ({\sim}\wedge\text{left})$$

$$\frac{\Gamma \Rightarrow \Delta, {\sim}\alpha}{\Gamma \Rightarrow \Delta, {\sim}(\alpha \wedge \beta)} \; ({\sim}\wedge\text{right1}) \qquad \frac{\Gamma \Rightarrow \Delta, {\sim}\beta}{\Gamma \Rightarrow \Delta, {\sim}(\alpha \wedge \beta)} \; ({\sim}\wedge\text{right2})$$

$$\frac{{\sim}\alpha, \Gamma \Rightarrow \Delta}{{\sim}(\alpha \vee \beta), \Gamma \Rightarrow \Delta} \; ({\sim}\vee\text{left1}) \qquad \frac{{\sim}\beta, \Gamma \Rightarrow \Delta}{{\sim}(\alpha \vee \beta), \Gamma \Rightarrow \Delta} \; ({\sim}\vee\text{left2})$$

$$\frac{\Gamma \Rightarrow \Delta, {\sim}\alpha \quad \Gamma \Rightarrow \Delta, {\sim}\beta}{\Gamma \Rightarrow \Delta, {\sim}(\alpha \vee \beta)} \; ({\sim}\vee\text{right})$$

$$\frac{{\sim}\alpha \Rightarrow \Diamond\Gamma, {\sim}\Box\Sigma}{{\sim}\Box\alpha \Rightarrow \Diamond\Gamma, {\sim}\Box\Sigma} \; ({\sim}\Box\text{left}^*) \qquad \frac{\Gamma \Rightarrow \Delta, {\sim}\alpha}{\Gamma \Rightarrow \Delta, {\sim}\Box\alpha} \; ({\sim}\Box\text{right})$$

$$\frac{{\sim}\alpha, \Gamma \Rightarrow \Delta}{{\sim}\Diamond\alpha, \Gamma \Rightarrow \Delta} \; ({\sim}\Diamond\text{left}) \qquad \frac{\Box\Gamma, {\sim}\Diamond\Sigma \Rightarrow {\sim}\alpha}{\Box\Gamma, {\sim}\Diamond\Sigma \Rightarrow {\sim}\Diamond\alpha} \; ({\sim}\Diamond\text{right}^*).$$

A sequent calculus pS4 for the positive fragment of the normal modal logic S4 is obtained from the $\{\wedge, \vee, \rightarrow\}$-fragment of CS4 by adding the inference rules of the form:

$$\frac{\alpha, \Gamma \Rightarrow \Delta}{\Box\alpha, \Gamma \Rightarrow \Delta} \; (\Box\text{left}) \qquad \frac{\Box\Gamma \Rightarrow \alpha}{\Box\Gamma \Rightarrow \Box\alpha} \; (\Box\text{right})$$

$$\frac{\alpha \Rightarrow \Diamond\Gamma}{\Diamond\alpha \Rightarrow \Diamond\Gamma} \; (\Diamond\text{left}) \qquad \frac{\Gamma \Rightarrow \Delta, \alpha}{\Gamma \Rightarrow \Delta, \Diamond\alpha} \; (\Diamond\text{right}).$$

Note that ($\Box$left), ($\Box$right), ($\Diamond$left), and ($\Diamond$right) are the modal inference rules from Ohnishi and Matsumoto's cut-free sequent calculus for S4 [Ohnishi and Matsumoto, 1957] and that pS4 enjoys cut-elimination. Note also that ($\Box$right) and ($\Diamond$left) are special cases of ($\Box$right*) and ($\Diamond$left*) for $\Sigma =$ the empty multiset. Finally, note that the exchange rules are omitted in CS4 and pS4, since we have agreed that the antecedents and succedents of the sequents in these systems are multisets.

The sequents of the form $\alpha \Rightarrow \alpha$ for any formula α are provable in cut-free CS4 and cut-free pS4. This fact can be shown by induction on α.

PROPOSITION 2. *The following sequents are provable in cut-free* CS4:

1. $\sim\sim\alpha \Leftrightarrow \alpha$,

2. $\sim(\alpha \wedge \beta) \Leftrightarrow \sim\alpha \vee \sim\beta$,

3. $\sim(\alpha \vee \beta) \Leftrightarrow \sim\alpha \wedge \sim\beta$,

4. $\sim(\alpha \to \beta) \Leftrightarrow \alpha \to \sim\beta$,

5. $\sim\Box\alpha \Leftrightarrow \Diamond\sim\alpha$,

6. $\sim\Diamond\alpha \Leftrightarrow \Box\sim\alpha$.

Proof. We show some cases.

(4):

$$\dfrac{\dfrac{\alpha \Rightarrow \alpha \quad \sim\beta \Rightarrow \sim\beta}{\alpha, \sim(\alpha \to \beta) \Rightarrow \sim\beta}(\sim\!\to\!\text{left})}{\sim(\alpha \to \beta) \Rightarrow \alpha \to \sim\beta}(\to\!\text{right}) \qquad \dfrac{\dfrac{\alpha \Rightarrow \alpha \quad \sim\beta \Rightarrow \sim\beta}{\alpha, \alpha \to \sim\beta \Rightarrow \sim\beta}(\to\!\text{left})}{\alpha \to \sim\beta \Rightarrow \sim(\alpha \to \beta)}(\sim\!\to\!\text{right}).$$

(5):

$$\dfrac{\dfrac{\sim\alpha \Rightarrow \sim\alpha}{\sim\alpha \Rightarrow \Diamond\sim\alpha}(\Diamond\text{right})}{\sim\Box\alpha \Rightarrow \Diamond\sim\alpha}(\sim\!\Box\text{left}^*) \qquad \dfrac{\dfrac{\sim\alpha \Rightarrow \sim\alpha}{\sim\alpha \Rightarrow \sim\Box\alpha}(\sim\!\Box\text{right})}{\Diamond\sim\alpha \Rightarrow \sim\Box\alpha}(\Diamond\text{left}^*).$$

∎

3 Cut-elimination and decidability

In the following, we introduce a translation of CS4 into pS4, and by using this translation, we show a theorem for *syntactically* embedding CS4 into pS4. The syntactical embedding theorem for CS4 into pS4 is used to show the cut-elimination and decidability theorems for CS4. A theorem for *semantically* embedding CS4 into pS4 will also be shown for giving the Kripke-completeness theorem for CS4.

DEFINITION 3. Let Φ be a non-empty set of propositional variables and Φ' be the set $\{p' \mid p \in \Phi\}$ of propositional variables. The language $\mathcal{L}_{\text{CS4}}$ (the set of formulas) of CS4 is defined using $\Phi, \sim, \to, \wedge, \vee, \Box$ and $\Diamond$. The language $\mathcal{L}$ of pS4 is obtained from $\mathcal{L}_{\text{CS4}}$ by adding Φ' and deleting $\sim$.

A mapping f from $\mathcal{L}_{\text{CS4}}$ to $\mathcal{L}$ is defined inductively by

1. for any $p \in \Phi$, $f(p) := p$ and $f(\sim p) := p' \in \Phi'$,

2. $f(\alpha \sharp \beta) := f(\alpha) \sharp f(\beta)$ where $\sharp \in \{\wedge, \vee, \to\}$,

3. $f(\sharp\alpha) := \sharp f(\alpha)$ where $\sharp \in \{\Box, \Diamond\}$,

4. $f(\sim\sim\alpha) := f(\alpha)$,

5. $f(\sim(\alpha \wedge \beta)) := f(\sim\alpha) \vee f(\sim\beta)$,

6. $f(\sim(\alpha \vee \beta)) := f(\sim\alpha) \wedge f(\sim\beta)$,
7. $f(\sim(\alpha \to \beta)) := f(\alpha) \to f(\sim\beta)$,
8. $f(\sim\Box\alpha) := \Diamond f(\sim\alpha)$,
9. $f(\sim\Diamond\alpha) := \Box f(\sim\alpha)$.

An expression $f(\Gamma)$ denotes the result of replacing every occurrence of a formula α in Γ by an occurrence of $f(\alpha)$.

THEOREM 4 (Syntactical embedding). *Let Γ and Δ be sets of formulas in $\mathcal{L}_{CS4}$, and f be the mapping defined in Definition 3. Then:*

1. $CS4 \vdash \Gamma \Rightarrow \Delta$ *iff* $pS4 \vdash f(\Gamma) \Rightarrow f(\Delta)$.

2. $CS4 - (cut) \vdash \Gamma \Rightarrow \Delta$ *iff* $pS4 - (cut) \vdash f(\Gamma) \Rightarrow f(\Delta)$.

Proof. Since the claim (2) can be obtained by a subproof of (1), we show only (1) in the following.

• ($\Longrightarrow$) : By induction on the proofs P of $\Gamma \Rightarrow \Delta$ in CS4. We distinguish the cases according to the last inference of P, and show some cases.

Case ($\sim p \Rightarrow \sim p$): The last inference of P is of the form: $\sim p \Rightarrow \sim p$. In this case, we obtain the required fact $pS4 \vdash f(\sim p) \Rightarrow f(\sim p)$, since $f(\sim p)$ coincides with the propositional variable p' by the definition of f.

Case ($\sim\to$left): The last inference of P is of the form:

$$\frac{\Gamma \Rightarrow \Delta, \alpha \quad \sim\beta, \Sigma \Rightarrow \Pi}{\sim(\alpha\to\beta), \Gamma, \Sigma \Rightarrow \Delta, \Pi} \ (\sim\to\text{left}).$$

By induction hypothesis, we have: $pS4 \vdash f(\Gamma) \Rightarrow f(\Delta), f(\alpha)$ and $pS4 \vdash f(\sim\beta), f(\Sigma) \Rightarrow f(\Pi)$. Then, we obtain the required fact:

$$\frac{f(\Gamma) \Rightarrow f(\Delta), f(\alpha) \quad f(\sim\beta), f(\Sigma) \Rightarrow f(\Pi)}{f(\alpha)\to f(\sim\beta), f(\Gamma), f(\Sigma) \Rightarrow f(\Delta), f(\Pi)} \ (\to\text{left})$$

where $f(\alpha)\to f(\sim\beta)$ coincides with $f(\sim(\alpha\to\beta))$ by the definition of f.

Case ($\sim\Diamond$right*): The last inference of P is of the form:

$$\frac{\Box\Gamma, \sim\Diamond\Sigma \Rightarrow \sim\alpha}{\Box\Gamma, \sim\Diamond\Sigma \Rightarrow \sim\Diamond\alpha} \ (\sim\Diamond\text{right}^*).$$

By induction hypothesis, we have: $pS4 \vdash f(\Box\Gamma), f(\sim\Diamond\Sigma) \Rightarrow f(\sim\alpha)$ where $f(\Box\Gamma)$ and $f(\sim\Diamond\Sigma)$ respectively coincide with $\Box f(\Gamma)$ and $\Box f(\sim\Sigma)$ by the definition of f. Then, we obtain:

$$\frac{\Box f(\Gamma), \Box f(\sim\Sigma) \Rightarrow f(\sim\alpha)}{\Box f(\Gamma), \Box f(\sim\Sigma) \Rightarrow \Box f(\sim\alpha)} \ (\Box\text{right})$$

where $\Box f(\sim\alpha)$ coincides with $f(\sim\Diamond\alpha)$ by the definition of f. Therefore we have the required fact: pS4 $\vdash f(\Box\Gamma), f(\sim\Diamond\Sigma) \Rightarrow f(\sim\Diamond\alpha)$.

Case ($\sim\Box$left*): The last inference of P is of the form:

$$\frac{\sim\alpha \Rightarrow \Diamond\Gamma, \sim\Box\Sigma}{\sim\Box\alpha \Rightarrow \Diamond\Gamma, \sim\Box\Sigma} \; (\sim\Box\text{left}^*).$$

By induction hypothesis, we have: pS4 $\vdash f(\sim\alpha) \Rightarrow f(\Diamond\Gamma), f(\sim\Box\Sigma)$ where $f(\Diamond\Gamma)$ and $f(\sim\Box\Sigma)$ respectively coincide with $\Diamond f(\Gamma)$ and $\Diamond f(\sim\Sigma)$ by the definition of f. Then, we obtain:

$$\frac{\vdots}{\begin{array}{c} f(\sim\alpha) \Rightarrow \Diamond f(\Gamma), \Diamond f(\sim\Sigma) \\ \hline \Diamond f(\sim\alpha) \Rightarrow \Diamond f(\Gamma), \Diamond f(\sim\Sigma) \end{array}} \; (\Diamond\text{left})$$

where $\Diamond f(\sim\alpha)$ coincides with $f(\sim\Box\alpha)$ by the definition of f. Therefore we have the required fact: pS4 $\vdash f(\sim\Box\alpha) \Rightarrow f(\Diamond\Gamma), f(\sim\Box\alpha)$.

- ($\Longleftarrow$) : By induction on the proofs Q of $f(\Gamma) \Rightarrow f(\Delta)$ in pS4. We distinguish the cases according to the last inference of Q, and show some cases.

Case (cut): The last inference of Q is of the form:

$$\frac{f(\Gamma_1) \Rightarrow f(\Delta_1), \beta \quad \beta, f(\Gamma_2) \Rightarrow f(\Delta_2)}{f(\Gamma_1), f(\Gamma_2) \Rightarrow f(\Delta_1), f(\Delta_2)} \; (\text{cut}).$$

Since β is in $\mathcal{L}$, we have the fact $\beta = f(\beta)$. This fact can be shown by induction on β. Then, by induction hypothesis, we have: CS4 $\vdash \Gamma_1 \Rightarrow \Delta_1, \beta$ and CS4 $\vdash \beta, \Gamma_2 \Rightarrow \Delta_2$. We then obtain the required fact: CS4 $\vdash \Gamma_1, \Gamma_2 \Rightarrow \Delta_1, \Delta_2$ by using (cut) in CS4.

Case ($\Box$right):

Subcase 1: The last inference of P is of the form:

$$\frac{f(\Box\Gamma), f(\sim\Diamond\Sigma) \Rightarrow f(\sim\alpha)}{f(\Box\Gamma), f(\sim\Diamond\Sigma) \Rightarrow f(\sim\Diamond\alpha)} \; (\Box\text{right})$$

where $f(\Box\Gamma), f(\sim\Diamond\Sigma)$ and $f(\sim\Diamond\alpha)$ respectively coincide with $\Box f(\Gamma), \Box f(\sim\Sigma)$ and $\Box f(\sim\alpha)$ by the definition of f. By induction hypothesis, we have: CS4 $\vdash \Box\Gamma, \sim\Diamond\Sigma \Rightarrow \sim\alpha$. Hence, we obtain the required fact:

$$\frac{\vdots}{\begin{array}{c} \Box\Gamma, \sim\Diamond\Sigma \Rightarrow \sim\alpha \\ \hline \Box\Gamma, \sim\Diamond\Sigma \Rightarrow \sim\Diamond\alpha \end{array}} \; (\sim\Diamond\text{right}^*)$$

Subcase 2: The last inference of P is of the form:

$$\frac{f(\Box\Gamma), f(\sim\Diamond\Sigma) \Rightarrow f(\alpha)}{f(\Box\Gamma), f(\sim\Diamond\Sigma) \Rightarrow f(\Box\alpha)} \; (\Box\text{right})$$

where $f(\Box\Gamma), f(\sim\Diamond\Sigma)$ and $f(\Box\alpha)$ respectively coincide with $\Box f(\Gamma), \Box f(\sim\Sigma)$ and $\Box f(\alpha)$ by the definition of f. This case can be shown in a similar way as in Subcase 1. ∎

THEOREM 5 (Cut-elimination). *The rule* (cut) *is admissible in cut-free* CS4.

Proof. Suppose CS4 $\vdash \Gamma \Rightarrow \Delta$. Then, we have pS4 $\vdash f(\Gamma) \Rightarrow f(\Delta)$ by Theorem 4 (1), and hence pS4 $-$ (cut) $\vdash f(\Gamma) \Rightarrow f(\Delta)$ by the cut-elimination theorem for pS4. By Theorem 4 (2), we obtain CS4 $-$ (cut) $\vdash \Gamma \Rightarrow \Delta$. ∎

THEOREM 6 (Decidability). CS4 *is decidable.*

Proof. By decidability of pS4, for each α, it is possible to decide if $f(\alpha)$ is pS4-provable. Then, by Theorem 4, CS4 is decidable. ∎

DEFINITION 7. Let $\sharp$ be a unary connective. A sequent calculus L is called *explosive* with respect to $\sharp$ if for any formulas α and β, the sequent $\alpha, \sharp\alpha \Rightarrow \beta$ is provable in L. It is called *paraconsistent* with respect to $\sharp$ if it is not explosive with respect to $\sharp$.

PROPOSITION 8 (Paraconsistency). CS4 *is paraconsistent with respect to* $\sim$.

Proof. Consider a sequent $p, \sim p \Rightarrow q$ where p and q are distinct propositional variables. The unprovability of this sequent is guaranteed by Theorem 5. ∎

4 Completeness

Firstly, we define the notion of a Kripke model for CS4.

DEFINITION 9. A structure $\langle M, R \rangle$ is called a Kripke frame if

1. M is a non-empty set,

2. R is a transitive and reflexive binary relation on M.

DEFINITION 10. *CS4-valuations* $\models^+$ and $\models^-$ on a Kripke frame $\langle M, R \rangle$ are mappings from the set Φ of all propositional variables to the power set 2^M of M. We will write $x \models^\star p$ for $x \in \models^\star (p)$ with $\star \in \{+, -\}$. These CS4-valuations $\models^+$ and $\models^-$ are extended to mappings from the set of all formulas to 2^M by

1. $x \models^+ \sim\alpha$ iff $x \models^- \alpha$,

2. $x \models^+ \alpha \wedge \beta$ iff $x \models^+ \alpha$ and $x \models^+ \beta$,

3. $x \models^+ \alpha \vee \beta$ iff $x \models^+ \alpha$ or $x \models^+ \beta$,

4. $x \models^+ \alpha \rightarrow \beta$ iff $x \models^+ \alpha$ implies $x \models^+ \beta$,

5. $x \models^+ \Box\alpha$ iff $\forall y \in M\ [xRy$ implies $y \models^+ \alpha]$,

6. $x \models^+ \Diamond\alpha$ iff $\exists y \in M\ [xRy$ and $y \models^+ \alpha]$,

7. $x \models^- \sim\alpha$ iff $x \models^+ \alpha$,

8. $x \models^- \alpha \wedge \beta$ iff $x \models^- \alpha$ or $x \models^- \beta$,

9. $x \models^- \alpha \vee \beta$ iff $x \models^- \alpha$ and $x \models^- \beta$,

10. $x \models^- \alpha \rightarrow \beta$ iff $x \models^+ \alpha$ implies $x \models^- \beta$,

11. $x \models^- \Box\alpha$ iff $\exists y \in M \ [xRy \text{ and } y \models^- \alpha]$,

12. $x \models^- \Diamond\alpha$ iff $\forall y \in M \ [xRy \text{ implies } y \models^- \alpha]$.

DEFINITION 11. A *CS4-Kripke model* is a structure $\langle M, R, \models^+, \models^- \rangle$ such that

1. $\langle M, R \rangle$ is a Kripke frame,

2. $\models^+$ and $\models^-$ are *CS4*-valuations on $\langle M, R \rangle$.

A formula α is *true* in a CS4-Kripke model $\langle M, R, \models^+, \models^- \rangle$ iff $x \models^+ \alpha$ for any $x \in M$, and is *CS4-valid* in a Kripke frame $\langle M, R \rangle$ iff it is true for every *CS4*-valuations $\models^+$ and $\models^-$ on the Kripke frame.

Next, we define the notion of a Kripke model for pS4.

DEFINITION 12. A *pS4-valuation* $\models$ on a Kripke frame $\langle M, R \rangle$ is a mapping from the set Φ of all propositional variables to the power set 2^M of M. We will write $x \models p$ for $x \in \models (p)$. The *pS4*-valuation $\models$ is extended to a mapping from the set of all formulas to 2^M by

1. $x \models \alpha \wedge \beta$ iff $x \models \alpha$ and $x \models \beta$,

2. $x \models \alpha \vee \beta$ iff $x \models \alpha$ or $x \models \beta$,

3. $x \models \alpha \to \beta$ iff $x \models \alpha$ implies $x \models \beta$,

4. $x \models \Box\alpha$ iff $\forall y \in M \ [xRy \text{ implies } y \models \alpha]$,

5. $x \models \Diamond\alpha$ iff $\exists y \in M \ [xRy \text{ and } y \models \alpha]$.

DEFINITION 13. A *pS4-Kripke model* is a structure $\langle M, R, \models \rangle$ such that

1. $\langle M, R \rangle$ is a Kripke frame,

2. $\models$ is a *pS4*-valuation on $\langle M, R \rangle$.

A formula α is *true* in a pS4-Kripke model $\langle M, R, \models \rangle$ iff $x \models \alpha$ for any $x \in M$, and is *pS4-valid* in a Kripke frame $\langle M, R \rangle$ iff it is true for every *pS4*-valuation $\models$ on the Kripke frame.

The following completeness theorem w.r.t. pS4-Kripke models is known: for any formula α, $S4 \vdash \Rightarrow \alpha$ iff α is pS4-valid.

LEMMA 14. *Let f be the mapping defined in Definition 3. For any* CS4-*Kripke model* $\langle M, R, \models^+, \models^- \rangle$, *we can construct a* pS4-*Kripke model* $\langle M, R, \models \rangle$ *such that for any formula α and any $x \in M$,*

1. $x \models^+ \alpha$ iff $x \models f(\alpha)$,

2. $x \models^- \alpha$ iff $x \models f(\sim\alpha)$.

Proof. Let Φ be a set of propositional variables and Φ' be the set $\{p' \mid p \in \Phi\}$ of propositional variables. Suppose that $\langle M, R, \models^+, \models^- \rangle$ is a CS4-Kripke model where $\models^+$ and $\models^-$ are mappings from Φ to 2^M. Suppose that $\langle M, R, \models \rangle$ is a pS4-Kripke model where $\models$ is a mapping from $\Phi \cup \Phi'$ to 2^M. Suppose moreover that these models satisfy the following conditions: for any $x \in M$ and any $p \in \Phi$,

1. $x \models^+ p$ iff $x \models p$,

2. $x \models^- p$ iff $x \models p'$.

Then, the claim of the lemma is proved by (simultaneous) induction on the complexity of α.

- Base step:

Case $\alpha \equiv p \in \Phi$: For 1, we obtain: $x \models^+ p$ iff $x \models p$ iff $x \models f(p)$ (by the definition of f). For 2, we obtain: $x \models^- p$ iff $x \models p'$ iff $x \models f(\sim p)$ (by the definition of f).

- Induction step:

Case $\alpha \equiv \sim\beta$: For 1, we obtain: $x \models^+ \sim\beta$ iff $x \models^- \beta$ iff $x \models f(\sim\beta)$ (by induction hypothesis for 2). For 2, we obtain: $x \models^- \sim\beta$ iff $x \models^+ \beta$ iff $x \models f(\beta)$ (by induction hypothesis for 1) iff $x \models f(\sim\sim\beta)$ (by the definition of f).

Case $\alpha \equiv \beta \wedge \gamma$: For 1, we obtain: $x \models^+ \beta \wedge \gamma$ iff $x \models^+ \beta$ and $x \models^+ \gamma$ iff $x \models f(\beta)$ and $x \models f(\gamma)$ (by induction hypothesis for 1) iff $x \models f(\beta) \wedge f(\gamma)$ iff $x \models f(\beta \wedge \gamma)$ (by the definition of f). For 2, we obtain: $x \models^- \beta \wedge \gamma$ iff $x \models^- \beta$ or $x \models^- \gamma$ iff $x \models f(\sim\beta)$ or $x \models f(\sim\gamma)$ (by induction hypothesis for 2) iff $x \models f(\sim\beta) \vee f(\sim\gamma)$ iff $x \models f(\sim(\beta \wedge \gamma))$ (by the definition of f).

Case $\alpha \equiv \beta \vee \gamma$: Similar to the above case.

Case $\alpha \equiv \beta \to \gamma$: For 1, we obtain: $x \models^+ \beta \to \gamma$ iff $x \models^+ \beta$ implies $x \models^+ \gamma$ iff $x \models f(\beta)$ implies $x \models f(\gamma)$ (by induction hypothesis for 1) iff $x \models f(\beta) \to f(\gamma)$ iff $x \models f(\beta \to \gamma)$ (by the definition of f). For 2, we obtain: $x \models^- \beta \to \gamma$ iff $x \models^+ \beta$ implies $x \models^- \gamma$ iff $x \models f(\beta)$ implies $x \models f(\sim\gamma)$ (by induction hypotheses for 1 and 2) iff $x \models f(\beta) \to f(\sim\gamma)$ iff $x \models f(\sim(\beta \to \gamma))$ (by the definition of f).

Case $\alpha \equiv \Box\beta$: For 1, we obtain: $x \models^+ \Box\beta$ iff $\forall y \in M[xRy$ implies $y \models^+ \beta]$ iff $\forall y \in M[xRy$ implies $y \models f(\beta)]$ (by induction hypothesis for 1) iff $x \models \Box f(\beta)$ iff $x \models f(\Box\beta)$ (by the definition of f). For 2, we obtain: $x \models^- \Box\beta$ iff $\exists y \in M[xRy$ and $y \models^- \beta]$ iff $\exists y \in M[xRy$ and $y \models f(\sim\beta)]$ (by induction hypothesis for 2) iff $x \models \Diamond f(\sim\beta)$ iff $x \models f(\sim\Box\beta)$ (by the definition of f).

Case $\alpha \equiv \Diamond\beta$: Similar to the above case. ∎

LEMMA 15. *Let f be the mapping defined in Definition 3. For any pS4-Kripke model $\langle M, R, \models \rangle$, we can construct a CS4-Kripke model $\langle M, R, \models^+, \models^- \rangle$ such that for any formula α and any $x \in M$,*

1. $x \models f(\alpha)$ iff $x \models^+ \alpha$,

2. $x \models f(\sim\alpha)$ iff $x \models^- \alpha$.

Proof. Similar to the proof of Lemma 14. ∎

THEOREM 16 (Semantical embedding). *Let f be the mapping defined in Definition 3. For any formula α,*

α *is* CS4-*valid iff* $f(\alpha)$ *is* pS4-*valid.*

Proof. By Lemmas 14 and 15. ∎

THEOREM 17 (Completeness). *For any formula α,*

CS4 $\vdash \Rightarrow \alpha$ *iff* α *is* CS4-*valid.*

Proof. CS4 $\vdash \Rightarrow \alpha$ iff pS4 $\vdash \Rightarrow f(\alpha)$ (by Theorem 4) iff $f(\alpha)$ is pS4-valid (by the Kripke completeness theorem for pS4) iff α is CS4-valid (by Theorem 16). ∎

5 Embedding into CS4

The language of C is obtained from that of CS4 by deleting $\Box$ and $\Diamond$. The same notation and terminology as for CS4 is used for C. An *intuitionistic sequent* for C is an expression of the form $\Gamma \Rightarrow \gamma$.

C is then defined as a Gentzen-type sequent calculus. For the inference rules of C, we use the same rule names as those of CS4.

DEFINITION 18 (C). The initial sequents of C are of the form: for any propositional variable p,

$$p \Rightarrow p \qquad \sim p \Rightarrow \sim p.$$

The inference rules of C are of the form:

$$\frac{\Gamma \Rightarrow \alpha \quad \alpha, \Sigma \Rightarrow \gamma}{\Gamma, \Sigma \Rightarrow \gamma} \text{ (cut)}$$

$$\frac{\Gamma \Rightarrow \gamma}{\alpha, \Gamma \Rightarrow \gamma} \text{ (we)} \qquad \frac{\alpha, \alpha, \Gamma \Rightarrow \gamma}{\alpha, \Gamma \Rightarrow \gamma} \text{ (co)}$$

$$\frac{\Gamma \Rightarrow \alpha \quad \beta, \Delta \Rightarrow \gamma}{\alpha \rightarrow \beta, \Gamma, \Delta \Rightarrow \gamma} \text{ ($\rightarrow$left)} \qquad \frac{\alpha, \Gamma \Rightarrow \beta}{\Gamma \Rightarrow \alpha \rightarrow \beta} \text{ ($\rightarrow$right)}$$

$$\frac{\alpha, \Gamma \Rightarrow \gamma}{\alpha \wedge \beta, \Gamma \Rightarrow \gamma} \text{ ($\wedge$left1)} \qquad \frac{\beta, \Gamma \Rightarrow \gamma}{\alpha \wedge \beta, \Gamma \Rightarrow \gamma} \text{ ($\wedge$left2)}$$

$$\frac{\Gamma \Rightarrow \alpha \quad \Gamma \Rightarrow \beta}{\Gamma \Rightarrow \alpha \wedge \beta} \text{ ($\wedge$right)} \qquad \frac{\alpha, \Gamma \Rightarrow \gamma \quad \beta, \Gamma \Rightarrow \gamma}{\alpha \vee \beta, \Gamma \Rightarrow \gamma} \text{ ($\vee$left)}$$

$$\frac{\Gamma \Rightarrow \alpha}{\Gamma \Rightarrow \alpha \vee \beta} \text{ ($\vee$right1)} \qquad \frac{\Gamma \Rightarrow \beta}{\Gamma \Rightarrow \alpha \vee \beta} \text{ ($\vee$right2)}$$

$$\frac{\alpha, \Gamma \Rightarrow \gamma}{\sim\sim\alpha, \Gamma \Rightarrow \gamma} \text{ ($\sim$left)} \qquad \frac{\Gamma \Rightarrow \alpha}{\Gamma \Rightarrow \sim\sim\alpha} \text{ ($\sim$right)}$$

$$\frac{\Gamma \Rightarrow \alpha \quad \sim\beta, \Delta \Rightarrow \gamma}{\sim(\alpha \rightarrow \beta), \Gamma, \Delta \Rightarrow \gamma} \text{ ($\sim\rightarrow$left)} \qquad \frac{\alpha, \Gamma \Rightarrow \sim\beta}{\Gamma \Rightarrow \sim(\alpha \rightarrow \beta)} \text{ ($\sim\rightarrow$right)}$$

$$\frac{\sim\alpha, \Gamma \Rightarrow \gamma \quad \sim\beta, \Gamma \Rightarrow \gamma}{\sim(\alpha \wedge \beta), \Gamma \Rightarrow \gamma} \text{ ($\sim\wedge$left)}$$

$$\frac{\Gamma \Rightarrow \sim\alpha}{\Gamma \Rightarrow \sim(\alpha \wedge \beta)} \ (\sim \wedge \text{ right1}) \qquad \frac{\Gamma \Rightarrow \sim\beta}{\Gamma \Rightarrow \sim(\alpha \wedge \beta)} \ (\sim \wedge \text{ right2})$$

$$\frac{\sim\alpha, \Gamma \Rightarrow \gamma}{\sim(\alpha \vee \beta), \Gamma \Rightarrow \gamma} \ (\sim \vee \text{ left1}) \qquad \frac{\sim\beta, \Gamma \Rightarrow \gamma}{\sim(\alpha \vee \beta), \Gamma \Rightarrow \gamma} \ (\sim \vee \text{ left2})$$

$$\frac{\Gamma \Rightarrow \sim\alpha \quad \Gamma \Rightarrow \sim\beta}{\Gamma \Rightarrow \sim(\alpha \vee \beta)} \ (\sim \vee \text{ right}).$$

The sequents of the form $\alpha \Rightarrow \alpha$ for any formula α are provable in cut-free C.

DEFINITION 19. *C-valuations* $\models^+$ and $\models^-$ on a Kripke frame $\langle M, R \rangle$ are mappings from the set Φ of propositional variables to the power set 2^M of M such that for any $\star \in \{+, -\}$, any $p \in \Phi$ and any $x, y \in M$, if $x \in \models^\star (p)$ and xRy, then $y \in \models^\star (p)$. We will write $x \models^\star p$ for $x \in \models^\star (p)$. These *C*-valuations $\models^+$ and $\models^-$ are extended to mappings from the set of all formulas to 2^M by

1. $x \models^+ \sim\alpha$ iff $x \models^- \alpha$,
2. $x \models^+ \alpha \rightarrow \beta$ iff $\forall y \in M \ [xRy \text{ and } y \models^+ \alpha \text{ imply } y \models^+ \beta]$,
3. $x \models^+ \alpha \wedge \beta$ iff $x \models^+ \alpha$ and $x \models^+ \beta$,
4. $x \models^+ \alpha \vee \beta$ iff $x \models^+ \alpha$ or $x \models^+ \beta$,
5. $x \models^- \sim\alpha$ iff $x \models^+ \alpha$,
6. $x \models^- \alpha \rightarrow \beta$ iff $\forall y \in M \ [xRy \text{ and } y \models^+ \alpha \text{ imply } y \models^- \beta]$,
7. $x \models^- \alpha \wedge \beta$ iff $x \models^- \alpha$ or $x \models^- \beta$,
8. $x \models^- \alpha \vee \beta$ iff $x \models^- \alpha$ and $x \models^- \beta$.

The following *hereditary condition* holds for $\models^\star$ with $\star \in \{+, -\}$: for any formula α and any $x, y \in M$, if $x \models^\star \alpha$ and xRy, then $y \models^\star \alpha$.

DEFINITION 20. A *C-Kripke model* is a structure $\langle M, R, \models^+, \models^- \rangle$ such that

1. $\langle M, R \rangle$ is a Kripke frame,
2. $\models^+$ and $\models^-$ are *C*-valuations on $\langle M, R \rangle$.

A formula α is *true* in a *C*-Kripke model $\langle M, R, \models^+, \models^- \rangle$ iff $x \models^+ \alpha$ for any $x \in M$, and is *C-valid* in a Kripke frame $\langle M, R \rangle$ iff it is true for every *C*-valuations $\models^+$ and $\models^-$ on the Kripke frame.

The following completeness theorem is known [Wansing, 2005]: for any formula α, $C \vdash \Rightarrow \alpha$ iff α is C-valid.

DEFINITION 21. Let $\mathcal{L}_C$ and $\mathcal{L}_{CS4}$ be the languages of C and CS4, respectively. A mapping g from $\mathcal{L}_C$ into $\mathcal{L}_{CS4}$ is defined inductively by

1. for any propositional variable p, $g(p) := \Box p$,
2. $g(\alpha \sharp \beta) := g(\alpha) \sharp g(\beta)$ where $\sharp \in \{\wedge, \vee\}$,

3. $g(\alpha \to \beta) := \Box(g(\alpha) \to g(\beta))$,

4. for any propositional variable p, $g(\sim p) := \Box \sim p$,

5. $g(\sim\sim\alpha) := g(\alpha)$,

6. $g(\sim(\alpha \land \beta)) := g(\sim\alpha) \lor g(\sim\beta)$,

7. $g(\sim(\alpha \lor \beta)) := g(\sim\alpha) \land g(\sim\beta)$,

8. $g(\sim(\alpha \to \beta)) := \Box(g(\alpha) \to g(\sim\beta))$.

Note that $\Diamond$ in $\mathcal{L}_{\text{CS4}}$ is not used in the definition of g.

LEMMA 22. *Let g be the mapping defined in Definition 21. For any formula α in $\mathcal{L}_C$, the following sequents are provable in CS4:*

$$\Box g(\alpha) \Leftrightarrow g(\alpha).$$

Proof. The case CS4 $\vdash \Box g(\alpha) \Rightarrow g(\alpha)$ is straightforward Thus, we show CS4 $\vdash g(\alpha) \Rightarrow \Box g(\alpha)$ by induction on α. We consider the following cases: $\alpha \equiv p$ (p is a propositional variable), $\alpha \equiv \beta \land \gamma$, $\alpha \equiv \beta \lor \gamma$, $\alpha \equiv \beta \to \gamma$, $\alpha \equiv \sim p$ (p is a propositional variable), $\alpha \equiv \sim\sim\beta$, $\alpha \equiv \sim(\beta \land \gamma)$, $\alpha \equiv \sim(\beta \lor \gamma)$ and $\alpha \equiv \sim(\beta \to \gamma)$. We show some cases below.

Case ($\alpha \equiv \sim p$ for any propositional variable p): We obtain the required fact:

$$\dfrac{\dfrac{\dfrac{\sim p \Rightarrow \sim p}{\Box \sim p \Rightarrow \sim p}\,(\Box\text{left})}{\Box \sim p \Rightarrow \Box \sim p}\,(\Box\text{right})}{\Box \sim p \Rightarrow \Box\Box \sim p}\,(\Box\text{right})$$

where $\Box \sim p \Rightarrow \Box\Box \sim p$ coincides with $g(\sim p) \Rightarrow \Box g(\sim p)$ by the definition of g.

Case ($\alpha \equiv \sim\sim\beta$): By induction hypothesis, we have: CS4 $\vdash g(\beta) \Rightarrow \Box g(\beta)$. Then, we obtain the required fact: CS4 $\vdash g(\sim\sim\beta) \Rightarrow \Box g(\sim\sim\beta)$ by the definition of g.

Case ($\alpha \equiv \sim(\beta \to \gamma)$): We obtain the required fact:

$$\dfrac{\dfrac{\dfrac{\vdots}{g(\beta) \to g(\sim\gamma) \Rightarrow g(\beta) \to g(\sim\gamma)}}{\dfrac{\Box(g(\beta) \to g(\sim\gamma)) \Rightarrow g(\beta) \to g(\sim\gamma)}{\dfrac{\Box(g(\beta) \to (\sim\gamma)) \Rightarrow \Box(g(\beta) \to g(\sim\gamma))}{\Box(g(\beta) \to g(\sim\gamma)) \Rightarrow \Box\Box(g(\beta) \to g(\sim\gamma))}\,(\Box\text{right})}\,(\Box\text{right})}\,(\Box\text{left})}$$

where $\Box(g(\beta) \to g(\sim\gamma))$ coincides with $g(\sim(\beta \to \gamma))$ by the definition of g.

Case ($\alpha \equiv \sim(\beta \land \gamma)$): By induction hypothesis, we have: CS4 $\vdash g(\sim\beta) \Rightarrow \Box g(\sim\beta)$ and CS4 $\vdash g(\sim\gamma) \Rightarrow \Box g(\sim\gamma)$. Then, we obtain: CS4 $\vdash g(\sim\beta) \lor g(\sim\gamma) \Rightarrow \Box g(\sim\beta) \lor \Box g(\sim\gamma)$ by using ($\lor$left) and ($\lor$right). Suppose that S is $g(\sim\beta) \lor g(\sim\gamma) \Rightarrow \Box g(\sim\beta) \lor \Box g(\sim\gamma)$. By using (cut), we then obtain the required fact:

$$\frac{\overset{\vdots}{S} \quad \frac{\Box g(\sim\beta) \vee \Box g(\sim\gamma) \Rightarrow \Box(\Box g(\sim\beta) \vee \Box g(\sim\gamma))}{g(\sim\beta) \vee g(\sim\gamma) \Rightarrow \Box(\Box g(\sim\beta) \vee \Box g(\sim\gamma))} \quad \frac{\overset{\vdots}{A} \qquad \overset{\vdots}{B}}{\Box(\Box g(\sim\beta) \vee \Box g(\sim\gamma)) \Rightarrow \Box(g(\sim\beta) \vee g(\sim\gamma))}}{g(\sim\beta) \vee g(\sim\gamma) \Rightarrow \Box(g(\sim\beta) \vee g(\sim\gamma))}$$

where $g(\sim\beta) \vee g(\sim\gamma)$ coincides with $g(\sim(\beta \wedge \gamma))$ by the definition of g, A is of the form:

$$\frac{\frac{\overset{\vdots}{\Box g(\sim\beta) \Rightarrow \Box g(\sim\beta)}}{\Box g(\sim\beta) \Rightarrow \Box g(\sim\beta) \vee \Box g(\sim\gamma)}}{\Box g(\sim\beta) \Rightarrow \Box(\Box g(\sim\beta) \vee \Box g(\sim\gamma))} \quad \frac{\frac{\overset{\vdots}{\Box g(\sim\gamma) \Rightarrow \Box g(\sim\gamma)}}{\Box g(\sim\gamma) \Rightarrow \Box g(\sim\beta) \vee \Box g(\sim\gamma)}}{\Box g(\sim\gamma) \Rightarrow \Box(\Box g(\sim\beta) \vee \Box g(\sim\gamma))}$$
$$\Box g(\sim\beta) \vee \Box g(\sim\gamma) \Rightarrow \Box(\Box g(\sim\beta) \vee \Box g(\sim\gamma))$$

and B is of the form:

$$\frac{\frac{\frac{\overset{\vdots}{g(\sim\beta) \Rightarrow g(\sim\beta)}}{\Box g(\sim\beta) \Rightarrow g(\sim\beta)}}{\Box g(\sim\beta) \Rightarrow g(\sim\beta) \vee g(\sim\gamma)} \quad \frac{\frac{\overset{\vdots}{g(\sim\gamma) \Rightarrow g(\sim\gamma)}}{\Box g(\sim\gamma) \Rightarrow g(\sim\gamma)}}{\Box g(\sim\gamma) \Rightarrow g(\sim\beta) \vee g(\sim\gamma)}}{\frac{\Box g(\sim\beta) \vee \Box g(\sim\gamma) \Rightarrow g(\sim\beta) \vee g(\sim\gamma)}{\frac{\Box(\Box g(\sim\beta) \vee \Box g(\sim\gamma)) \Rightarrow g(\sim\beta) \vee g(\sim\gamma)}{\Box(\Box g(\sim\beta) \vee \Box g(\sim\gamma)) \Rightarrow \Box(g(\sim\beta) \vee g(\sim\gamma))}}}.$$

∎

LEMMA 23. *Let g be the mapping defined in Definition 21. For any intuitionistic sequent $\Gamma \Rightarrow \gamma$ in $\mathcal{L}_C$,*

if $C \vdash \Gamma \Rightarrow \gamma$, then $CS4 \vdash g(\Gamma) \Rightarrow g(\gamma)$.

Proof. By induction on the proofs P of $\Gamma \Rightarrow \gamma$ in C. We distinguish the cases according to the last inference of P. We show some cases.

Case ($\sim p \Rightarrow \sim p$): The last inference of P is of the form: $\sim p \Rightarrow \sim p$ for any propositional variable p. We obtain the required fact:

$$\frac{\frac{\sim p \Rightarrow \sim p}{\Box \sim p \Rightarrow \sim p} \;(\Box \text{left})}{\Box \sim p \Rightarrow \Box \sim p} \;(\Box \text{right})$$

where $\Box \sim p$ coincides with $g(\sim p)$ by the definition of g.

Case ($\sim \to$right): The last inference of P is of the form:

$$\frac{\alpha, \Gamma \Rightarrow \sim\beta}{\Gamma \Rightarrow \sim(\alpha \to \beta)} \;(\sim \to \text{right}).$$

By induction hypothesis, we have: CS4 $\vdash g(\alpha), g(\Gamma) \Rightarrow g(\sim\beta)$. Then, we obtain:

$$\vdots$$
$$\dfrac{g(\alpha), g(\Gamma) \Rightarrow g(\sim\beta)}{g(\Gamma) \Rightarrow g(\alpha){\rightarrow}g(\sim\beta)} \; (\rightarrow\text{right})$$
$$\vdots \; (\Box\text{left})$$
$$\dfrac{\Box g(\Gamma) \Rightarrow g(\alpha){\rightarrow}g(\sim\beta)}{\Box g(\Gamma) \Rightarrow \Box(g(\alpha){\rightarrow}g(\sim\beta))} \; (\Box\text{right})$$

where $\Box(g(\alpha){\rightarrow}g(\sim\beta))$ coincides with $g(\sim(\alpha{\rightarrow}\beta))$ by the definition of g. By Lemma 22, we have: CS4 $\vdash g(\pi) \Leftrightarrow \Box g(\pi)$ for any $\pi \in \Gamma$. Then, we obtain the required fact: CS4 $\vdash g(\Gamma) \Rightarrow g(\sim(\alpha{\rightarrow}\beta))$ by applying (cut) to $g(\pi) \Rightarrow \Box g(\pi)$ ($\pi \in \Gamma$) and $\Box g(\Gamma) \Rightarrow g(\sim(\alpha{\rightarrow}\beta))$, repeatedly.

Case ($\sim\rightarrow$left): The last inference of P is of the form:

$$\dfrac{\Gamma \Rightarrow \alpha \quad \sim\beta, \Delta \Rightarrow \gamma}{\sim(\alpha{\rightarrow}\beta), \Gamma, \Delta \Rightarrow \gamma} \; (\sim\rightarrow\text{left}).$$

By induction hypothesis, we have: CS4 $\vdash g(\Gamma) \Rightarrow g(\alpha)$ and CS4 $\vdash g(\sim\beta), g(\Delta) \Rightarrow g(\gamma)$. Then, we obtain the required fact:

$$\vdots \qquad \vdots$$
$$\dfrac{g(\Gamma) \Rightarrow g(\alpha) \quad g(\sim\beta), g(\Delta) \Rightarrow g(\gamma)}{g(\alpha){\rightarrow}g(\sim\beta), g(\Gamma), g(\Delta) \Rightarrow g(\gamma)} \; (\rightarrow\text{left})$$
$$\dfrac{}{\Box(g(\alpha){\rightarrow}g(\sim\beta)), g(\Gamma), g(\Delta) \Rightarrow g(\gamma)} \; (\Box\text{left})$$

where $\Box(g(\alpha){\rightarrow}g(\sim\beta))$ coincides with $g(\sim(\alpha{\rightarrow}\beta))$ by the definition of g.

Case ($\sim\sim$left): The last inference of P is of the form:

$$\dfrac{\alpha, \Gamma \Rightarrow \gamma}{\sim\sim\alpha, \Gamma \Rightarrow \gamma} \; (\sim\sim\text{left}).$$

By induction hypothesis, we have: CS4 $\vdash g(\alpha), g(\Gamma) \Rightarrow g(\gamma)$. Then, we obtain the required fact: CS4 $\vdash g(\sim\sim\alpha), g(\Gamma) \Rightarrow g(\gamma)$ by the definition of g. ∎

LEMMA 24. *Let g be the mapping defined in Definition 21. For any formula α in $\mathcal{L}_C$,*

if CS4 $\vdash \Rightarrow g(\alpha)$, then C $\vdash \Rightarrow \alpha$.

Proof. We show: if $\Rightarrow \alpha$ is not provable in C, then $\Rightarrow g(\alpha)$ is not provable in CS4. Suppose that $\Rightarrow \alpha$ is not provable in C. By the Kripke-completeness theorem for C, there exists a C-Kripke model $\langle M, R, \models^+, \models^- \rangle$ such that α is not true in $\langle M, R, \models^+, \models^- \rangle$. Suppose that $x_0 \models^+ \alpha$ does not hold for $x_0 \in M$. We then define a CS4-Kripke model $\langle M, R, \models^+_{\text{CS4}}, \models^-_{\text{CS4}} \rangle$ by:

1. for any propositional variable p, $x \models^+_{\text{CS4}} p$ iff $x \models^+ p$,

2. for any propositional variable p, $x \models^-_{\text{CS4}} p$ iff $x \models^- p$.

Claim: For any $\alpha \in \mathcal{L}_C$ and any $x \in M$,

1. $x \models^+_{CS4} g(\alpha)$ iff $x \models^+ \alpha$,
2. $x \models^+_{CS4} g(\sim\alpha)$ iff $x \models^- \alpha$.

By this claim (which will be proved) and the assumption ($x_0 \models^+ \alpha$ does not hold), we obtain the fact that $x_0 \models^+_{CS4} g(\alpha)$ does not hold. Therefore $\Rightarrow g(\alpha)$ is not provable in CS4 by Theorem 17.

We now prove the claim by (simultaneous) induction on α. We show some cases.

Case ($\alpha \equiv p$ for any propositional variable p): For 1, we obtain:

$$x \models^+ p$$

iff $\forall y \in M$ [xRy implies $y \models^+ p$] (by the hereditary condition)

iff $\forall y \in M$ [xRy implies $y \models^+_{CS4} p$]

iff $x \models^+_{CS4} \Box p$

iff $x \models^+_{CS4} g(p)$ (by the definition of g).

For 2, we obtain:

$$x \models^- p$$

iff $\forall y \in M$ [xRy implies $y \models^- p$] (by the hereditary condition)

iff $\forall y \in M$ [xRy implies $y \models^-_{CS4} p$]

iff $\forall y \in M$ [xRy implies $y \models^+_{CS4} \sim p$]

iff $x \models^+_{CS4} \Box \sim p$

iff $x \models^+_{CS4} g(\sim p)$ (by the definition of g).

Case ($\alpha \equiv \beta \to \gamma$): For 1, we obtain:

$$x \models^+ \beta \to \gamma$$

iff $\forall y \in M$ [xRy and $y \models^+ \beta$ imply $y \models^+ \gamma$]

iff $\forall y \in M$ [xRy and $y \models^+_{CS4} g(\beta)$ imply $y \models^+_{CS4} g(\gamma)$] (by induction hypothesis for 1)

iff $x \models^+_{CS4} \Box(g(\beta) \to g(\gamma))$

iff $x \models^+_{CS4} g(\beta \to \gamma)$ (by the definition of g).

For 2, we obtain:

$$x \models^- \beta \to \gamma$$

iff $\forall y \in M$ [xRy and $y \models^+ \beta$ imply $y \models^- \gamma$]

iff $\forall y \in M$ [xRy and $y \models^+_{\text{CS4}} g(\beta)$ imply $y \models^+_{\text{CS4}} g(\sim\gamma)$] (by induction hypotheses for 1 and 2)

iff $x \models^+_{\text{CS4}} \Box(g(\beta) \to g(\sim\gamma))$

iff $x \models^+_{\text{CS4}} g(\sim(\beta \to \gamma))$ (by the definition of g).

Case ($\alpha \equiv \sim\beta$): For 1, we have: $x \models^+ \sim\beta$ iff $x \models^- \beta$ iff $x \models^+_{\text{CS4}} g(\sim\beta)$ (by induction hypothesis for 2). For 2, we have: $x \models^- \sim\beta$ iff $x \models^+ \beta$ iff $x \models^+_{\text{CS4}} g(\beta)$ (by induction hypothesis for 1) iff $x \models^+_{\text{CS4}} g(\sim\sim\beta)$ (by the definition of g). ∎

THEOREM 25 (Embedding). *Let g be the mapping defined in Definition 21. For any formula α in $\mathcal{L}_\text{C}$,*

$$\text{C} \vdash \Rightarrow \alpha \ \textit{iff}\ \text{CS4} \vdash \Rightarrow g(\alpha).$$

Proof. By Lemmas 23 and 24. ∎

A similar embedding of C into the logic BS4, which is a modal extension of Belnap and Dunn's four-valued logic of first-degree entailment, is proved in [Odintsov and Wansing, 2010].

6 Embedding into CS4^{d-}

In Ohnishi and Matsumoto's cut-free sequent calculus for the modal logic S4 [Ohnishi and Matsumoto, 1957], $\Box\alpha$ is an abbreviation of $\neg\Diamond\neg\alpha$ (where $\neg$ is classical negation). Their rules for $\Box$ and $\Diamond$ "work in isolation but not when they are combined" [Kripke, 1963, p. 91]. Whereas in their system the equivalence between $\Box\alpha$ and $\neg\Diamond\neg\alpha$ and between $\Diamond\alpha$ and $\neg\Box\neg A$ cannot be proved, in CS4 the syntactic duality between $\Box$ and $\Diamond$ with respect to $\sim$ is provable. For a strong negation $\sim$, this duality seems to be natural, since in the first-order versions of Nelson's constructive logics with strong negation (see, e.g., [Odintsov and Wansing, 2003; Odintsov and Wansing, 2004]) the universal and the existential quantifiers also are duals of each other with respect to $\sim$. Nevertheless, constructive modal logics with strong negation but without syntactic duality between $\Box$ and $\Diamond$ have been investigated as well, cf. [Odintsov and Wansing, 2003; Odintsov and Wansing, 2004]. In particular, in [Odintsov and Wansing, 2003] a very natural constructive description logic $\mathcal{CALC}^{\mathbf{N4}}$ without syntactical duality is defined by a translation from the language of constructive description logic into the language of constructive predicate logic. An additional motivation for considering a subsystem of CS4 in which the syntactic duality between $\Box$ and $\Diamond$ fails to hold comes with the observation that $\Diamond$ in $\mathcal{L}_\text{CS4}$ is not used in the specification of the translation function g from Definition 21.

DEFINITION 26. The sequent calculus CS4^{d-} is defined as the result of (i) adding for any $\mathcal{L}_\text{CS4}$-formula α initial sequents of the form

$$\sim\Box\alpha \Rightarrow \sim\Box\alpha \qquad \sim\Diamond\alpha \Rightarrow \sim\Diamond\alpha,$$

(ii) removing from CS4 the rules ($\sim\Box$left*), ($\sim\Box$right), ($\sim\Diamond$left), and ($\sim\Diamond$right*) and (iii) at the same time replacing in CS4 the rule ($\Box$right*) by ($\Box$right) and the rule ($\Diamond$left*) by ($\Diamond$left).

The language $\mathcal{L}_{\text{CS4}^{d-}}$ (the set of all formulas of CS4^{d-}) is the same as $\mathcal{L}_{\text{CS4}}$. The superscript d indicates that the modal operators satisfy semantic duality in the sense that $\Box$ and $\Diamond$ are interpreted as a restricted universal, respectively existential quantifier with respect to *the same* binary accessibility relation, and the superscript $^-$ indicates that the syntactic duality between $\Box$ and $\Diamond$ fails, cf. [Odintsov and Wansing, 2004]. The set $\Box\mathcal{L}_{\text{CS4}}$ is defined as $\{\Box\gamma \mid \gamma \in \mathcal{L}_{\text{CS4}}\}$, and $\Diamond\mathcal{L}_{\text{CS4}} := \{\Diamond\gamma \mid \gamma \in \mathcal{L}_{\text{CS4}}\}$.

DEFINITION 27. CS4^{d-}-valuations $\models^+$ and $\models^-$ on a Kripke frame $\langle M, R \rangle$ are defined exactly as CS4-valuations on $\langle M, R \rangle$, except that we consider functions v^+ from Φ to 2^M and v^- from $\Phi \cup \Box\mathcal{L}_{\text{CS4}} \cup \Diamond\mathcal{L}_{\text{CS4}}$ to 2^M, where Φ is the set of all propositional variables, and postulate that for every $p \in \Phi$, $x \models^+ p$ iff $x \in v^+(p)$, $x \models^- p$ iff $x \in v^-(p)$, and:

11.' $x \models^- \Box\alpha$ iff $x \in v^-(\Box\alpha)$,

12.' $x \models^- \Diamond\alpha$ iff $x \in v^-(\Diamond\alpha)$.

DEFINITION 28. A *CS4^{d-}-Kripke model* is a structure $\langle M, R, \models^+, \models^- \rangle$ such that

1. $\langle M, R \rangle$ is a Kripke frame,

2. $\models^+$ and $\models^-$ are *CS4^{d-}*-valuations on $\langle M, R \rangle$.

An $\mathcal{L}_{\text{CS4}}$-formula α is *true* in a *CS4^{d-}*-Kripke model $\langle M, R, \models^+, \models^- \rangle$ iff $x \models^+ \alpha$ for any $x \in M$, and is *CS4^{d-}-valid* in a Kripke frame $\langle M, R \rangle$ iff it is true for every *CS4^{d-}*-valuations $\models^+$ and $\models^-$ on $\langle M, R \rangle$.

DEFINITION 29. Let the sets of (new) propositional variables Φ'' and Φ''' be defined as follows:

$$\Phi'' := \{p_\alpha \mid \Box\alpha \in \mathcal{L}_{\text{CS4}}\} \text{ and } \Phi''' := \{q_\alpha \mid \Diamond\alpha \in \mathcal{L}_{\text{CS4}}\}$$

and let the language of pS4 now comprise as propositional letters in addition to the elements from Φ and Φ' also the elements from Φ'' and Φ'''. The mapping f' from $\mathcal{L}_{\text{CS4}}$ to the language of pS4 is defined exactly as the mapping f from Definition 3, except that:

8.' $f'(\sim\Box\alpha) := p_\alpha$,

9.' $f'(\sim\Diamond\alpha) := q_\alpha$.

THEOREM 30 (Syntactical embedding). *Let Γ and Δ be sets of formulas in $\mathcal{L}_{\text{CS4}}$, and f' be the mapping defined in Definition 29. Then:*

1. *CS4^{d-} $\vdash \Gamma \Rightarrow \Delta$ iff pS4 $\vdash f'(\Gamma) \Rightarrow f'(\Delta)$.*

2. *CS4^{d-} $-$ (cut) $\vdash \Gamma \Rightarrow \Delta$ iff pS4 $-$ (cut) $\vdash f'(\Gamma) \Rightarrow f'(\Delta)$.*

Proof. Since the claim (2) can be obtained by a subproof of (1), it is enough to consider (1). All cases of the induction on proofs in CS4^{d-} and proofs in pS4 are covered already by the proof of Theorem 4. In particular, the distinction between two subcases of Case ($\Box$right) in the induction on proofs in pS4 is superfluous due to the difference between f and f'. ∎

As in the case of CS4 we obtain cut-elimination and decidability for $CS4^{d-}$.

THEOREM 31 (Cut-elimination). *The rule* (cut) *is admissible in cut-free* $CS4^{d-}$.

THEOREM 32 (Decidability). $CS4^{d-}$ *is decidable*.

LEMMA 33. *Let f' be the mapping defined in Definition 29. For any $CS4^{d-}$-Kripke model $\langle M, R, \models^+, \models^- \rangle$, we can construct a pS4-Kripke model $\langle M, R, \models \rangle$ such that for any formula α and any $x \in M$,*

1. $x \models^+ \alpha$ *iff* $x \models f'(\alpha)$,

2. $x \models^- \alpha$ *iff* $x \models f'(\sim\alpha)$.

Proof. Let Φ be a set of propositional variables and Φ' be the set $\{p' \mid p \in \Phi\}$ of propositional variables, and let again $\Phi'' := \{p_\alpha \mid \Box\alpha \in \mathcal{L}_{CS4}\}$ and $\Phi''' := \{q_\alpha \mid \Diamond\alpha \in \mathcal{L}_{CS4}\}$. Suppose that $\langle M, R, \models^+, \models^- \rangle$ is a $CS4^{d-}$-Kripke model where v^+ is a mapping from Φ to 2^M and v^- is a mapping from $\Phi \cup \Box\mathcal{L}_{CS4} \cup \Diamond\mathcal{L}_{CS4}$ to 2^M. Suppose that $\langle M, R, \models \rangle$ is a pS4-Kripke model where $\models$ is a mapping from $\Phi \cup \Phi' \cup \Phi'' \cup \Phi'''$ to 2^M. Suppose moreover that these models satisfy the following conditions: for any $x \in M$ and any $p \in \Phi$, $p_\alpha \in \Phi''$, and $q_\alpha \in \Phi'''$,

1. $x \models^+ p$ iff $x \models p$,

2. $x \models^- p$ iff $x \models p'$,

3. $x \models^- \Box\alpha$ iff $x \models p_\alpha$,

4. $x \models^- \Diamond\alpha$ iff $x \models q_\alpha$.

The claim of the lemma is proved by (simultaneous) induction on the complexity of α. We consider the subcases not covered by the proof of Lemma 14.

Case $\alpha \equiv \Box\beta$, Claim 2: We have $x \models^- \Box\beta$ iff $x \models p_\beta$ iff $x \models f'(\sim\Box\beta)$.

Case $\alpha \equiv \Diamond\beta$, Claim 2: Analogous to the previous subcase. ∎

LEMMA 34. *Let f' be the mapping defined in Definition 29. For any pS4-Kripke model $\langle M, R, \models \rangle$, we can construct a $CS4^{d-}$-Kripke model $\langle M, R, \models^+, \models^- \rangle$ such that for any formula α and any $x \in M$,*

1. $x \models f'(\alpha)$ *iff* $x \models^+ \alpha$,

2. $x \models f'(\sim\alpha)$ *iff* $x \models^- \alpha$.

Proof. Similar to the proof of Lemma 33. ∎

THEOREM 35 (Semantical embedding). *Let again f' be the mapping defined in Definition 29. For any formula α,*

α *is* $CS4^{d-}$*-valid iff* $f'(\alpha)$ *is pS4-valid*.

Proof. By Lemmas 33 and 34. ∎

THEOREM 36 (Completeness). *For any formula α,*

$$\text{CS4}^{d-} \vdash \Rightarrow \alpha \text{ iff } \alpha \text{ is CS4}^{d-}\text{-valid.}$$

Proof. Analogous to the proof of Theorem 17. ∎

The next two lemmas for CS4^{d-} can be proved in complete analogy to Lemmas 22 and 23 for CS4.

LEMMA 37. *Let g be the mapping defined in Definition 21. For any $\mathcal{L}_C$-formula α, the following sequents are provable in CS4^{d-}:*

$$\Box g(\alpha) \Leftrightarrow g(\alpha).$$

LEMMA 38. *Let g be the mapping defined in Definition 21. For any intuitionistic sequent $\Gamma \Rightarrow \gamma$ in $\mathcal{L}_C$,*

if $C \vdash \Gamma \Rightarrow \gamma$, then $\text{CS4}^{d-} \vdash g(\Gamma) \Rightarrow g(\gamma)$.

LEMMA 39. *Let g be the mapping defined in Definition 21. For any $\mathcal{L}_C$-formula α,*

if $\text{CS4} \vdash \Rightarrow g(\alpha)$, then $C \vdash \Rightarrow \alpha$.

Proof. The only essential difference to the proof of Lemma 24 is the definition of the CS4^{d-}-valuation $\models^-$ in a CS4^{d-}-Kripke countermodel. Namely, it has to be remarked that the definition of $\models^-$ for formulas from the sets $\Box \mathcal{L}_{\text{CS4}}$ and $\Diamond \mathcal{L}_{\text{CS4}}$ is arbitrary. ∎

THEOREM 40 (Embedding). *Let g be the mapping defined in Definition 21. For any formula α in $\mathcal{L}_C$,*

$$C \vdash \Rightarrow \alpha \text{ iff } \text{CS4}^{d-} \vdash \Rightarrow g(\alpha).$$

Proof. By Lemmas 38 and 39. ∎

If both syntactic and semantic duality between $\Box$ and $\Diamond$ in CS4 are given up, we obtain the logic CS4^-.

DEFINITION 41. A Kripke frame for CS4^- is a structure $\langle M, R_\Box, R_\Diamond \rangle$, where $R_\Box$ and $R_\Diamond$ are binary, reflexive and transitive relations on the non-empty set M.

DEFINITION 42. CS4^--valuations $\models^+$ and $\models^-$ on a Kripke frame $\langle M, R_\Box, R_\Diamond \rangle$ for CS4^- are defined exactly as CS4^{d-}-valuations on a Kripke frame except that:

5.' $x \models^+ \Box \alpha$ iff $\forall y \in M \ [xR_\Box y \text{ implies } y \models^+ \alpha]$,

6.' $x \models^+ \Diamond \alpha$ iff $\exists y \in M \ [xR_\Diamond y \text{ and } y \models^+ \alpha]$.

The notions of a CS4$^-$-Kripke model, truth of an $\mathcal{L}_{\text{CS4}}$-formula α in a CS4$^-$-Kripke model, and CS4$^-$-validity in a Kripke frame for CS4$^-$ are defined in analogy to the corresponding notions for CS4^{d-}. Note that the formula $\Box\alpha \to \Diamond\alpha$ is CS4-valid and CS4^{d-}-valid in any Kripke frame, but it is not the case that $\Box\alpha \to \Diamond\alpha$ is CS4$^-$-valid in any Kripke frame for CS4$^-$.

We close this paper with the statement of an open problem, namely to define a cut-free sequent calculus for the semantically defined logic CS4$^-$.

Acknowledgments. We would like to thank Sergei P. Odintsov for his valuable comments. N. Kamide was partially supported by the Ministry of Education, Science, Sports and Culture, Grant-in-Aid for Young Scientists, 20700015.

BIBLIOGRAPHY

[Almukdad and Nelson, 1984] A. Almukdad and D. Nelson. Constructible falsity and inexact predicates. *Journal of Symbolic Logic*, 49:231–233, 1984.

[Angell, 1962] R. Angell. A propositional logics with subjunctive conditionals. *Journal of Symbolic Logic*, 27:327–343, 1962.

[Gödel, 1933] K. Gödel. Eine interpretation des intuitionistischen aussagenkalküls. *Ergebnisse eines mathematischen Kolloquiums*, 4:39–40, 1933.

[Gurevich, 1977] Y. Gurevich. Intuitionistic logic with strong negation. *Studia Logica*, 36:49–59, 1977.

[Kamide and Wansing, 2010] N. Kamide and H. Wansing. Symmetric and dual paraconsistent logics. *Logic and Logical Philosophy*, 19:7–30, 2010.

[Kripke, 1963] S. Kripke. Semantical analysis of modal logic i: Normal modal propositional calculi. *Zeitschrift für mathematische Logik und Grundlagen der Mathematik*, 11:3–16, 1963.

[McCall, 1966] S. McCall. Connexive implication. *Journal of Symbolic Logic*, 31:415–433, 1966.

[Nelson, 1949] D. Nelson. Constructible falsity. *Journal of Symbolic Logic*, 14:16–26, 1949.

[Odintsov and Wansing, 2003] S. Odintsov and H. Wansing. Inconsistency-tolerant description logic. motivation and basic systems. In V. Hendricks and J. Malinowski, editors, *Trends in Logic, 50 Years of Studia Logica*, pages 301–335, Dordrecht, 2003. Kluwer Academic Publishers.

[Odintsov and Wansing, 2004] S. Odintsov and H. Wansing. Constructive predicate logic and constructive modal logic. formal duality versus semantical duality. In V. Hendricks et al., editor, *First-Order Logic Revisited*, pages 269–286, Berlin, 2004. Logos Verlag.

[Odintsov and Wansing, 2010] S. Odintsov and H. Wansing. Modal logics with belnapian truth values. *Journal of Applied Non-Classical Logics*, 20:279–301, 2010.

[Ohnishi and Matsumoto, 1957] M. Ohnishi and K. Matsumoto. Gentzen method in modal calculi. *Osaka Mathematical Journal*, 9:113–130, 1957.

[Rautenberg, 1979] W. Rautenberg. *Klassische und nicht-klassische Aussagenlogik*. Vieweg, 1979.

[Vorob'ev, 1952] N.N. Vorob'ev. A constructive propositional calculus with strong negation (in russian). *Doklady Akademii Nauk SSR*, 85:465–468, 1952.

[Wansing, 2005] H. Wansing. Connexive modal logic. *Advances in Modal Logic*, 5:367–385, 2005.

[Wansing, 2010] H. Wansing. Connexive logic. In E.N. Zalta, editor, *The Stanford Encyclopedia of Philosophy (Fall 2010 Edition)*, URL = ⟨http://plato.stanford.edu/archives/fall2010/entries/logic-connexive/⟩, 2010.

Norihiro Kamide Waseda
Institute for Advanced Study
Waseda University
Tokyo, Japan
E-mail: drnkamide08@kpd.biglobe.ne.jp

Heinrich Wansing
Department of Philosophy II
Ruhr University Bochum
Bochum, Germany
E-mail: Heinrich.Wansing@rub.de

Basic Constructive Modality

VALERIA DE PAIVA AND EIKE RITTER

ABSTRACT. The benefits of the extended Curry-Howard correspondence relating the simply typed lambda-calculus to proofs of intuitionistic propositional logic and to appropriate classes of categories that model the calculus are widely known. In this paper we show an analogous correspondence between a simple constructive modal logic CK (with both necessity □ and possibility ◊ operators) and a lambda-calculus with modality constructors. Then we investigate classes of categorical models for this logic. Parallel work for constructive S4 (CS4) has appeared before in [Bierman and de Paiva, 2000; Alechina et al., 2001]. The work on the basic system CK has appeared initially with co-authors Bellin and Ritter in the conference Methods for the Modalities [Bellin et al., 2001]. Since then the technical work has been improved by [Kakutani, 2007] and taken to a different, higher-order categorical setting by Ritter and myself. Here we expound on the logical significance of the earlier work.

Preface

It is a great honor to contribute to this volume celebrating Walter Carnielli's work. Professor Carnielli is a source of inspiration and support for people who believe that logic in all its manifestations, pure and applied, is an engine for change and improvement in the sciences and in technology – maybe even in society and in the social sciences, if we, logicians, can get there. With honor comes duty and I have worried about which kind of the technical work would be suitable for this celebration. I have decided to write on "constructive modal logics". Constructive (or intuitionistic, I will use the terms interchangeably) modal logics are modal logics based on a constructive perspective of the world. Constructivists believe in producing witnesses for existential statements; in knowing, given a disjunction $A \lor B$ that one of the disjuncts really holds and that you should know which one is that. They distrust excluded middles and double-negations, which look a bit like "magic" and they really like the notion of implication as a form of internalization of the inferencing process. Constructivists ought to be allowed their own modal logics.

Personally I have been working on constructive modal logics for a long while. I started doing it when I realized that formally the system S4 is just like Linear Logic, the subject of my thesis. Yes, I know that historically this is "back-to-front", the rules for modal S4 must have been in Girard's mind when he conceived Linear Logic, but this was the order that made me interested in modal logic. Work on this formal similarity between the systems with Gavin Bierman eventually became [Bierman and de Paiva, 2000]. Then I discovered the affordances (and intricacies) of formal, explicit substitutions and the preliminary work on a dual system for intuitionistic and (S4) modal logic, called DIML for Dual and Intuitionistic Modal logic (joint with Neil Ghani and Eike Ritter[Ghani et al., 1998]) came to light. By then I was truly bitten by the bug: I wanted to see how far we can push the frontiers of the Curry-Howard correspondence for modal systems. But I wanted my Curry-Howard correspondence to be a categorical one, that is, I wanted "triangles" of maps relating logics, their type-theoretical formulations and their (equivalent) categorical semantics formulations. With the help of Natasha Alechina, Eike Ritter and Michael Mendler I wrote about the relationship between categorical semantics and possible world semantics in [Alechina et al., 2001], but we barely scratched the surface of the question. Paying attention to the philosophical tradition that considered K (named after Kripke[Kripke, 1963]) the basic system for normal modal logic, Gianluigi Bellin, Eike Ritter and myself applied to this system our basic Natural Deduction intuitions in [Bellin et al., 2001].This is the work we discuss here.

Meanwhile I have been helping to organize a collection of workshops on "Intuitionistic Modal Logics and Applications (IMLA)", hoping to get philosophers and computer scientists to share their insights on the big quest for a Curry-Howard-Lawvere[1] correspondence for intuitionistic Modal Logic in general. The IMLA workshops started in 1999, as part of the Federated Logic Conference (FLoC1999) and the fifth installment has just happened as part of the Congress of Logic, Methodology and Philosophy of Science in Nancy, France, 2011. Associated with the IMLA workshops there have been journal special volumes published as [Fairtlough et al., 2001; Goré et al., 2004; de Paiva and Pientka, 2011].

But work on intuitionistic modal logics is still very much in construction, the quest is just beginning, we have some pieces of the puzzle in place, but much remains to be done. It is also a work where more philosophical intuition is required. Mathematics alone can only go so far and hence this is a work where we could do with help from logicians that are well-versed in philosophical questions. This brings us back to Walter Carnielli and his ability of, not only doing first-rate work on his own, straddling philosophical and mathematical fields of expertise, but also of congregating and organizing other logicians to work on interesting problems. Thus this work is dedicated to Walter in the hope that he will like the project of Curry-Howard-Lawvere correspondences for constructive Modal Logic and he will use some of his uncanny abilities to further it. Happy Birthday, Walter!

1 Introduction

Modal logic is arguably the logic formalism most used in Computer Science. The *explicit* logic formalism most used, as classical logic is used implicitly everywhere. Modal Logic in its several variants, e.g. epistemic logic, temporal logic, description logics, probabilistic logic, etc.. have found compelling applications in Artificial Intelligence, Knowledge Representation, Verification, Software En-

---

[1]Some people would call it a Curry-Howard-Lambek correspondence and this would be a good name too.

gineering, etc. The work of the Coalgebraic Modal school, for example as described in [Cirstea et al., 2007], makes a convincing case for modal logic in general and the coalgebraic approach in particular. But while the coalgebraic approach is encompassing, it is not the whole story on the categorical way to modal logic. The coalgebraic approach is based on semantics of modal logics in terms of relational structures and can be seen as following the tradition of categorical 'model' theory. Here we are interested in the categorical 'proof' theory approach to modal logic.

Methods of categorical proof theory are useful both in explaining logic systems and in providing us with possible implementations and applications of those systems. Our guiding intuitions come from the Curry-Howard interpretation and its uses as foundations for Functional Programming. As usual, the categorical proof theory approach to modal logic is less developed than the corresponding modal theoretical one. Our aim is to help balance the issue by providing a basic, but fundamental piece of the proof theoretical approach in detail.

The most basic classical modal logic is usually taken to be system K (after Kripke), where one has two operators $\Box$ (necessity) and $\Diamond$ (possibility) satisfying the axiom $\Box(A \to B) \to \Box A \to \Box B$ and the necessitation rule. The operator $\Diamond$ is usually taken to be defined in terms of $\Box$ and negation, as $\Diamond A = \neg(\Box(\neg A))$. But categorical proof theory is more transparent over a constructive basis, so we need to have independent definitions for $\Box$ and $\Diamond$ and, as traditional when constructivizing concepts, one is faced with multiple possibilities.

2 Which basic constructive modal logic?

It is traditional to face a 'plurality' problem when constructivizing notions. Usually a single notion in classical mathematics gives rise to several possible notions when using constructive logic. When confronted with the problem of defining the intuitionistic or constructive modal system corresponding to the classical modal system K many different systems present themselves. Rather than choosing one such system and calling it 'the' constructive system corresponding to K modal logic, we prefer to discuss briefly two such systems and then then concentrate in the one we prefer.

The first constructive system we discuss corresponding to K was described (together with a whole framework of other constructive modal logics) in Simpson's thesis[Simpson, 1994]. This system (called IK for intuitionistic K) had independently being proposed earlier by Fisher-Servi and others and satisfies many properties that one might expect of a basic constructive modal logic. These include non-interdefinable modal operators $\Box$ (for necessity) and $\Diamond$ (for possibility). The logical basis of the system is intuitionistic propositional logic (IPL) and adding to the independently defined modalities the law of the excluded middle takes us back to the the traditional system K. In addition the system satisfies a *disjunctive* as well as an *existence* property, as characteristic of intuitionistic logic. One possible axiomatization of the system is presented below.

As it happens in the classical system K, the necessity operator distributes over conjunctions, while symmetrically the possibility operator distributes over disjunctions. The constants follow the same pattern.

$$\Box(A \wedge B) \iff \Box A \wedge \Box B$$
$$\Box(\top) \iff \top$$
$$\Diamond(A \vee B) \iff \Diamond A \vee \Diamond B$$
$$\Diamond(\bot) \iff \bot$$

The system IK comes from the strong semantic intuition of possible worlds, where we say that $\Box A$ holds in the current world, if A holds in all the worlds accessible from it. And symmetrically, $\Diamond A$ holds in the current world if there exists one world accessible from the current world where A holds.

The second system, called CK (from constructive K comes from proof-theoretical intuitions provided by Natural Deduction formulations of logic. It was described i [Bellin et al., 2001], following the adaptation of Prawitz's suggestions in his seminal work [Prawitz, 1965] for $S4_l$ to K. Like the previous system IK we have an intuitionistic propositional basis and non-interdefinable operators for necessity and possibility. Like IK the system satisfies a disjunctive as well as an existence property, as characteristic of intuitionistic logic. Unlike IK this system does not warrant distribution of the possibility operator over disjunction. The problem is that in a naive Natural Deduction environment one cannot see how to deduce $\Diamond A \vee \Diamond B$ from $\Diamond(A \vee B)$ and hence this distribution (algebraically very appealing) is left out of the system.

A main advantage of system CK is that it is easy to define terms in a lambda-calculus corresponding to the operations of CK, as done in [Bellin et al., 2001]. The rules repeated below are not as symmetric as in constructive versions of S4. The introduction of *commuting conversion* rules, necessary to expose β-redexes, as in the case of CS4 make the term calculus not as streamlined as we would like it to be. However, this is a natural logic to consider, if one is determined to push the frontiers of the Curry-Howard Isomorphism, as far as they will go.

The modal system CK fewer symmetries than system CS4 make it harder to conform to the Natural Deduction requirements of introduction and elimination pairs of rules, each pair defining a single connective, each pair satisfying local reduction rules. Thus we are faced with a problem of lack of structure, which makes the modelling of CK more challenging than the modelling of CS4. Meanwhile the system IK has more algebraic structure, but as it is usually presented does not lend itself well to categorical modelling. The problem is that the kinds of judgements we are modelling (M is a term of type A and A is related to A' via an accessibility relation) are quite different. Despite the lack of overall best system, in this particular work, we prefer system CK to system IK, as the categorical modelling is our immediate goal.

3 The System CK

We aim at a propositional system that, like classical or intuitionistic logic, can be presented either as an axiomatic system, or a sequent calculus or a Natural Deduction system and such that these different presentations are proved 'equivalent', in the sense of, at least, proving the same theorems. (Whether each proof in one system can be transformed or not in a proof of the other system is a harder case, left for future work.) We discuss these different formalisms for our chosen basic constructive necessity system CK.

3.1 Sequent Calculus and Axiomatic System

The sequent calculus rules and Hilbert-style axioms for the system CK are relatively uncontroversial and well-known.

To define a sequent calculus for CK we add to the sequent calculus rules for intuitionistic propositional logic (IPL) two rules. One rule for the necessity modal operator ($\Box$) and a similar rule for the modality of possibility ($\Diamond$).

$$\frac{\Gamma \vdash B}{\Box\Gamma \vdash \Box B}\text{-}Box \qquad \frac{\Gamma, A \vdash B}{\Box\Gamma, \Diamond A \vdash \Diamond B}\text{-}Diamond$$

These rules do double-duty as they work as both left and right introduction rules for the necessity $\Box$ and the possibility $\Diamond$ modal operators. These rules are also slightly awkward in that they are not strictly left or right rules and the rule for $\Diamond$ already mentions the $\Box$ operator. However, the rules are sufficient to prove the necessary syntactic theorems and they do provide us with a cut elimination theorem, as shown, for instance in (in a more complicated form) by Wisejekera in [Wijesekera, 1990].

Similarly to the sequent calculus, we can add to any axiomatization of propositional Intuitionistic Logic (IPL) the following three axioms:

$$\Box(A \to B) \to (\Box A \to \Box B)$$

$$\Box(A \to B) \to (\Diamond A \to \Diamond B)$$

$$\Box A \times \Diamond B \to \Diamond(A \times B)$$

and the Necessitation Rule:

$$\frac{\vdash B}{\vdash \Box B}$$

to obtain an axiomatization of CK. Other axiomatizations are possible, but do not shed much light on the essence of the system. Wijesekera shows that the sequent calculus above corresponds to the axiomatic formulation given by axioms for intuitionistic logic, plus axiom

$$\Box(A \to B) \to (\Box A \to \Box B)$$

together with rules for Modus Ponens and Necessitation:

$$\frac{\vdash A \to B \quad \vdash A}{\vdash B}\text{-}MP \qquad \frac{\vdash A}{\vdash \Box A}\text{-}Nec$$

He then proves a Craig interpolation theorem for his system, one of the usual consequences of syntactic cut-elimination.

Wisejekera also produces Kripke, algebraic and topological semantics for this calculus. From our "wish list" for logical systems a natural deduction formulation and a categorical semantics are missing. These we proceed to discuss, in turn.

Prawitz in his classic monograph on Natural Deduction only discusses natural deduction formulations for the basic modal logics S4 and S5. As Bull and Segerberg note in their survey "Basic Modal Logic" in page 27

> [..]It has proved difficult to extend this sort of analysis [the Natural Deduction one] to the great multitude of other systems of modal logic.

According to Bull and Segerberg, some logicians tried to blame the intensional character of modal logic for the poor fit between modal logic and Gentzen proof theoretical methods. But intuitionistic logic is also intensional and Gentzen methods work like a treat for it. Another suggestion is that there is an unreasonable proliferation of modal logics and "natural" deduction methods would only apply to "natural" enough logics. Granting for the moment that this may be the case, it is still strange that sequent calculus systems have been devised for a whole family of modal logics, while natural deduction formulations only exist, traditionally, for S4 and S5.

The problem is easy to see: if one wants to think of the sequent calculus rule for $\Box$ above as applying to natural deduction derivation-trees we have that a tree

$$A_1, A_2, \ldots, A_k$$
$$\vdots \pi$$
$$B$$

has to be transformed into a tree of shape somewhat like

$$\Box A_1, \Box A_2, \ldots, \Box A_k$$
$$\vdots \pi^*$$
$$\frac{B}{\Box B}$$

but while transforming the conclusion of a natural deduction derivation tree is perfectly acceptable, modifying its premisses is not allowed, usually. Moreover, if modifying of premisses was allowed, we would still have a problem as the inference

$$\frac{\Box A}{A}$$

perhaps used to get from the 'new' hypotheses $\Box A_i$ back to the given ones A_i, is valid in S4, but not in K, the system we want to model.

In the following discussion, as in the previous one ([Bellin et al., 2001]) in which this is based, we present three different solutions for this problem. Neither of the solutions is completely satisfactory, for reasons we discuss in section 6. We feel that discussing the problem and its partial solutions is worthwhile though, hoping that someone will produce better solutions.

4 Natural Deduction for CK?

The first "solution" has existed for quite a while. In the 50's Fitch proposed a variant of Gentzen's Natural Deduction, which for classical or intuitionistic logic seems simply a notational variant: one writes linear derivations instead of tree-like derivations. But Fitch-style Natural Deduction and Gentzen-style Natural Deduction are further apart than one might expect. While at the level of classical (or intuitionistic) logic the differences seem only notational, when modal operators are considered the gap seems to widen. Fitting's 1983 monograph "Proof Methods for Modal and Intuitionistic Logics" considers several Fitch-style natural deduction systems for different modal systems built from system K plus axioms. More recently Borghuis (in his doctoral thesis) has developed some of the Fitch-ND systems to devise a Curry-Howard "proofs-as-types"[2] interpretation for several systems of modal logic.

A second solution to the problem can be seen as a corollary of work of Gianluigi Bellin done in the early 80's[Bellin, 1985]. We recap that work briefly and extend it, both with a term assignment system and a categorical model.

Finally a third solution, inspired by work on systems where the context is divided in zones, such as Girard's LU or Plotkin and Barber's DILL[Barber, 1996] is presented. Relating these different solutions is left for future work[3].

Because of our emphasis in categorical modelling, we will concentrate in solutions two and three above. Putting it bluntly I do not know how to extract terms and a categorical semantics from a Fitch-like Natural Deduction formulation of CK. As we will see in sequel, the terms and categorical semantics we do develop still have some issues in need of clarification.

Summarizing: Our aim in this note is two-fold. First we want to develop a term calculus with usual syntactic properties (strong normalization, subject reduction, confluence) for a Gentzen-like (as opposed to Fitch-like) Natural Deduction version of modal system K. Second we would like to extend (Borghuis and our own) work on natural deduction and term assignment systems, with the further correspondence between typed λ-calculus and category theory, that is usually referred as the "extended" (or categorical) or Curry-Howard-Lawvere correspondence.

4.1 Fitch-style Modal Natural Deduction

Borghuis work ([Borghuis, 1994; Borghuis, 1998]) is based on Pure Type Systems, Barendregt's beautiful systematization of work on several (higher-order) lambda-calculi. His framework is very expressive, but since we are not interested (at least in this note) on higher-order systems, some of the complications of Borghuis system can be avoided.

The main idea of Fitch-style ND proofs is that to prove an implication, say $A \to B$, we go into what is called a *subordinate proof* where we assume that

---

[2]Borghuis system seems strange in that, despite the fact that his basic systems are all constructive, the modal logics on top are classical. This choice seems dictated more by tradition (Barendregt's cube is constructive, while most work on modal logic is classical) than by application suitability.

[3]Recent other approaches (Vigano, Gabbay, Martini& Masini) that tie in the semantics of modal logics (in terms of possible worlds) with their syntactic presentation have been devised. We will not have much to say about this line of work in this paper.

the antecedent of the proposed implication, A is true. If we can derive B under this assumption, then we can discharge the assumption, by exiting the subordinate proof (which is confusedly also called a *box*) and adding $A \to B$ to the original proof. One is also allowed to reiterate any formulas from the main proof into a subordinate proof, this simply corresponds to many uses of the same assumption. This device makes it possible to write proofs in a linear order, but does not change the intuitive meaning of the implication rule.

Having invented this box for dealing with implication, it is now natural to invent a different kind of box to deal with the modality $\Box$. Fitch adds a new kind of subordinate proof, a *strict subordinate proof* to his system in his book[Fitch, 1952]. A strict subordinate proof requires no hypothesis, and more importantly "reiteration" in a strict subordinate proof is restricted to formulas of a certain form. For the logic we are interested here, system K, one is only allowed to reiterate formulas of the form $\Box A$, ie modal formulas and their reiteration appears in the subordinate proof without the $\Box$ operator., ie as simply A. This is the so-called *K-import rule*. A formula imported into a strict subordinate proof does not count as a hypothesis of that proof.

To export proofs from the box, we have to be a bit more careful: a conclusion A can be exported if it was derived by means of a *categorical* strict subordinate proof. (Note that this categorical has nothing to do with Category Theory.) A categorical strict subordinate proof has to satisfy:

1. All of its assumptions have been discharged;

2. the conclusion lies directly inside the modal interval;

3. there are no nested subordinate proofs that are still "open".

We want to simplify Borghuis type theory to deal simply with the *propositional* modal logic K. We also want a *constructive* propositional basis, so we shall not use encodings of operators. We have ground types, implications $A \to A$, conjunctions $A \wedge A$, and modal types $\Box A$. We omit disjunctions and falsum, for the time being.

For modal Fitch-style natural deduction we need to attribute a degree of nestedness to each formula in each proof. In a non-modal Fitch setting the degree of a formula is simply the number of hypotheses at that stage of the proof. In a modal setting the degree of a formula is a pair of natural numbers (m, h) where m is the modal depth of the formula and h is the number of hypotheses at that stage of the proof.

The following is a simplification (no higher-order logic) of Borghuis system. The type theory has judgements

$$\Gamma_k | \ldots | \Gamma_1 | \Gamma_0 \vdash M : A$$

meaning that (logically) M is a proof of A where a context has k-many compartments, where k is the maximum modal depth of formulas in the derivation. We have the following raw terms:

$$M ::= x \mid \lambda a : A.M \mid MM \mid \mathsf{Box}(M) \mid \mathsf{Unbox}(M)$$

The typing rules are as follows

$$\overline{\Gamma|\Gamma_0, x{:}\,A \vdash x{:}\,A}$$

$$\frac{\Gamma|\Gamma_0, a{:}\,A \vdash M{:}\,B}{\Gamma|\Gamma_0 \vdash \lambda a{:}\,M{:}\,A \to B} \qquad \frac{\Gamma|\Gamma_0 \vdash M{:}\,A \to B \quad \Gamma|\Gamma_0 \vdash N{:}\,A}{\Gamma|\Gamma_0 \vdash MN{:}\,B}$$

$$\frac{\Gamma|\Gamma_0 \vdash M{:}\,\Box A}{\Gamma|\Gamma_0|_- \vdash \mathtt{Unbox}(M){:}\,A} \qquad \frac{\Gamma|_- \vdash M{:}\,A}{\Gamma \vdash \mathtt{Box}(M){:}\,\Box A}$$

Usual desirable meta-theoretical results hold for this calculus. In particular Borghuis proved:

THEOREM 1 (Borghuis). *Subject Reduction, Strong Normalization (SN) and the Church-Rosser property hold for the calculus above.*

The proof detailed in Borghuis thesis proceeds by defining an erasing mapping from the modal lambda-calculus to the simply typed λ-calculus and proving properties of this erasing.

Fitch-style Natural Deduction does not emphasize (as much as Gentzen-style natural deduction) the importance of the normalization procedure. Clearly one can translate from Fitch-ND proofs to Gentzen-ND proofs, normalize them and translate back into Fitch-style. But it would be interesting to see whether we could say something about the normalization procedure staying in the Fitch-style system throughout. More importantly, one can translate Fitch-ND proofs into Gentzen-ND ones and then provide categorical semantics for the Gentzen-style system, but a more direct route would be preferred. Especially because the Fitch-style does seem to accommodate more variation in the logical system.

4.2 Gentzen-style Modal Natural Deduction

We now present a second Natural Deduction calculus for the intuitionistic modal system CK. This calculus can be seen as revisiting Bellin's ideas in [Bellin, 1985], except that we simplify our basis to a *constructive* one and we add terms to the natural deduction system, to obtain a Curry-Howard isomorphism[4]. We have judgements of the form

$$\Gamma \vdash M{:}\,A$$

meaning that (logically) M is a proof of the Natural Deduction sequent $\Gamma \vdash A$. There are usual rules for conjunctions and implications. There is a single rule for the modality, which requires some explanation. The introduction rule for necessity, which in sequent calculus is the usual

$$\frac{\Gamma \vdash B}{\Box \Gamma \vdash \Box B}$$

---

[4]Note that the main goal of the paper[Bellin, 1985] was to deal with the Gödel-Löb provability logic GL, the work on K was only necessary background to it.

when looked at in tree-shape, requires that

$$\begin{array}{c} A_1, A_2, \ldots, A_k \\ \vdots \pi \\ B \end{array}$$

is transformed into a tree like

$$\dfrac{\begin{array}{cc} & A_1, A_2, \ldots, A_k \\ & \vdots \pi \\ \Box A_1, \Box A_2, \ldots, \Box A_k & B \end{array}}{\Box B}$$

where $A_1, \ldots, A_k$ have all been closed for substitutions. The only place in this tree where substitutions can occur, is on top of the boxed A_i, ie $\Box A_i$. This shape of rule, which was fairly sensible in the S4-case (as $\Box A_i$ does imply A_i) is less obvious here, but still reasonable. Thus the rule in Martin-Loef ND looks like:

$$\dfrac{\Delta \vdash \vec{N} \colon \vec{\Box A} \quad x_1 \colon A_1, x_2 \colon A_2, \ldots, x_k \colon A_k \vdash M \colon B}{\Delta \vdash \texttt{Box}\ M\ \texttt{with}\ \vec{N}\ \texttt{for}\ \vec{x} \colon \Box B}$$

where $\vec{\Box A} = \Box A_1, \Box A_2, \ldots, \Box A_k$ and $\vec{N}$ means $N_1, \ldots, N_k$, so $\Delta \vdash \vec{N} \colon \vec{\Box A}$ really means a collections of derivations $\Delta \vdash N_1 \colon \Box A_1, \ldots, \Delta \vdash N_k \colon \Box A_k$.

Note that this introduction rule for $\Box$ is mixed, it already mentions the $\Box A_i$ that we are defining how to introduce, a bad characteristic of modal Natural Deduction, which is similar to the S4 case.

This calculus is equivalent to its sequent and axiomatic formulations above and satisfies normalization, which should have been an easy corollary of work in [Bellin, 1985]).

But getting the right orientation for the equality rules given by the normalization process can be confusing. As it turns out, our original reduction rules in [Bellin et al., 2001] were sloppy and needed to be tightened up. The necessary corrections were made by Kakutani[Kakutani, 2007], who proves, all the required results for the $\Box$-fragment.

THEOREM 2 (Kakutani). *Subject Reduction, Strong Normalization (SN) and the Church-Rosser property hold for the $\Box$-fragment of the calculus above.*

The system satisfies the subject reduction property, the system is strongly normalizing, the system is confluent and the system has the subformula property. Kakutani also provides call-by-name and call-by-value versions of the calculus, and a continuation passing style (CPS) transformation from the call-by-value version to the the call-by-name one, proved sound and complete. Details can be found in [Kakutani, 2007].

4.3 Dual-context Modal Natural Deduction

The third version of a Natural Deduction calculus for CK follows a pattern exploited recently by Girard (LU), Miller, Barber and Plotkin (DILL) amongst others [Ghani et al., 1998]: contexts are divided into two zones, one where assumptions have modal ($\Box$) types and the other where assumptions have the usual (intuitionistic) types. We have judgements of the form

$$\Gamma|\Delta \vdash M : A$$

meaning that (logically) M is a proof of the sequent $\Box\Gamma, \Delta \vdash A$.

DEFINITION 3. Consider the calculus DK, with the following raw terms:
$M ::= x \mid a \mid \lambda a{:}A.M \mid MM \mid \texttt{Box } M \texttt{ with } \vec{N} \texttt{ for } \vec{a} \mid \texttt{Unbox } M \texttt{ for } \Box x \texttt{ in } M$

The typing rules are the usual typing rules for the simply typed λ-calculus plus the following rules for the modality $\Box$:

$$\overline{\Gamma, x{:}A|\Delta \vdash x{:}\Box A}$$

$$\frac{\Gamma|_{-} \vdash \vec{N}{:}\Box\vec{A} \quad _{-}|\vec{a}{:}\vec{A} \vdash M{:}A}{\Gamma|_{-} \vdash \texttt{Box } M \texttt{ with } \vec{N} \texttt{ for } \vec{a}{:}\Box A} \qquad \frac{\Gamma|\Delta \vdash M{:}\Box A \quad \Gamma, x{:}A|\Delta \vdash N{:}B}{\Gamma|\Delta \vdash \texttt{Unbox } M \texttt{ for } \Box x \texttt{ in } N{:}B}$$

where by $\vec{a}{:}\vec{A}$ we mean $a_1{:}A_1, a_2{:}A_2, \ldots a_n{:}A_n$.

Note that unlike our previous work on CS4 here the right-hand side of the context does not mean without a box. Actually the left-side means with at least one box, possibly more. So there might be boxed formulas on the right-hand side of the context.

We can built in all the necessary substitutions in the rule:

$$\frac{\Gamma|_{-} \vdash \vec{N}{:}\Box\vec{A} \quad _{-}|\vec{a}{:}\vec{A} \vdash M{:}A}{\Gamma|_{-} \vdash \texttt{Box } M \texttt{ with } \vec{N} \texttt{ for } \vec{a}{:}\Box A}$$

The rule for $\Box$-elimination is a bit of a trick to get terms that look like introduction and elimination. This elimination rule can be seen as explaining how to move formulae away from the modal side of the context.

To obtain a reduction calculus we introduce the obvious reduction rules. We consider only β-rules and commuting conversions at the moment. We have two β-rules, namely:

$$(\lambda a{:}A.M)N \rightsquigarrow M[N/a]$$
$$\texttt{Unbox } (\texttt{Box } M \texttt{ with } \vec{N} \texttt{ for } \vec{a}) \texttt{ for } \Box x \texttt{ in } R \rightsquigarrow R[\texttt{Box } M \texttt{ with } \vec{N} \texttt{ for } \vec{a}/x]$$

and the following commuting conversions:

$$(\texttt{Unbox } M \texttt{ for } \Box x \texttt{ in } N)R \rightsquigarrow \texttt{Unbox } M \texttt{ for } \Box x \texttt{ in } NR$$

$$\texttt{Unbox } (\texttt{Unbox } M \texttt{ for } \Box x \texttt{ in } N) \texttt{ for } y \texttt{ in } R \rightsquigarrow$$
$$\texttt{Unbox } M \texttt{ for } \Box x \texttt{ in Unbox } N \texttt{ for } y \texttt{ in } R$$

$$\texttt{Box } M \texttt{ with } \vec{N}, \texttt{Unbox } M \texttt{ for } \Box x \texttt{ in } N, \vec{R} \texttt{ for } \vec{b}, a, \vec{a} \rightsquigarrow$$
$$\texttt{Unbox } M \texttt{ for } \Box x \texttt{ in Box } M \texttt{ with } \vec{N}, N, \vec{R} \texttt{ for } \vec{b}, a, \vec{a}$$

While the β-conversion rules are clearly necessary, we are still working on (newer) categorical models that explain why the commuting conversions are adequate [Ritter and de Paiva, 2012].

We show subject reduction, strong normalisation and confluence for the box-fragment system DK discussed in this section.

PROPOSITION 4 (Subject Reduction). *Assume that term M has type A in the context $\Gamma|\Delta$, $\Gamma|\Delta \vdash M: A$ and that M reduces to N using the rules above $M \rightsquigarrow N$. Then also N has type A, $\Gamma|\Delta \vdash N: A$.*

For strong normalisation we define a translation from our calculus into the dual calculus for CS4, DIML[Ghani et al., 1998], which preserves reductions. This translation is given in the following definition.

DEFINITION 5. For each term M define a DIML-term $(M)^D$ by induction over the structure of M as follows:

$$\begin{aligned}
(a)^D &= a \\
(x)^D &= \Box x \\
(\lambda a.M)^D &= \lambda a.(M)^D \\
(MN)^D &= (M)^D(N)^D \\
(\text{Box } M \text{ with } \vec{N} \text{ for } \vec{a})^D &= \text{let } (\vec{N})^D \text{ be } \Box\vec{x} \text{ in } \Box(M)^D[\vec{x}/\vec{a}] \\
(\text{Unbox } M \text{ for } \Box x \text{ in } N)^D &= \text{let } (M)^D \text{ be } \Box x \text{ in } (N)^D
\end{aligned}$$

This translation preserves typing:

LEMMA 6. *Assume $\Gamma|\Delta \vdash M: A$ in the $\mathbf{IK}_\Box$ calculus. Then $\Gamma|\Delta \vdash (M)^D: A$ in DIML.*

Proof. Easy induction over the definition of $(M)^D$. ∎

Strong normalisation follows now directly:

THEOREM 7. *The $\mathbf{IK}_\Box$ calculus is strongly normalising.*

Proof. One shows that if $M \rightsquigarrow N$, then $(M)^D \rightsquigarrow^+ (N)^D$. But DIML is strongly normalising. Hence $\nu(M) \leq \nu((M)^D) < \infty$, where $\nu(M)$ is the length of the longest reduction sequence of M if M is strongly normalising and is ∞ otherwise. ∎

Confluence cannot be inferred directly from confluence of DIML but needs to be shown separately.

THEOREM 8. *The $\mathbf{IK}_\Box$ calculus is confluent.*

Proof. As we have already shown strong normalisation it suffices to show local confluence. It is now possible to see that all critical pairs can be completed. ∎

So we invented a more complicated syntax, proved the meta-theoretical results we usually need for lambda calculi for it, but we have not shown, yet, that this is really equivalent to the system $\mathsf{CK}_\Box$ that we started from. For the equivalence proof one can use a Hilbert-style presentation with the **K** axiom $\Box(A \to B) \to \Box A \to \Box B$ together with the necessitation inference rule

$$\frac{\_|\_ \vdash M\colon A}{\_|\_ \vdash \Box M\colon \Box A}$$

(i) Assume $\Gamma|\Delta \vdash M\colon A$. Then there exists a system K derivation of $\Box\Gamma \to \Delta \to A$.

(ii) Assume there exists a system K derivation of $\Gamma \to A$. Then there exists a term M such that $\Gamma \vdash M\colon A$.

Proof.

(i) The only interesting case is the $\Box I$-rule. For this, we have the following derivation:

$$\frac{\dfrac{\Delta \to A}{\Box(\Delta \to A)}\ necessitation \qquad \Box(\Delta \to A) \to \Box\Delta \to \Box A}{\Box\Delta \to \Box A}\ MP$$

and now use modus ponens again.

(ii) For the other direction, necessitation is obvious from the $\Box I$-rule with empty left-hand side and empty set of terms $\vec{M}$.

Now observe for a start that we have terms $x\colon A|\_ \vdash x\colon \Box A$ and for any term N in $\Gamma, x\colon A|\Delta \vdash N\colon B$ we have $\Gamma|\Delta, a\colon \Box A \vdash \mathtt{Unbox}\ a\ \mathtt{for}\ \Box x\ \mathtt{in}\ N\colon B$. Hence if we are asked to construct a term $\_|\Box\Gamma, \Delta \vdash M\colon A$ it suffices to give a term $\Gamma|\Delta \vdash M'\colon A$. Hence for the axiom it suffices to give a term $x\colon A \to B, y\colon A|\_ \vdash M\colon \Box B$. Indeed, we have

$$x\colon A \to B, y\colon A|\_ \vdash \mathtt{Unbox}\ a, b\ \mathtt{for}\ x, y\ \mathtt{in}\ Box(ab)\colon \Box B$$

∎

5 Categorical Models

Categorical models distinguish between different proofs of the same formula. A category consists of objects, which model the propositional variables, and for every two objects A and B each morphism in the category from A to B, corresponds to a proof of B using A as hypothesis.

Cartesian closed categories are the categorical models for (disjunction-free) intuitionistic propositional logic. For a careful explanation the reader should consult [Lambek and Scott, 1985]: here we just outline the intuitions. Conjunction is modelled by cartesian products, a suitable generalisation of the products in Heyting algebras. Logically the definition of a product says that a proof of a conjunction $A \wedge B$ from C corresponds to a proof of A (from C) and a proof of B (from C). The usual logical relationship between conjunction and implication

$$A \wedge B \longrightarrow C \text{ if and only if } A \longrightarrow (B \to C)$$

is modelled by an adjunction and it defines categorically the implication connective. That is we require that for any two objects B and C there is an object $B \to C$ such that there is a bijection between morphisms from $A \wedge B$ to C and morphisms from A to $B \to C$. Disjunctions could be modelled by co-products, again a suitable generalisation of the sums of Heyting algebras, but since they cause a few problems, we prefer to restrict our language. True and false are modelled by the empty product (called a terminal object) and empty co-product (the initial object), respectively. Finally negation, as traditional in constructive logic, is modelled as implication into falsum.

Categorical models for the Gentzen-system presented CK are not problematic: clearly we need a cartesian closed category to model implications and conjunctions and a *monoidal* endofunctor to model the modality $\Box$. The monoidicity of the functor corresponds to the modelling of the K characteristic axiom $\Box(A \to B) \to \Box A \to \Box B$.

In previous work [Bierman and de Paiva, 2000] it was shown that to model the S4 necessity $\Box$ operator one needs a *monoidal comonad*. Here we have less structure and hence only the monoidal endofunctor remains.

DEFINITION 9. A $\mathsf{CK}_\Box$-category consists of a cartesian closed category $\mathcal{C}$, together with a monoidal endofunctor $\Box\colon \mathcal{C} \longrightarrow \mathcal{C}$. The functor being monoidal means the existence of a natural transformation with components: $m_{A,B}\colon \Box A \times \Box B \to \Box(A \times B)$ and of a morphism $m_1\colon 1 \to \Box 1$ satisfying the well-known commuting diagrams below.

The soundness theorem shows in detail how the categorical semantics models the modal logic.

THEOREM 10 (Soundness). *Let $\mathcal{C}$ be any $\mathsf{CK}_\Box$-category. Then there is a canonical interpretation $[\![\_]\!]$ of $\mathsf{CK}_\Box$ in $\mathcal{C}$ such that*

- *a formula A is mapped to an object $[\![A]\!]$ of $\mathcal{C}$;*
- *a natural deduction proof ψ of B using formulae $A_1, \ldots, A_n$ as hypotheses is mapped to a morphism $[\![\psi]\!]$ from $[\![A_1]\!] \times \cdots \times [\![A_n]\!]$ to $[\![B]\!]$;*
- *each two natural deduction proofs ψ and ψ of B using formulae $A_1, \ldots, A_n$ as hypotheses which are equal (modulo normalisation of proofs) are mapped to the same morphism, in other words $[\![\phi]\!] = [\![\psi]\!]$.*

Proof. We use an induction over the structure of natural deduction proofs. We describe the modality rule. Consider a proof ψ

$$\cfrac{\begin{array}{c}\Gamma_1\\ \vdots\, \phi_1\\ \Box A_1\end{array} \quad \cdots \quad \begin{array}{c}\Gamma_n\\ \vdots\, \phi_n\\ \Box A_n\end{array} \quad \begin{array}{c}[A_1 \cdots A_n]\\ \vdots\, \phi\\ B\end{array}}{\Box B}\,\Box_\mathcal{I}$$

By induction hypothesis, let $f_1, \ldots, f_n, f$ be the interpretation of $\phi_1, \ldots, \phi_n, \phi$ respectively. Then the interpretation of ψ is

$$(\Box f) \circ m_{A_1,\ldots,A_n} \circ (f_1 \times \cdots \times f_n)$$

where $m_{A_1,\ldots,A_n}$ is inductively defined by

$$m_{A_1,\ldots,A_{m-1},A_m} = m_{A_1\times\cdots\times A_{m-1},A_m} \circ (m_{A_1,\ldots,A_{m-1}}) \times id_{A_m}$$

We omit the routine verification that the desired equalities hold. ∎

A trivial degenerate example of an $\mathsf{CK}_\square$-category consists of taking any ccc, say Sets for example and considering the identity functor (as a monoidal functor) on it. Less trivial, but still degenerate models are Heyting algebras (the poset version of a bi-ccc) together with a closure operator.

To prove categorical completeness we use a term model construction.

THEOREM 11 (Categorical Completeness).

(i) *There exists a $\mathsf{CK}_\square$-category such that all morphisms are interpretations of natural deduction proofs.*

(ii) *If the interpretation of two natural deduction proofs is equal in all $\mathsf{CK}_\square$-categories, then the two proofs are equal modulo proof-normalisation in natural deduction.*

Proof. We show both statements by constructing an $\mathsf{CK}_\square$-category $\mathcal{C}$ out of the natural deduction proofs. We give here only the morphisms, and omit the verification that the required equalities between proofs hold. We write a natural deduction proof

as $A \vdash B$. The objects of the category are formulae, and a morphism between A and B is a proof of B using A as a hypothesis. The identity morphism is the basic axiom $A \vdash A$, and composition is given by cut. The bi-cartesian closed structure of $\mathcal{C}$ follows in the usual way from the conjunction, disjunction and implication in intuitionistic logic.

The $\square$-modality gives rise to a monoidal functor. The functor $\square$ sends an object A to $\square A$ and a morphism $f\colon A, B \vdash C$ to the morphism $\square f\colon \square A, \square B \vdash \square C$, obtained by applying the $\square\mathcal{I}$-rule. Note that the monoidicity of the functor is only used to glue together the proofs of the several boxed assumptions into the original proof: if $f\colon A_1, A_2, A_3 \vdash B$, then we need

$$m_{A_1,A_2,A_3}\colon \square A_1 \times \square A_2 \times \square A_3 \to \square(A_1 \times A_2 \times A_3)$$

This category $\mathcal{C}$ shows now the claim: Assume an equation between proofs holds in all $\mathsf{CK}_\square$-categories. Because $\mathcal{C}$ is a $\mathsf{CK}_\square$-category, it holds in $\mathcal{C}$. But equality in $\mathcal{C}$ is equality between natural deduction proofs, hence the two proofs are equal.

∎

6 Conclusions and more work

We have discussed several constructive versions of a basic modal system K, including Simpson's IK, Borghuis' Fitch-style calculus K_b, constructive CK (the system inspired by Prawitz's $S4_1$) and the dual context calculus $\Box$-only fragment DK.

What is bad about this proliferation of calculi and models? For the necessity only fragment, we have almost as good a situation as for constructive S4. We have parallel systems and we have subject reduction, normalization and confluence for the parallel lambda-calculi. We have categorical models easy to state and prove correct, but these do not constrain us much. There are many cartesian closed categories with monoidal endofunctors around, but none stands out as being particularly informative. Maybe this is right, the system is really this weak, we need to check its applications (and/or extensions) to see the usefulness of the categorical semantics. Or maybe we are missing one essential part in the modelling. In any case, it would be nice to see a more mathematically-independent model, a less syntax-derived construction.

One big positive point, that we have not described here, is that we now have a categorical model that works directly for the system DK. This is important as the calculus DK is structurally similar to most of the systems produced by Pfenning and his collaborators. Until recently we could only provide categorical semantics for those systems by first translating them into single contexts calculi. These new 'fibrational models' fit well with the modelling of more expressive logics, but require more sophisticated tools, reason why they need a separate manuscript.

What is good about this proliferation of type theories and models for basic constructive modal logic? Well, the classic system K is still one of the most basic modal logics extant and that fact that we can do (more than one) type theory for it, and produce more than one kind of categorical model for it is very good indeed and needs more exposure. The 'choices' between calculi mostly come down to the possibility-fragment and the modelling of that lags behind. Modelling $\Diamond$s in the style of CK is easy, we have terms and categorical models for this system. But we do not have a developed dual contexts syntax nor do we have fibrational categorical models for it. And we do not have categorical models for Simpson-style calculi either. Given the work on intuitionistic hybrid systems [Braüner and de Paiva, 2006] this ought to be feasible, but we are not there, yet.

BIBLIOGRAPHY

[Alechina et al., 2001] Natasha Alechina, Michael Mendler, Valeria de Paiva, and Eike Ritter. Categorical and Kripke Semantics for Constructive Modal Logics. In *Computer Science Logic (CSL01), Paris*, 2001.

[Barber, 1996] A. Barber. *Dual Intuitionistic Linear Logic*. The Laboratory for Foundations of Computer Science, School of Informatics at the University of Edinburgh, ECS-LFCS-96-347, 1996.

[Bellin et al., 2001] Gianluigi Bellin, Valeria De Paiva, and Eike Ritter. Extended Curry-Howard correspondence for a basic constructive modal logic. In *Proceedings of Methods for the Modalities (M4M)*. CSLI, 2001.

[Bellin, 1985] G. Bellin. A system of natural deduction for GL. *Theoria, 2: 89–114*, 1985.

[Bierman and de Paiva, 2000] Gavin M. Bierman and Valeria de Paiva. On an intuitionistic modal logic. *Studia Logica*, 65(3):383–416, 2000.

[Boolos, 1993] G. Boolos. *The Logic of Provability*. Cambridge University Press, 1993.

[Borghuis, 1994] T. J. Borghuis. *Coming to Terms with Modal Logic: On the interpretation of modalities in typed λ-calculus*. PhD thesis, Technical University of Eindhoven, 1994.

[Borghuis, 1998] Tijn Borghuis. Modal pure type systems. *Journal of Logic, Language and Information*, 7(3):265-296, 1998.

[Braüner and de Paiva, 2006] T. Braüner and V. de Paiva. Intuitionistic hybrid logic. *Journal of Applied Logic*, 4(3):231–255, 2006.

[Cirstea et al., 2007] Corina Cirstea, Alexander Kurz, Dirk Pattinson, Lutz Schroeder, and Yde Venema. Modal Logics are Coalgebraic. *Electronic Workshops in Computing, The British Computer Society*, 2007.

[de Paiva and Pientka, 2011] Valeria de Paiva and Brigitte Pientka. Intuitionistic Modal Logic and Applications. *Information and Computation*, Volume 209, Issue 12:1435–1538, 2011.

[Fairtlough et al., 2001] M. Fairtlough, M. Mendler, and E. Moggi. Modalities in Type Theory. *Mathematical Structures in Computer Science*, 11(4), 2001.

[Fitch, 1952] F. Fitch. *Symbolic Logic: An Introduction*. The Ronald press Co., New York, 1952.

[Gentzen, 1969] G. Gentzen. On the relation between intuitionist and classical arithmetic (1933). In M.E. Szabo, editor, *The collected papers of Gerhard Gentzen*, pages 53–67, Amsterdam, 1969. North-Holland.

[Ghani et al., 1998] N. Ghani, V. Paiva, and E. Ritter. Explicit substitutions for constructive necessity. *Automata, Languages and Programming*, page 743, 1998.

[Gödel, 1986] K. Gödel. An interpretation of the intuitionistic propositional calculus (1933f). In S. Feferman, J.W. Dawson Jr., S.C. Kleene, G.H. Moore, R.M. Solovay, and J. van Heijenoort, editors, *K. Gödel's Collected Works: Volume I: Publications 1929-1936*, pages 301–302, Oxford, 1986. Oxford University Press.

[Goré et al., 2004] Rajeev Goré, Valeria de Paiva, and Michael Mendler. Intuitionistic Modal Logic and Application. *Journal of Logic and Computation*, 14(4), 2004.

[Kakutani, 2007] Y. Kakutani. Calculi for intuitionistic normal modal logic. In *Proceedings of Programming and Programming Languages*, pages 234–248, 2007.

[Kripke, 1963] S. Kripke. Semantical Analysis of Modal Logic I. Normal Proposicional Calculi. *Zeitschrift fur mathematische Logik und Grundlagen der Mathematik*, 9:67–96, 1963.

[Kripke, 1965] S. Kripke. Semantical Analysis of Modal Logic II. Non-Normal Modal Propositional Calculi. In J. W. Addison, L. Henkin, and A. Tarski, editors, *The Theory of Models (Proceedings of the 1963 International Symposium at Berkeley)*, pages 206–220, Amsterdam, 1965. North-Holland.

[Lambek and Scott, 1985] J. Lambek and Ph. J. Scott. *Introduction to Higher-Order Categorical Logic*. Cambridge University Press, 1985.

[Lemmon, 1957] E. J Lemmon. New Foundations for Lewis Modal Systems. *The Journal of Symbolic Logic*, 22(2):176–186, 1957.

[Lewis and Langford, 1932] C. I. Lewis and C. H. Langford. *Symbolic Logic*. Century, 1932.

[Prawitz, 1965] D. Prawitz. *Natural Deduction: A Proof-Theoretic Study*. Almqvist and Wiksell, 1965.

[Ritter and de Paiva, 2012] Eike Ritter and Valeria de Paiva. *Fibrational Models of Modal Logic*. University of Birmingham, UK, 2012.

[Simpson, 1994] A.K. Simpson. *The Proof Theory and Semantics of Intuitionistic Modal Logic*. PhD thesis, University of Edinburgh, 1994.

[Wijesekera, 1990] D Wijesekera. Constructive modal logic I. *Annals of Pure and Applied Logic*, 50:271–301, 1990.

Valeria de Paiva
Senior Research Scientist
Rearden Commerce
Foster City, CA, USA
E-mail: valeria.depaiva@gmail.com

Eike Ritter
Lecturer
School of Computer Science
University of Birmingham
Birmingham, UK
E-mail: E.Ritter@cs.bham.ac.uk

On the theory of Dynamic Sets

MANUEL FIDEL AND MARTÍN FIGALLO

ABSTRACT. Classical sets contain all elements that fulfill their defining property. Dynamic sets, on the other hand, are defined by a property but they only contain those elements which have been explicitely added by accepted methods. Besides, it is possible to eliminate elements if this is considered necessary. All this makes that a given dynamic set varies over time in spite of the fact that the property that defines it never change. Classical sets are a particular case of dynamic sets.

This concept constitute a more general model that results to be a suitable tool for many situations in which classical sets are not adequate. For example, in Computer Science and Foundations of Mathematics.

On the other hand, Alchourrón, Gänderfors and Makinson introduced in [Alchourrón, 1993] a new kind of logic presently known as "Belief Revision" (see also [Alchourrón *et al.*, 1985], [Alchourrón and Makinson, 1981] and [Alchourrón and Makinson, 1982]). This model has been widely spread as AGM in honor of its authors. This theory made a great impact on logicians, epistemologists and A.I. scientists.

In the last few years, there have been presented a lot of theories in order to formalize this kind of reasoning. In many occasions, this is not consistent with the actual notion of set. But, the task of finding an appropriate mathematical framework for the study of these logics have not been easy; and we think that this is due to the fact that the notion of set established in Zermelo-Fraenkel axiomatic theory is not the appropriate one to deal with them.

In this paper, we present an informal introduction to the axiomatic of **DS**, as well as, we exhibit some of its more important properties. In particular, we show that Russell paradox can be easily avoided (or incorporated) in **DS**.

1 Introduction

The paradoxes of set-theory which appeared about the turn of 20th century showed that the quasi-constructive procedure of Cantor's set-theory had to be restricted in some way, and thus an axiomatic determination of the restriction became imperative. After the initial shock of the antinomies, the stress was laid on restricting the concept of set axiomatically in such a way that the well–known paradoxes were eliminated and new ones were not to be expected. This was done mainly by Zermelo.

Thus, the axiomatic method for developing set theory was used by the first time and it showed to be the fundamental method for mathematics in the sense that every mathematical theory must be presented axiomatically in order to be developed rigorously.

Set theory was one of the biggest revolutions in the field of mathematics, this is due mainly to the fact that every pure mathematical notion can be defined in terms of sets and usual mathematical theorems can be demonstrated from the postulates of set theory and the specific axioms of the branch of mathematics in which the work is being developed. In a few words, we can say that the set theory is the basement of ordinary mathematics.

The set theory, as it was thought by Cantor, was not axiomatic and therefore it constituted a naive theory. Later, the appearance of paradoxes led to the first axiomatization, the Zermelos' one in 1908. The Cantorian work was based on two main principles: *extensionality*: if two sets have the same elements then, they are equal; and *separation:* every property determines a set composed only by the elements that verify that property. As it is well–known, using classical logic (and many others) as base logic, the separation principle leads to the famous Russell paradox. More paradoxes were also discovered by Curry, Moh Shaw-Kwei, Cantor, Burali-Forti among others.

Preserving Classical Logic as the base logic, with or without equality, the only possible solution was to restrict the separation principle and, since the resulting system is in general too weak, add new axioms to guarantee the existence of the union–set, the power–set, and so on. This is the case of the Zermelo-Fraenkel system (**ZF**).

For those who are not acquainted with the paradoxes of the set–theory, separation without limitations seems to be acceptable. Every property determines a set constituted by the objects (elements) that verify it and only by them. A property (or collection of properties) forms the *intention* of a concept and the collection of objects that satisfy it forms its *extension* (see [da Costa et al., 1998]). Since the birth of logic, in times of Aristotle, it was accepted as an absolute truth that every concept had endowed with one extension and one intention. But, there are (were) logicians that procured laws to regulate the relation between intention and extension. Well–known traditional manuals of logic, like Liar's, associate an operation of definition of the concepts to intention and an operation of division to extension.

But this notion of set, introduced by Cantor and refined by Zermelo, is very restrictive in many aspects. Intuitively, a set represents a static collection of objects. Nevertheless, in many circumstances of real life collections of objects that change over time must be modeled. When a set is defined using the notion of Cantor, it remains invariant in the course of the considerations. In many situations, this is not consistent with the actual notion of set.

For instance, if we consider the collection of *"all the inhabitants of the city X"*, this collection is not a set in the sense of Cantor since its elements vary in the course of time. Indeed, with every birth a new element is added to the set and with every death an element must be deleted from it.

2 Motivations and state of knowledge

Computer Science has provided us with an innumerable amount of examples of sets, all with a common characteristic: they are not static entities but dynamic ones that changes in the course of the time. Some of them are the following:

- When the set x stand for a Data Base we have in mind a dynamic set.

- In Imperatives Programming Languages, every variable can be thought as a dynamic singleton.

- Many applications require insertion and suppression of registers during the execution. Searches performed on this sets are known as *dynamic searches*. For instance, an Airline Reservations System.

Another weak side of the Cantor's notion is that if we consider a determined property then, we "magically" have the set of all elements that verify that property. Again, there exist numerous examples in which this does not happen. For instance, consider the proposition *"x is a planet where there exist intelligent life"*: it is clear that this set has at least one element, but nothing more can be said about it.

Finally, let us consider a principal motivation to consider the notion of dynamic set. Alchourrón, Gänderfors and Makinson introduced in [Alchourrón, 1993] a new kind of logic presently known as "Belief Revision" . This model has been widely spread as AGM in honor of its authors. This theory made a great impact on logicians, epistemologists and A.I. scientists. In the last few years, there have been presented a lot of theories in order to formalize this kind of reasoning. In many occasions, this is not consistent with the actual notion of set. But, the task of finding an appropriate mathematical framework for the study of these logics have not been easy; and we think that this is due to the fact that the notion of set established in

Zermelo-Fraenkel axiomatic theory is not the appropriate one to deal with them.

Dynamic sets, that is to say, sets with the property that their elements can vary over time have been consider previously by different authors. For example, in [Baruah et al.,], the authors present an algorithm to schedule the access to a single shared resource by a dynamic set of tasks that compete for it.

In [Steere, 1995] and [Steere and Satyanarayanan, 1994] dynamic sets are consider as an Abstract Database (ADB), and it is shown how this structure can offer substantial benefits in the solutions of problems like the raise of latency in I/O operations as dominant factor in the performance of different applications.

In Computer Science, the notion of Dynamic Set has been widely used, but in many cases, this was done without exhibit a precise definition for them. To the best of our knowledge, the only definition of dynamic set that has been set is in *Dynamic Sets and Their Application in VDM* by S. Liu and J. McDermid (see [Liu and McDermid,]).

Vienna Development Method (**VDM**) is both a method and a language to build software specifications whose behaviour are dynamic processes. However, since **VDM** is based in classical set theory, it presents many problems. The most serious of them is the lack of a well–defined mechanism for structuring. For this reason, the authors defined dynamic sets.

This presentation limits itself to consider a specific set of objects called universal set X. Besides, it is used the concept of "time interval":

A time interval T is an interval of real numbers $[t_1, t_2]$, $[t_1, t_2)$, $(t_1, t_2]$ or (t_1, t_2), where t_1 and t_2 are real numbers, $t_1 < t_2$. Then,

DEFINITION 1. A dynamic set A of elements of X over a time interval T is a function $A : T \to \mathcal{P}(X)$. ($\mathcal{P}(X) = \{Y : Y \subseteq X\}$). Besides, if A is a dynamic set in a time interval T then, it will be noted A^T.

By $\mathcal{P}_T(X)$ it is represented the collection of all dynamic sets of elements of X in the time interval T.

If $A^T(t_1) = A^T(t_2)$ for all $t_1, t_2 \in T$, the dynamic set A^T represents a classical set. On the other hand, the belonging is defined as follows:

DEFINITION 2. Let A^T be a dynamic set and $x \in X$. Then, it is said that x is a member of A^T, noted $x \in A^T$, if and only if $\exists_{t \in T} \cdot x \in A^T(t)$.

Besides,

DEFINITION 3. Let $A^T, B^T \in \mathcal{P}_T(X)$. Then, $A^T = B^T$ if and only if $\forall_{t \in T} \cdot A^T(t) = B^T(t)$.

DEFINITION 4. Let $A^T, B^T, C^T \in \mathcal{P}_T(X)$. Then, there are defined the following operations in $\mathcal{P}_T(X)$.

$A^T \cap B^T = C^T$ if and only if $\forall_{t \in T} \cdot A^T(t) \cap B^T(t) = C^T(t)$.

$A^T \cup B^T = C^T$ if and only if $\forall_{t \in T} \cdot A^T(t) \cup B^T(t) = C^T(t)$

$A^T \setminus B^T = C^T$ if and only if $\forall_{t \in T} \cdot A^T(t) \setminus B^T(t) = C^T(t)$

$\mathcal{C} A^T = C^T$ if and only if $\forall_{t \in T} \cdot \mathcal{C} A^T(t) = C^T(t)$

On the other hand, the studies on belief revision have a relatively recent origin. They started with the works of the philosophers William Harper (1976) and Issac Levi in the 1970s. But, it was a decade later when revision beliefs was established as a research area with the work of Carlos Alchourrón, Peter Gänderfors and David Makinson whose model became standard in the area and known as AGM in honor to the authors (see [Alchourrón and Makinson, 1981] and [Alchourrón and Makinson, 1982]). These works present a confluence of previous works by Gänderfors, by the one side, and by Alchourrón and Makinson, by the other. In the last few years numerous theories have been presented in order to formalize this kind of reasoning.

Our purpose in this paper is, working in an acceptable" way for mathematics, to investigate alternative choices that allow us to apply the methods of the algebraic logic to the development of this logics, and particularly, to find a semantic for them. To do so, we shall introduce a new theory of sets that we also named *Theory of Dynamic Sets* (**DS** for short) that intend to overcome the limitations of classical set–theory just mentioned.

In this new theory, every set will be defined 'by an *intention* and an *extension*. The intention will be determined by a property, a predicate; and the extension by any collection of objects that verifies this property.

The main difference between this new theory and **Z** or **ZF** is that the extension of a given set must not necessarily be the collection of all the objects that verify the property but just the collection of all objects that we want to belong to it.

We shall use the book *Elementos de teoria paraconsistente de conjuntos*, Coleção CLE-Unicamp (1998), of N.C.A. da Costa, J.-Y.Beziau y O.A.S. Bueno as a reference text (see also [da Costa, 1963], [da Costa, 1964] and [da Costa, 1997]).

3 The Theory of Change: Belief Revision

The simplest theory of change and the more well–known is the *expansion* where a new proposition (axiom), that could be inconsistent with the theory A is added to A (an element is added to a set). This new set is next closed by the consequence logic operator.

There exist, of course, many other theories of change, whose logics were less understood. One of them is the *contraction*, where a proposition p that already was in the theory is refuted (rejected). When A is a code of norms this process is known by lawyers as the *derogation* of p from A. The main problem of this form of theory of change is determine those propositions of A that should be rejected along with p in such a way that the contracted theory is closed under the operator of consequence.

Another class of theory of change is the *revision*, where a proposition p, inconsistent with a given theory A, is added to A under the requirement that the revised theory is consistent and closed under the operator of consequence. In a legal context, this kind of theory of change is known as *amendment*.

Gänderfors developed postulates of an equational nature in order to capture the basic properties of this processes. Also, he showed that the process of revision can be reduced to contraction via the so–called *identity of Levi*.

Let A be a theory and x a proposition, if we denote $A \dot{-} x$ the contraction of A by x, then the revision of A by x, denoted by $A \dot{+} x$, can be defined:

(1) $\quad A \dot{+} x = \mathbf{Cn}((A \dot{-} \neg x) \cup \{x\})$

where $\mathbf{Cn}$ is the logical operator of consequence.

Gänderfors' postulates for contraction

(C1) $A \dot{-} x$ is a theory, for all theory A (closure).

(C2) $A \dot{-} \neg x \subset A$.

(C3) If $x \notin \mathbf{Cn}(A)$ then $A \dot{-} \neg x = A$.

(C4) If $x \notin \mathbf{Cn}(\emptyset)$ then $x \notin \mathbf{Cn}(A \dot{-} \neg x)$.

(C5) If $\mathbf{Cn}(\{x\}) = \mathbf{Cn}(\{y\})$ then $A \dot{-} \neg x = A \dot{-} \neg y$.

(C6) $A \subset \mathbf{Cn}((A \dot{-} \neg x) \cup \{x\})$, for all theory A.

Gänderfors' postulates for revision

(C1) $A \dot{+} x$ is a theory, for all theory A (closure).

(C2) $x \in A \dot{+} x$.

(C3) If $\neg x \notin \mathbf{Cn}(A)$ then, $A \dot{+} x = \mathbf{Cn}(A \cup \{x\})$.

(C4) If $\neg x \notin \mathbf{Cn}(\emptyset)$ then, $A \dot{+} x$ is consistent under $\mathbf{Cn}$.

(C5) If $\mathbf{Cn}(\{x\}) = \mathbf{Cn}(\{y\})$ then, $A \dot{+} \neg x = A \dot{+} \neg y$.

(C6) $(A \dot{+} x) \cap A = A \dot{-} x$, for all theory A.

4 Dynamic Sets. A naive presentation

Intuitively, a dynamic set A (or D-set) will be any pair $A = (\phi(x), E)$, where $\phi(x)$ is a predicate with free variable x and E is a collection of elements that satisfy the predicate ϕ. We shall say that is the *intention* of the dynamic set A and the collection of elements E is the *extension* of A.

EXAMPLE 5. Let $\phi(x)$ be the predicate "x is a natural number". Then, the following are dynamic sets:

$$(\phi(x), \emptyset),$$
$$(\phi(x), \{1\}),$$
$$(\phi(x), \{1, 2\}),$$

In this theory it will be necessary to distinguish two kinds of belonging. Let $A = (\phi(x), E)$ a dynamic set, the expression $x \in_e A$, read "*x belongs to the extension of A*" (or "*x extensionally belongs to A*"), means that x is one of the elements of E. On the other hand, an object x intentionally belongs to A iff $\phi(x)$ is verified, i.e, $x \in_i A$ iff $\phi(x)$.
Then, the membership relation $\in$ is defined as follows.

DEFINITION 6. We shall say that a belongs to the dynamic set A, noted $a \in A$, iff $a \in_i A$ and $a \in_e A$ hold simultaneously.

Action of addition and elimination of an element

Let $A = (\phi_A, A)$ be a D-set and x an element such that $x \in_i A$ then, we shall accept the next axiom.

If $x \notin_e A$ and the action $A \dotplus x$ is performed then, $x \in_e A$.

The expression $A \dotplus x$ will be read "x is added to A". Let us remark that if $x \notin A$ and x intentionally belongs to A then, if we added x to A we shall have $x \in A$.

EXAMPLE 7. Let $\phi(x)$ be the predicate consider in the previous example and let $A = (\phi(x), \emptyset)$. If we perform the action $A \dotplus 1$ then, we shall have $A = (\phi(x), \{1\})$.

Analogously, we accept the following axiom.

If $x \in A$ and the action $A \dotminus x$ is performed then, $x \notin_e A$.

The expression $A \dotminus x$ will be read "x is eliminated from A". That is to say, if x belongs to A then, x belongs both intentionally and extensionally to A. Then, if we perform the action $A \dotminus x$ we shall have that x extensionally no longer belongs to A and therefore it will not belong to A.

EXAMPLE 8. If $A = (\phi(x), \{1, 2\})$ and we perform $A \dotminus 1$ then, we shall have $A = (\phi(x), \{1\})$.

4.1 A preliminary formalization for DS

Our logical frame could be a discrete temporal logic or any logic of discrete steps where in each step the relations between the different actors may change. To formalize this theory we shall use a first order language enriched with additional formulas that we shall call actions. The symbols of the language will be:

$$\in, \in_e, \in_i, =, =_e, =_i, \vee, \wedge, \rightarrow, \leftrightarrow, \neg, \forall, \exists, \dotplus, \dotminus.$$

together with a denumerable collection of variables that will be denoted by $a, b, c, \ldots, x, y, z$.

- The expressions $x \in y$, $x \in_e y$, $x \in_i y$ will be read, "x belongs to y", "x extensionally belongs to a y" and "x intentionally belongs to y", respectively.

- The formulas $x \dot{+} y$ and $x \dot{-} y$ are called actions and will be read "x is added to y" and "x is eliminated from y", respectively.

Axioms

Extensionality Axioms

[**DS1**] $\forall x \forall y (\forall z (z \in_e x \leftrightarrow z \in_e y) \to (x =_e y))$

[**DS2**] $\forall x \forall y (\forall z (z \in_i x \leftrightarrow z \in_i y) \to (x =_i y))$

[**DS3**] $\forall x \forall y (x = y \leftrightarrow (x =_e y \land x =_i y))$

Specification Axioms

[**DS4**] $\forall x \forall z (z \in x \leftrightarrow (z \in_e x \land z \in_i x))$,

[**DS5**] $\exists x (\forall z (z \in_i x \leftrightarrow \phi(z) \land \forall y (y \notin_e x))$, where ϕ is a formula with free variable z.

The set x will be represented by the pair $(\phi, \emptyset_e)$ and we shall say that the first and second component of the pair is the *intention* and *extension* of x, respectively. Axiom scheme [**DS5**] state that for every predicate ϕ there exists the set x whose intention is given by ϕ and the extension is empty.

Axiom for the addition of an element

[**DS6**] $\forall x \forall y (((x \neq y) \land (y \in_i x) \land (x \dot{+} y)) \to y \in_e x)$

For instance, let $\phi(z)$ be a formula with free variable z and x the set whose existence is guaranteed by [**DS5**]. If $(x \neq a) \land (a \in_i x)$ and we perform the action $x \dot{+} a$, by [**DS6**], we have $a \in_e x$. Then, x will be noted $x = (\phi(z), \{a\})$.

Axiom for the elimination of an element

[**DS7**] $\forall x \forall y (((x \in y) \land (x \dot{-} y)) \to y \notin_e x)$

REMARK 9.

(i) An element extensionally belongs to a given set x if and only if it was explicitly added to x in some moment.

(ii) Taking into account the above remark every set in this theory is finite.

From the axioms is not difficult to prove the following results.

THEOREM 10.

(i) $\vdash \exists x \forall (y \notin_e x)$.

(ii) $\vdash \exists x \forall (y \notin_i x)$.

(iii) $\vdash \exists x \forall (y \notin x)$ \hfill (Existence of the null-set).

(iv) $\vdash \forall x \forall y \exists z (\forall t (t = x \lor t = y) \leftrightarrow t \in z)$ \hfill (Axiom of Pairing).

(v) $\vdash \forall x_1 \ldots \forall x_n \exists z (\forall t (t =_i x_1 \lor \ldots \lor t =_i x_2) \leftrightarrow t \in_i z)$
(existence of intentionally finite sets).

(viii) $\vdash \forall z \exists x (\forall t (\exists y (y \in_i z \land t \in_i y)) \to t \in_i x)$
(Axiom of intentional sum-set).

(ix) $\vdash \exists y \forall x (x \in_e y \leftrightarrow (x \in_e a \lor x \in_e b))$, a and b finites.
(Axiom of extensional sum-set)

(x) $\vdash \exists y \forall x (x \in_e y \leftrightarrow (x \in_e a \land x \in_e b))$, a and b finites.
(Extensional intersection of finite sets)

REMARK 11. We shall note

(i) $x \subseteq_i y$ iff $\forall t (t \in_i x \to t \in_i y)$.

(ii) $x \subseteq_e y$ iff $\forall t (t \in_e x \to t \in_e y)$.

(iii) $x \subseteq y$ iff $x \subseteq_i y \land x \subseteq_e y$.

(iv) $\emptyset$ the set whose existence is guaranteed by Theorem (iii).

(v) $\mathcal{U}$ the set whose intention is given by the formula $z = z$ and whose existence is guaranteed by Axiom [**DS5**].

DEFINITION 12. Let x and y be D-sets. Then,

$x \cup_i y =_{def} (z \in_i x \lor z \in_i y, \emptyset_e)$

$x \cap_i y =_{def} (z \in_i x \land z \in_i y, \emptyset_e)$

$\mathcal{C}_i x =_{def} (z \notin_i x, \emptyset_e)$

THEOREM 13.

(xi) $\vdash \forall x \exists y (\forall t (t \subseteq_i x \rightarrow t \in_i y))$. (Axiom of intentional Power-Set)

Besides, it holds:

THEOREM 14.

(xii) $\vdash \forall x \forall y \forall z (x \cup_i (y \cup_i z) = (x \cup_i y) \cup_i z)$.

(xiii) $\vdash \forall x \forall y (x \cup_i y = y \cup_i x)$

(xiv) $\vdash \forall x (x = x \cup_i x)$

(xv) $\vdash \forall x \forall y \forall z (x \cap_i (y \cap_i z) = (x \cap_i y) \cap_i z)$

(xvi) $\vdash \forall x \forall y (x \cap_i y = y \cap_i x)$

(xvii) $\vdash \forall x (x = x \cap_i x)$

(xviii) $\vdash \forall x \forall y (x = x \cap_i (y \cup_i x))$

(xix) $\vdash \forall x \forall y \forall z (x \cup_i (y \cap_i z) = (x \cup_i y) \cap_i (x \cup_i z))$

(xx) $\vdash \forall x (\emptyset \cap_i x = \emptyset)$

(xxi) $\vdash \forall x (\mathcal{U} \cup_i x = \mathcal{U})$

(xxii) $\vdash \forall x (\mathcal{C}_i x \cap_i x = \emptyset)$

(xxiii) $\vdash \forall x (\mathcal{C}_i x \cup_i x = \mathcal{U})$

For every D-set x we shall note $\mathcal{P}_i(x)$ the set whose existence is guaranteed by Theorem 13. Then,

THEOREM 15. For every D-set x, the structure $\langle \mathcal{P}_i(x), \cup_i, \cap_i, \mathcal{C}_i, \emptyset, \mathcal{U} \rangle$ is a Boolean algebra.

Analogously, it holds:

THEOREM 16.

(xxiv) $\vdash \forall x \forall y \forall z (x \cup_e (y \cup_e z) =_e (x \cup_e y) \cup_e z)$.

(xxv) $\vdash \forall x \forall y (x \cup_e y =_e y \cup_e x)$

(xxvi) $\vdash \forall x (x =_e x \cup_e x)$

(xxvii) $\vdash \forall x \forall y \forall z (x \cap_e (y \cap_e z) =_e (x \cap_e y) \cap_e z)$

(xxviii) $\vdash \forall x \forall y (x \cap_e y =_e y \cap_e x)$

(xxix) $\vdash \forall x (x =_e x \cap_e x)$

(xxx) $\vdash \forall x \forall y (x =_e x \cap_e (y \cup_e x))$

(xxxi) $\vdash \forall x \forall y \forall z (x \cup_e (y \cap_e z) =_e (x \cup_e y) \cap_e (x \cup_e z))$

(xxxii) $\vdash \forall x (\emptyset \cap_e x =_e \emptyset)$

Then,

THEOREM 17. *The structure* $\langle \mathcal{U}, \cup_e, \cap_e \rangle$ *is a finite distributive lattice.*

5 Conclusions

In this presentation we have introduced the concept of dynamic set, some of its applications as well as part of its possible development. The later could present various alternatives since regular operations between dynamic sets can be defined in many different ways, making the theory more complex.

It is worth mentioning that **DS** is inconsistent. However, the Russell set r defined by the formula $\phi(z) \equiv z \notin z$ leads to no contradiction as long as r is not added to r, i.e., if we do not perform the action $r \dot{+} r$.

It will remain as a future work to show that a good semantic to AGM (with classical base) could be based in Boolean Algebras with filters that would be dynamic sets. This will imply a lot of new difficulties to overcome.

BIBLIOGRAPHY

[Alchourrón and Makinson, 1981] C. Alchourrón and D. Makinson. Hierarchies of regulations and their logic. In R. Hilpinen, D. Reidel, and Dordrecht, editors, *New Studies in Deontic Logic: Norms, Actions, and the Foundations of Ethics*, pages 125–148. 1981.

[Alchourrón and Makinson, 1982] C. Alchourrón and D. Makinson. On the logic of theory change: Contraction functions and their associated revision functions. *Theoria*, 48:14–37, 1982.

[Alchourrón et al., 1985] C. Alchourrón, P. Gardenförs, and D. Makinson. On the logic of theory change: Partial meet functions for contractions and revisions. *Journal of Symbolic Logic*, 50:510–530, 1985.

[Alchourrón, 1993] C. Alchourrón. *Philosophical Foundations of Deontic Logic and the Logic Defeasible Conditionals*. Meyer, J.J.Ch. & Wieringa, R.J. (eds.). Deontic Logic in Computer Science. Chichester: Wiley; 43-84., 1993.

[Baruah et al.,] Baruah, Gehrke, Paxton, Stoica, Abdel-Wahab, and Jeffay. Fair on-line scheduling of a dynamic set of tasks on a single resource. *Preprint submitted to Elsevier Science*.

[da Costa et al., 1998] N.C.A. da Costa, J.-Y.Beziau, and O.A.S. Bueno. *Elementos de teoria paraconsistente de conjuntos*. CLE-Unicamp, Campinas, 1998.

[da Costa, 1963] N.C.A. da Costa. Calculs propositionnels pour les systèmes formels inconsistants. *Comptes Rendus de lAcademie des Sciences de Paris*, 257:3790–3793, 1963.

[da Costa, 1964] N.C.A. da Costa. Calculs propositionnels pour les systèmes formels inconsistants. *Comptes Rendus de lAcademie des Sciences de Paris*, 258:27–29, 1964.

[da Costa, 1997] N.C.A. da Costa. *Logiques classiques et non classiques*. Masson, Paris, 1997.

[Halpern, 1998] J. Y. Halpern. Set-theoretic completeness for epistemic and conditional logic. *Ann. Math. Artificial Intelligence*, 26:1–27, 1998.

[Liu and McDermid,] S. Liu and J. McDermid. Dynamic sets and their application in vdm. *University of York. (Preprint)*.

[Steere and Satyanarayanan, 1994] D. Steere and M. Satyanarayanan. A case for dynamic sets in operating systems. *School of Computer Science. Carnegie Mellon University. Pittsburgh, PA 15213. Preprint.*, 1994.

[Steere, 1995] D. Steere. Using dynamic sets to speed search in word wide information systems. *School of Computer Science. Carnegie Mellon University. Pittsburgh, PA 15213. Preprint.*, 1995.

Manuel Fidel
Departamento de Ciencias e Ingeniería de la Computación
Universidad Nacional del Sur
Bahía Blanca, Argentina
E-mail: mmfidel@gmail.com

Martín Figallo
Departamento de Matemática
Universidad Nacional del Sur
Bahía Blanca, Argentina
E-mail: figallomartin@gmail.com

Truth-value gaps and the minimalist conception of truth

OSWALDO CHATEAUBRIAND

ABSTRACT. Although in my book *Logical Forms* I raise several objections to the basic ideas of the minimalist conception of truth, as formulated by Frege[1] and by Tarski[2], I did not discuss more recent formulations. Since I think the main problem for the minimalist conception of truth is the problem of truth-value gaps, I will raise some questions about Paul Horwich's influential discussion of this problem, especially in connection with his definition of falsity in sections 26-27 of *Truth*.

1 Preliminary considerations

When dealing with propositions, or sentences[3], that are neither true nor false it is important to realize that logical equivalence, interpreted as mutual logical consequence, does not imply material equivalence, interpreted as sameness of truth-value. That is, whereas the logical equivalence of two propositions rules out the possibility of one proposition being true and the other not true, it does not rule out the possibility of one proposition being false and the other truth-valueless. This is an interesting phenomenon, usually unnoticed, and is relevant to Tarski's truth schema, among other things.[4]

Tarski formulates the truth schema for sentences of a given language as:

(T) X is true $\leftrightarrow S$,

where instances of (T) are obtained replacing 'S' by a sentence of the language in question, and replacing 'X' by a designator of that sentence. It is easy to see that for any instance of (T), the two sides are logically equivalent, but if we take a sentence which is neither true nor false, then the left-hand side is false

---
[1][Chateaubriand, 2001], pp. 80-81.
[2][Chateaubriand, 2001], pp. 232-240.
[3]I am not going to be very careful with the separation of propositions from sentences because I do not think it will affect my arguments.
[4]An interesting case is that of definite descriptions, which I discuss in [Chateaubriand, 2001], chapter 3 and in (2002). For example, a sentence of the form '*the F is G*' in which the description '*the F*' does not denote, is truth-valueless according to [Frege, 1960] analysis, and is false according to [Russell, 1905] analysis. From this it is usually inferred that the two analyses are not logically equivalent, but if logical equivalence is characterized as mutual logical consequence, the inference is incorrect. For, if '*the F is G*' is true in Frege's analysis, then '*the F*' must denote a unique thing that is G, and Russell's analysis is also true. Conversely, if Russell's analysis is true, then there must be a unique thing that is F and is G, making Frege's analysis true.

and the right-hand side is truth-valueless.

Thus, if we agree with Frege that a sentence (or thought) containing a non-denoting name is truth-valueless[5], then an instance of (T) such as

(1) 'Sherlock Holmes is tall' is true ↔ Sherlock Holmes is tall,

is also truth-valueless, even though the left-hand side is logically equivalent to the right-hand side. The reason for the truth-value gap is that the left-hand side is false whereas the right-hand side is neither true nor false—but the truth of either of them guarantees the truth of the other; i.e., they are logical consequences of each other.

Evidently, the same happens for the schema

(E) ⟨p⟩ is true ↔ p,

which Horwich uses to define truth. He is quite right, therefore, in holding that truth-value gaps cannot be admitted in his minimalist account of truth (p. 76). In the next few sections I will discuss his account of falsity and his strategies for avoiding truth-value gaps and non-denoting terms.

2 The characterization of falsity

Horwich characterizes falsity as absence of truth, either as

(2*) ⟨p⟩ is false ↔ not[⟨p⟩ is true]

or, equivalently, as

(2**) ⟨p⟩ is false ↔ not p.

On p. 77 he gives three main reasons in favor of this definition.

The first reason is that since together with (E) the definition of falsity yields

(3) ⟨p⟩ is not true and not false → not p & not not p,

"we cannot claim of some proposition that it has no truth-value, for that would imply a contradiction."

The second reason is that "[t]he account reflects our pre-theoretical intuition that if a proposition is *not* true then it is false, and that if something is *not* the case then the claim that it *is* the case would be false."

---

[5][Frege, 1960], pp. 32-33 in the original pagination.

The third reason is that "[n]o reasonably plausible alternative characterization of falsity is able to accommodate these features of the concept."

I will discuss these reasons below, but first it is necessary to make some remarks about negation.

3 Predicate negation and sentential negation

In ordinary language, and in predicate logic, we use negation both as predicate negation—as in 'John is not tall'—and as sentential negation—as in 'it is not the case that John is tall'. Since to say it is not the case that John is tall actually means *it is not true* that John is tall, sentential negation also operates on a predicate; namely, the predicate 'is true'.

If a sentence is neither true nor false, then its predicate negation is also neither true nor false. Thus, if we hold that the sentence

(4) Sherlock Holmes is tall

is neither true nor false, because the name 'Sherlock Holmes' does not denote, then we should also hold, for the same reason, that the sentence

(5) Sherlock Holmes is not tall

is neither true nor false.

The sentential negation 'it is not true that', on the other hand, always gives a truth-value, because if a sentence is neither true nor false, then it is not true, and its sentential negation is true. Thus, as opposed to the predicate negation (5), which is neither true nor false, the sentential negation

(6) It is not true that Sherlock Holmes is tall

is true. And, of course, the sentential negation of a true sentence is false and the sentential negation of a false sentence is true.

It is also important to notice that the affirmation 'it is the case that'—i.e., 'it is true that'—also has the feature of always giving a truth-value. For, if a sentence is neither true nor false, then to say that it is true is false. Thus, in the case of our sentence (4), the sentence

(7) It is true that Sherlock Holmes is tall

is false. And if a sentence is either true or false, then to say that it is true yields the same truth-value as the sentence itself.

Let us now consider Horwich's reasons in favor of the definition of falsity.

4 The reasons

If we remember that falsity is defined as the absence of truth, and make appropriate replacements in (3), we obtain

(3') ⟨p⟩ is not true and not not true → ⟨p⟩ is not true & it is not true that ⟨p⟩ is not true,

which is actually a tautology. It is quite true that the consequent is a contradiction, but so is the antecedent if 'false' means 'not true'.

With respect to the second reason, it is not at all clear to me that what is not true is false is a pre-theoretical intuition of ours. I think there are many of us whose pre-theoretical intuition is that there are sentences, or propositions, which are neither true nor false. But, in any case, the second part of Horwich's argument in support of his position is based on an equivocation. For, as I pointed out above, if a proposition is not true, then to say that it is true is false, independently of whether it is neither true nor false, or it is false.

I agree with the third reason, although I do not think that the features Horwich wants to preserve are in fact features of the concept of falsity. [6]

5 Vacuous terms

Since non-denoting terms are usually taken to be an obvious source of truth-value gaps, Horwich confronts this issue in section 27. He gives the following argument that he attributes to Russell:[7]

> ...an atomic proposition entails that the referents of its singular terms exist: *a is F* entails that *a exists*; and in that case, it is natural to allow that if it is false that *a exists* then it is false that *a is F*. Therefore atomic propositions containing vacuous singular terms may very plausibly be regarded as *false* and dont call for truth-value gaps.

This argument raises some issues related to my discussion in section 1 above.

Let us interpret 'entails' in the sense of logical consequence, and let us think of '*a is F*' and '*a exists*' as sentence forms. For a sentence of the form

(8) *a is F*

---

[6] Russell also insists on defining falsity as absence of truth, both in (1905) and in (1957), where he is especially insistent on the point. I discuss his argument in [Chateaubriand, 2001], p. 131.

[7] Horwich refers to [Russell, 1905], but I am not sure which specific argument he has in mind. In [Chateaubriand, 2005a] I argue that most of the arguments in "On Denoting" are fallacious for reasons deriving from the issues discussed in sections 1 and 3 above.

to be true, the name a must denote something that is F, and for a sentence of the form

(9) a exists

to be true the name a must denote. Thus, consider (4). For this sentence to be true the name 'Sherlock Holmes' must denote something that is tall. And, evidently, if this is so, then the name 'Sherlock Holmes' must denote. Hence, the sentence

(10) Sherlock Holmes exists

is a logical consequence of (4).

In fact, a sentence of the form (9) can *never* be false, in a strict sense of falsity, but can only be not true. Hence, Horwich's inference that if (9) is false, then (8) is false, is only an inference that if (9) is not true, then (8) is not true; and this is perfectly compatible with (8) being neither true nor false.[8] Horwich's argument presupposes his own definition of falsity in order to argue for it.

But Horwich supplements his argument with an appeal to Russell's theory of descriptions and to Quine's technique for eliminating singular terms "to produce logical forms that are free of singular terms." He argues that, "[f]or example,

(11) Everyone has an ancestor from Atlantis,

which contains the empty name 'Atlantis', becomes

(11*) There is a place with the property of *being-Atlantis*, and everyone has an ancestor from there,

which is uncontroversially false."[9]

There are two main problems with this strategy. One is that the treatment of singular terms in Russell' theory of descriptions is quite questionable. This is a controversial issue, which I have discussed in several places.[10] The second problem is that I do not think Quine's technique of eliminating singular terms by introducing *ad hoc* predicates such as 'is-Pegasus', or 'pegasizes', or 'is-Atlantis', or 'atlantisizes', is justified.

A large part of the problem can be formulated as the question: What is a predicate? Is *anything* that behaves syntactically as a predicate a predicate?

---

[8]For an expanded version of this discussion see [Chateaubriand, 2001], pp. 110-114.
[9]I have changed his numbering from (10) to (11).
[10]See [Chateaubriand, 2001], chapter 3, [Chateaubriand, 2002], and [Chateaubriand, 2005a].

Is 'mimsy' a predicate? (As in Lewis Carroll's "mimsy were the borogroves"?) And even if we agree to say that it is a syntactic predicate, can we say that it is true of or false of anything? Is it either true or false that something is mimsy? Is Quine mimsy?

When Quine discusses his elimination technique in "On What There Is" ([Quine, 1953], p. 8) he says:

> In order thus to subsume a one-word name or alleged name such as 'Pegasus' under Russell's theory of description, we must, of course, be able first to translate the word into a description. But this is no real restriction. If the notion of Pegasus had been so obscure or so basic a one that no pat translation into a descriptive phrase had offered itself along familiar lines, we could still have availed ourselves of the following artificial and trivial-seeming device; we could have appealed to the *ex hypothesi* unanalyzable, irreducible attribute of *being Pegasus*, adopting, for its expression, the verb 'is-Pegasus', or 'pegasizes'. The noun 'Pegasus' itself could then be treated as derivative, and identified after all with a description: 'the thing that is-Pegasus', 'the thing that pegasizes'.

If, as Quine says, the notion of Pegasus had been very obscure, what then are the grounds for holding that 'is-Pegasus', or 'pegasizes', are *predicates*? What determines whether these alleged predicates apply to, or do not apply to, a thing? [11]

My view is that in order for something to be a predicate, in a significant semantic sense, and not merely in a purely syntactic sense, it must have conditions of applicability, and that if there are no such conditions, then sentences involving the alleged predicate are neither true nor false.[12] Not only singular terms may fail to refer and produce truth-value gaps, but also general terms may fail to refer and produce truth-value gaps. I hold, therefore, there are no such properties as *being-Pegasus* or *being-Atlantis*.[13] Of course, if a name refers, then there is a corresponding predicate and property, because the conditions of applicability are given by the reference of the name. The predicate 'is-Quine' applies to something if, and only if, that thing is Quine. But obviously it would not do to say that the predicate 'is-Pegasus' applies to something if, and only if, that thing is Pegasus, or to say that the predicate 'is-Atlantis' applies to something if, and only if, that thing is Atlantis.

---

[11] This discussion of Quine's strategy derives from [Chateaubriand, 2005b], p. 243.

[12] The conditions of applicability need not be precisely specified—and for most ordinary predicates they are not—but there must be *some* conditions.

[13] I derive this point from [Kripke, 1980] discussion of vacuous terms such as 'Sherlock Holmes' and 'unicorn', and expand on it in ([Chateaubriand, 2001], pp. 381-385). See also my discussion of predicates and properties in [Chateaubriand, 2004] and [Chateaubriand, 2007].

My overall conclusion, therefore, is that the strategies suggested by Horwich in sections 26-27 of *Truth* do not succeed in avoiding the problem of truth-value gaps.

BIBLIOGRAPHY

[Chateaubriand, 2001] O. Chateaubriand. *Logical Forms. Part I: Truth and Description*, volume 34 of *Coleção CLE*. CLE-UNICAMP, 2001.
[Chateaubriand, 2002] O. Chateaubriand. Descriptions: Frege and Russell combined. *Synthese*, 130:213–226, 2002.
[Chateaubriand, 2004] O. Chateaubriand. Negation and negative properties: reply to Richard Vallée. *Manuscrito*, 27:235–242, 2004.
[Chateaubriand, 2005a] O. Chateaubriand. Deconstructing "On Denoting". In B. Linsky and G. Imaguire, editors, *On Denoting: 1905-2005*, pages 361–380, Munich, 2005. Philosophia Verlag.
[Chateaubriand, 2005b] O. Chateaubriand. *Logical Forms. Part II: Logic, Language, and Knowledge*, volume 42 of *Coleção CLE*. CLE-UNICAMP, 2005.
[Chateaubriand, 2007] O. Chateaubriand. The truth of thoughts: variations on Fregean themes. *Grazer Philosophische Studien*, 75:199–215, 2007.
[Frege, 1960] G Frege. On sense and reference. In P. Geach and M. Black, editors, *Translations from the Philosophical Writings of Gottlob Frege*, pages 56–78, Oxford, 1960. Oxford University Press.
[Horwich, 1998] P. Horwich. *Truth*. Oxford University Press, 2nd edition, 1998.
[Kripke, 1980] S. Kripke. *Naming and Necessity*. Harvard University Press, 1980.
[Quine, 1953] W.V. Quine. *From a Logical Point of View*. Harvard University Press, 1953.
[Russell, 1905] B. Russell. On Denoting. *Mind*, 14:479–493, 1905.
[Russell, 1957] B. Russell. Mr Strawson on Referring. *Mind*, 66:385–389, 1957.

Oswaldo Chateaubriand
Department of Philosophy
Pontifical Catholic University of Rio de Janeiro (PUC-Rio)
Rio de Janeiro, Brazil
E-mail: oswaldo@puc-rio.br

On Skolem Functions in Proof Theory[1]

JAAKKO HINTIKKA

ABSTRACT. In the usual proof theory, the order of applications of different rules is restricted by the requirement that rules of inference apply only to the dominant quantifier or connective. This requirement can be liberalized, among other things by replacing existential-force quantifiers by Skolem functions. Existential instantiation then becomes substitution of a constant term for an argument variable of a Skolem function. This liberalizes the order of existential instantiations among other rules and among themselves.

1 The importance of being a Skolem function

Skolem functions are the machinery that keeps the logic of quantification going. This is conspicuous in model theory where they are literally the truth-makers: A quantificational sentence is true iff there exists a full set of its Skolem functions. This paper is calculated to illustrate their uses also in proof theory.

The important role of Skolem functions is rooted in the meaning of quantifiers. The only way in which the dependence of a variable y on another variable x can be expressed in a usual first-order logic is by the formal dependence of its quantifier $(Q_2 y)$ on the quantifier $(Q_1 x)$ to which the other variable is bound. What Skolem functions do is to codify those dependence relations for existential quantifiers that depend on universal ones. (Formulas are throughout this paper taken to be in their negation normal form.) Their doing so puts them in a position to take over the job of existential quantifiers. This can be spelled out by defining the Skolemized variant of a given first-order formula (in a negation normal form) F. It is obtained from F by replacing (one by one from inside out) each existential subformula $(\exists x)G[x]$ by $G[g(y_1, y_2, \ldots)]$. Here g is a new function constant, different for different existential quantifiers, and $(\forall y_1), (\forall y_2), \ldots$, are all the universal quantifiers on which $(\exists x)$ depends. Individual constants instantiating existential quantifiers are considered as constant Skolem functions. Since a sentence and its Skolemized version contains different symbols, they are not deductively equivalent. But it is obvious that they are interpretationally equivalent and behave in the same way in logic.

The transformation of all sentences into the Skolemized form amounts to a kind of quantifier elimination, for after the elimination of existential quantifiers, the universal quantifiers could all be thought of as sentence-initial and hence represented in the rest of the sentence only via their variables.

---
[1] This paper was written when Jaakko Hintikka was a Distinguished Visiting Fellow of the Collegium for Advanced Studies of the University of Helsinki. He was assisted by Antti Kylänpää

2 Notation

In this paper, a first-order language with identity will be considered, with its usual formation rules. Its nonlogical constants may contain (besides individual constants) either predicates only or else also function symbols. All formulas discussed are assumed to be in a negation normal form. That is to say, the only propositional connectives are &, ∨, ∼, and all negation signs are prefixed to atomic formulas or identities.

The usual game-theoretical semantics is assumed throughout. In order to apply this semantics we need the notion of informational dependence. A quantifier $(Q_2 y)$ depends on another one $(Q_1 x)$ iff it occurs in the formal scope of $(Q_1 x)$. A quantifier (Qx) prefixed to a subformula $F[x]$ also depends on individual constants occurring in $F[x]$.

These notions can be extended to independence friendly (IF) logic by allowing an existential quantifier $(\exists y)$ to be independent of a universal quantifier within whose formal $(\forall x)$ scope it occurs by writing it $(\exists y/\forall x)$. Likewise $(\exists y/b)$ shows that $(\exists y)$ is independent of the constant b.

The dependence of $(\exists y)$ on $(\forall x)$ is also shown by the fact that x is one of the arguments of the Skolem function replacing $(\exists y)$. Likewise a constant b is one of the arguments of such a Skolem function iff $(\exists y)$ depends on b (in the given context). (Constants are also Skolem functions, viz. constant ones.)

A term is defined in the usual way:

(i) An individual constant is a term.

(ii) An individual variable is a term.

(iii) If f is a function symbol with k argument and $t_1, t_2, \ldots, t_k$ are terms, then $f(t_1, t_2, \ldots, t_k)$ is a term.

A constant term is one without any individual variables.

3 Disproofs as frustrated model constructions

Proof theory operates in its most usual version with Gentzen-type sequence calculi. Proofs in such calculi are mirror images of proofs using a corresponding Beth-type tableau. Alternatively, one can use the equivalent tree method. (For this method, see e.g. [Smullyan, 1963].) These methods bring out more clearly the model-theoretical import of formal proofs and are therefore more intuitive than sequence calculi. In this paper the tree method is used. In it, a proof of a formula S is a disproof of ∼S. And a disproof of a set $\{S_1, S_2, \ldots\}$ of formulas can be thought of as an inevitably frustrated attempt to describe (a description of) a model in which they are all true. As such a description, we can use model sets defined in the usual way. Such sets of formulas can literally serve as their own models. For the precise sense in which they do so, see [Hintikka, 1973], chapters 1-2, and [Hintikka, 2007].

A model set must be consistent in the sense that two closed formulas of the form F, ∼F must not occur in it. Also, no sentence $(a \neq a)$ may occur in a model set. A model set must also satisfy certain closure conditions. The disproof rules are rules of eliminating one by violations of these closure conditions, in other words, for constructing a model set.

These construction rules constitute the proof-theoretical subject of this paper. The propositional rules can be formulated as follows with λ being a state in the construction:

(R.&)   If $(F_1 \& F_2) \in \lambda$, add F_1 and F_2 to λ

(R.∨)   If $(F_1 \vee F_2) \in \lambda$, split this construction into two alternatives branches λ_1, λ_2 with $F_1 \in \lambda_1$, $F_2 \in \lambda_2$

(R.=)   If $F[a] \in \lambda$, $(a = b) \in \lambda$ or $(b = a) \in \lambda$, add $F[b]$ to λ

It is well known that these rules (together with the rules for quantifiers explained later) define a complete disproof procedure for the usual first-order logic. Since the law of excluded middle holds in the usual first-order logic, it also give a complete proof procedure.

In independence-friendly (IF) logic, tertium non datur does not hold, and accordingly we do not obtain there a complete proof procedure. But in virtue of the compactness of IF logic, we still obtain a complete disproof procedure. In the rest of this paper the term "proof" is used also of disproofs; unless the contrary is explicitly indicated, especially when proof rules are concerned. (For IF logic, see [Hintikka, 1996], [Mann et al., 2011].

4 Rules for quantifiers

The crucial rules are the ones governing quantified expressions. They are in the usual approach essentially instantiation rules. The only minor complications are occasioned by the presence of functions.

(R.E)   (Existential instantiation) If a sentence $(\exists x)F[x]$ is in λ, one may add to λ a sentence $F[b]$ where b is a new individual constant.

This rule can be restricted to cases where there is no sentence of the form $F[t]$ in λ, where t is any constant term.

(R.A)   (Universal instantiation) If a sentence $(\forall x)G[x]$ occurs in λ, one may add $G[t]$ to λ, where t is a constant term formed from function symbols and constant terms occurring in the members of λ.

5 Operations on subformulas

The tree method rules (model set building rules) apply only to the logical constants with the largest scope, for instance only to formula-initial quantifiers. They operate by introducing substitution-instances of subformulas of earlier formulas. In order to apply logical rules to subformulas, we have to wait until this subformula-introducing process has brought them up to the surface, that is, made them entire formulas (putative members of a model set). This is an unnecessary restriction that among other drawbacks complicates proof theory needlessly. It can be relaxed by allowing logical rules to be applied also to conjuncts and not only to members of sets of formulas. What is needed in the first place is merely a new notational perspective. The alternative stages of model set construction can be conceptualized as disjunctions of conjunctions instead of alternative sets of formulas. This leaves old rules intact, mutatis mutandis, but it now allows their application to conjunctions instead of sets of formulas. The old terms like "branch" and "construction" can likewise be used still.

The generalized rules may be formulated as follows:

(R.∨)*   If $(F_1 \vee F_2)$ occurs in a conjunction
$$(F_1 \vee F_2)\&G,$$
then this conjunction may be replaced by
$$(F_1\&G) \vee (F_2\&G).$$

(R.E)*   An existential formula $(\exists x)F[x]$ occurring in a context may be replaced in that context by $F[g(y_1, y_2, \ldots)]$.

Here g is a new function constant and $(\forall y_1), (\forall y_2), \ldots$ are all the universal quantifiers on which $(\exists x)$ depends in the given context.

(R.A)*   A universal formula $(\forall x)F[x]$ occurring in a context may be replaced by $(\forall x)(F[x]\&F[t])$.

Here t is a term formed from (i) function terms occurring in the same branch; (ii) constant terms in the same branch; (iii) variables bound to quantifiers in the same branch.

A few comments are in order. In original (dis)proofs, the category (iii) of ingredients of the substituted term t were not used in rule (R.A)*. Proofs obeying this restriction are called conventional as distinguished from liberated ones. The new rule is nevertheless obviously valid, and the restriction therefore unnecessary. Removing it helps to liberate the order of rule applications.

The liberated rule (R.E)* is what allows Skolemization. It is obviously valid and yet has a great deal of deductive power. This is because the applicability of (R.A)* in a given context depends only on the supply of available function constants and constant terms.

Other rules must also be adjusted to the new perspective and notation.

(R. =)*   If $F[a]$ occurs within the scope of $(a = b)$ or $(b = a)$ it can be replaced by $(F[a]\&F[b])$.

Here the scope of $(a = b)$ in $((a = b)\&B_1\&B_2\&\ldots)$ is $(B_1\&B_2\&\ldots)$. The formulas $F[a]$ and $F[b]$ are like each other except for an exchange of a and b at one or several of their occurrences.

We can also generalize the contradiction rule.

(R.∼)*   If a conjunction contains closed conjuncts of the form $F, \sim F$, the nearest disjunction of which the conjunction is a part can be replaced by $(F\&\sim F)$.

With these liberalizations, it is seen that certain other syntactical transformations change only the order of the application of different rules in a disproof without affecting anything else. Among them there are the following:

(i) Permutation and association for $\&, \vee$.

(ii) Distribution laws for $\&, \vee$.

(iii) Distribution law for $(\forall x)$ over conjunction and for $(\exists x)$ over disjunction.

(iv) Changing the order of $(\forall x)$ and $\&$ or of $(\exists x)$ and $\vee$.

(v) Interchange of $((\exists x)F[x]\&G)$ and $(\exists x)(F[x]\&G)$ as well as $((\forall x)F[x] \vee G)$ and $(\forall x)(F[x] \vee G)$ if it does not occur in G.

6 Skolemized proofs

A further simplification is obtained by Skolemizing all the initial formulas of a tree method disproof by $(R.E)$. Proofs of this kind are called Skolemized proofs. Then in the rest of the proof (construction) all the existential quantifiers disappear (together with variables bound to them) and all the universal quantifiers can be thought of as being moved to the beginning of a conjunction of all the members of the model set approximation so far reached. The rules for existential quantifiers become redundant, and the only rule (needed) for quantifiers is $(R.A)$ with a fixed supply of function constants and the Skolem functions. In the following, this requirement is relaxed somewhat by allowing also composite functions of the earlier ones into the formation of t in $(R.A)^*$. Then we can simplify $(R.A)^*$ by stipulating that t be of the form $f(t_1, t_2, \ldots)$. Where f is a function formed from the given initial functions plus Skolem functions, and $t_1, t_2, \ldots$ are either (i) constant terms occurring in the same branch or (ii) variables bound to quantifiers in the given context.

Now new constant terms are introduced only by applications of $(R.A)^*$ with all terms $t_1, t_2, \ldots$ constant. The rule $(R.A)^*$ nevertheless allows us in certain circumstances to postpone the introduction of a constant term (and perhaps to avoid it altogether) by a suitable nonconstant (variable-dependent) term. This can be illustrated by a simple example.

7 An example

The liberation of the order of applications of different rules and especially the liberation of the order in which constant terms are introduced is facilitated by having suitable nonconstant terms to do the work of a constant term before it has actually been introduced. This and some of the other aspects of the functioning of Skolem functions is illustrated by the following example. In it, a certain proof of a formula (i.e. a disproof of its negation) is presented in three different frameworks, first in the traditional tree method form and then in a Skolemized but conventional format and finally in a liberated version. As such an example, consider the usual disproof of the negation of

$$(\exists x)(A(x) \supset (\forall y)A(y))$$

This negation is

(7.1.1) $(\forall x)(A(x)\&(\exists y)\sim A(y))$

In a traditional tree method, an argument might run as follows

(7.1.2) $A(a)\&(\exists y)\sim A(y)$ (By $(R.A)$ from (7.1.1). It can be assumed trivially that a is present in λ.)
(7.1.3) $(\exists y)\sim A(y)$ (By $(R.\&)$ from (7.1.2))
(7.1.4) $\sim A(b)$ (By $(R.E)$ from (7.1.3))
(7.1.5) $A(b)\&(\exists y)\sim A(y)$ (By $(R.A)$ from (7.1.1))
(7.1.6) $A(b)$ (By $(R.\&)$ from (7.1.5))

In a conventional Skolemized version of the tree method, an analogous proof might run as follows:

(7.2.1) $(\forall x)(A(x)\&\sim A(f(x)))$ (By Skolemization)
(7.2.2) $A(a)\&\sim A(f(a))$ (By $(R.A)$ from (7.2.1) with a as constant Skolem function
(7.2.3) $A(f(a))\&\sim A(f(f(a)))$ (By $(R.A)$ from (7.2.1))
(7.2.4) $\sim A(f(a))$ (By $(R.\&)$ from (7.2.2))
(7.2.5) $A(f(a))$ (By $(R.\&)$ from (7.2.3))

Again, two constant terms a and $f(a)$ (corresponding to b in (7.1.4)) are used. A liberated version might run in the following way:

(7.3.1) $(\forall x)(A(x)\&\sim A(f(x)))$
(By Skolemization)
(7.3.2) $(\forall x)((A(x)\&\sim A(f(x)))\&(A(f(x))\&\sim A(f(f(x)))))$
(By $(R.A)^*$ from (7.3.1)
(7.3.3) $(\forall x)(\sim A(f(x))\&A(f(x)))\&(\forall x)(A(x)\&\sim A(f(f(x))))$
(By reordering the conjuncts and distributing $(\forall x)$)
(7.3.4) $(\forall x)(\sim A(f(x))\&A(f(x)))$
(By $(R.\&)$ from (7.3.3))
(7.3.5) $\sim A(f(a))\&A(f(a))$
(By $(R.A)$ from (7.3.4))

In this proof, only one constant term, viz. $f(a)$, is used. The role of the other constant term a in (7.2.1)-(7.2.5) is played by the variable x. But this variable need not be instantiated in the proof (7.3.1)-(7.3.4). There is no counterpart to (7.2.2) there. This line of thought will be used later in the proof of the Theorem 4.

8 Rule-ordering questions

One of the main concerns of proof theory is the order of applications of different logical rules in proofs. It is now seen that the use of Skolem functions, together with the possibility of applying rules to subformulas, has the effect of relaxing this order greatly. After the initial replacement of existential quantifiers in terms of Skolem functions, the only quantifier rule we need is $(R.A)^*$. By examining it, we can immediately see interesting results.

THEOREM 1. *Constant terms in a completed liberated Skolemized proof can be introduced in any order. Changing their order of introduction does not affect the intervening rule applications, except for their order. (Some steps may become unnecessary).*

Proof. Consider two constant terms t_1 and t_2 introduced by two respective applications of $(R.A)^*$. If these applications are in different branches, then it is seen from $(R.A)^*$ that they depend only on their respective branches and hence can be made independently of each other.

If the applications are in the same branch, we can assume that t_1 precedes t_2, and that t_1 may therefore be an ingredient of t_2. But in terms our of which t_1 is built are available as ingredients of t_2, which therefore can be formed directly, independently of t_1. ■

COROLLARY 2. *All the introductions of all constant terms can be done before the application of other rules. Hence any proof can be turned into a conventional one by changing the order of rule applications.*

The possibility of introducing constant terms in an arbitrary order has a clear intuitive meaning in terms of the idea of attempted proofs as model constructions, individual by individual. What it means in this context is that the individuals constituting the fragment of a model at any stage of construction could have been introduced in any order.

9 The branching structure of proofs

A formulation of proof in Skolemized form also helps to obtain an overview on their branching structure. The basic idea is made intuitive by the semantical meaning of disproofs as frustrated model constructions. These constructions are in computer-theoretical terms nondeterministic computations. Accordingly the main theoretical and practical questions here concern the determinants of the length of branches, the determinants of branching, and the relation between the two.

In a model set construction, the number of steps in a given branch is called its overall length. The number of constant terms employed in the branch is called its combinational length. The length of the longest branch in a construction (tree proof) is called the overall length of the proof and the combinatorial length of its combinationally longest branch is called its combinatorial length.

It is easily seen that the combinational length of a branch in a construction (tree proof) is the same as the combinatorial length of the corresponding liberated proof. The first main fact here is:

THEOREM 3. *The overall length of a branch is at most a polynomial function of its combinatorial length.*

Proof. All the sentences in a branch are substitution-instances with respect to constant terms of subformulas of the initial sentences. The number of constant terms is the length l of the branch. If the number of different branches in all the subformulas of S is n, then at most l^n different substitution-instances can be formed. ■

As to the relation of the length of the branches of a disproof to their number, the interpretation of proofs as countermodel constructions suggests an important idea. The basic parameter is the number k of individuals already introduced into a branch. These individuals are represented by k constant terms. If we then introduce yet another k individual, we have to consider how it is related to each of the old ones. If in each case we have at least two different possibilities to consider, we have to consider at least 2^k different possibilities for which we need at least 2^k different branches. In brief, the number of branches is an exponential function of their length.

This idea has a precise formulation in

THEOREM 4. *If the combinatorially shortest disproof of S has the combinatorial overall length k, then it has at least 2^k different branches.*

Proving this theorem seems at first rather simple.

Consider, for the purpose, the combinatorially longest branch of the combinatorially shortest proof. Let the step in which the last constant c_b term is introduced take us from $(\forall x)F[x]$ to

(1) $F[g(c_1, c_2, \ldots, c_{k-1})]$.

Here $g(c_1, c_2, \ldots, c_{k-1}) = c_k$ and $c_1, c_2, \ldots, c_{k-1}$ are all the previously introduced constant terms. In the rest of the proof no new constant terms are introduced.

Since we are considering a single branch, F can be assumed to be a conjunction or negated or unnegated atomic formulas or identities. In the latter case, since c_k is a new constant, adding (1) creates an inconsistency only if $F[x]$ does likewise. Hence there is no need to introduce c_k. Any previous constant can do the same job.

Let us assume, then, that $F[g_k]$ is of the form

(2) $(\forall x)(G_1[x] \vee G_2[x] \vee \ldots \vee G_m[x])$.

The cases with multiple universal quantifiers can be dealt with in the same way. Hence we can assume that $m > 1$.

The only way in which the presence of (2) can open the way of completing the proof is via substitutions of $c_1, c_2, \ldots, c_{k-1}$ for x in (2). For no new constants are introduced after c_k.

Hence the addition of the following conjunction will create a contradiction:

(3) $(G_1[c_1] \vee G_2[c_1] \vee \ldots \vee G_m[c_1])$&
$(G_1[c_2] \vee G_2[c_2] \vee \ldots \vee G_m[c_2])$&
$\ldots$ &
$(G_1[c_{k-1}] \vee G_2[c_{k-1}] \vee \ldots \vee G_m[c_{k-1}])$.

If all the old constants $c_1, c_2, \ldots, c_{k-1}$ are indispensable in (3) we have here in effect a disjunction of m^{k-1} disjuncts and here at least m^{k-1} branches, as was asserted.

Assume, then, that one of the conjuncts in (3) is dispensable, say the first one. Then we can modify the proof as indicated in Theorem 1 and assume that c_1 is introduced only after c_k. But the only use of c_1 after c_k was introduced in the first conjunct of (3). If this conjunct is dispensable, we obtain a proof in which c_1 was not used at all, in other words a combinatorially shorter proof than the combinatorially shortest one. Hence all of the $c_1, c_2, \ldots, c_{k-1}$ are indispensable and the exponentiality of the number of branches follows.

Theorem 4 can be taken to establish the most basic fact about the relation of the length and the number of different branches in a logical (dis)proof procedure.

10 Application

According to [Hintikka, 2012] and [Hintikka, 2010], any computation can be represented as a cut-free deduction with a one-one correspondence of constant terms. Hence the above results concerning formal proofs have automatically analogs for computations. In particular, the conclusion about this connection between the length of branches and their number holds also for nondeterministic computations. For instance, if the length of the branches of a nondeterministic computation is polynomial, such a computation cannot be turned into linear computation of polynomial length since the number of branches is exponential.

BIBLIOGRAPHY

[Hintikka, 1973] J. Hintikka. *Logic, Language-Games and Information*. Clarendon Press, 1973.

[Hintikka, 1996] J. Hintikka. *Principles of Mathematics Revisited*. Cambridge University Press, 1996.

[Hintikka, 2007] J. Hintikka. Hilbert was an axiomatist, not a formalist. In S. Pihlstrom, P. Raatikainen, and M. Sintonen, editors, *Approaching Truth: Essays in Honour of Ilkka Niiniluoto*, pages 33–47, London, 2007. College Publications.

[Hintikka, 2010] J. Hintikka. Does logic count? deductive logic as a general theory of computation. In L. Magnani, W. Carnielli, and C. Pizzi, editors, *Model-Based Reasoning in Science and Technology: Abduction, Logic, and Computational Discovery*, pages 265–274. Springer, 2010.

[Hintikka, 2012] J. Hintikka. Logic in a theory of computation. *Newsletter of the APA Committee on Philosophy and Computers*, forthcoming, 2012.

[Mann et al., 2011] A. Mann, G. Sandu, and M. Sevenster. *Independence-friendly logic: A game-theoretical approach*. Cambridge University Press, 2011.

[Smullyan, 1963] R. Smullyan. *First-Order Logic*. Springer, 1963.

Jaakko Hintikka
Department of Philosophy
Boston University
Boston, USA
Email: hintikka@bu.edu

www.ingramcontent.com/pod-product-compliance
Lightning Source LLC
Chambersburg PA
CBHW071056230426
43666CB00009B/1726